Beyond Popular Science

Beyond Popular Science

Exploring Curious Phenomena

David H. Silver
david@remiza.ai

With illustrations by
Jess DG Hayes

OpenBook Publishers

BEYOND POPULAR SCIENCE
David H. Silver

https://www.openbookpublishers.com

Digital material and resources associated with this volume are available at https://doi.org/10.11647/OBP.0526#resources

ISBN Paperback:	978-1-80511-877-0
ISBN Hardback:	978-1-80511-878-7
ISBN Digital (PDF):	978-1-80511-879-4
ISBN HTML:	978-1-80511-881-7
ISBN Digital ebook (epub):	978-1-80511-880-0
DOI:	10.11647/OBP.0526

Cover image by Enny Silver and David H. Silver
Cover design by Jeevanjot Kaur Nagpal

Introduction

This book is not a popular science book. It is not a textbook. It is not an academic book. It is not even a chimera of the above.

It does share some goals with the three: to inspire wonder (like a popular science book), to include some rigour (like textbooks), and to introduce readers to phenomena that might challenge their understanding (as academic works often achieve).

As with some combinations, such as a sushi-pizza restaurant, it excels at none. The main exposition isn't long enough for full understanding, the technical part is often too abstract or detailed to follow, and despite claims of rigour, von Neumann and Morgenstern's observation that "there is no point in using exact methods where there is no clarity in the concepts and issues to which they are to be applied" (*Theory of Games*, 1944) fits too well.

But if my plan works, you'll get the appetite to leave this sushi-pizza diner. Maybe to a textbook because you're intrigued. Maybe to Wikipedia or blogs for more context about this fantastical world.

This book contains fifty stories, each structured to guide readers from the intuitive to the profound:

Backdrop Each chapter begins with concise background—the people, circumstances, and discoveries behind the phenomenon.

Phenomenon Description The phenomenon is described in straightforward terms, avoiding sensational language for clear, accurate explanations.

Hardcore Analysis For readers ready to dive deeper, the third section in each chapter provides rigorous academic analysis. Here, the mathematical and technical underpinnings of the phenomenon are laid bare, complete with equations, references, and detailed derivations. This section is unapologetically tough. As with references in a scientific article, this section is not required for the reader to grasp the main ideas, but it does provide scaffolding, justifies the claims presented earlier in the chapter, and offers readers the tools to validate the claims, explore further, or simply appreciate that simplified accounts are built on layers of rigour.

Each chapter also opens with a short quotation. I have tried to track down original sources instead of repeating popular misattributions, and when possible I use the original language rather than a familiar translation. A few attributions remain imperfect—one Leonardo da Vinci line defeated my search—but I'm content with the overall accuracy.

A Few Disclaimers: The book contains errors ranging from typos to wrong equations. Please report them, and be forgiving of mistakes. While precision is unrealizable, this serves as a more accurate guide to reality than popular science expositions. All chapters can be read independently. The **essence** is accessible to anyone, mostly in the chapter summaries. Some chapters are extremely mathematical and may not appeal to unfamiliar readers. With those caveats in mind, the stories begin.

Contents

Prologue

This project began as a small 'flight magazine' I put together for my family before a long transatlantic trip. I started filling it with notes I had collected over the years—questions that seem simple on the surface but are, on inspection, scientifically intricate and mathematically deep.

After drafting a few pieces—why gold is yellow (relativity) and why apples fall from trees (time dilation)—it became clear that this was more than a pamphlet. Throughout the day, I found myself recalling phenomena I had once wondered about, and the list of topics kept growing.

Anyone who knows me knows that every so often I get excited about some new piece of science and insist on explaining it, regardless of the time or place. It turned out there were many such topics. One thing led to another; more pages were written, and this became a real book.

I owe special thanks to everyone who supported the early stages of this project—both personally and through the Kickstarter campaign—and helped to make the first printing possible. I am especially grateful to my wife, Enny, for her unwavering support, patience, and love. I am also thankful to my kids, who patiently sat through countless explanations of physics and mathematics.

This book returns to the roots of scientific wonder, combining accessible explanations with rigorous mathematical foundations. Unlike contemporary science communication that oversimplifies or sensationalises, it highlights the beauty of science as it truly is: both elegant and complex. The focus is understanding, not just exposure.

Too often, modern science communicators rely on a 'laugh track' approach—telling readers how they should feel ('this is mind-blowing!') instead of letting wonder arise naturally from the ideas. This cheapens the experience, as though science requires manufactured excitement. Science doesn't need exaggeration; its wonder is self-evident to those who explore it properly.

I must apologise if my enthusiasm and flair are not easy to convey in this medium. But I assure you: even reading portions of this book should leave you feeling that our universe is more fantastical than any Tolkien creation. The effects we observe in the natural world work in wondrous ways—relativity and quantum mechanics are stranger than fiction, with more sorcerous underlying complexity than any mythological chant.

When a ray of sunlight hits your eyes and you move, the cascade of events is magnificent: trillions of quantum field excitations coordinate changes across millions of city-scale structures. Cells with politics and defence protocols—ribosomes pounding out translations like factories, mitochondria running proton gradients as power plants, immune patrols scanning for invaders, membranes running checkpoints and visa systems—trillions of these cities operating in parallel just to send an impulse down the spinal cord.

Every molecule performs Hamiltonian plays, issuing redistribution orders to orbitals. Atoms are not little spheres but dense arrangements of nuclei with surrounding clouds, and the protons and neutrons in those nuclei are bound states of three colourful quarks, constantly borrowing energy from the vacuum, exchanging gluons trillions of times per second, stitching colour fields so tight that the binding energy exceeds the sum of the parts, generating most of the mass that weighs the body down, mass that resists acceleration, mass that makes clocks tick more slowly.

Layered above, enzymes catalyse reactions in femtoseconds, metabolic pathways route energy into ATP, blood pumps uphill against gravity, oxygen convoys carried by haemoglobin through capillary labyrinths. Neurons fire spikes, action potentials race along axons, vesicles dump neurotransmitters into synapses, inhibitory and excitatory votes cast trillions of times per second, motor cortex computes commands, motoneurons release acetylcholine into neuromuscular junctions, muscle fibres flood with calcium, actin and myosin filaments slide, sarcomeres shorten, tendons tug, bones shift, and the person moves.

And the photon itself was generated in a star's core where the weak nuclear force converts protons to neutrons after quantum tunnelling through an energy barrier, then trapped in plasma for a million years in random-walk collisions, finally escaping the surface and flying straight for minutes across the vacuum (zero time from the photon's point of view), striking your retina and flipping rhodopsin from cis to trans—a femtosecond molecular rearrangement amplified into a millisecond spike.

The cascade from subnuclear quark fields to stellar photon journeys to cellular cities to muscular contraction all chained together so that when you think 'I should move,' your body shifts in space and every layer of physics and biology has fired in unison to make it happen.

This must be less mundane than any grumpy villain that can fly forks around telekinetically. Maybe after reading a few chapters you will agree with this claim even more.

All topics in this book have personal stories behind them—I remember how I learned about them. *I hope I can infect you with some of that excitement.*

The goal is to respect the reader's intelligence and curiosity. Whether discussing topological insulators, the mechanics of atomic clocks, or the subtleties of time dilation, these chapters present science as it is: demanding, rewarding, and truly inspiring.

This book counteracts oversimplified science communication. Science isn't slogans or easy answers—its complexity is a feature to celebrate. Understanding takes effort, but transforms fleeting curiosity into lasting enlightenment.

If you're ready to explore science in its full intellectual glory, I invite you to turn the page.

Relatively Yellow

Energy–Mass Equivalence: A box of mass M and length L emits a light pulse of energy E from its left wall. The pulse carries momentum $p = E/c$, so to conserve momentum the box recoils leftward with speed $v \approx E/(Mc)$ (for $v \ll c$). The pulse reaches the right wall in time $t \approx L/c$, during which the box drifts left by $\Delta x = vt = \frac{EL}{Mc^2}$. After absorption the box is again at rest, but its location has shifted—seemingly moving the system's centre of energy despite no external forces. To prevent any net shift, the emission must reduce the box's rest mass at the left wall by ΔM (which reappears at the right upon absorption). This relocation shifts the system's centre of energy rightward by $\Delta x' = \frac{\Delta M}{M} L$. Setting $\Delta x' = \Delta x$ gives $\frac{\Delta M}{M} L = \frac{EL}{Mc^2}$, hence $\Delta M = E/c^2$.

Relativistic Energy and Momentum: In the middle-right panel, a massive particle crosses the box; both rest mass and motion contribute. Lorentz symmetry packages energy and momentum into the four-momentum $P^\mu = (E/c, p_x, p_y, p_z)$. Its Minkowski norm is invariant: $P^\mu P_\mu = (E/c)^2 - p^2 = m^2c^2$. Therefore $E^2 = m^2c^4 + p^2c^2$ (or, with $c = 1$, $E^2 = m^2 + p^2$).

Why Gold Is Yellow: In the final panel, blue light is absorbed at the surface of a gold atom. Relativistic contraction of inner orbitals (due to high-velocity 6s electrons) shifts energy levels. This narrows the 5d–6s gap, bringing blue transitions into range. The missing blues tint the reflection yellow. Gold is yellow because relativity bends its atomic spectrum.

Relatively Yellow

The yellow colour of gold requires relativistic quantum mechanics to explain, unlike silver's silvery appearance. Electrons in gold atoms reach 58% of light speed, causing changes in the 6s and 5d orbitals. This shifts absorption to blue wavelengths, resulting in the reflection of yellow-red light. Similar relativistic effects explain mercury's liquid state and platinum's white appearance. These everyday properties demonstrate how modern physics manifests in macroscopic observations.

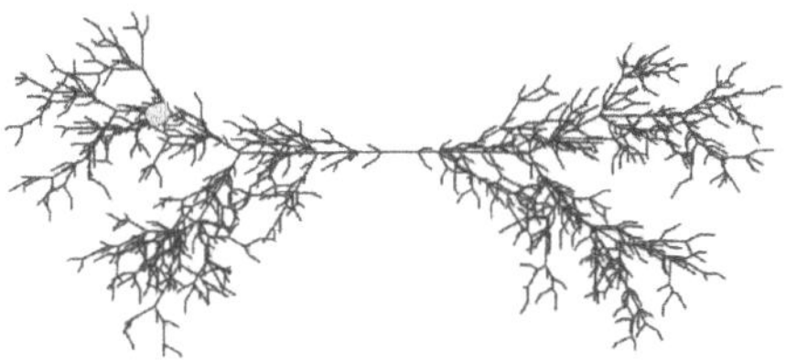

Quantum Numbers ∘ Gold's Yellow Colour ∘ Relativistic Orbital Contraction ∘ $v \sim Z\alpha c$ Scaling ∘ 5d-6s Energy Reversal ∘ Blue Light Absorption ∘ Dirac vs Schrödinger ∘ Heavy Atom Effects ∘ Inert Pair Effect ∘ Mercury Liquid State ∘ Relativistic Chemistry

"All that is gold does not glitter,
not all those who wander are lost;
the old that is strong does not wither,
deep roots are not reached by the frost."
— J. R. R. Tolkien, 1954

"All that glitters may not be gold,
but at least it contains free electrons."
— John Desmond Bernal, 1962

Relatively Yellow

Arnold Sommerfeld's 1916 work on relativistic extensions to the Bohr model laid the groundwork for understanding how high nuclear charge alters electronic structure in heavy atoms. In 1940, A. O. Williams improved the Hartree self-consistent field method by incorporating the Dirac equation, demonstrating relativistic corrections in Cu^+ and quantifying the spin–orbit splitting—the splitting of degenerate energy levels due to coupling between an electron's spin and its orbital motion.

David Francis Mayers extended this work in 1957, identifying that electrons in heavy atoms, travelling at significant fractions of the speed of light, experience orbital contraction.

Boyd, Larson, and Waber (1973) expanded upon this by demonstrating the relativistic expansion of certain d orbitals, emphasising the intricate interplay between electron velocity and orbital behaviour. Kenneth S. Pitzer's 1971 research marked a turning point, showing that mercury's unusually low melting point could be attributed to these relativistic effects. Later in the 1970s, Pekka Pyykkö and Jean-Pierre Desclaux carried the idea further, using theoretical methods to connect mercury's liquid state and gold's distinct colouration to changes in orbital energies brought about by relativistic corrections.

By the early 1980s, X-ray photoelectron spectroscopy offered direct experimental confirmation of these phenomena, although Lennart Norrby noted in 1991 that such insights still struggled to gain widespread inclusion in general chemistry curricula.

Atoms consist of a dense nucleus, composed of protons and neutrons, surrounded by electrons that do not follow classical trajectories. Instead of moving in well-defined orbits similar to planets around a star, electrons in atoms are described by orbitals: spatial distributions derived from the quantum mechanical wavefunction that give the probability of finding the electron in a given region.

These orbitals emerge as solutions to the Schrödinger equation (Schrödinger, 1926) applied to the Coulomb potential created by the positively charged nucleus. Each solution is characterised by a discrete set of parameters—the quantum numbers—which determine the electron's energy, spatial distribution, and angular properties. There are four such quantum numbers:

- The **principal quantum number** $n = 1, 2, 3, \ldots$ determines the energy level and average radial extent of the orbital. Higher n corresponds to greater distance from the nucleus and higher energy.
- The **angular momentum quantum number** $\ell = 0, 1, \ldots, n-1$ determines the orbital's shape. Values of ℓ are labelled spectroscopically as s ($\ell = 0$), p ($\ell = 1$), d ($\ell = 2$), f ($\ell = 3$), and continue alphabetically. For example, s orbitals are spherically symmetric, while p orbitals have a nodal plane and a dumbbell-like shape.
- The **magnetic quantum number** $m = -\ell, \ldots, +\ell$ defines the orientation of the orbital in space relative to a chosen axis (typically the z-axis).

 https://doi.org/10.11647/OBP.0526.01

- The **spin quantum number** $s = \pm\frac{1}{2}$ captures a quantum property with no classical analogue which behaves like an angular magnetic moment.

No two electrons in a single atom may occupy the same quantum state. This constraint—known as the Pauli exclusion principle (Pauli, 1925)—means that each combination (n, ℓ, m, s) can be occupied by at most one electron. For example, the 1s orbital can host two electrons: one with spin up and one with spin down.

As electrons are added to an atom, they fill available orbitals according to energy minimization principles. This leads to the well-known electron filling sequence:

$$1\text{s},\ 2\text{s},\ 2\text{p},\ 3\text{s},\ 3\text{p},\ 4\text{s},\ 3\text{d},\ 4\text{p},\ 5\text{s},\ \dots .$$

This order does not follow a strict progression in n, due to interactions such as shielding and penetration. Inner electrons partially screen the nucleus, reducing the effective nuclear charge experienced by outer electrons. Orbitals with the same n but different ℓ values can therefore have different energies.

In hydrogen-like atoms (single electron, full nuclear charge), all orbitals with the same n are degenerate—they have the same energy regardless of ℓ. This symmetry is broken in multielectron atoms, where electron–electron repulsion and the shape of orbitals result in energy level splitting.

Orbital shapes are determined by both radial and angular components. The radial part depends on n and ℓ, while the angular part (controlled by m and ℓ) determines nodal planes and symmetry axes. For instance, a 3d orbital has two angular nodes and occupies a region shaped like a cloverleaf.

The periodic table reflects these principles. Each row corresponds roughly to a value of n, and each column—especially among main-group elements—reflects the number and configuration of valence electrons, those in the outermost shell. Elements with similar valence configurations (e.g., noble gases, alkali metals) exhibit analogous chemical properties.

The colour and optical appearance of a material are determined by how it interacts with light. Light is an electromagnetic wave, characterised by its wavelength, λ. When light strikes a material, some wavelengths are absorbed, while others are reflected or transmitted. The observed colour corresponds to the reflected portion of the spectrum. For instance, a substance that absorbs blue light and reflects red and green will appear yellow. The mechanism behind absorption is electronic: photons transfer their energy to electrons, promoting them from lower to higher energy states.

This promotion requires the photon's energy to match the gap between electronic states. By the Planck-Einstein relation, $E = hc/\lambda$, shorter wavelengths carry higher energies. When the energy gap between orbitals aligns with a photon's energy, that wavelength is absorbed. These absorptions determine the material's colour.

Now let's talk relativity. Special relativity (Einstein, 1905) describes the behaviour of physical systems at speeds approaching the speed of light c. Its core principle is that measurements of time, length, and mass depend on the observer's inertial frame. In Minkowski spacetime, c is not merely a large speed—it is the natural speed scale that defines the causal structure. Just as angles are bounded by 2π radians or percentages by 100%, velocities in spacetime are bounded by c. This is a geometric constraint, not a practical limitation.

For a particle with rest mass m_0, the total energy increases with velocity according to:

$$E = \frac{m_0 c^2}{\sqrt{1 - v^2/c^2}}.$$

As $v \to c$, the ratio v^2/c^2 approaches 1, causing the denominator $\sqrt{1 - v^2/c^2}$ to shrink toward zero, and the energy diverges. The formula reflects that c marks the boundary of physically realisable velocities.

In practice, even modest fractions of c can result in measurable relativistic energy corrections. For example, at $v/c = 0.6$, the denominator becomes $\sqrt{1 - 0.36} = 0.8$, so the total energy is increased by a factor of $1/0.8 = 1.25$ above the rest energy. These corrections are small for everyday objects but become substantial for subatomic particles, especially in high-energy regimes such as atomic orbitals in heavy atoms.

When considering electrons in atoms, however, the notion of velocity requires care. Electrons do not follow classical trajectories—they are described by wavefunctions, and dynamical quantities like momentum are represented by operators. Nevertheless, expectation values of these operators yield characteristic speeds that determine whether relativistic corrections matter.

The uncertainty principle (Heisenberg, 1927) connects localisation to momentum: electrons confined near a high-Z nucleus must carry large momentum components. In the Schrödinger equation, the kinetic energy term penalises sharp spatial confinement, while the Coulomb potential favours proximity to the nucleus. The balance determines the orbital's size and shape. For heavy atoms, where electrons are drawn tightly inward, the resulting momenta reach kinetic energies where relativistic corrections become crucial.

To see why, we can first estimate electron speeds using the standard (non-relativistic) framework, then check whether relativistic corrections are needed. In quantum mechanics, the momentum operator $\hat{p} = -i\hbar\nabla$ extracts how rapidly the wavefunction oscillates in space—sharper oscillations correspond to higher momentum. The expectation value $\langle p \rangle$ gives the average momentum, and dividing by the electron mass yields a characteristic velocity: $v \sim \langle p \rangle / m_e$. If this estimate approaches a significant fraction of c, the non-relativistic treatment breaks down and must be replaced by relativistic quantum mechanics.

A useful estimate for these characteristic speeds comes from hydrogen-like atoms—idealised atoms with a single electron and full nuclear charge. The expectation value scales as:

$$\langle v \rangle \sim Z\alpha c,$$

where Z is the atomic number and $\alpha \approx \frac{1}{137}$ is the fine-structure constant. Electrons are drawn more tightly to higher-Z nuclei and thus move faster. (A more rigorous treatment uses $\langle v^2 \rangle$ or the full Dirac equation to properly account for relativistic kinematics.) As Z increases, relativistic energy corrections become non-negligible.

These relativistic effects modify the quantum mechanical equations that describe bound states. The Schrödinger equation must be replaced or augmented by relativistic formulations such as the Dirac equation (Dirac, 1928). These corrections modify the energies and shapes of orbitals. The resulting deviations from non-relativistic predictions grow with atomic number and are especially pronounced for the inner (core) electrons in heavy elements.

For gold with $Z = 79$, the 1s electrons reach velocities around $v \approx 0.58c$—more than half the speed of light! These extreme speeds require relativistic treatment. The resulting orbital contractions and energy shifts cascade through all electron shells, ultimately affecting even the outermost electrons responsible for optical properties.

In most metals, optical behaviour is dominated by conduction electrons. These electrons are not localised to individual atoms but move freely through the crystal, occupying partially filled energy bands. Because these conduction bands are broad and continuous, they allow electrons to respond uniformly to incoming electromagnetic waves. As a result, nearly all visible wavelengths are reflected equally, and the metal appears silvery or white. This is why typical metals, such as aluminium, iron, or silver, lack colour: their optical response is effectively achromatic.

However, when deeper (non-conduction) bands lie close to the Fermi level (the highest occupied energy level at absolute zero), interband transitions become possible. In this case, photons can excite electrons from filled lower bands (such as d-bands) into the conduction band when their energy matches the band gap. If this gap lies within the visible range, the metal absorbs certain wavelengths and reflects the rest, producing colour.

In gold and other heavy elements, relativistic effects shift the energies of these bands. s-orbitals contract because they have zero angular momentum and can penetrate directly through the nuclear centre, experiencing the full relativistic effects. In contrast, d-orbitals have angular momentum that keeps them away from the nucleus via a centrifugal barrier, reducing relativistic corrections. This differential contraction reduces the energy difference between d and s bands and brings the d-to-s gap into the visible spectrum.

In silver ($Z = 47$), relativistic effects are minor. The 4d band lies well below the Fermi level, and interband transitions require photon energies above the visible range. As a result, silver reflects nearly all visible light uniformly and appears bright white.

In gold ($Z = 79$), relativistic contraction of the 6s orbital lowers the conduction band, while expansion and destabilisation of the 5d orbitals raises the valence band. The gap between them narrows to become smaller than the non-relativistic prediction of around 3.7 eV: $E_{5\text{d}\to 6\text{s}} \approx 2.4$ eV, which corresponds to an absorption wavelength of: $\lambda \approx \frac{1240 \text{ eV}\cdot\text{nm}}{2.4 \text{ eV}} \approx$ 520 nm.

This lies in the blue region of the spectrum. Because the 5d and 6s bands are broad, interband transitions occur across a range of energies. The absorption is not narrow, but spread slightly, selectively removing blue shades. The result is gold's distinctive yellow colour, enriched in red and green wavelengths.

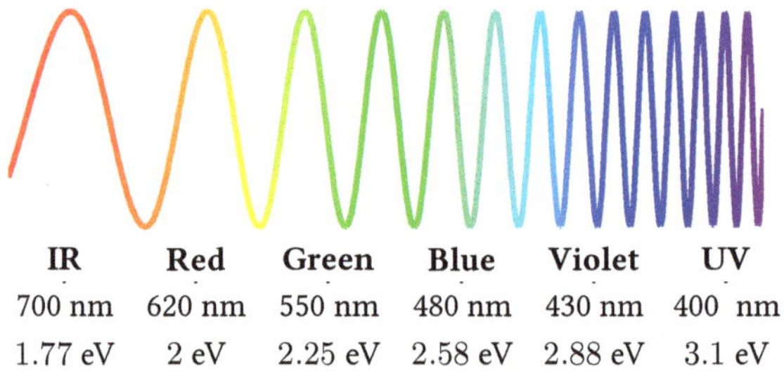

Platinum ($Z = 78$) also experiences strong relativistic shifts, but its partially filled $5d^9$ configuration and broader interband spacing push absorption into the ultraviolet. Thus, platinum reflects visible light uniformly and appears silvery-white.

Copper ($Z = 29$), though far lighter, has a naturally narrow d–s gap of about 2.1 eV even without relativistic effects. This gap also falls within the visible range and leads to selective absorption of blue-green shades, producing its reddish hue.

Mercury ($Z = 80$) reveals a different consequence of relativistic orbital modifications. The 6s orbital contracts so strongly that its electrons become tightly bound and chemically inert. This reduces orbital overlap and weakens metallic bonding. The result is a low cohesive energy and a low melting point for a metal. Mercury remains liquid at room temperature–a phase behaviour that non-relativistic models cannot reproduce.

Low-Limit Theories

Usually in physical sciences, there are simplifications to precise theories that are applicable when one 'zooms out.' For example, at low velocities, the Galilean formulation suffices without relativistic corrections. In low mass scenarios, Newtonian gravity replaces general relativity. These **'low-limit theories'** have well-defined domains of validity characterised by dimensionless parameters—velocity ratio (v/c), gravitational potential (GM/rc^2), or wavelength ratio (λ/L). When these parameters approach unity, the simplified theories break down.

In mechanics, one typically uses classical physics, and in chemistry, simplistic atomic models (electrons 'orbiting' a nucleus) are adequate for most applications. **This is why it is remarkable that the colour of gold requires relativistic corrections**. The relativistic effects on gold's electron orbitals cause it to absorb blue light and appear yellow, whereas non-relativistic predictions would yield a silvery appearance like its periodic table neighbours.

It is rare to need special relativity for macroscopic phenomena we encounter daily. This makes gold particularly noteworthy—it represents one of the few cases where a common macroscopic property (colour) can be correctly predicted only by including relativistic effects.

Relativistic Quantum Chemistry and the Colour of Gold

Quantum Mechanical Origin of High Electron Velocities

Electrons in atoms are described by wavefunctions obeying the Schrödinger equation:

$$\left[-\frac{\hbar^2}{2m}\nabla^2 - \frac{Ze^2}{r}\right]\psi = E\psi.$$

The kinetic term penalises localization, while the Coulomb term favours proximity to the nucleus. Their balance is constrained by the uncertainty principle: $\Delta x \cdot \Delta p \gtrsim \hbar$.

Strong nuclear attraction forces wavefunction localization near $r = 0$, requiring large momentum components and thus high typical velocities.

In nonrelativistic quantum mechanics, velocity is an operator:

$$\hat{v} = \hat{p}/m, \quad \hat{p} = -i\hbar\nabla.$$

For Dirac hydrogen-like ions, the characteristic electron speed scale for the innermost shells is $v_{\text{char}} \sim Z\alpha c$, where $\alpha \approx 1/137$. For gold ($Z = 79$), this yields $v_{\text{char}} \approx 0.58c$, indicating that inner electrons have characteristic speeds that are a significant fraction of c.

Relativistic Orbital Contraction

For Dirac hydrogenic s states, a factor $\sqrt{1-(Z\alpha)^2}$ appears in the energy and radial functions; this is often re-expressed as an effective 'Lorentz factor' $\gamma = 1/\sqrt{1-(Z\alpha)^2}$ and used as a measure of relativistic contraction. As Z increases, γ increases from 1, indicating stronger relativistic effects. Relativistic corrections cause s and $p_{1/2}$ orbitals to contract relative to the non-relativistic case, whereas d and f orbitals become more diffuse due to the different angular behaviour and spin–orbit structure. This contraction has consequences for interband transitions and optical properties.

Electronic Transitions and Optical Properties

Gold's configuration is $[Xe]4f^{14}5d^{10}6s^1$. Relativistic 6s contraction and 5d expansion reduce the 5d–conduction-band gap to roughly:

$$\Delta E \sim 2.3\text{–}2.5\,\text{eV} \quad \Rightarrow \quad \lambda \sim 500\text{–}540\,\text{nm},$$

in the green–blue region of the visible spectrum.

In solids, these transitions span energy bands rather than discrete levels. Finite band widths and electron lifetimes broaden the absorption, leading to selective attenuation of blue light and reflection of red/green—the physical basis for gold's colour.

Other Relativistic Effects in Heavy Elements

- **Platinum (Silvery-White)**: In Pt ($5d^9 6s^1$), relativistic effects also contract 6s and modify the 5d manifold, but the detailed band filling and d-band position keep the main interband onset in the ultraviolet. As a result, the reflectivity is practically flat across the visible, so platinum appears silvery-white.
- **Mercury (Liquid)**: Relativistic contraction of the $6s^2$ shell in Hg lowers and localises these electrons, reducing 6s–6s overlap and narrowing the 6s band. Metallic bonding is therefore unusually weak, and dispersion (van der Waals–type) interactions play a larger role compared to typical metals. This weakened cohesion explains mercury's anomalously low melting point of -38.8°C.

References:

Williams, A. O. (1940). A Relativistic Self-Consistent Field for Cu^+. *Phys. Rev.*, **58**, 723.

Mayers, D. F. (1957). Relativistic Self-Consistent Field Calculation for Mercury. *Proc. R. Soc. Lond. A*, **241**, 93.

Norrby, L. J. (1991). Why Is Mercury Liquid? Or, why do relativistic effects not get into chemistry textbooks? *J. Chem. Educ.*, **68**, 110.

Pyykkö, P., Desclaux, J. P. (1979). Relativity and the periodic system of elements. *Acc. Chem. Res.*, **12** (8), 276–281.

Dark Energies Are Pushing Us Apart

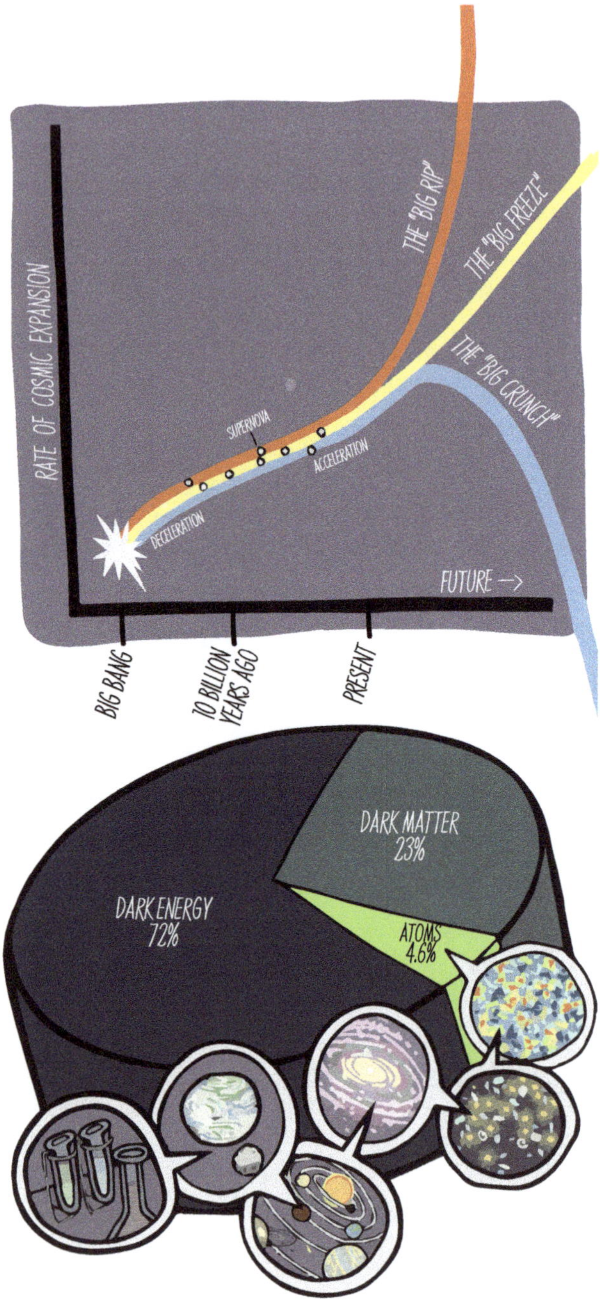

Top (Cosmic Expansion Scenarios): The universe's rate of expansion is shown across time, beginning with an early deceleration due to gravity, followed by the observed acceleration linked to dark energy. Possible future outcomes are indicated: the *Big Crunch* (re-collapse), the *Big Freeze* (endless cooling expansion), and the *Big Rip* (runaway acceleration tearing structures apart). Supernova data points provide the evidence anchoring this curve.
Bottom (Cosmic Energy Composition): The universe's mass-energy content is divided into dark energy (72%), dark matter (23%), and atoms (4.6%), with small additional contributions from neutrinos and photons. Within this narrow slice of atomic matter we can zoom in further: the large-scale structures of the cosmos → stars and galaxies → planetary systems → laboratory beakers. This exponentially dropping sequence highlights how minuscule the portion we directly observe is compared with the vast domain over which the same physical laws are successfully extrapolated.

Dark Energies Are Pushing Us Apart

Observations of distant supernovae in the 1990s revealed that the universe's expansion is accelerating, contradicting earlier models that predicted gravity would slow cosmic expansion. This acceleration requires dark energy, an unknown component comprising about 70% of the universe's energy density. Dark energy counteracts gravity at cosmological scales, manifesting as either a cosmological constant in Einstein's equations or a dynamical field.

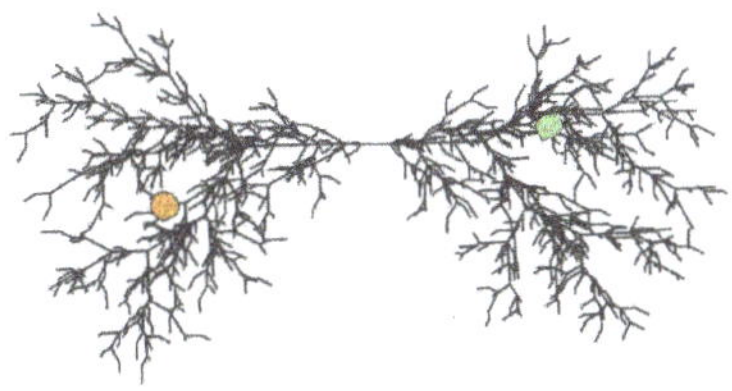

TYPE IA SUPERNOVAE ∘ STANDARD CANDLES ∘ HUBBLE DIAGRAM ∘ COSMIC ACCELERATION DISCOVERY ∘ DARK ENERGY ∘ COSMOLOGICAL CONSTANT Λ ∘ DISTANCE MODULUS ∘ REDSHIFT-LUMINOSITY ∘ ΛCDM MODEL ∘ CMB CONFIRMATION ∘ 68% DARK ENERGY

"The simplest calculation involves summing the quantum mechanical zero-point energies of all the fields known in Nature. This gives an answer about 120 orders of magnitude higher than the upper limits on Λ set by cosmological observations. This is probably the worst theoretical prediction in the history of physics! Nobody knows how to make sense of this result. Some physics mechanism must exist that makes the cosmological constant very small."

— Hobson, Efstathiou & Lasenby, 2006

Dark Energies Are Pushing Us Apart

In the late 1990s, two independent collaborations studying distant Type Ia supernovae—the Supernova Cosmology Project led by Saul Perlmutter and the High-Z Supernova Search Team led by Adam Riess and Brian Schmidt—reported that these stellar explosions were fainter than anticipated. Their apparent brightness suggested they were farther away than a purely decelerating cosmological model would indicate. This unexpected outcome signalled an accelerating rate of cosmic expansion.

Early theoretical groundwork on cosmic expansion stemmed from solutions to Einstein's field equations by Alexander Friedmann in the 1920s and Georges Lemaître in the 1930s. Although Einstein initially introduced a cosmological constant Λ to force a static universe, Hubble's observations of galactic recession (1929) established expansion as a fact. Decades later, detailed measurements of supernova luminosities confirmed that this expansion was not slowing down, but accelerating.

Recognition of this discovery came in 2011, when the Nobel Prize in Physics was awarded to Perlmutter, Riess, and Schmidt for uncovering cosmic acceleration. Additional evidence arose from measurements of cosmic microwave background anisotropies and large-scale galaxy distributions, all converging on the conclusion that the expansion of the universe is not merely continuing but speeding up.

Stars are thermonuclear engines balanced between gravitational contraction and fusion pressure. This pressure is sustained by nuclear fusion: the conversion of light elements into heavier ones, accompanied by energy release. Low-mass stars fuse hydrogen through the proton-proton chain; massive stars use the CNO (Carbon-Nitrogen-Oxygen) cycle (Bethe, 1939) at their higher core temperatures.

Stellar evolution proceeds as nuclear fuel depletes. Once hydrogen fusion subsides, gravity compresses the core, raising the temperature and enabling new fusion stages. Helium is converted into carbon and oxygen, followed by carbon, neon, oxygen, and silicon burning in increasingly rapid succession. These stages produce successively heavier elements up to iron. Fusion beyond iron is endothermic, and no further energy can be extracted from nuclear reactions.

Stellar fate depends on initial mass. Stars below ~ 8 solar masses cannot fuse heavy elements. Their cores contract into white dwarfs composed of carbon and oxygen, while their outer layers are ejected as planetary nebulae. White dwarfs are stabilised by electron degeneracy pressure up to the Chandrasekhar limit (Chandrasekhar, 1931), $M_{\text{Ch}} \approx 1.4 M_{\odot}$ (the symbol $\odot$ denotes the Sun).

Heavier stars burn through all stages to iron cores. The core collapses in milliseconds when fusion ceases. Infalling material rebounds off the dense core and triggers an explosion that ejects the outer layers. This core-collapse supernova leaves behind a neutron star or, if massive enough, a black hole.

Supernovae are classified according to their spectral features. Type II supernovae retain strong hydrogen lines and arise from stars that preserve their outer hydrogen envelopes.

 https://doi.org/10.11647/OBP.0526.02

Type Ib and Ic supernovae lack hydrogen and, in the latter case, helium signatures, indicating extensive pre-explosion mass loss. These core-collapse events produce irregular light curves and depend sensitively on progenitor properties.

Type Ia supernovae originate from a different mechanism. These events occur when a white dwarf in a binary system accretes matter from a companion and approaches the Chandrasekhar mass. Carbon and oxygen ignite under degenerate conditions, completely disrupting the white dwarf. Unlike core-collapse events, Type Ia supernovae leave no remnant. Their uniformity leads to a characteristic light curve with a distinctive rise and decay pattern, allowing for empirical standardisation and making them cosmological distance indicators.

Type Ia supernovae serve as standard candles with intrinsic luminosity L inferred after empirical correction. Observers measure apparent flux F, related by $F = L/(4\pi d_L^2)$ where d_L is luminosity distance. Astronomers use magnitudes: apparent magnitude m minus absolute magnitude M (at 10 parsecs, where one parsec is approximately 3.26 light-years and 1 Mpc = 10^6 parsecs) gives the distance modulus $\mu = m - M = 5\log_{10}(d_L/\mathrm{Mpc}) + 25$. After correcting for light curve shape (brighter events fade slower) and extinction, standardised magnitudes yield cosmological information.

Each supernova provides two observables: a redshift z and a distance modulus $\mu(z)$. The redshift is defined by $1 + z = \frac{\lambda_{\mathrm{obs}}}{\lambda_{\mathrm{emit}}}$ where λ_{emit} is the rest-frame emission wavelength and λ_{obs} is the observed wavelength. At cosmological scales this reflects the changing scale factor. The set of points (z, μ) defines a Hubble diagram, which records the history of cosmic expansion.

In the 1990s, two major collaborations used wide-field imaging and spectroscopy to discover Type Ia supernovae out to $z \sim 1$, coordinating ground-based and Hubble telescopes. Each event was monitored across bands, with light curves compared to templates and corrected for extinction and host properties.

Distant supernovae were fainter than expected (Riess et al., 1998) for a decelerating universe. The effect persisted across all filters, instruments, and analysis methods. Cross-calibration against low-redshift samples and repeated observations supported the result. The implication was that the universe had expanded more during the photons' travel than gravitational deceleration would allow, requiring an expansion rate that increases with time.

Cosmic evolution follows the scale factor $a(t)$—expansion occurs when $a(t)$ increases, acceleration when $\ddot{a}(t) > 0$. Matter and radiation cause deceleration (positive density, non-negative pressure), so acceleration requires a component with sufficiently negative pressure. Supernovae revealed this directly: $d_L(z)$ increases faster with redshift than expected for deceleration, following purely from empirical brightness-distance relations.

The cosmic microwave background (CMB) confirmed this: temperature fluctuations indicate flat geometry, but matter provides only one-third of the required density. The remainder must be a non-clustering, smooth component. Large-scale clustering observations, including baryon acoustic oscillations, support this picture (Eisenstein et al., 2005).

Type Ia supernovae first detected acceleration; CMB and structure surveys confirmed it. The converging evidence strongly favours an accelerating expansion history.

The possibility of a component that modifies expansion on large scales had been considered decades earlier. Einstein added Λ to his field equations in 1917 to construct a static universe, with repulsion balancing gravity. This static solution was unstable and soon discarded after Hubble's 1929 observations demonstrated that galaxies are receding from one another. The cosmological constant was removed from most models and became a historical footnote.

Its mathematical form, however, remained valid. Λ represents a constant energy density that does not dilute as the universe expands. In general relativity, it contributes with a sign and magnitude that counteract the deceleration caused by matter. In quantum field theory, the vacuum itself carries an energy density that enters the gravitational equations in the same way. Estimates of this vacuum energy, however, exceed observed values by many orders of magnitude, creating a mismatch that remains unresolved.

When supernovae revealed acceleration, the cosmological constant was reintroduced. Its inclusion adjusts the predicted luminosity-distance curve to align with observations.

'Dark energy' drives cosmic acceleration with behaviour distinct from matter or radiation—it doesn't cluster and maintains constant energy density as space expands. In its simplest form, it corresponds to Λ, though models allow time variation via scalar fields or modified gravity. Observationally, it's defined by producing acceleration: models excluding it fail to match supernovae, CMB patterns, and clustering measurements.

The inference of cosmic acceleration relies on systematic measurement of distance and redshift. Each supernova provides a data point (z, μ) on the Hubble diagram, building one of the most important experimental datasets for cosmological models.

Surveys use low-redshift controls and Cepheid calibration (standard candles) to eliminate systematics. Large surveys (SNLS, SDSS-II, DES) compiled thousands of supernovae to $z \sim 1.5$, confirming acceleration—the Hubble diagram curvature cannot be replicated by matter-only models. Independent measurements from baryon acoustic oscillations serve as standard rulers across redshift, further constraining the expansion history.

Combined data constrain acceleration's evolution. Future surveys—including the Vera Rubin LSST (first data 2025) and the Nancy Grace Roman Space Telescope (expected 2027)—aim to refine measurements of $d_L(z)$ to sub-percent precision, distinguishing between a true cosmological constant and time-varying models of dark energy.

The cosmological constant Λ provides the simplest model for dark energy. Within general relativity, cosmic expansion follows the Friedmann equations (Friedmann, 1922) governing the scale factor $a(t)$—comoving separations grow as $a(t)$ increases. It can be shown that acceleration depends on energy density and pressure: ordinary matter and radiation (non-negative pressure) cause deceleration, while components with sufficiently negative pressure drive acceleration. The cosmological constant corresponds to uniform energy density with negative pressure equal to its energy density, acting as a repulsive gravitational source that accelerates expansion.

Combining Λ with cold dark matter—non-relativistic, weakly interacting matter that clusters gravitationally—produces ΛCDM, the standard cosmological model. It successfully describes a wide range of observations, placing the universe's current energy budget at

68% dark energy, 27% dark matter, and 5% ordinary matter—consistent with supernovae, CMB, and structure observations.

High-redshift supernovae confirm a transition from an earlier decelerating phase, when matter dominated the energy budget, to the current accelerating phase driven by dark energy. This transition occurred roughly when the universe was half its current age, when the energy densities of matter and dark energy became comparable.

The physical nature of dark energy remains unknown. The observed value of the cosmological constant is extraordinarily small compared to theoretical expectations from quantum field theory. Quantum mechanics predicts that even empty space should possess a vacuum energy density, arising from zero-point fluctuations of all possible fields. Yet this predicted vacuum energy exceeds the observed dark energy density by approximately 120 orders of magnitude—a discrepancy so severe it has been called 'the worst theoretical prediction in the history of physics' (Hobson, Efstathiou and Lasenby, 2006, p. xi; see also the chapter epigraph)

This cosmological constant problem exposes an incomplete understanding of quantum gravity. General relativity couples to absolute energy density, not just differences. The vacuum energy cannot simply be subtracted away without affecting the geometry of space and time. The resolution may require entirely new physics beyond the Standard Model or a radical revision of our understanding of gravity.

How Much Can Be Measured

Scientific understanding is constrained by what can be observed. In cosmology, measurements rely on electromagnetic radiation, primarily in a few accessible bands of the spectrum. Telescopes capture only a fraction of the sky at any given time. Most stars, galaxies, and intergalactic matter are never directly observed, and yet, from this limited sample, consistent physical laws have been extracted.

This is a serendipitous outcome! Sampling 0.0...01% of the universe's contents has been sufficient to uncover mathematical relationships that describe its large-scale behaviour. The same formalism that governs an apple's descent also predicts planetary motion, galaxy trajectories, and the expansion of space.

The precision is remarkable: measuring a falling spoon on Earth can predict planetary orbits in distant galaxies with 99% accuracy. Adding the considerable effort of developing and testing general relativity gains the remaining 0.99% precision needed for GPS satellites and gravitational wave detectors. Yet despite this extraordinary success in describing gravity and motion across cosmic scales, 95% of the total energy content in the universe remains completely unaccounted for!

These components dominate the dynamics. The success of mathematical modelling in organising what is accessible illustrates both the remarkable reach of inference from incomplete information and the humbling limits of our cosmic perspective.

The Custodians of One

The monastery on Tethra-9 had no name, it needed none. Carved from igneous glass in a moon's crust whose atmosphere had boiled off in the Proxima Flare, it housed monks who had transcended insignia, watching null-entropy with infinite patience.

Commander Rafe Lin arrived in *Shibboleth*, a vacuum-energy cruiser extracting work from the ground state itself. Its hull was reinforced against gradients that could collapse neutron stars into geometric points—an abomination, even to its builders.

The monks observed his approach with studied indifference. Seven levels down through recursive dimensional barriers: a featureless black sphere hovering in defiance of every conservation law except the one that mattered.

The Monad. A magnetic monopole—not merely a particle, but the *arithmetic* upon which particle physics balanced its ledgers. Without it, electric charge would be as arbitrary as poetry, gauge symmetries unravelling like a bad proof.

"Extraction protocol, immediate," Rafe commanded, his voice carrying the authority of those who were never disobeyed.

The eldest monk spoke with the gentleness reserved for those about to discover the difference between locally and globally defined: "Commander, your vessel maintains structural integrity only because spacetime agrees to be continuous. Remove the boundary condition, and agreement becomes... negotiable."

The containment field collapsed. The sphere shifted through the space of *definitions* themselves—somewhere, the Riemann hypothesis lost a prime, fundamental constants performed a probabilistic quadrille.

Shibboleth discovered its existence had been predicated on several no-longer-valid assumptions. The ship lost coherence with the dignity of a well-posed problem becoming ill-defined. Rafe learned, too late, that identity is a function of existence.

The monks reactivated containment with the practised efficiency of librarians re-shelving infinity. The sphere returned to its fixed point. The universe exhaled. No rescue missions were authorised, for there was nothing left to rescue but mathematics.

And the cosmos, whole once more, continued its stately rotation, because $qg = \frac{n\hbar}{2}$ only holds if $\oint_{\partial V} \mathbf{B} \cdot d\mathbf{A} \neq 0$ *somewhere.*

The Cosmological Constant Problem: Quantum Vacuum Energy vs Observations

Quantum Field Theory Prediction

In quantum field theory, even empty space possesses energy due to zero-point fluctuations. For a free massless scalar field $\phi(x, t)$ with Hamiltonian,

$$H = (1/2) \int \left[\pi^2(x) + |\nabla\phi(x)|^2\right] d^3x,$$

Upon canonical quantization, the vacuum expectation value becomes:

$$\langle 0|H|0\rangle = (1/2) \int d^3k/(2\pi)^3 \cdot \hbar\omega_k,$$

where $\omega_k = c|\mathbf{k}|$. This yields:

$$\langle 0|H|0\rangle \propto \int_0^\infty k^3 dk,$$

which diverges as k^4 at high momentum, requiring a cutoff.

Planck-Scale Cutoff

Assuming quantum field theory remains valid up to the Planck energy scale:

$$k_{\max} = M_{\text{Planck}}c/\hbar = \sqrt{c^3/(\hbar G)}.$$

The vacuum energy density becomes:

$$\begin{aligned}\rho_{\text{vac}}^{\text{theory}} &= \hbar c/(16\pi^2) \cdot k_{\max}^4 \\ &= \hbar c/(16\pi^2) \cdot (c^6/(\hbar^2 G^2)) \\ &= c^7/(16\pi^2 \hbar G^2) \sim 10^{76}\ \text{GeV}^4.\end{aligned}$$

Observational Constraints

Cosmological observations from supernovae, CMB, and large-scale structure constrain the dark energy density to:

$$\rho_{\text{DE}}^{\text{obs}} = \rho_{\text{crit}}\Omega_\Lambda \approx (3H_0^2/(8\pi G)) \times 0.68,$$

where $H_0 \approx 70$ km/s/Mpc. Converting to natural units:

$$\rho_{\text{DE}}^{\text{obs}} \approx 10^{-47}\ \text{GeV}^4.$$

The Discrepancy

The ratio of theoretical prediction to observational constraint:

$$\rho_{\text{vac}}^{\text{theory}}/\rho_{\text{DE}}^{\text{obs}} \sim 10^{76}/10^{-47} = 10^{123}.$$

This represents the largest mismatch between theory and observation in physics history. (Depending on cutoff choices and degrees of freedom, values from 10^{118} to 10^{123} are quoted in the literature.)

The Fine-Tuning Problem

Unlike other physics areas where only energy differences matter, general relativity couples directly to absolute energy density through Einstein's field equations:

$$G_{\mu\nu} + \Lambda g_{\mu\nu} = (8\pi G/c^4)T_{\mu\nu},$$

where $\Lambda = (8\pi G/c^2)\rho_{\text{vac}}$.

If quantum vacuum energy contributed at the predicted level, it would drive exponential expansion so rapid that structure formation would be impossible. The observed value requires either:

1. Extraordinary cancellation reducing vacuum energy by 120 orders of magnitude.
2. New physics beyond the Standard Model altering vacuum structure.
3. Modification of general relativity at cosmological scales.

No proposed solution has gained broad acceptance, making this one of the most pressing problems in theoretical physics.

References:

Weinberg, S. (1989). The cosmological constant problem. *Rev. Mod. Phys.*, 61, 1.

Carroll, S. M. (2001). The cosmological constant. *Living Rev. Relativ.*, 4, 1.

Padmanabhan, T. (2003). Cosmological constant—the weight of the vacuum. *Phys. Rep.*, 380, 235.

An Axiom of Your Choice

Hilbert's Hotel with Infinite Buses: An infinite number of buses, each carrying infinitely many passengers, arrives at a hotel with countably infinite rooms. To accommodate everyone, each passenger n on bus m is assigned to room $(p_n)^m$, where p_n is the nth prime number. No two passengers ever collide, since distinct prime powers are never equal: for $p \neq q$, there are no integers k, ℓ such that $p^k = q^\ell$.
Banach-Tarski as a Card Game: Imagine a deck with cards A, B, A^{-1}, B^{-1}, where you can make any pile of cards, under one rule: A and A^{-1} cancel, B and B^{-1} cancel (the cards disappear if they are next to one another). Each card represents a rigid rotation: A, A^{-1} rotate around one axis; B, B^{-1} around another. In the diagram, blue and red are inverses, as are green and yellow.
Four players each take all reduced sequences (no cancelling cards neighbour each other) beginning with a specific card, forming four disjoint point sets: $S_A, S_{A^{-1}}, S_B, S_{B^{-1}}$.
Now prepend an A to every sequence in $S_{A^{-1}}$, practically rotating the whole set, forming $A \cdot S_{A^{-1}}$. This transforms sequences that began with A^{-1} into sequences that no longer do. For instance, $A^{-1}BA \mapsto BA$, and $A^{-1} \mapsto$ identity. The union $S_A \cup A \cdot S_{A^{-1}}$ now covers all sequences—forming one full copy of the sphere. Similarly, $S_B \cup B \cdot S_{B^{-1}}$ forms a second.

An Axiom of Your Choice

The Banach–Tarski paradox shows that a solid sphere can be partitioned into finitely many disjoint pieces and, using only rigid motions, reassembled into two spheres identical to the original. This construction depends on the Axiom of Choice and the existence of non-measurable sets, whose behaviour diverges from intuitions about volume. While not physically realisable, the result illustrates how certain set-theoretic assumptions allow decompositions that defy standard notions of size and conservation.

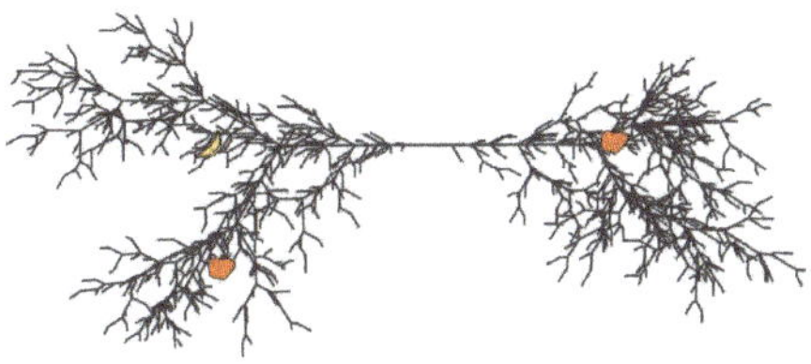

Axiom of Choice ∘ Non-Measurable Sets ∘ Sphere Paradox ∘ Free Group F_2 ∘ Orbit Decomposition ∘ Rigid Motion Invariance ∘ Countable Additivity ∘ von Neumann Naturals ∘ Hilbert's Hotel ∘ Volume Non-Conservation ∘ Mathematical Constructivism

„Das ist nicht Mathematik, das ist Theologie!“
(“This is not mathematics, this is theology!”)
— Paul Gordan (1888), protesting Hilbert's use of non-constructive existence proofs

„... auch die Theologie hat ihre Verdienste.“
(“... even theology has its merits.”)
— David Hilbert's reply, defending non-constructive reasoning

An Axiom of Your Choice

In the second half of the nineteenth century, questions about the foundations of analysis led mathematicians to examine the basic assumptions underlying number, function, and space. Dedekind formalised the real numbers via cuts, while Weierstrass removed appeals to geometric intuition from calculus. At the same time, mathematicians encountered pathologies—such as functions continuous everywhere but differentiable nowhere, or nowhere-dense sets of positive measure—that challenged classical notions of size and shape. These developments revealed that intuitive notions of length, area, and convergence required formal clarification, especially in the context of infinite processes.

In the late nineteenth century, Georg Cantor changed mathematics with his work on infinite sets, laying the groundwork for new perspectives on measure and cardinalities. In 1905, Giuseppe Vitali introduced the first example of a non-measurable set, suggesting that some subsets of $\mathbb{R}^n$ cannot be assigned an intuitive notion of size. Building on this foundation, Felix Hausdorff presented a paradox in 1914, showing that a sphere could be decomposed in a way that hinted at even more surprising outcomes.

A decade later, in 1924, Stefan Banach and Alfred Tarski formulated it further in a concrete result, known as the Banach–Tarski paradox. They demonstrated that a solid ball in three-dimensional space could be split into a finite number of pieces and then reassembled, through rigid motions, into two full copies of the original in a process that requires the Axiom of Choice, introduced two decades prior (Zermelo, 1904) and still a subject of research. Though it does not apply to physical objects, the Banach–Tarski paradox remains a powerful example of how set-theoretic assumptions can lead to unexpected and extraordinary results in geometry.

A mathematical system begins with a specification of its elements: which objects exist, which operations are defined on them, and which relations must hold. These specifications are encoded in the system's axioms. An axiom is a formal assumption that serves as the foundation for the system. Within that system, no statement can be derived unless it is implied by the axioms in conjunction with the rules of logical inference.

Once a set of axioms is fixed, all further reasoning, including definitions, proofs, and theorems, must proceed within the structure they determine. The consistency and character of the system depend entirely on these initial choices.

Different axiomatic systems describe different mathematical worlds. In one system, every set may have a well-ordering (every non-empty subset has a least element with respect to the order). In another, it may not. In one geometry, parallel lines exist; in another, they do not.

Even familiar objects depend on axiomatic choices. One way to see this is through the von Neumann construction of the natural numbers (von Neumann, 1923) inside set theory. Think of sets as 'bags' that can hold distinct objects: duplicates vanish, so $\{\text{dog}, \text{dog}, \text{cat}\} = \{\text{dog}, \text{cat}\}$. The union of two bags (denoted by $A \cup B$) merges their contents, ignoring duplicates. With this picture, the naturals are nested bags, because all you have in this

 https://doi.org/10.11647/OBP.0526.03

universe are empty bags. Mark an empty bag as $\{\}$.

$$0 := \{\}, \quad 1 := \{0\} = \{\{\}\}, \quad 2 := \{0, 1\} = \{\{\}, \{\{\}\}\}, \quad 3 := \{0, 1, 2\}, \ \dots .$$

Notice the pattern. Zero is an empty bag. One is a bag containing the empty bag. Two is a bag containing both zero and one. Three contains zero, one, and two. Each number n is the bag containing all smaller numbers. This means $m \in n$ (the set m is an element of the set n) exactly when $m < n$. The natural numbers are the smallest set that satisfies the axioms.

Addition is defined recursively: $m + 0 = m$ and $m + S(n) = S(m + n)$, where $S(n) = n \cup \{n\}$ is the successor. Since n is the set of all smaller numbers, $S(n)$ contains all those smaller numbers plus the set n itself as a new element. This makes even the simplest arithmetic fact a theorem rather than a definition. For instance, one proves that $1 + 1 = 2$. Since $1 + 1 = 1 + S(0) = S(1 + 0) = S(1) = 2$, the familiar statement is established within the axioms.

This illustrates how set theory provides foundations for all mathematics. Relations such as $<$ become sets of ordered pairs, and axioms govern the construction of increasingly abstract objects. Some axioms describe intuitive operations such as forming power sets, while others assert the existence of entities that cannot be explicitly constructed, such as inaccessible cardinals or non-measurable sets.

The Axiom of Choice is one such axiom (Zermelo, 1904). It asserts that for any collection of non-empty sets, there exists a function that selects exactly one element from each set. In finite cases, such selections can be written down explicitly or proved to exist using elementary methods. In infinite settings, this is not always possible. For example, consider an infinite collection of drawers, each containing a left and a right shoe. A rule such as 'choose the right shoe' provides a well-defined selection and does not require the Axiom of Choice. But if each drawer contains a pair of identical socks with no distinguishing features, then no explicit rule can be formulated. The existence of a function that selects one sock from each drawer in this case depends on accepting the Axiom of Choice.

Let us now move down the ladder of abstraction, from axioms to the notion of size. Understanding how we measure things—length, area, volume—will be important for seeing why the Banach-Tarski paradox is so surprising.

How do we assign 'size' to things? In everyday life, it's intuitive—a room's area is its length times width, and if you combine two non-overlapping rooms, their total area is simply the sum of their individual areas.

Measure theory formalises this intuition. A *measure* μ is a function that assigns a non-negative real number to certain subsets of a space. The central requirement is **additivity**: if two disjoint measurable sets A and B are combined, then their measure is the sum of the measures of the two sets: $\mu(A \cup B) = \mu(A) + \mu(B)$ (provided that $A \cap B = \emptyset$, i.e., they share no elements). This principle extends to infinite collections: if a set is decomposed into countably infinite disjoint measurable subsets $\{A_i\}_{i=1}^{\infty}$, then $\mu\left(\bigcup_{i=1}^{\infty} A_i\right) = \sum_{i=1}^{\infty} \mu(A_i)$. This is called **countable additivity**.

Not every subset can be assigned a measure. Some sets resist consistent size assignment—they are *non-measurable.* The existence of non-measurable sets depends on accepting axioms such as the Axiom of Choice.

In familiar settings, the standard measure corresponds to area in the 2D space $\mathbb{R}^2$ or volume in the 3D space $\mathbb{R}^3$. But even in these cases, not all subsets are measurable.

When measuring spatial objects, **invariance under rigid motions** is required. Translating or rotating a measurable set leaves its measure unchanged. Rigid motions preserve distances and angles; they don't stretch, tear, or compress. This reflects the expectation that volume is an intrinsic property—not dependent on where the object is or how it's oriented.

Non-measurable sets lead to paradoxical results, including the Banach–Tarski paradox (Banach & Tarski, 1924). The paradox states that a 3-dimensional ball can be partitioned into five disjoint subsets, which can then be recombined—using only rigid motions—into two balls congruent to the original. This result demonstrates the failure of volume to be preserved when applied to non-measurable sets. In physical systems such as a stone or a fluid body, each component contributes additively to the whole. The Banach–Tarski construction defines a setting where this principle fails and volume is not additive.

Consider Hilbert's Hotel—an infinite hotel (Hilbert, 1926) with rooms numbered $1, 2, 3, \ldots$, all occupied. To accommodate one new guest, shift everyone from room n to room $n+1$, freeing room 1. To accommodate infinitely many new guests, move everyone from room n to room $2n$, freeing all odd-numbered rooms.

The Banach–Tarski construction can be presented as a combinatorial game. Consider a deck with four types of cards: A, B, A^{-1}, and B^{-1}. These cards can be arranged in sequences, with one rule: if A and A^{-1} appear next to each other, they annihilate. The same applies to B and B^{-1}. A sequence like $ABA^{-1}B$ is stable, but a sequence like $B^{-1}BA$ immediately reduces to just A. The set of all irreducible sequences forms the free group F_2—all sequences of the cards A and B, with the rule that A and A^{-1} annihilate, and B and B^{-1} annihilate.

Now here's where the cards become geometric transformations. Each card corresponds to a rotation of a sphere: A rotates by an irrational angle (e.g., $\sqrt{2}$ degrees) around the Z-axis, A^{-1} rotates back by the same angle. Similarly, B rotates by an irrational angle around the X-axis, B^{-1} rotates back.

Why irrational angles? Because they ensure that no finite sequence of these rotations will ever bring a point back to exactly where it started (unless all the cards cancel out).

When we fix two rotations A and B, we can build arbitrary sequences of them and their inverses to move points around the sphere. Pick a point p on the sphere. Apply every possible sequence of A, A^{-1}, B, and B^{-1} to p. The resulting collection of points is called the orbit of p under this group of rotations.

Two points lie in the same orbit if one can be turned into the other by some sequence of these rotations. Orbits are disjoint. A point belongs to exactly one orbit, because sequences of rotations either connect two points or they don't. The union of all orbits is the whole sphere, so the orbits form a partition.

Because A and B are chosen carefully (rotations about different axes by irrational angles), the group they generate is free and each orbit is infinite, spreading densely across the sphere.

At this point the Axiom of Choice enters. From each orbit, pick a single representative point. Call the set of these chosen representatives R. Every other point on the sphere can be written uniquely as $g \cdot r$, where g is some sequence of rotations and $r \in R$ is the representative of its orbit. In other words, the entire sphere is recovered by 'shuffling' the representatives through all possible rotation sequences.

Now comes the partitioning that makes the paradox work. Four players will divide the entire sphere among themselves by each taking the representatives R and applying only certain rotation sequences:

- *Player 1*: Gets all points reached by sequences starting with card A (avoiding immediate A^{-1} cancellation). This creates the point set S_A.
- *Player 2*: Gets all points from sequences starting with A^{-1} (avoiding A), creating $S_{A^{-1}}$.
- *Player 3*: Gets all points from sequences starting with B (avoiding B^{-1}), creating S_B.
- *Player 4*: Gets all points from sequences starting with B^{-1} (avoiding B), creating $S_{B^{-1}}$.

Every possible rotation sequence must start with one of these four cards (A, A^{-1}, B, or B^{-1}). This means the four sets are disjoint—no point belongs to two different players—and together they cover the entire sphere (except for fixed points on rotation axes, which are handled separately).

First Sphere Reconstruction:

Player 2 gives their entire collection $S_{A^{-1}}$ to Player 1. Player 1 then rotates every point in $S_{A^{-1}}$ by applying rotation A before each sequence, creating the new set $A \cdot S_{A^{-1}}$.

When you rotate a point reached by sequence $A^{-1}BA$ using rotation A, you get $A \cdot (A^{-1}BA) = (AA^{-1})BA = BA$—the A and A^{-1} cancel. Similarly, rotating the point from sequence A^{-1} gives $A \cdot A^{-1} =$ identity (the original representative point).

Player 1 now has S_A (all points reached by sequences starting with A) and $A \cdot S_{A^{-1}}$ (all points reached by sequences *not* starting with A). Together, these form a complete sphere!

Meanwhile, Players 3 and 4 perform the same trick. Player 4 gives $S_{B^{-1}}$ to Player 3, who rotates it by B to get $B \cdot S_{B^{-1}}$. Combining S_B and $B \cdot S_{B^{-1}}$ gives another complete sphere.

We started with four disjoint pieces that made up one sphere. By rotating two of those pieces, we reassembled them into two identical spheres. No points were added or removed—only rearranged. That is the heart of the paradox.

This paradoxical outcome depends on selecting representative points from each orbit, which requires the Axiom of Choice. The resulting sets are non-measurable—they cannot be assigned consistent volumes that preserve both countable additivity and invariance under rotation. The non-measurability arises from the paradoxical structure of the free group's action on the sphere.

This construction illustrates the core ideas of the Banach–Tarski paradox. The full theorem decomposes a solid three-dimensional ball into five pieces, which can then be rotated into two balls identical to the original. The essential ingredients—free groups, orbit decompositions, the Axiom of Choice, and non-measurable sets—remain the same.

The Axiom of Choice occupies a unique position in modern mathematics. Gödel proved that if ZF (the Zermelo–Fraenkel axioms) (Gödel, 1938) is consistent, then so is ZFC (ZF plus Choice). Cohen (1963) later proved that ZF with the negation of Choice is also consistent, establishing that Choice is independent of ZF—neither provable nor disprovable from the other axioms. This independence was revolutionary. Classical theorems across mathematics depend on it—"every vector space has a basis" and "the product of compact spaces is compact" are each equivalent to accepting Choice.

Before dismissing Choice to avoid the paradox, consider that its absence permits equally frightening results. When you partition a set into disjoint subsets, selecting one representative from each part typically creates an injection from the index set into the original—the number of parts cannot exceed the number of elements. Without Choice, this intuition fails. In ZF alone, it is consistent that a partition $X = \bigsqcup_{i \in I} B_i$ exists with no injection $I \hookrightarrow X$, so the number of parts $|I|$ can exceed $|X|$. A set can be split into more parts than it has elements. Rejecting Choice trades one counterintuitive result for another.

Mathematics versus Reality

This chapter forces a distinction between mathematical and physical reasoning. The Banach–Tarski construction is not a paradox in the sense of contradiction or physical impossibility, but it is a good example of the consequences of adopting the Axiom of Choice, showing that intuitive notions such as volume are not preserved across all set decompositions. The result is clean and formally sound, yet incompatible with empirical modelling. That gap—between internally consistent mathematics and physically grounded expectation—illustrates the conceptual boundaries explored throughout this book. As with other chapters that emphasise when simplifications fail (relativity in gold, curvature in gravity, topology in voting), this example shows that what appears absurd may instead be a well-posed feature of a chosen formal system.

For the following trick we will require the axiom of choice

Mathematical Foundations of the Banach-Tarski Paradox

Overview. The Banach-Tarski paradox states that a solid ball in $\mathbb{R}^3$ can be partitioned into finitely many disjoint subsets, and, using only rigid motions, reassembled into two identical copies of the original. The result depends on the Axiom of Choice, the existence of a free subgroup of SO(3), and the absence of a finitely additive rotation-invariant measure on all subsets of the sphere.

Construction Outline

- **Group-Theoretic Basis.** The free group $F_2 = \langle a, b \rangle$ is equidecomposable with two disjoint copies of itself under left multiplication. This violates the property of amenability, which forbids such duplications.
- **Geometric Embedding.** Rotations S and T in SO(3), each by angle $\theta = \arccos(1/3)$ around orthogonal axes, generate a subgroup isomorphic to F_2. No nontrivial reduced word in these rotations fixes any point off the axes.
- **Removing Fixed Points.** The set D of all points fixed by nontrivial group elements is countable. One can remove D from the sphere by applying a suitable rotation to each of its images.
- **Orbit Decomposition.** The action of F_2 on $S^2 \setminus D$ partitions it into orbits. The Axiom of Choice selects one representative per orbit to form a set M. Then $S^2 \setminus D = \bigsqcup_{g \in F_2} gM$.
- **Duplication via Equidecomposition.** Since $F_2 = A \sqcup B$ with $A \cong F_2$ and $B \cong F_2$, define bijections $\phi_i : F_i \to F_2$, and extend them to $S^2 \setminus D$ via $x = gm \mapsto \phi_i(g)m$. This yields two disjoint subsets each equidecomposable with the whole.
- **Lifting to the Ball.** The unit ball is viewed as concentric spherical shells. The paradoxical decomposition is applied to each shell simultaneously. The centre is treated separately.

Amenability and Non-Amenability. A group is amenable if there exists a finitely additive, invariant probability measure on all its subsets. Free groups on two or more generators are non-amenable: they admit no such measure. This failure is what allows equidecomposition of a set with two disjoint isometric copies of itself. The Banach-Tarski paradox is a geometric manifestation of non-amenability.

Cayley Graph Interpretation. The Cayley graph of F_2 is a 4-regular tree. Each branch corresponds to a left coset, and the entire group acts by translation. It is somewhat easy to see that a careful merger of two opposite branches of the tree will result in a tree with the same structure as the original full tree, in a fractal way.

Rotations Generating the Free Group. Define $R = \frac{1}{3}\begin{bmatrix} 1 & -2\sqrt{2} \\ 2\sqrt{2} & 1 \end{bmatrix}$. Then:

$$S = \begin{bmatrix} R & 0 \\ 0 & 1 \end{bmatrix}, \quad T = \begin{bmatrix} 1 & 0 \\ 0 & R \end{bmatrix}.$$

Any reduced word W in S, T, and their inverses maps $(1,0,0)$ to a vector of the form $\left(\frac{a}{3^n}, \frac{\sqrt{b}}{3^n}, \frac{c}{3^n}\right)$, with $b > 0$, confirming that the group acts freely on the orbit of $(1,0,0)$.

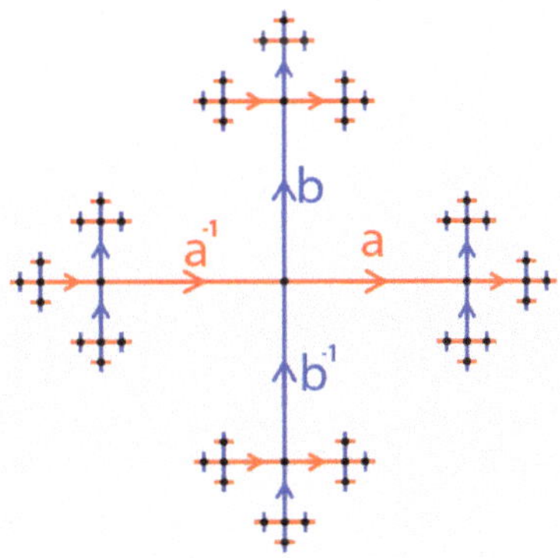

Figure: Cayley graph of F_2, visualised as a tree with no cycles and uniform branching.

References:

Banach, S. and Tarski, A. (1924). *Fund. Math.* **6**
Wu, A. (2008). *The Banach-Tarski Paradox.*

Think Outside the Wire

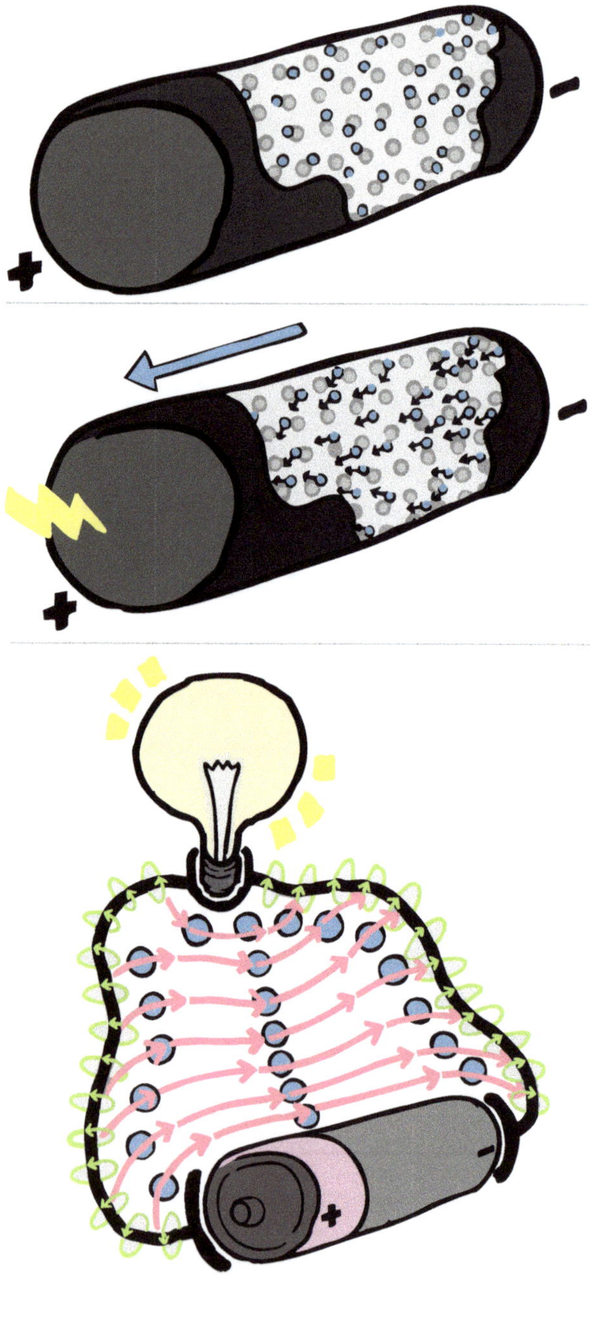

Top (Thermal Motion): Thermal motion in equilibrium. Even with no applied voltage, conduction electrons within the wire exhibit rapid random motion (thermal velocities on the order of 10^5 to 10^6 m/s), but with no net directional flow—resulting in zero macroscopic current.
Middle (Net Drift): Net drift under an applied electric field. Applying a voltage creates an electric field along the wire, introducing a slight statistical bias in electron velocities. This produces a slow net drift (typically fractions of a millimetre per second), superimposed on the much faster thermal motion. Despite the minuscule drift speed, the circuit responds almost instantly to changes in voltage, as the electromagnetic field propagates at near light speed.
Bottom (Energy Flow): Macroscopic current and surrounding fields. The collective electron drift constitutes a measurable current, which, according to Ampère's Law, is accompanied by a magnetic field encircling the wire (as shown by the green loops and the right-hand rule). Together, the electric field driving the current and the magnetic field generated by it produce a nonzero Poynting vector $\vec{S} = \vec{E} \times \vec{H}$ oriented along the wire. This indicates that energy flow occurs through the electromagnetic field in the space surrounding the conductor—not within the conductor itself. The drift of electrons ensures charge continuity but plays no role in determining the speed of energy delivery.

Think Outside the Wire

Electrical energy travels primarily through electromagnetic fields surrounding conductors, not through the movement of electrons in wires. While electrons drift at millimetres per second, energy transfer occurs near light speed through the Poynting vector ($S = E \times H$), which describes energy flow perpendicular to both electric and magnetic fields. This field-based transmission explains why circuits respond almost instantly despite slow electron movement, as opposed to the common misconception that electricity flows like water through pipes.

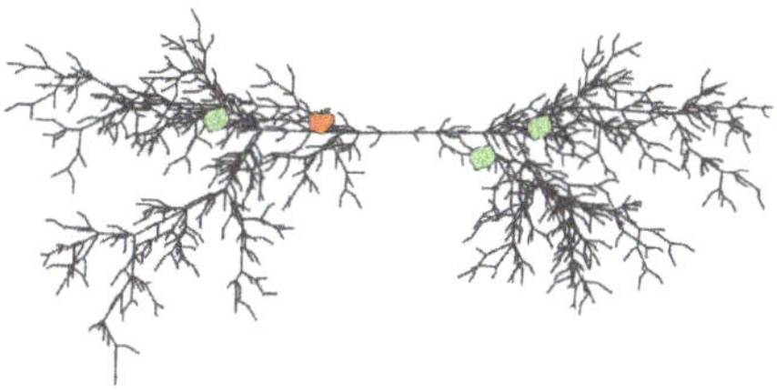

Poynting Vector **S** ◦ Energy in Fields ◦ Maxwell's Displacement Current ◦ Drift Velocity Slowness ◦ Drude Model Failures ◦ Debye Phonons ◦ Sommerfeld Quantum Gas ◦ Boundary Conditions ◦ Wire as Field Guide ◦ Veritasium Misconception ◦ Field vs Charge Flow

"Electricity is actually made up of extremely tiny particles called electrons, that you cannot see with the naked eye unless you have been drinking."

— Dave Barry, 1998

Think Outside the Wire

Early investigations linking electricity and magnetism began with Hans Christian Ørsted's 1820 observation that a current-carrying wire deflected a nearby compass needle, demonstrating that electric currents generate magnetic effects. Shortly thereafter, André-Marie Ampère quantified these interactions, paving the way for a unified framework. Michael Faraday's idea of lines of force emphasised that fields permeate space and mediate electrical phenomena.

In the 1860s, James Clerk Maxwell brought together these concepts, formulating a concise set of equations that governs how changing electric and magnetic fields propagate as electromagnetic waves. This discovery challenged earlier assumptions that electrical energy was confined to wires alone. John Henry Poynting then introduced the Poynting vector in 1884, clarifying how electromagnetic energy flows through the space surrounding conductors.

At the same time, Oliver Heaviside simplified Maxwell's equations into the modern vector calculus form, making them more accessible for engineers and physicists. His insights led to a better understanding of power transmission, highlighting that energy is not carried by the motion of electrons in a wire but by the surrounding electromagnetic fields. This change in perspective would later prove critical in the development of radio transmission, telecommunication systems, and waveguide theory.

Nonetheless, the older 'current as fluid in a pipe' analogy persisted in basic electrical education well into the twentieth century. Only with the advent of high-frequency engineering and transmission line theory did the role of electromagnetic fields become widely acknowledged in practical applications. Today, the principles laid down by Ørsted, Ampère, Faraday, Maxwell, Poynting, and Heaviside form the foundation of modern electromagnetism, from power grids to fibre-optic communication.

Electric energy is often described as flowing through wires, like water through a pipe. This analogy is common but misleading. It treats energy as a substance carried by electrons moving from source to device. But electrons inside a conductor move slowly. Their average drift velocity under a typical voltage is only a few millimetres per second. What moves quickly is not the charge, but the disturbance in the electromagnetic field that propagates through space. A better analogy, if still inaccurate, is a wave travelling across the surface of a pond: the water does not move forward, but the wave does. In the same way, electric energy is transmitted by the wave-like interaction of fields, not by the displacement of material particles.

This effect is clear in static electricity: when materials are rubbed together, electrons transfer but no current flows, yet a strong electric field appears in surrounding space that can exert forces and store energy. Similarly, when a switch closes in a circuit, devices respond almost instantly because electromagnetic fields establish throughout the geometry simultaneously. A voltage sets up an electric field $\mathbf{E}$ along the wire, current produces a magnetic field $\mathbf{B}$ encircling it, and these fields extend beyond the conductor's surface, occupying surrounding space and determining energy's path.

This behaviour is formalised in Maxwell's equations (Maxwell, 1865). Before Maxwell, electric and magnetic phenomena were treated separately. Electric fields originated from static charges, and magnetic fields from moving charges, that is, from currents. These laws worked well for static situations but failed in time-varying regimes. Maxwell identified a critical gap in Ampère's law. According to its original form, a magnetic field was produced only by conduction current. But this led to contradictions in cases where the electric field changed in time but no actual current flowed, such as inside a capacitor during charging. Maxwell resolved the inconsistency by introducing the concept of displacement current, the idea that a changing electric field $\partial\mathbf{E}/\partial t$ acts like a current, generating a magnetic field even in the absence of moving charge.

This single correction was momentous. The displacement current term transformed a collection of separate electromagnetic laws into a mathematically consistent system of equations. This completed system contained a wave equation with a definite propagation speed: $c = 1/\sqrt{\varepsilon_0\mu_0}$. When Maxwell calculated this speed using known electromagnetic constants, it matched the measured speed of light. This indicated that light itself was an electromagnetic phenomenon. However, this consistency exposed an incompatibility: Maxwell's equations predicted that light travels at the same speed in all reference frames, while Newton's mechanics (and common sense) required speeds to transform according to Galilean relativity—that if you move away from a light source at half the speed of light, you should see the light move away from you at half the speed. The consistency that Maxwell achieved revealed an incompatibility at the heart of classical physics—one that would not be resolved until Einstein's special relativity.

The completed equations describe how electric and magnetic fields sustain each other: a changing electric field $\mathbf{E}$ generates a magnetic field $\mathbf{B}$, and a changing $\mathbf{B}$ regenerates $\mathbf{E}$. This mutual coupling produces wave propagation even in the absence of charge or current, with perpendicular oscillating fields that constitute the electromagnetic energy. To compute energy flow, the magnetic field $\mathbf{B}$ must be normalised by the vacuum permeability: $\mathbf{H} = \mathbf{B}/\mu_0$. This ensures that both $\mathbf{E}$ and $\mathbf{H}$ are expressed in compatible units when evaluating energy flux. The Poynting vector, $\mathbf{S} = \mathbf{E} \times \mathbf{H}$, describes the instantaneous direction and intensity of electromagnetic energy transport. It has units of power per unit area ($\mathrm{W/m^2}$) and is always perpendicular to both fields.

In high-energy conventions one often sets $c = 1$ and $\varepsilon_0 = \mu_0 = 1$, in which case $\mathbf{B}$ and $\mathbf{H}$ share units and coincide in vacuum.

Near a long straight wire, $\mathbf{B}$ encircles the wire azimuthally, while surface charges establish an $\mathbf{E}$ field with a component along the wire; in coaxial or two-wire transmission lines, $\mathbf{E}$ is predominantly transverse (radial). In all cases, the Poynting vector $\mathbf{S} = \mathbf{E} \times \mathbf{H}$ points along the direction of power flow in the surrounding space, not within the conductor. The conductors establish boundary conditions that constrain and guide the fields; the energy transfer occurs in the fields occupying the space around the conductors.

Now for the role of electrons. Before electrons were understood as discrete particles, early models of electricity imagined it as a continuous substance flowing through wires, like water through a pipe. In this mechanical analogy, the conductor acted as a passive conduit, and the electric current was treated as an invisible, uniform fluid. The observable effects of voltage and current were attributed to the movement of this substance through the wire. While this model offered some intuition for the flow of charge, it could not account for material differences or thermal effects, lacking any microscopic description of matter that would allow calculation of materials' behaviour under applied voltage.

In 1900, Paul Drude introduced a kinetic theory of conduction that treated electrons as classical particles moving freely between instantaneous collisions with heavy, stationary ions in a metallic lattice. Under an applied electric field, the electrons acquired a small net drift velocity superimposed on their thermal motion, giving rise to a steady current. This model successfully reproduced Ohm's law of resistance and introduced the concept of mean free path. However, it relied on Maxwell–Boltzmann statistics and treated electrons as distinguishable particles in thermal equilibrium. These assumptions, though reasonable for a dilute gas, led to contradictions when applied to dense electron systems in metals.

Multiple contradictions arose. Classical theory predicted each electron would contribute $\frac{3}{2}k_B$ (where k_B is the Boltzmann constant), but calorimetry showed values over a hundred times smaller, indicating most electrons couldn't gain thermal energy. According to the Wiedemann–Franz law, the ratio $L = \kappa/(\sigma T)$ of thermal conductivity κ to electrical conductivity σ at temperature T should be constant ($\approx 2 \times 10^{-8}$ W·Ω/K^2), but experiments showed temperature variation, implying different transport mechanisms. Furthermore, Drude's model failed to explain electrostatic shielding—the blocking of external electric fields inside conductive enclosures—because it treated electrons as isolated particles.

Further contradictions came from temperature-dependent resistivity. Drude's model predicted that resistivity should increase linearly with temperature due to more frequent electron–ion collisions. In practice, resistivity curves showed deviations from linearity, especially at low temperatures where resistance often plateaued or decreased. High-purity metals with large crystalline domains exhibited behaviour that depended sensitively on defect density, lattice structure, and impurity concentration. These features played no role in Drude's theory, which treated the lattice as a uniform background. The observed dependence on details suggested that new scattering mechanisms and quantum restrictions were at play.

In 1912, Peter Debye addressed the failures of the Drude model by incorporating lattice dynamics into the theory of conduction. Instead of treating ions as fixed scattering centres, he modelled them as thermally vibrating masses whose motion becomes increasingly pronounced with temperature. These vibrations were treated as quantized normal modes—modernly termed phonons—which represent collective oscillations of the atomic lattice. Unlike localised particle collisions, phonons describe delocalized, wave-like excitations that span the crystal and interact coherently with conduction electrons. As the temperature rises, the number and amplitude of accessible phonon modes increase, leading to more frequent electron–phonon collisions and higher resistivity. This introduced a temperature-dependent scattering mechanism that aligned more closely with observed trends in metallic resistance.

The phonon model explained resistivity behaviour: linear growth at high temperatures due to increased phonon population, and saturation at low temperatures where phonon modes freeze out. High-purity metals showed stronger temperature effects since electron–phonon scattering dominated over impurity scattering.

Debye also introduced electrostatic screening. Conduction electrons collectively redistribute to cancel external fields within a characteristic Debye length (typically nanometres). This explained perfect shielding in conductors and redefined them as collectively responsive media. In metals, electrostatic screening is governed by the dense electron gas and characterised by the Thomas–Fermi screening length (Thomas, 1927; Fermi, 1927) (typically on the order of an ångström). The Debye length is appropriate for dilute plasmas and electrolytes; in conductors, free electrons rearrange to cancel external fields within this much shorter scale, producing near-perfect shielding.

While Debye focused on the quantized behaviour of the lattice, in 1928, Arnold Sommerfeld turned to the electron gas itself. Debye's model resolved key thermal anomalies by treating lattice vibrations as phonons and introducing collective screening, but it still relied on classical statistics for the electrons. Sommerfeld's contribution was to replace the classical electron gas with a quantum one, governed by the Pauli exclusion principle (Pauli, 1925). In this revised model, electrons occupy discrete quantum states and fill all available levels up to the Fermi energy (the highest occupied level at absolute zero). Only those near this surface can change state when a weak external field is applied. This restriction explains why most electrons do not contribute to conduction or heat capacity, despite their large individual velocities. It also accounts for the small but nonzero electronic heat capacity and the weak temperature dependence of conductivity in pure metals. Sommerfeld's approach completed the redefinition of conduction: not as thermal drift through a static lattice, but as the quantum response of a filled electron sea to external perturbation.

The Sommerfeld model resolved the longstanding discrepancy in the electronic heat capacity. Classical theories assumed that all conduction electrons share thermal energy, leading to a heat capacity proportional to temperature and electron count. Measurements showed a smaller contribution, growing linearly with temperature but with a suppressed coefficient. Sommerfeld explained this through quantum mechanics: only electrons within a narrow energy window around the Fermi level can absorb energy and transition to higher states. The rest are blocked by the exclusion principle. This result matched calorimetric data and clarified why the electronic contribution vanishes at low temperature, while the lattice contribution remains governed by phonon dynamics.

The same explanation also accounts for the weak temperature dependence of conductivity. Because only a small fraction of electrons near the Fermi surface can shift momentum under an applied field, the number of active carriers remains nearly constant as temperature changes. Scattering rates still vary—especially due to phonons—but the carrier population does not. This also improved the theoretical form of the Wiedemann–Franz law. By combining quantum statistics for the electron gas with Debye's treatment of the lattice, the temperature scaling of both thermal and electrical conductivity was derived with the correct proportionality constant. Sommerfeld's model provided a foundation for the thermal and electrical behaviour of metals across temperature regimes.

Although each electron near the Fermi surface contributes to conduction, the resulting motion is slow. The net velocity acquired from an applied electric field is called the drift velocity. It is given by $v_d = I/(nAe)$, where I is the current, n is the charge carrier density, A is the cross-sectional area of the conductor, and e is the elementary charge. For typical metals carrying macroscopic currents, this drift speed is on the order of a fraction of a millimetre per second. Despite the vast number of electrons involved, their collective motion results in a current that builds slowly and transports charge gradually along the wire. The slowness of this process is an outcome of the Fermi-level restriction and the small imbalance imposed by weak electric fields.

At the macroscopic level, conductors don't carry energy—they impose boundary conditions on electromagnetic fields. Free charge ensures the electric field **E** vanishes inside conductors, shaping fields outside and fixing their orientation. Current sets the magnetic field **B** in surrounding space, with wire geometry anchoring the field configuration. This role becomes explicit in guided-wave systems: waveguides and coaxial cables confine fields by geometry, allowing energy to flow through space between conductors as modes determined by Maxwell's equations. In contrast, radiative systems such as antennas lack boundaries, so fields spread outward and energy disperses.

Energy Beyond the Wire: The Veritasium Debate

A popular Veritasium video brought renewed attention to electromagnetic energy flow in circuits. It emphasised that energy resides in the fields surrounding conductors rather than inside them, and that the Poynting vector describes this flow. The example of a bulb positioned one metre from a battery created the impression that power reaches the bulb directly through the air because the bulb responds after roughly $1\,\mathrm{m}/c$.

The spatial layout in the demonstration was misleading. The bulb was one metre away in physical space, but electrically it was located many metres along the conductor path. The early response is driven by the local electromagnetic disturbance that propagates at the speed of light through the conductor–air geometry near the bulb. This response does not indicate direct energy transfer across the one-metre air gap and does not imply that geometric proximity determines power flow.

Electromagnetic energy is guided by boundary conditions set by the conductors. The Poynting vector follows the field configuration enforced by the circuit geometry rather than the shortest spatial route between battery and load. A resistive load receives sustained power only when the fields correspond to a continuous conductive path that supports current and consistent boundary conditions.

A useful check is to imagine the return path cut far away so that the two ends of the wire are separated by 0.1 light-seconds. For the first 0.1 s after the switch is closed, the bulb cannot distinguish between an intact loop and a broken one. The local electromagnetic field around the bulb is the same in both cases because the information about the distant break has not yet arrived. The early behaviour reflects only the nearby field adjustment. The long-time behaviour, where sustained power transfer either occurs or fails, depends on whether the global circuit is continuous.

Electromagnetic Energy Flow Outside Wires

Maxwell's equations reveal that *electromagnetic fields*, rather than moving electrons, transport energy. In free space, Maxwell's equations in SI units are:

$$\nabla \cdot \mathbf{E} = \frac{\rho}{\varepsilon_0}, \qquad \nabla \cdot \mathbf{B} = 0,$$
$$\nabla \times \mathbf{E} = -\frac{\partial \mathbf{B}}{\partial t}, \qquad \nabla \times \mathbf{B} = \mu_0 \mathbf{J} + \mu_0 \varepsilon_0 \frac{\partial \mathbf{E}}{\partial t},$$

where **E** and **B** are the electric and magnetic fields, ρ is charge density, and **J** is current density. The *Poynting vector*, defined as

$$\mathbf{S} = \mathbf{E} \times \mathbf{H},$$

describes the direction and magnitude of energy flow, where $\mathbf{H} = \mathbf{B}/\mu_0$ in vacuum. In a typical circuit, **S** is concentrated in the space around conductors, not within them, showing that fields, not electron drift, convey energy.

Near a Conductor: Shaping the Fields

Wires carry charges that generate **E** and **B**, but the power flux **S** remains predominantly outside. Applied voltage establishes the electric field **E**, current generates the magnetic field **B** and auxiliary field **H**, and their cross product $\mathbf{E} \times \mathbf{H}$ directs energy flow outside the conductor. Conductors constrain and guide the fields, enabling controlled power transfer with minimal radiation losses.

Electrons Are Slow

The speed of electrons in a wire, known as the *drift velocity*, is given by $v_d = I/nqA$, where I is the current, n is the number density of free electrons in the conductor, q is the elementary charge, and A is the cross-sectional area of the wire.

Example. For a copper wire of cross-sectional area $A = 1\,\text{mm}^2 = 10^{-6}\,\text{m}^2$, carrying a current of $I = 3\,\text{A}$, and using:

$$n \approx 10^{29}\,\text{electrons/m}^3, \quad q \approx 1.6 \times 10^{-19}\,\text{C},$$

the drift velocity is:

$$v_d \approx \frac{3}{(10^{29})(1.6 \times 10^{-19})(10^{-6})} \approx 1.9 \times 10^{-4}\,\text{m/s}.$$

Bonus Section: Maxwell's Equations in 4D Differential Forms (Warning: Jargon Ahead)

A remarkably elegant formulation uses differential forms in four-dimensional spacetime. Instead of treating electric and magnetic fields separately, one defines the field-strength 2-form F from a potential 1-form A: $F = \mathrm{d}A$. Maxwell's equations in vacuum then reduce to:

$$\mathrm{d}F = 0, \quad \mathrm{d}(\star F) = \mu_0 J,$$

where $\star F$ is the Hodge dual of F, and J is the 3-form representing charge and current density, Gauss's law. These equations encapsulate:

- $\mathrm{d}F = 0$: Magnetic fields are divergence-free (no monopoles), and electric fields induce magnetic circulation.
- $\mathrm{d}(\star F) = \mu_0 J$: Charge and current generate fields, unifying Gauss's law and Ampère's law.

This formulation emphasises that electromagnetic phenomena, including the Poynting vector, arise naturally from spacetime geometry rather than as separate electric and magnetic field concepts in three-dimensional space.

While defining these objects and proving their properties can take months, the payoff is nice: results like Maxwell's equations emerge from simple geometric principles. Other results, such as the generalised Stokes's theorem, follow with similar elegance.

References:

Feynman, R. P., Leighton, R. B., and Sands, M. (1964). *The Feynman Lectures on Physics.*

Baez, J. C., and Muniain, J. P. (1994). *Gauge Fields, Knots and Gravity.*

A Circle of PIE

Indo-European Language Family Tree: This phylogenetic reconstruction illustrates the hierarchical relationships among Indo-European languages, flowing from Proto-Indo-European (PIE) on the left to modern languages on the right. The tree demonstrates the systematic branching described by the comparative method, where shared innovations define intermediate nodes and regular sound correspondences link ancestral forms to their descendants.

The diagram directly supports the etymological analysis presented in this chapter. The PIE root *k^{w}ék^{w}los ('wheel, circle') appears across multiple branches with predictable phonological transformations: Greek κύκλος (*kyklos*) via labiovelar to velar shift, Sanskrit चक्र (*chakra*) through labiovelar palatalisation, and English 'wheel' via Grimm's Law (*k^{w} > hw). Each pathway reflects the systematic sound changes that characterise individual language families.

Major branches are colour-coded: Germanic (red) encompasses English, German, and Scandinavian languages; Celtic (green) includes Irish and Welsh; Italic (purple) covers Latin and its descendants; Balto-Slavic (blue) spans Russian, Polish, and Baltic languages; Indo-Iranian (yellow) includes Hindi, Persian, and related languages. The tree reveals morphological and phonological spread through subgroups.

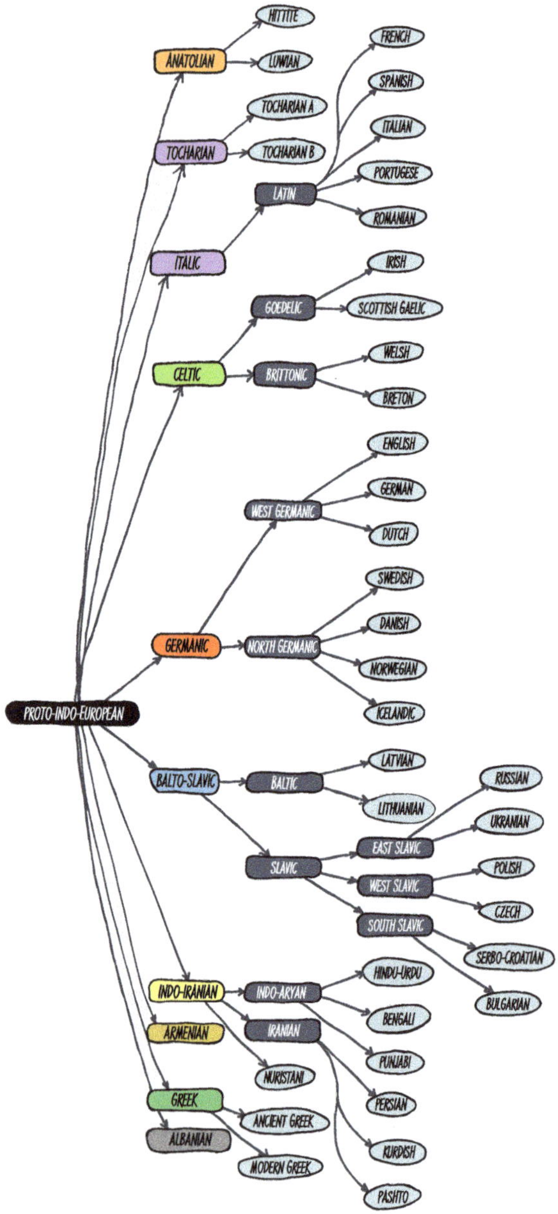

A Circle of PIE

The terms 'wheel' and 'cycle' (but not circle!) derive from Proto-Indo-European *k^{w}ék^{w}los despite their phonetic dissimilarity in modern languages. Regular sound shifts transformed this root differently in Germanic and Hellenic branches through documented phonological processes. These linguistic patterns preserve evidence of Bronze Age terminology and illustrate consistent patterns of language change. Comparative methods identify these transformations through sound correspondences across Indo-European languages.

Proto-Indo-European ◦ *K^{W}ÉK^{W}LOS Wheel ◦ *D^{h}UGH$_{2}$TḗR Daughter ◦ Comparative Method ◦ Sound Correspondences ◦ False Cognates ◦ Reduplication & Circularity ◦ Hebrew *G-L-L* Root ◦ וחוזר חלילה Mystery ◦ High-Retention Vocabulary ◦ Linguistic Reconstruction

> "English isn't a language, it's three languages stacked on top of each other wearing a trenchcoat."
>
> — Various Attributions

A Circle of PIE

The systematic study of language families emerged in the eighteenth and nineteenth centuries, when philologists began identifying regular correspondences between phonemes, grammatical structures, and syntactic patterns across geographically distant languages. Sir William Jones's 1786 observation that Sanskrit, Greek, and Latin exhibited similarities unlikely to arise by chance laid the foundation for reconstructing their common ancestor: *Proto-Indo-European (PIE)*, a prehistoric language hypothesised to have been spoken around 3000–4000 BCE in the Pontic–Caspian steppe.

Franz Bopp advanced the field by developing the *comparative method*, a procedure for recovering unattested forms through systematic analysis of sound correspondences and inflectional morphology. August Schleicher introduced genealogical tree diagrams to represent linguistic divergence (Schleicher, 1861)—a model that remains standard today. These methods enabled reconstruction of PIE roots with high consistency, revealing shared grammatical principles across its descendants.

The PIE daughter branches—Indo-Iranian, Hellenic, Italic, Celtic, Germanic, Balto-Slavic, and Anatolian, among others—developed distinct phonologies while preserving identifiable ancestral features. For example, the PIE voiced aspirated stop $*b^h$ appears as *bh* in Sanskrit, *f* in Latin, and *b* in English. Such rules apply across lexicon and morphology, enabling broad reconstruction. Grimm's Law (Grimm, 1822) captured the phonetic shifts distinguishing Proto-Germanic from other Indo-European languages, accounting for correspondences such as Latin *pater*, Greek *patēr*, Sanskrit *pitṛ́*, and English 'father.'

Though no written PIE record survives, its form emerges from consistent patterns in attested ancient languages including Hittite, Old Church Slavonic, and Old Persian. Core vocabulary—kinship terms, natural elements, agriculture, and tools—resists borrowing and anchors the comparative framework. PIE reconstruction also illuminates semantic evolution. Many roots generate both concrete and abstract derivatives across daughter languages. Motion, time, and cyclical processes often yield terms spanning physical action, ritual practice, and philosophical speculation.

The reconstruction of ancestral languages depends on regularities that persist across phonological evolution. Sound change occurs according to consistent patterns that affect entire grammatical systems (Osthoff & Brugmann, 1878). These patterns allow historical inference to proceed by rule-governed comparison. When daughter languages show aligned differences in equivalent words, the shape of the ancestral form can often be inferred with precision.

Historical linguistics considers shared morphological and syntactic structures as indicators of genealogical descent. Correspondences in case endings, agreement systems, and word order provide signals of common ancestry. These features must appear across lexical items to count as evidence for descent. Chance resemblance or cultural borrowing cannot produce such system-wide alignment.

Proto-Indo-European (PIE) is the name assigned to the unattested language reconstructed from parallels among Indo-European languages. Its existence is inferred from regularities in grammar and phonology shared by Sanskrit, Ancient Greek, Latin, Hittite, Old Church Slavonic, and others. These languages exhibit consistent transformations that converge on reconstructed PIE forms, reflecting their shared ancestry.

The comparative method identifies sound correspondences (Bopp, 1816) that link descendant languages to a shared root. For example, Latin *f*, Sanskrit *bh*, and English *b* align in inherited words, implying a common source consonant in PIE. Such correspondences must be supported by examples across word families. Once established, they allow reconstruction of ancestral forms that conform to a phonological system.

Phonological transformations affect all levels of morphology, including declensions, conjugations, and derivational patterns. These shifts are governed by well-defined constraints such as syllable structure, stress placement, and adjacent sounds. A given transformation applies across the lexicon once its domain is defined. This internal consistency permits reconstructions that are testable.

Kinship terms, natural elements, and tools form high-retention vocabulary with cross-linguistic stability (Swadesh, 1955). These words resist borrowing, undergo regular phonological change, and remain semantically intact across time scales. They serve as indicators of shared ancestry in historical reconstruction.

The PIE root *$\mathrm{d^hugh_2t\acute{\bar{e}}r}$, meaning "daughter," provides one of the clearest examples of stability across Indo-European languages. Despite phonological divergence, the kinship meaning is retained with consistency from Vedic Sanskrit to modern English.

In Sanskrit, the form is दुहितृ (*duhitṛ*), preserving both the root and the feminine suffix. Ancient Greek gives θυγάτηρ (*thygatēr*)—where the initial aspirated dental is retained. Latin replaced the expected cognate of PIE *$\mathrm{d^hugh_2t\acute{\bar{e}}r}$ with *filia*, from an entirely different root meaning 'suckling' (related to *filius* for 'son'). In Gothic, the reflex is *dauhtar*, leading to Old English *dohtor* and eventually modern English 'daughter.'

These forms are an example of predictable sound changes. The PIE voiced aspirated dental *$\mathrm{d^h}$ becomes *th* in Greek and *d* in Germanic. The laryngeal *$\mathrm{h_2}$ affects surrounding vowels and often disappears. Preservation of suffixes and semantic continuity reinforce the reconstruction's accuracy.

A second example, drawn from material instead of kinship vocabulary, illustrates the principles of retention and transformation. The PIE root *$\mathrm{k^w\acute{e}k^wlos}$, meaning 'wheel' or 'circle,' illustrates how this root produced enduring derivatives across Indo-European languages. Across language families, this root led to distinct yet semantically linked terms for circularity and motion.

In Greek, κύκλος (*kyklos*) retained the meanings of 'circle' and 'wheel,' later influencing Latin *cyclus* and English 'cycle.' Sanskrit preserved the root as चक्र (*chakra*), initially referring to a physical wheel, and later extended to cycles in philosophical and spiritual contexts. In Proto-Germanic, the root evolved into *hweulą, producing Old English *hweol*, Middle English *whele*, and modern English 'wheel.'

Phonological shifts altered the surface form, but the meaning remained around rotation and recurrence. Latin *colere*, meaning 'to cultivate' or 'to tend,' may derive from *$\mathrm{k^wel}$-

('to turn')—though this etymology remains speculative—generating *cultus* ('ritual care') and eventually 'cult.' A related case is Latin *circulus*, a diminutive of *circus* ('ring'), which became English 'circle' via Old French *cercle*. Although derived from a separate PIE root (*sker-, 'to bend, turn'), its semantic parallel to κύκλος reflects linguistic convergence.

In Semitic languages, comparable formations are found. Hebrew גלגל (*galgal*), meaning 'wheel' or 'rolling object,' derives from the root *g-l-l*, which denotes circular motion. Related terms include גל (*gal*, 'wave'), גללים (*galalim*, 'dung pellets'), and גולגולת (*gulgoleth*, 'skull'). The reduplication in *galgal* superficially resembles the PIE form *k^{w}ék^{w}los (k^{w}e-k^{w}l-os), but the similarity is incidental, as these forms derive from distinct morphological systems.

Reduplication as a strategy for emphasising repetition or motion appears independently across language families. Its presence in both Indo-European and Semitic systems points to a broader cross-linguistic pattern. The recurrence of phonemes, as in גלגל and *k^{w}ék^{w}los, reinforces the idea of circularity through sound. Just as the wheel itself arose independently in different cultures, linguistic forms encoding rotation also emerged separately.

The comparative method excludes false cognates. Consider English 'day' and Latin *dies*—both refer to a twenty-four-hour period and share similar sounds, yet they derive from entirely unrelated PIE roots.

English 'day' traces back through Old English *dæg* to PIE *d^{h}eg^{wh}-, meaning 'to burn' or 'to be hot.' The semantic connection runs from the heat of daylight to the daylight period itself. In contrast, Latin *dies* descends from PIE *dyéws, meaning 'sky' or 'to shine,' related to *deus* ('god') and Sanskrit *dyáus* ('sky, heaven'). Both roots metaphorically extended to 'day,' but through independent pathways.

The comparative method distinguishes such cases by requiring regular sound correspondences across word families. English 'day' follows Germanic sound laws: PIE *d^{h} regularly becomes *d* in English, and *g^{wh} becomes *g* (later weakened to zero). Latin *dies*, however, shows the expected Latin treatment of PIE *dy: the sequence becomes *di-* in Latin, as seen in *Iovis* (Jupiter) from *dyēws.

Had these words been genuine cognates, we would expect to find the same root appearing across Romance and Germanic languages with parallel semantic development. Instead, we find that other Germanic languages show the *d^{h}eg^{wh}- root (German *Tag*, Dutch *dag*), while Romance languages consistently reflect *dyéws (French *jour* from Latin *diurnus*, Spanish *día*). The pattern confirms separate origins despite surface similarity.

Another case involves English 'much' and Spanish 'mucho'—words that are nearly identical in sound and meaning yet stem from unrelated PIE roots. English 'much' derives from Old English *micel* and ultimately PIE *méǵh$_{2}$s ('great'), whose Latin cognate is *magnus*. Spanish 'mucho,' however, comes from Latin *multus* ('many'), which traces to PIE *mel- ('strong'). The superficial resemblance results from convergent phonological development: Germanic *k > English *ch*, and Latin consonant clusters *-lt-* > Spanish *ch*. The true English cognate of 'mucho' would be a derivative of *magnus*, while the Spanish cognate of 'much' appears in *más* (from Latin *magis*).

This systematic approach prevents the method from accepting coincidental resemblances, borrowings, or parallel semantic developments. True cognates must satisfy constraints simultaneously: sound correspondences, morphological patterns, and semantic plausibility across the language family.

A Personal Encounter

At thirteen, I spent lunch breaks calling the Academy of the Hebrew Language from a pay phone, taking advantage of their public consultation hours. One question preoccupied me: the meaning of the Hebrew phrase וחוזר חלילה (*vechozer chalilah*). The expression denotes endless repetition, 'again and again' or 'in a cycle,' yet the word חלילה (*chalilah*) also means 'God forbid.' Why would a phrase about recurrence contain a word implying prohibition?

The Academy asked for two weeks to investigate. When I called again, they proposed three hypotheses. One traced it to חלל (*chalal*, 'void'), implying unboundedness. Another derived it from חול (*chol*, 'sand'), whose accumulation metaphorically signals continuity. A final suggestion pointed to חליל (*chalil*, 'flute'), possibly named for its cylindrical form.

None of these answers resolved my curiosity. Years later, I encountered $^*k^w\acute{e}k^w los$ and its descendants: κύκλος (*kyklos*), चक्र (*chakra*), 'cycle,' and 'wheel.' I remembered that call. Across language families, words for turning often imply repetition.

Sound plays a role. Words like חלילה (*chalilah*) and גלגל (*galgal*) echo themselves, as do Greek ροίζος (*rhoizos*, 'whirring noise') and other motion-related terms. Reduplication strengthens the perception of rotation. Whether וחוזר חלילה originated independently or reflects a linguistic universal, it illustrates a principle: what turns, returns.

Years later, at the Technion, I worked in an EEG lab run by Professor Hillel Pratt, a scientist of towering breadth and compassion. Our conversations drifted across neuroscience, etymology, and Aramaic grammar. For over a couple of years, we debated word origins.

Then one day I learned he was a sitting member of the Academy of the Hebrew Language. He had never mentioned it. He let the ideas speak for themselves. Technically, he was *always* right.

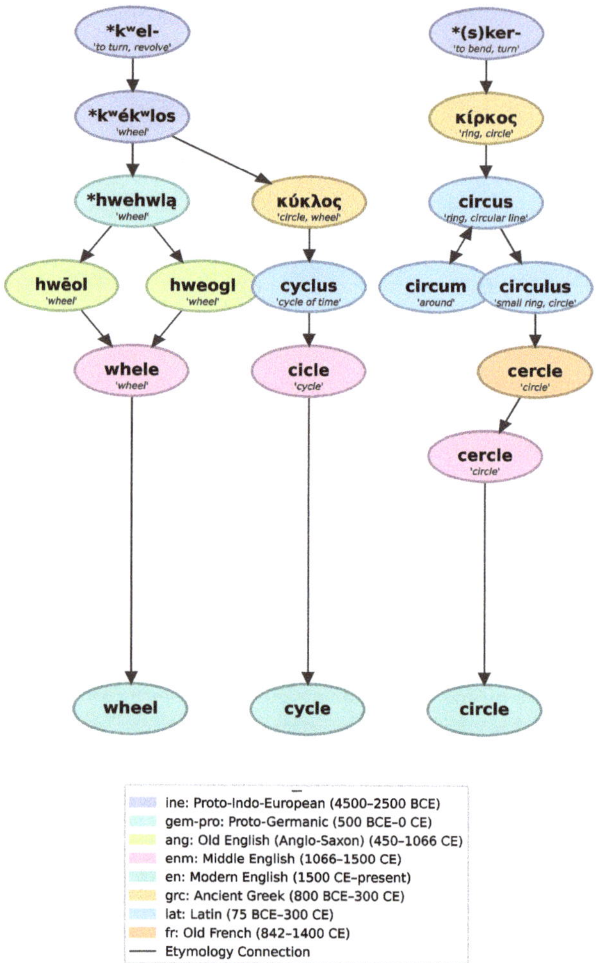

Figure by author. Etymological relationships in the Indo-European family showing the evolution of the PIE root *kʷékʷlos ('wheel, circle') across major language branches. The diagram illustrates systematic phonological transformations: de-labialization in Latin (*colere*) versus preservation as *qu* (e.g., *equus, quattuor*), palatalization in Sanskrit (*chakra*), contextual neutralisation in Greek (*kyklos*), and fricativization through Grimm's Law in Germanic (*wheel*). Colour coding distinguishes language families while maintaining visual clarity of the genealogical relationships that underlie comparative reconstruction.

The Phonological Evolution of Labiovelars in the Indo-European Descendants of *kʷékʷlos

Proto-Indo-European (PIE) contained labiovelar stops (*kʷ, *gʷ, *gʷʰ) with simultaneous velar closure and labialization, contrasting with plain velars (*k, *g, *gʰ) and palatalized velars (*ḱ, *ǵ, *ǵʰ). The PIE root *kʷékʷlos, a reduplicated form of *kʷel- ('to turn'), underwent systematic shifts across branches.

In Greek, By Mycenaean Greek (c. 1400 BCE, attested in Linear B, the earliest known form of the Greek writing system), labiovelars were still distinguished. Later Greek neutralised them context-dependently: *kʷ became t before front vowels (πέντε 'five' < PIE *pénkʷe), p in many environments (λείπω 'I leave' < PIE *leikʷ-), and k in certain contexts:

$$\kappa\kappa\lambda o\varsigma \text{ (\textit{kyklos}) < PIE } *k^w k^w los.$$

The labiovelars were thus fully neutralised, with reflexes depending on phonological environment.

In Sanskrit, labiovelars merged with palatals before front vowels, so *kʷ became च (*c*, [t͡ʃ]):

$$\text{(\textit{chakra}) < PIE } *k^w k^w los.$$

This is part of a broader Indo-Iranian shift where labiovelars fronted or merged with palatals.

In Latin, *kʷ was generally preserved as *qu* in most environments (*quis, quo, equus, aqua, quattuor*). However, certain roots show regular de-labialization to *c*-, notably the *kʷel- family: PIE *kʷel- → Latin *colere* ('to cultivate'), *incola* ('inhabitant'). This reflects dissimilation: the labial element of *kʷ is lost before a rounded vowel in the following syllable. In contrast, *circulus* ('circle'), from Greek *kirkos* ('ring'), derives from PIE *(s)ker- 'to turn, bend'—a distinct root without labiovelars.

In Proto-Germanic, Grimm's Law altered the stop system: *kʷ → *hw. Thus, *kʷékʷlos became *hweulą (Proto-Germanic), which evolved into Old English *hwēol*, Middle English *whele*, and Modern English *wheel*.

Summary: Greek neutralised labiovelars context-dependently (*kʷ > t/p/k), Sanskrit palatalized (*kʷ > c → चक्र), Latin generally preserved *qu* but de-labialized in certain roots like *colere*, and Germanic fricativized via Grimm's Law (*kʷ > hw → *wheel*).

These transformations illustrate how a single PIE labiovelar stop produced diverse reflexes across Indo-European languages, shaping words that remain etymologically linked despite significant phonetic divergence.

References:

Fortson, B. (2010). *Indo-European Language and Culture: An Introduction.* Wiley-Blackwell.

Ringe, D. (2006). *From Proto-Indo-European to Proto-Germanic.* Oxford University Press.

Online Etymology Dictionary: `https://www.etymonline.com/`

The Apple Falls the Slowest from the Tree

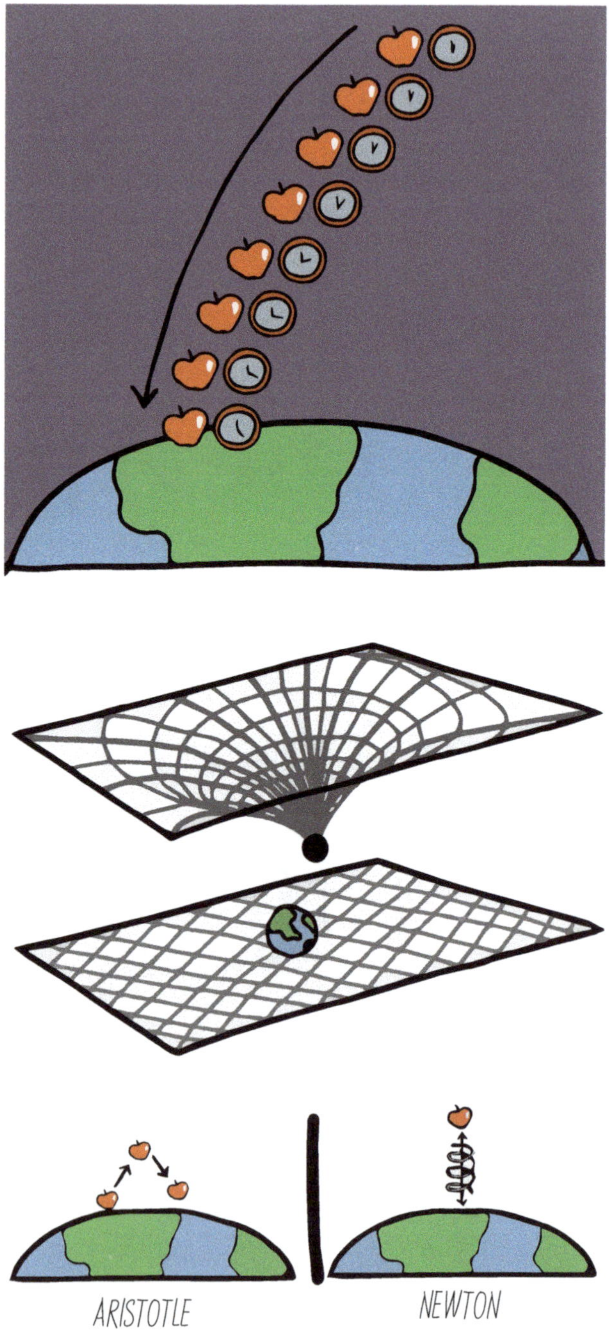

Top (GR—Falling Toward Slow Time): An apple follows a curved free-fall trajectory toward Earth. Along the path, it is shown accompanied by clocks, which tick progressively slower as they get closer to Earth. This is an insight relativity: free-falling objects move toward regions where proper time is maximised—in other words, where the time component of the metric, g_{00}, decreases. The apple is not being pulled by a force, but slides through spacetime toward slower time.
Middle (Misleading Space Bending): Two 'rubber sheet' grid diagrams are shown. The top depicts an exaggerated funnel-shaped distortion caused by a black mass—a common pop-science depiction of gravity. The bottom shows the Earth resting on a seemingly flat grid, emphasising that for small masses such as Earth, spatial curvature is negligible. What matters is the distortion of time, not space. Gravity in this regime is governed almost exclusively by the g_{00} gradient.
Bottom (Aristotle to Newton): Left: Aristotle's model shows apples falling back to Earth because they 'belong' there—matter returning to its natural place. Right: Newton's model depicts a single apple being pulled downward by an invisible gravitational force, with spring lines suggesting mutual attraction between masses. This is Newton's universal law of gravitation—a force acting at a distance.

The Apple Falls the Slowest from the Tree

General relativity formulates gravity as spacetime curvature where time and space metrics are affected by mass and energy. Contrary to the depiction of gravity as bending space, as in the rubber sheet visualisations, in cases in which masses are small (the Earth, for example) it is the gradient in time's rate that creates gravitational attraction, guiding objects towards regions of slower time. The reason an apple is falling is not because it is affected by force radiated by Earth, nor is it due to the curvature of space, but rather because it is following the shortest path through curved time.

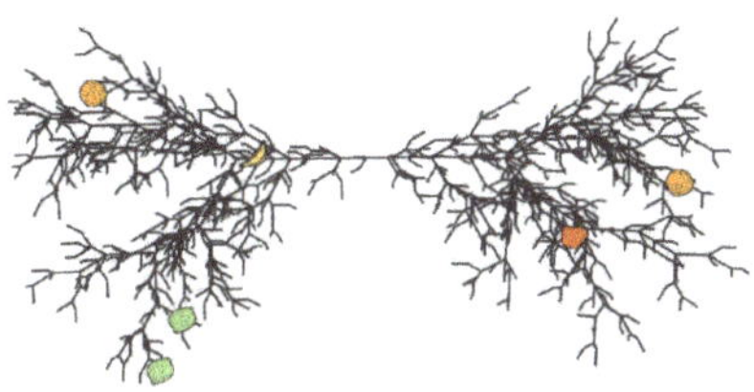

EQUIVALENCE PRINCIPLES ∘ FREE FALL AS GEODESIC ∘ SPACETIME METRIC $g_{\mu\nu}$ ∘ GRAVITATIONAL TIME DILATION ∘ g_{00} COMPONENT ∘ FALLING INTO SLOWER TIME ∘ LOCAL INERTIAL FRAMES ∘ TIDAL EFFECTS ∘ EINSTEIN'S ELEVATOR ∘ PROPER TIME MAXIMISATION ∘ GPS CLOCK CORRECTIONS

"Spacetime tells matter how to move;
matter tells spacetime how to curve."
— John Archibald Wheeler, 1962

"We have ways of making people talk...
by giving them fresh apple slices."
— Kenneth Parcell, 2010

The Apple Falls the Slowest from the Tree

Isaac Newton's 1687 *Principia* described gravity as a force acting at a distance between masses, accurately predicting the behaviour of falling objects and planetary orbits. This framework dominated physics for more than two centuries. Yet as astronomical measurements grew more precise, small but persistent anomalies emerged—most notably the unexplained excess precession of Mercury's orbit, which Newtonian mechanics could not account for.

In 1915, Albert Einstein introduced general relativity, reframing gravity as the curvature of spacetime. This insight came from Einstein's realisation that the equivalence of gravitational and inertial mass was no coincidence. His theory predicted phenomena beyond Newton's reach: time would flow differently in gravitational fields, and light would bend twice as much as predicted by Newton when passing massive bodies.

The name 'general' relativity captures the theory's ambitious scope—it applies to all observers, whether accelerating, rotating, or in gravitational fields, generalising Einstein's 1905 special relativity which was limited to inertial frames.

Experimental confirmation came in 1919, with Arthur Eddington's expedition to observe a solar eclipse. They found starlight bending around the Sun consistent with Einstein's prediction within experimental uncertainties—instantly making the German physicist a global celebrity. Later experiments provided increasingly precise validation: the 1959 Pound–Rebka experiment detected gravitational redshift using gamma rays in a Harvard tower; Gravity Probe A (1976) launched a hydrogen maser clock on a rocket to confirm time dilation with altitude; and by 1980, ground-based cesium clocks could measure these effects with exquisite precision.

By the late twentieth century, relativistic effects had become engineering concerns—GPS satellites must continuously correct for both special and general relativistic temporal effects to maintain metre-level positioning accuracy.

The ultimate confirmation came a century after Einstein's publication. In 2015, LIGO detected gravitational waves from two black holes spiralling together over a billion light-years away. This observation, followed by dozens more including neutron star collisions, validated Einstein's theory in extreme gravitational regimes.

General relativity remains one of physics's most tested theories, validated from subatomic to cosmological scales.

The weak equivalence principle states that all objects follow identical free-fall trajectories when released from the same point–whether lead or feathers. Formulated by Galileo and tested for centuries, it shows gravitational acceleration is and tested for centuries, it shows gravitational acceleration is independent of mass, charge, or composition. This universality implies that gravitational (how strongly gravity pulls on an object) and inertial mass (the object's resistance to acceleration) are proportional–a coincidence in Newtonian physics that inspired Einstein to reinterpret gravity geometrically (Einstein, 1915), as motion through curved spacetime.

 https://doi.org/10.11647/OBP.0526.06

The strong equivalence principle generalises this: in any local region of spacetime, gravitational effects can be eliminated by choosing a freely falling reference frame. Within such a frame, spacetime curvature becomes negligible. The geometry flattens to first order, and all physical processes proceed as they would in the absence of gravitation. Mechanical systems evolve according to Newton's laws, electromagnetic fields obey Maxwell's equations in vacuum (Maxwell, 1865), and the motion of particles is governed by the inertial character of special relativity.

While spacetime may be globally curved, it admits neighbourhoods indistinguishable from flat Minkowski space. No experiment in a small free-falling laboratory can detect gravity.

Einstein's elevator thought experiment illustrates this: in a free-falling elevator, released objects float weightlessly and light travels straight. No experiment inside can detect the external gravitational field. Mechanical pendulums, electromagnetic resonators, and radioactive decay rates all behave as in gravity-free space. Free fall and inertial motion are locally identical when tidal effects are negligible—that is, when the variation in gravitational field strength across the laboratory is too small to measure. Over larger regions, these tidal effects become detectable as objects at different positions experience slightly different accelerations, causing initially parallel trajectories to converge or diverge.

The spacetime metric is the mathematical object that defines how distances and time intervals are measured. It quantifies the separation between nearby events.

In flat spacetime, the metric is constant (Minkowski, 1908): $ds^2 = -c^2dt^2+dx^2+dy^2+dz^2$. Time gets a negative coefficient $-c^2$ while space coordinates get positive coefficients, meaning the shortest distance between two points is when space separation is minimised and time separation is maximised. In curved spacetime where gravity is present, the metric tensor $g_{\mu\nu}$ varies from point to point. This variation determines how clocks tick at different locations and how distances are measured—it is what we experience as gravity.

These variations manifest in observable ways. Clocks at different gravitational potentials accumulate time at different rates. Initially parallel free-falling trajectories converge or diverge. Tidal effects—the differential forces that stretch objects toward massive bodies and compress them perpendicular to that direction—arise because gravitational field strength varies with position. On Earth, these variations cause ocean tides as the Moon pulls more strongly on the near side than the far side. In extreme cases near black holes, tidal forces can tear objects apart or spaghettify them. When exchanging signals between different altitudes, identical clocks emit pulses at regular intervals but receivers measure changed intervals: upward light is redshifted, downward light is blueshifted.

A local inertial frame has no proper acceleration and follows special-relativistic laws—light travels straight and clocks tick uniformly. The metric matches flat spacetime at a point, with deviations appearing only at second order in displacement. But when comparing such frames at different locations, clocks at different altitudes tick at different rates, transported vectors fail to align, and no single coordinate system makes the metric flat everywhere. The curvature is part of the metric's second derivatives.

A helpful analogy comes from curved surfaces. Imagine two travellers walking north from the equator along different lines of longitude. Their paths begin parallel and appear straight locally, yet they eventually converge at the pole. The meeting point is not due to any force between them but to the geometry. In spacetime, freely falling objects can similarly start with zero relative velocity and later converge or diverge, not because of an interaction, but because the metric changes from point to point.

The metric tensor contains ten independent components in four dimensions. Each component encodes different aspects of spacetime geometry. The spatial components g_{ij} govern how distances are measured and produce effects such as gravitational lensing–light bending around massive objects. The time-space cross terms g_{0i} appear when spacetime itself rotates, as around spinning black holes, causing frame dragging (Lense & Thirring, 1918). But near Earth's surface, where rotation is negligible and velocities remain small compared to light, one component dominates all others in determining motion: g_{00}.

The metric component g_{00} encodes how proper time flows for stationary observers. When g_{00} varies with position, identical coordinate intervals correspond to different amounts of proper time–this is gravitational time dilation. Atomic clocks aboard aeroplanes, satellites, and mountaintops have confirmed these predictions with precision essential for GPS accuracy. While spatial curvature produces phenomena such as gravitational lensing and tidal forces, the gradient of g_{00} determines both the rate at which clocks tick and the direction objects fall.

This connection between time and motion is the essence of gravity. In general relativity, freely falling objects follow geodesics–paths that extremize proper time through spacetime. The insight lies in the metric signature: time enters with an opposite sign to the spatial components $ds^2 = -c^2dt^2 + dx^2 + dy^2 + dz^2$. This opposite sign means that minimising spatial separation and maximising temporal duration both contribute to extremizing the spacetime interval in the same direction. A straight line in space is the shortest spatial path; a geodesic in spacetime is the longest proper time. Near Earth's surface, these geodesics curve downward in space precisely because proper time accumulates more slowly at lower altitudes–moving toward slower clocks maximises the proper time experienced along the path. The 'force' we attribute to gravity is actually the spatial projection of motion along these curved spacetime paths.

Consider an apple hanging from a tree. While attached to the branch, it resists its natural geodesic. Once the stem breaks, the apple enters free fall. As it descends, the changing gradient of g_{00} continuously adjusts its trajectory toward regions where time passes more slowly. This curvature in the apple's spacetime path manifests as what we perceive as gravitational acceleration. The metric itself, encoding how time flows differently at each point in space, tells matter how to move.

Falling Into Slower Time

The realisation that gravity arises from differences in the rate of time's passage, rather than from any applied force, marks a sharp shift in our description of nature. This perspective is mathematically well-defined and supported by high-precision experiments, but it remains difficult to visualise and understand intuitively.

The rubber sheet analogy shows mass distorting a surface, with objects curving toward the indentation. While this conveys how mass alters geometry, it emphasises spatial curvature. In general relativity, gravitational motion comes primarily from varying proper time—time flows more slowly deeper in gravitational fields. This temporal gradient, not spatial curvature, governs free fall.

This is, for me, a *fantastic* observation. An apple falls from a tree not because it is pulled downward, but because time flows slightly faster at the top of the tree than at the bottom. Once released, the apple is no longer constrained. It follows a trajectory through spacetime that maximises the amount of proper time experienced along the way. The curve of this path is determined by how the clock rate changes with altitude.

In everyday reasoning, we often think of objects as taking the shortest route through space. But in general relativity, freely falling objects follow the most direct path through spacetime as a whole, which is about minimising spatial distance while maximising proper time, until even the distinction between space and time becomes irrelevant (as in the vicinity of a black hole which we will discuss in Chapter 23).

A race against time.

Exercise: Deriving the Schwarzschild Metric

Objective: Derive the Schwarzschild metric, which describes spacetime outside a spherically symmetric, non-rotating, uncharged mass M.
Assumptions: Spherical symmetry, static spacetime, vacuum solution ($R_{\mu\nu} = 0$), and the Newtonian limit at large r.

1. **General Form of the Metric.** A spherically symmetric, static spacetime is described by the most general metric:

$$ds^2 = -A(r)c^2dt^2 + B(r)dr^2 + r^2d\Omega^2,$$

where $d\Omega^2 = d\theta^2 + \sin^2\theta d\phi^2$, and $A(r)$, $B(r)$ are functions of r only. *(Why only r? Symmetry! Time-independence and spherical symmetry ensure that metric components cannot depend on t, θ, or ϕ).*

2. **Newtonian Limit.** In the weak-field, slow-motion limit, the metric must reduce to the Newtonian potential: $g_{00} \approx -(1 + 2\Phi/c^2)$, where $\Phi = -GM/r$ is the Newtonian potential of a point mass M. This implies $A(r) = 1 - 2GM/rc^2$. Define the **Schwarzschild radius** $r_s = 2GM/c^2$, so $A(r) = 1 - r_s/r$.

3. **Key Assumption: Relationship Between $A(r)$ and $B(r)$.** Instead of solving Einstein's field equations explicitly, we assume that $B(r) = 1/A(r)$. *Why?*
   a) **Birkhoff's Theorem:** The unique spherically symmetric vacuum solution must be static and match the Schwarzschild metric.
   b) **Energy Conservation in Radial Free-Fall:** If an object falls radially inward, the proper time and coordinate time should be related in a way that matches Newtonian conservation of energy in the weak-field limit.

   This assumption gives:

$$B(r) = \frac{1}{1 - r_s/r}.$$

4. **The Schwarzschild Metric.** Substituting $A(r) = 1 - r_s/r$ and $B(r) = (1 - r_s/r)^{-1}$ into the general metric:

$$ds^2 = -\left(1 - r_s/r\right)c^2dt^2 + \frac{dr^2}{1 - r_s/r} + r^2d\Omega^2.$$

5. **Physical Consequences:**
   a) **Event Horizon ($r = r_s$):** The metric component g_{00} vanishes and g_{rr} diverges. This is a coordinate singularity, marking the event horizon of a black hole.
   b) **Inside the Schwarzschild Radius ($r < r_s$):** The roles of space and time effectively switch, leading to an inevitable collapse toward $r = 0$.
   c) **Observational Predictions:** The Schwarzschild metric predicts gravitational time dilation, light bending, and the precession of planetary orbits (as seen in Mercury).

Gravity as Curved Time: The Time-Only Ansatz

General relativity describes gravity through the curvature of spacetime. But which part of spacetime does the work? Rather than beginning with the full Schwarzschild solution, we follow a more direct route: assume space is perfectly flat and ask whether variations in the rate of time alone can reproduce Newtonian gravity. The answer is yes—and the derivation shows why spatial curvature is irrelevant for everyday gravitational phenomena.

The Ansatz: Flat Space, Variable Time

We assume Euclidean spatial geometry but allow the rate at which clocks tick to depend on position. This gives a spacetime interval of the form:

$$ds^2 = -f(\mathbf{x})\, c^2\, dt^2 + dx^2 + dy^2 + dz^2,$$

where $f(\mathbf{x})$ is a static function encoding how fast a clock runs at location $\mathbf{x}$. In the absence of any gravitational source, $f = 1$ everywhere and we recover the flat Minkowski metric of special relativity.

Connecting f to the Newtonian Potential

The Newtonian gravitational potential outside a mass M is $\Phi(\mathbf{x}) = -GM/r$, where $r = |\mathbf{x}|$. We set:

$$f(\mathbf{x}) = 1 + \frac{2\Phi(\mathbf{x})}{c^2}.$$

Since $\Phi < 0$ near a mass, we have $f < 1$: clocks deeper in the gravitational well tick slower. The dimensionless ratio $|\Phi|/c^2$ measures the strength of the effect. At Earth's surface, $GM/(c^2 R_\oplus) \approx 7 \times 10^{-10}$—a tiny perturbation, but one with real consequences.

Proper Time and the Geodesic Equation

A freely falling particle follows a geodesic—the path that maximises proper time between two events. The proper time along a worldline is:

$$d\tau^2 = f(\mathbf{x})\, dt^2 - \frac{1}{c^2}(dx^2 + dy^2 + dz^2).$$

For a particle moving slowly ($v \ll c$), the spatial terms are negligible compared to the time term, and $d\tau \approx \sqrt{f}\, dt$. The particle's four-velocity is dominated by $u^0 \approx c/\sqrt{f}$, with spatial components $u^i \ll u^0$.

The geodesic equation $\ddot{x}^\mu + \Gamma^\mu_{\alpha\beta}\, \dot{x}^\alpha \dot{x}^\beta = 0$ then simplifies. Because spatial velocities are small, only the Γ^i_{00} Christoffel symbol matters. For our diagonal metric with flat spatial part $\Gamma^i_{00} = -\frac{1}{2}\,\delta^{ij}\,\frac{\partial f}{\partial x^j} = -\frac{1}{c^2}\,\frac{\partial \Phi}{\partial x^i}$.

Recovering Newton's Second Law

Substituting into the spatial geodesic equation gives the coordinate acceleration:

$$\frac{d^2 x^i}{dt^2} \approx -c^2\, \Gamma^i_{00} = -\frac{\partial \Phi}{\partial x^i}.$$

In vector form: $\mathbf{a} = -\nabla\Phi$. For a point mass, $\Phi = -GM/r$ yields $\mathbf{a} = -GM\hat{\mathbf{r}}/r^2$, which is Newton's inverse-square law. No spatial curvature was needed at any step. The gravitational acceleration arises entirely from the spatial gradient of the clock-rate function f.

How Small Is the Spatial Curvature Correction?

In the full Schwarzschild solution, the spatial metric component $g_{rr} = (1 - 2GM/c^2 r)^{-1}$ also deviates from unity. This introduces an additional Christoffel symbol $\Gamma^r_{rr} \approx -GM/(c^2 r^2)$, which contributes to the geodesic equation via the term $-\Gamma^r_{rr}(v^r)^2$. For an object falling from rest near Earth's surface, $v^2 \sim 2GM/R_\oplus$, so the spatial curvature contribution to acceleration is of order:

$$\frac{a_{\text{spatial}}}{a_{\text{temporal}}} \sim \frac{v^2}{c^2} \approx 2\frac{GM}{c^2 R_\oplus} \approx 1.4 \times 10^{-9}.$$

The spatial geometry of general relativity contributes less than two parts per billion to the gravitational acceleration of a falling apple. For all non-relativistic experience, gravity is an effect of temporal curvature alone.

References:

Misner, C. W., Thorne, K. S., and Wheeler, J. A. (1973). *Gravitation.* (Section 17.4: Newton's Theory in the Language of Spacetime Curvature).

I thank Prof. A. Winkler for suggesting this presentation.

A Complex (Projective) Billiard Game

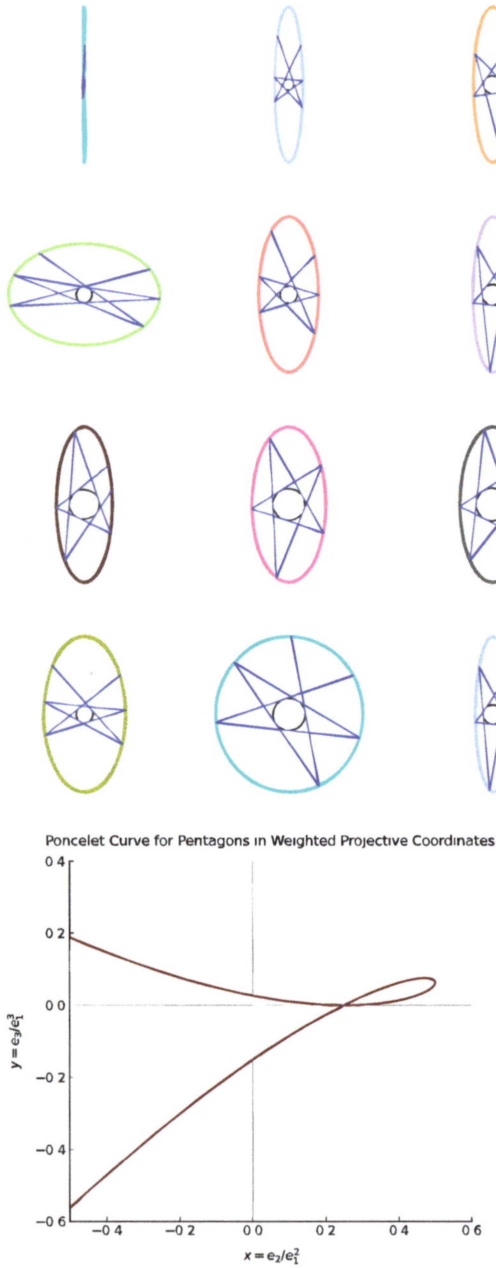

Top (Poncelet Trajectories): Each subplot in the top section displays a pair of conics: the outer circle $x^2 + y^2 = 1$ and an inner ellipse of the form $\left(\frac{x}{a}\right)^2 + \left(\frac{y}{b}\right)^2 = 1$, where (a, b) are derived from sampled Cayley invariants (x, y). A five-step Poncelet trajectory is drawn in blue, starting from a fixed angle and iteratively constructing tangents from the circle to the ellipse. If the polygon closes after five steps, the pair lies on the Poncelet curve.
Bottom (Poncelet Curve): The curve shown is the Poncelet curve associated with pentagonal (5-periodic) Poncelet polygons inscribed in one conic and circumscribed around another. Each point on this curve corresponds to a set of conic pairs for which a closed 5-gon exists that satisfies Poncelet's closure condition. The coordinates (x, y) represent algebraic invariants of the conic pair, specifically $x = e_2/e_1^2$ and $y = e_3/e_1^3$, where e_k are the elementary symmetric functions of the characteristic multipliers derived from the conic configuration. A point on the curve means that a 5-periodic polygon can be inscribed and circumscribed for that particular combination of invariants. For more, see the excellent blog of Oliver Nash at http://oliver-nash.org/2018/07/08/poring-over-poncelet/index.html.

A Complex (Projective) Billiard Game

Poncelet's Porism describes an unexpected property of billiard trajectories between two nested ellipses: if one path returns to its starting point after a finite number of bounces, then all starting points generate periodic trajectories with the same number of bounces. This geometric result connects to elliptic curves and measure-preserving dynamical systems, exemplifying how problems in distinct fields reduce to the same equations through appropriate frameworks.

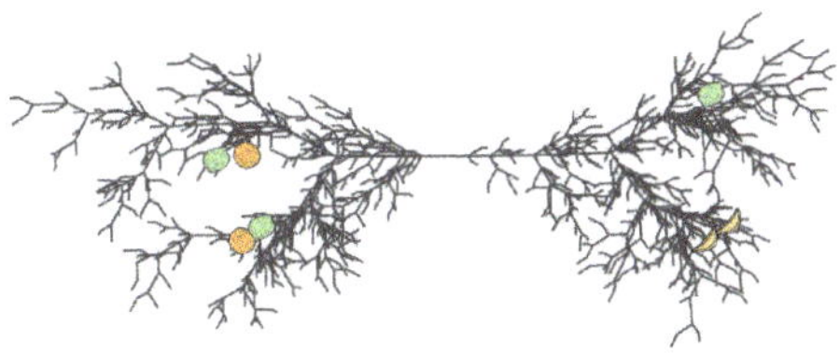

Elliptical Billiards ∘ Poncelet's Porism ∘ Nested Ellipses ∘ Poncelet Map ∘ Invariant Weighted Measure ∘ Rotation Number ∘ Rational vs Irrational ∘ Benford's Law ∘ Leading Digit Distribution ∘ Gelfand Framework ∘ Dynamical Proof

« La mathématique est l'art de donner le même nom à des choses différentes. »

("Mathematics is the art of giving the same name to different things.")

— Henri Poincaré, 1908

"Algebra is the offer made by the devil to the mathematician.
The devil says: I will give you this powerful machine,
it will answer any question you like.
All you need to do is give me your soul:
give up geometry and you will have this marvellous machine."

— Michael Atiyah, 2001

A Complex (Projective) Billiard Game

In the early nineteenth century, Jean-Victor Poncelet (1788–1867) pioneered projective geometry by examining how shapes transform under projection. Around 1822, he introduced the concept now known as Poncelet's Porism, demonstrating that a closed polygon can be inscribed in one conic and circumscribed about another conic, provided it exists once for a given number of sides. This insight spurred an intense study of conic sections, with mathematicians such as Carl Gustav Jacob Jacobi and Arthur Cayley extending Poncelet's results to explore more algebraic properties of these curves.

Over the mid to late nineteenth century, researchers recognised a link between geometric theorems such as Poncelet's Porism and physical billiard trajectories. Elliptical billiards, in particular, drew interest when it was observed that the classical reflection law led to periodic paths that echoed Poncelet's closure conditions. By the turn of the twentieth century, these geometric investigations began intertwining with the nascent field of algebraic geometry, revealing that repeated reflections could be described by equations resembling elliptic or hyperelliptic curves.

The rules for the motion of billiards are straightforward. A ball moves in a straight line until it meets a boundary—where it reflects according to the law of geometric optics: the angle of incidence equals the angle of reflection. On a rectangular table, this produces familiar trajectories. Some repeat periodically, some trace diagonals, and others fill up regions with a regular structure. The behaviour is fully determined by the shape of the table and the reflection rule.

When the boundary is not polygonal but curved, the situation changes. An ellipse introduces a new constraint. By definition, the sum of distances from any point on the ellipse to two fixed points, the foci, remains constant. Combined with the reflection law, this implies an optical property: any ray emanating from one focus reflects off the boundary and passes through the other. This follows from the fact that, at the point of reflection, the tangent line to the ellipse bisects the angle formed between the incoming and outgoing paths. That is, the segment from the first focus to the boundary, and then from the boundary to the second focus, meets the boundary at equal angles on either side. The configuration is symmetric, and the total path length is stationary with respect to small variations.

Most trajectories, however, do not begin at a focus. A generic path reflects from arbitrary points on the boundary. Some paths return to their origin after a finite number of reflections; others do not. Some fill annular regions densely, never repeating. The classification of such trajectories depends on both the initial direction and the shape of the boundary.

Now consider two nested, smooth, closed curves, an outer ellipse and an inner ellipse lying entirely within it. Imagine a polygon whose vertices lie on the outer ellipse and whose sides are tangent to the inner ellipse. When the two ellipses are confocal, such a polygon is exactly a closed billiard path— he reflection law at the outer boundary is equivalent to tangency with the confocal caustic. But the construction makes sense for any nested pair: each segment connects two points on the outer ellipse while remaining tangent to the inner one. If such a polygon exists and closes after n steps, it defines a Poncelet polygon.

The question is whether such a configuration is rare. If one closed polygon exists, does that imply anything about other starting points? Does the system admit only a single orbit, or does the existence of one periodic path imply a rule?

Poncelet's Porism answers affirmatively. If there exists a single n-gon inscribed in the outer ellipse and tangent to the inner one, then for every point on the outer ellipse there exists such an n-gon. The ellipse is foliated by periodic trajectories of the same type. The existence of one closed polygon implies the existence of an infinite family, each differing only by a rotation of the starting point.

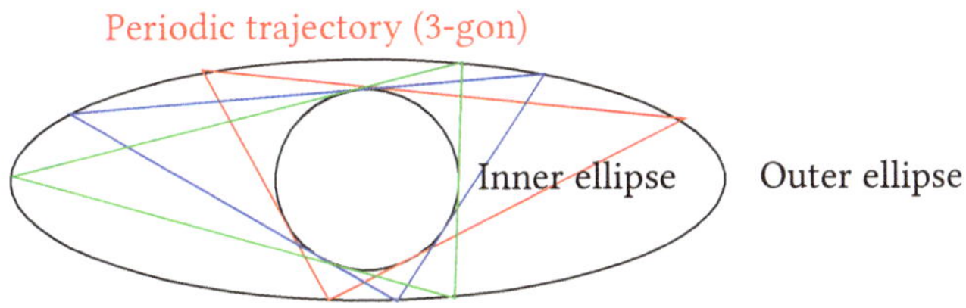

Poncelet's Porism: If one closed trajectory exists, then infinitely many exist.

The geometric construction of Poncelet polygons can be reinterpreted as a dynamical process. Fix two nested ellipses. Choose a point on the outer ellipse and draw a line tangent to the inner ellipse, continuing the segment until it intersects the outer ellipse again. Repeat this procedure: from each new intersection, draw the unique line tangent to the inner ellipse and find its next point of contact with the outer ellipse. The result is a discrete sequence of points on the outer ellipse, each determined from its predecessor. This iteration defines a map from the outer ellipse to itself.

This map—commonly referred to as the Poncelet map—sends a point on the outer ellipse to the next vertex of the corresponding Poncelet polygon.

(Side note: A *porism* is a byproduct of a theorem, usually a corollary which implies infinitely many solutions. Which reminds me of the joke by mathematician Jerry Bona, in which he described the equivalence of the Axiom of Choice (Zermelo, 1904), Zorn's Lemma, and the Well-Ordering Theorem as follows: *"The Axiom of Choice is obviously true, the Well-Ordering Theorem is obviously false, and Zorn's Lemma—who can tell?"* These statements are mathematically equivalent, yet their perceived plausibility varies widely.)

Poncelet originally proved his porism in 1822 using projective geometry and the theory of conics. Later approaches employed algebraic geometry, treating the problem as a question about curves in complex projective space. The configuration of two nested conics defines an elliptic curve, and the closure condition corresponds to torsion points on this curve. These methods, while powerful, require sophisticated machinery from algebraic geometry and complex analysis.

The dynamical systems approach sidesteps this complexity entirely. The question becomes: what kind of transformation is the Poncelet map?

It turns out that this map preserves a specific measure on the ellipse: a notion of size for subsets that remains unchanged under iteration. This measure is not uniform arc length but rather arc length weighted by the distance from each point to its tangency point on the inner ellipse. To see why, first apply an affine transformation to make the inner ellipse into a circle (this preserves tangency and incidence). Now consider nearby points p and p' on the outer ellipse, and their images $T(p)$ and $T(p')$. The tangent chords from these points form two similar triangles; taking the limit as $p' \to p$ and using $\sin\theta \sim \theta$ for small arcs gives $ds_1/ds = |p_1 m|/|pm|$, where m is the tangency point on the chord from p. Since the inner ellipse is now a circle, the two tangent segments from p_1 to it have equal length: $|p_1 m| = \rho(T(p))$. Thus $ds/\rho(p) = ds_1/\rho(T(p))$, and the weighted measure $d\mu = ds/\rho$ is invariant.

This measure is absolutely continuous with respect to arc length (it has a smooth, positive density) and is non-atomic (individual points have zero measure). The existence of such an invariant measure is a special feature of conic geometry and is not generic for arbitrary smooth curves.

A foundational result in ergodic theory—related to the classification of circle homeomorphisms—states that any orientation-preserving homeomorphism of a circle that preserves a finite, non-atomic, absolutely continuous measure must be topologically conjugate to a rigid rotation (Poincaré, 1885). 'Conjugate' here means there exists a homeomorphism φ of the circle such that $\varphi \circ T \circ \varphi^{-1}$ is a pure rotation $R_\alpha : \theta \mapsto \theta + \alpha \mod 2\pi$. The map and the rotation have identical dynamics, just expressed in different coordinates.

The conjugacy parameter α, called the rotation number, is an invariant of the map. It measures the average rate of angular advance per iteration and determines the orbit structure completely. If α is a rational multiple of 2π, say $\alpha = 2\pi p/q$ in lowest terms, then every orbit is periodic with period q. The circle decomposes into congruent periodic orbits, each forming a q-gon in the original ellipse geometry. If α is an irrational multiple of 2π, then no orbit closes. Instead, every orbit is dense: it passes arbitrarily close to every point on the ellipse, filling the curve uniformly in the limit. This is equidistribution, a consequence of Weyl's equidistribution theorem for irrational rotations (Weyl, 1916).

Poncelet's Porism now follows as a corollary. The existence or non-existence of a closed n-gon depends only on whether the rotation number is rational or irrational—a property intrinsic to the pair of ellipses. If one closed polygon exists, the rotation number is rational, and therefore all starting points produce closed polygons with the same number of sides. If no closed polygon exists, the rotation number is irrational, and no starting point produces one.

This same approach appears in problems involving the distribution of leading digits in exponential sequences. The following analysis concerns the frequency of initial digits in powers of integers, governed by irrational rotations on the unit interval.

Consider the powers of 2: $2^1 = 2$, $2^5 = 32$, $2^{10} = 1024$, $2^{53} = 9\,007\,199\,254\,740\,992$. Some begin with 1, others with 3, 9, etc. The digits appear irregularly, but over time, a pattern emerges. The frequency of each digit stabilises and matches a specific distribution.

The explanation is logarithmic. Write $2^n = 10^{n \log_{10} 2}$. The number of digits in 2^n grows roughly linearly in n, and its leading digit depends only on the decimal part (denoted by $\{x\}$) of $n \log_{10} 2$. As n increases, the sequence $\{n \log_{10} 2\}$ fills the interval $[0, 1)$ evenly, because

$\log_{10} 2$ is irrational. This means that 2^n is equally likely to appear in any logarithmic subinterval of a given order of magnitude.

A number begins with digit d if its logarithm lies between $\log_{10} d$ and $\log_{10}(d+1)$. So the proportion of terms 2^n that begin with digit d approaches

$$\log_{10}\left(1+\frac{1}{d}\right).$$

This is Benford's Law (Benford, 1938). It predicts that 1 appears as the leading digit about 30% of the time, while 9 appears less than 5% of the time. The same reasoning applies to any base-b sequence where $\log_{10} b$ is irrational. The decimal parts of $n \log_{10} b$ become evenly spread, and the digit frequencies converge to the same logarithmic formula.

Benford's Law extends far beyond powers of integers. Real-world datasets that span multiple orders of magnitude—river lengths, city populations, physical constants, file sizes, mountain heights—often follow the same logarithmic distribution. The key requirement is that the data range widely without artificial constraints or preferred scales.

When values are distributed across many orders of magnitude, they tend to be spread evenly on a logarithmic scale rather than a linear one. This occurs naturally when data arise from multiplicative processes or when no particular scale is privileged (so it should be invariant to multiplication by a constant). In such cases, the probability of a value falling into the interval that starts with digit d is proportional to the logarithmic length of that interval: $\log_{10}(1+1/d)$. This geometric fact, combined with scale invariance, produces Benford's distribution without requiring randomness or special assumptions. This property makes Benford's Law useful for fraud detection, where artificial datasets often deviate from the expected pattern.

Benford's Law does not apply universally. It fails when numbers are tightly clustered, rounded, or constrained by conventions, such as ID numbers, product prices, or phone records. But when applicable, it can serve as a diagnostic tool. Large deviations from the expected digit frequencies may indicate fabricated or manipulated data.

Forensic accountants have used Benford's Law to uncover anomalies in financial statements, including during the Enron scandal. Tax authorities apply it to detect suspicious filings. In the 2009 Iranian presidential election, some analysts claimed that statistical tests based on Benford's Law identified irregularities in reported vote counts. In each case, observed digit distributions diverged significantly from the logarithmic baseline, suggesting artificial data generation.

This formulation, noted by Gelfand, is identical to the classification of the Poncelet map. In both cases, a rotation acts on a compact one-dimensional space, and the long-term behaviour of orbits (whether periodic or equidistributed) depends only on the rationality of a single parameter. For powers of b, this parameter is $\log_{10} b$; for the Poncelet construction, it is the angular step induced by tangency.

Good Teachers and Invariant Measures

I fell in love with mathematics during my undergraduate studies, particularly through a set theory course taught by Professor Amos Nevo. The material was foundational, focusing on logic, sets, and proofs. Nevo consistently highlighted how abstract structures recur across fields, enhancing its significance. Even when teaching introductory content, he pointed toward broader connections: between algebraic symmetries, analysis, and geometry.

Later, in a seminar on dynamical systems (we were a group of students who tried to enroll in any course Amos offered), he asked me to present a proof of Poncelet's Porism. I began working through the classical approach: projective geometry, dual conics, and elliptic curves. After several weeks of effort, I came to him with the outline. He laughed and said that wasn't the proof he had in mind.

Instead, he pointed me to a short argument we had already studied in class, one based on the existence of an invariant measure. From that, topological conjugacy implies that the Poncelet map behaves like a rigid rotation. Rational rotation number implies periodicity. The porism drops out almost immediately.

The Poncelet map is a beautiful object and exemplifies how abstract tools from dynamical systems, such as invariant measures and topological conjugacy, can be seen doing the heavy lifting on a classical geometry problem. It was also one of the best seminars I ever took for accelerating mathematical maturity.

Mike's understanding of Snell's law made him unbeatable,
but considerably slowed down the game

Poncelet for Two Ellipses

Let C and D be two smooth, strictly convex, nested ellipses in the plane, with $D \subset \text{int}(C)$. The *Poncelet map* $T : C \to C$ is defined as follows: for a point $p \in C$, draw the line through p tangent to the inner ellipse D (choosing one of the two tangent directions consistently and smoothly along C); let $T(p)$ be the second point of intersection of this line with C. This consistent choice defines an orientation-preserving homeomorphism of C.

We show that T preserves a natural measure and is topologically conjugate to a circle rotation. This yields a complete classification of the dynamics of T, and with it, a proof of Poncelet's closure theorem.

Affine Reduction and the Invariant Measure To analyze T, we apply two affine reductions. First, by dilating along one axis of D, we may assume the inner ellipse D is a circle; this preserves incidence and tangency, so the Poncelet map is unchanged. Now suppose C is given by

$$x^2/a^2 + y^2/b^2 = 1, \quad \text{with } a > b > 0.$$

Let $U(x, y) = (x/a, y/b)$, sending C to the unit circle $\widetilde{C}$ and D to an ellipse $\widetilde{D}$. Let $\tilde{s}$ denote arc-length on $\widetilde{C}$. For $p \in C$, let m be the tangency point of the chord $p\,T(p)$ with D, and let m_1 be the tangency point of the next chord $T(p)\,T^2(p)$. Set $\rho(p) := |p - m|$. The invariant measure is $d\mu := d\tilde{s}\,/\,\rho$, mixing arc-length on $\widetilde{C}$ with the tangent-segment distance in the original (circular-D) coordinates.

Invariance of the Measure. For nearby $\tilde{p}, \tilde{p}' \in \widetilde{C}$, the chords $\tilde{p}\tilde{p}_1$ and $\tilde{p}'\tilde{p}'_1$ meet at a point $\tilde{n}$, forming similar triangles. As $\tilde{p}' \to \tilde{p}$, $\tilde{n} \to \widetilde{m}$ and chord ratios on $\widetilde{C}$ converge to arc-length ratios ($\sin\theta/\theta \to 1$), so

$$\frac{d\tilde{s}_1}{d\tilde{s}} = \frac{|\tilde{p}_1\widetilde{m}|}{|\tilde{p}\widetilde{m}|} = \frac{|p_1 m|}{|pm|},$$

where the second equality holds because U is linear and preserves distance ratios. Since D is a circle, the two tangent segments from p_1 to D are equal: $|p_1 m| = |p_1 m_1| = \rho(T(p))$.

Hence

$$\frac{d\tilde{s}_1}{d\tilde{s}} = \frac{\rho(T(p))}{\rho(p)}, \quad \text{i.e.,} \quad \frac{d\tilde{s}}{\rho(p)} = \frac{d\tilde{s}_1}{\rho(T(p))},$$

and $d\mu = d\tilde{s}\,/\,\rho$ is T-invariant.

Topological Conjugacy to a Circle Rotation An orientation-preserving circle homeomorphism with a finite non-atomic invariant measure positive on every non-empty open arc has no wandering intervals (a result due to Poincaré); therefore its Poincaré semiconjugacy to a rotation is a conjugacy.

Our measure μ satisfies: **Finite:** Since C and D are smooth and strictly convex with $D \subset \text{int}(C)$, $\tilde{\rho}$ is continuous on the compact set C' and strictly positive, so it has a positive infimum. Thus $\int_{C'} d\tilde{s}/\tilde{\rho} < \infty$. **Non-atomic:** points have zero measure. **Positive on arcs:** $\tilde{\rho}$ is bounded above on any arc, so $1/\tilde{\rho}$ is bounded below by a positive constant, ensuring open arcs have positive measure.

Thus, there exists a homeomorphism $\varphi : C \to S^1$ and $\alpha \in [0, 1)$ such that: $\varphi \circ T \circ \varphi^{-1} = R_\alpha$, where $R_\alpha(\theta) = \theta + \alpha \mod 1$

The number α, called the *rotation number* of T, quantifies the average angular displacement per iteration. Lift the circle map to $F : \mathbb{R} \to \mathbb{R}$ in an angular coordinate; then $\alpha := \lim_{n\to\infty} (F^n(\theta) - \theta)/n$ exists and is independent of θ. The rotation number of T is $\alpha \bmod 1$.

The rotation number classifies the dynamics:

- If $\alpha \in \mathbb{Q}$, write $\alpha = m/n$ in lowest terms. Then all orbits are periodic with period n. Every point on C traces a closed n-gon tangent to D.
- If $\alpha \notin \mathbb{Q}$, then no orbit is periodic. The sequence $p, T(p), T^2(p), \ldots$ becomes dense in C, and no Poncelet polygon closes.

References:

Leopold Flatto, *Poncelet's Theorem.* Mathematical Surveys and Monographs, Vol. 56. American Mathematical Society, 2009 (beautiful book!)
For visualization, see bit.ly/poporism.

Mind the Gap

Top (Prime Distribution): Natural numbers 1–25 with primes (red) and composites (teal), showing that twin prime pairs like (3,5), (5,7), (11,13), (17,19).

Second (Gap Sizes): Consecutive primes with gap sizes marked by blue arrows, from the minimal gap of 2 to larger separations such as 8 between 23 and 31.

Third (Gap Analysis): Scatter plot of prime gaps $p_{n+1} - p_n$ versus p_n for the first 10,000 primes, showing that most gaps are small with rare large exceptions. Reference lines mark key bounds: twin primes (gap = 2), Polymath8's refined bound (246), and Zhang's original breakthrough (70 million). Infinitely many gaps stay below some fixed bound, guaranteeing points always appear below these lines.

Note the key distinction: for small gaps we seek absolute bounds (fixed numbers such as 246), while for large gaps the bounds are functions of p_n that grow as primes get larger.

Bottom (Maximum Gaps): How large can prime gaps become? This shows the largest observed gaps (light blue) compared to theoretical predictions. The red lines show upper bound conjectures for maximum gap size, while coloured lower bounds prove that gaps must occasionally be large. The scatter demonstrates that while most gaps are small relative to the local prime density, some gaps are much larger than the typical spacing in their region.

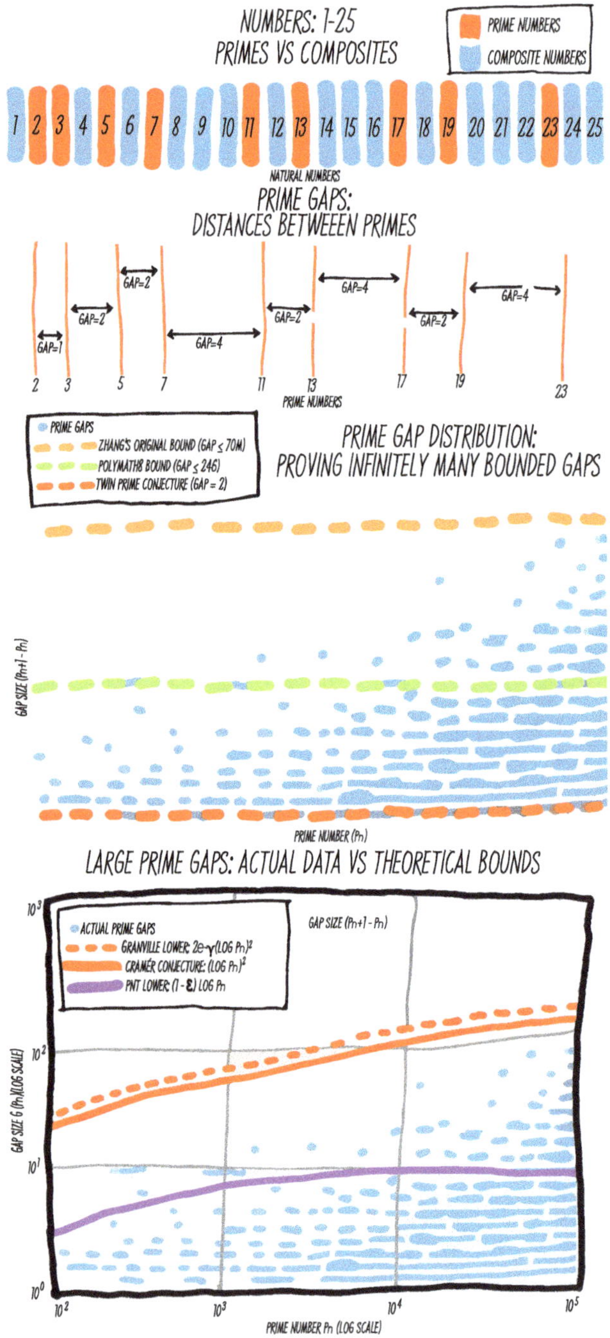

Mind the Gap

In 2013, an unaffiliated Yitang Zhang proved there exists a finite bound B (initially 70,000,000) such that infinitely many prime pairs differ by at most B. While prime gaps can grow arbitrarily large, this breakthrough showed they cannot drift apart arbitrarily far. The Polymath8 collaboration subsequently reduced this bound to a few hundred. Zhang's approach combined distribution properties of primes in arithmetic progressions with an advanced sieving technique, resolving a fundamental question about number patterns while falling short of proving the Twin Prime Conjecture that infinitely many primes differ by exactly 2.

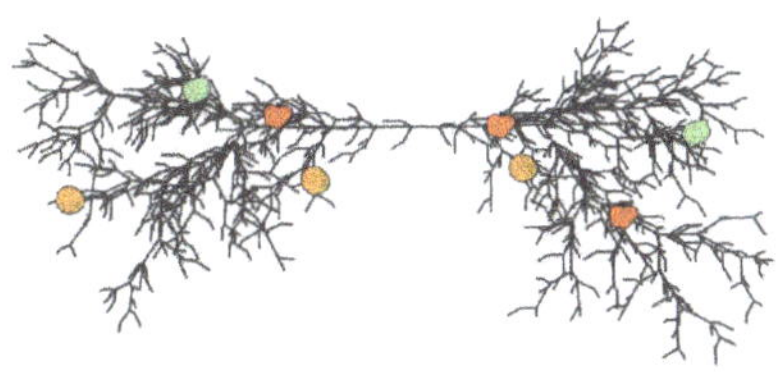

Prime Gaps ◦ Twin Prime Conjecture ◦ Zhang's Bounded Gaps ◦ 70 Million Bound ◦ Weighted Sieve Method ◦ Polymath8 Collaboration ◦ Maynard Simplification ◦ Green-Tao Progressions ◦ Cramér's Conjecture ◦ Riemann Hypothesis ◦ Zhang's Trajectory

"If I were to awaken after having slept
for a thousand years,
my first question would be:
has the Riemann Hypothesis been proven?"
— David Hilbert, 1900

Mind the Gap

In the third century BCE, Euclid proved that there are infinitely many prime numbers. His argument, based on contradiction, became one of the earliest and most enduring examples of a general mathematical method. The search for primes—and for patterns among them—soon followed. Eratosthenes introduced a sieve procedure (an algorithm that filters out composites by eliminating multiples) for enumerating primes. By the time of Diophantus primes were already recognised as foundational to arithmetic.

In the eighteenth century, Euler showed that the sum of reciprocals of primes diverges (a stronger quantitative refinement of Euclid's theorem that, in particular, implies infinitude), and he introduced analytic tools that connected primes to infinite products and logarithmic identities. This initiated the study of prime distribution through analytic functions.

In 1859, Bernhard Riemann introduced the zeta function into number theory (a complex analytic function encoding prime information via its Euler product) and conjectured that all its nontrivial zeros lie on the critical line. This hypothesis remains unproven. Riemann's formulation marked the beginning of analytic number theory—a field that uses tools from complex analysis to study the distribution and density of primes. G. H. Hardy and others developed this perspective further in the early twentieth century.

The study of prime gaps took a more technical turn when Viggo Brun introduced sieve methods in the 1910s (combinatorial procedures for bounding the count of integers with prescribed divisibility). Brun proved that the sum of reciprocals of twin primes converges, implying their overall scarcity, even if they might be infinite in number. Later refinements by Selberg and Bombieri led to the Bombieri–Vinogradov theorem (an average-case version of the Generalised Riemann Hypothesis for arithmetic progressions), which became central to modern sieve theory.

In the early 2000s, Goldston, Pintz, and Yıldırım (GPY) introduced a method for bounding small gaps between primes using weighted sums over admissible tuples (integer patterns that avoid local divisibility obstructions—for example, $0, 2, 4$ is not admissible, since modulo 3 it covers all residue classes, whereas $0, 2, 6$ is admissible). Their work showed that if primes are sufficiently regular in arithmetic progressions, then bounded gaps should follow. The approach relied on conjectural input—notably the Elliott–Halberstam conjecture (a proposed uniformity result for primes in arithmetic sequences).

In 2013, Yitang Zhang proved that there are infinitely many pairs of primes separated by at most 70 million. His argument used a modified version of the GPY sieve. This was the first proof that bounded prime gaps occur infinitely often, without relying on unproven conjectures.

Prime numbers are integers greater than 1 that have no positive divisors other than 1 and themselves. They form the multiplicative building blocks of arithmetic. The first few primes, 2, 3, 5, 7, 11, 13, 17, 19, 23, 29, occur without any obvious pattern. Their spacing varies.

 https://doi.org/10.11647/OBP.0526.08

As numbers increase, primes become less frequent, because they have more possible factors. The Prime Number Theorem formalises this observation (Hadamard, 1896; de la Vallée Poussin, 1896): the number of primes less than x grows like $x/\log x$. This gives an average spacing between primes near x of about $\log x$, but does not constrain individual gaps.

The differences between consecutive primes can be small or large. Pairs such as (3, 5), (5, 7), (11, 13), (17, 19), and (29, 31) each differ by 2. The smallest prime gap of 2 recurs often at the start of the number line. By contrast, the primes 370,261 and 370,373 are separated by 112, with no primes between them. For any given n, there exist consecutive primes with a gap larger than n. One construction uses the sequence $(n+1)!+2, (n+1)!+3, \ldots, (n+1)!+(n+1)$, which yields n consecutive composite numbers, hence a gap of at least n between the bounding primes.

What remains unknown is whether small gaps, such as a fixed difference of 2, occur infinitely often. The Twin Prime Conjecture asserts that there are infinitely many primes p such that $p+2$ is also prime. This remains one of the major conjectures in number theory.

In 2013, Yitang Zhang proved that there exists a constant B such that infinitely many pairs of primes differ by at most B (as opposed to exactly a fixed gap such as 2 in the twin prime conjecture). His original bound was $B < 70{,}000{,}000$—while this does not resolve the twin prime conjecture, it proves that small prime gaps occur infinitely often.

Zhang's method extended work by Goldston, Pintz, and Yıldırım. He combined improved estimates on the distribution of primes in arithmetic progressions with a weighted sieve construction that amplified configurations where primes appear close together. This yielded a finite bound on the gap size that recurs infinitely often.

Following Zhang's proof, the Polymath8 collaboration reduced the bound from 70 million to below 250 through analytic refinements. Later on, James Maynard introduced a simplified sieve method that removed the need for strong distributional estimates and extended the technique to detect many primes within bounded intervals.

Zhang's result drew attention for its mathematical content and the circumstances of its discovery—after completing his doctorate, he spent years outside academic mathematics, with no permanent university position and limited research output. His proof was written and submitted independently, lacking collaborators or institutional support. The publication of his result led to rapid follow-up work, large-scale collaboration, and the re-entry of a long-standing problem into the mathematical mainstream.

While Zhang's work addressed bounded gaps, the opposite question—how large prime gaps can become—has also attracted intense study. Let $G(x)$ be the largest gap between consecutive primes less than x. It is known that $G(x)$ increases faster than $\log x$, which is the average spacing predicted by the Prime Number Theorem. A classical result due to Paul Erdős shows that $G(x)$ exceeds a constant multiple of $\log x$ times another slowly growing function. This means that although most prime gaps are relatively small, unusually large gaps must still occur infinitely often. The best known upper bounds on $G(x)$ remain far from matching the lower bounds. Some of the strongest predictions, such as Cramér's conjecture, suggest that the maximal gap should grow no faster than $\log^2 x$, but this has not been proven.

Alongside these increasingly long gaps, primes can also appear in patterns. In 2004, Ben Green and Terence Tao proved that the primes contain arbitrarily long arithmetic progressions. For any integer k, there exists a sequence of the form $p, p+d, p+2d, \ldots, p+(k-1)d$ in which all terms are prime. The length k can be taken as large as desired. Although such progressions become rarer as k increases, the result shows that they never stop appearing.

The Green–Tao theorem uses tools from ergodic theory and additive combinatorics—it begins by approximating the set of primes using related sequences whose behaviour is easier to control. A transference principle then carries results from these surrogate sequences back to the primes themselves. The original result was later extended by Green and Ziegler to cover polynomial progressions, such as $p, p+q, p+4q, p+9q, p+16q$ in which the differences between terms follow a fixed polynomial pattern and all terms are again required to be prime.

Many questions regarding the distribution of primes, including the spacing between them and the occurrence of patterned arrangements, remain unresolved. Some of these questions cannot be settled with current methods because their answers depend on an open conjecture in complex analysis and number theory. This conjecture is known as the Riemann Hypothesis (Riemann, 1859).

The Riemann Hypothesis (RH) concerns a function called the Riemann zeta function. This function is initially defined as a sum over positive integers, $\zeta(s) = \sum_{n=1}^{\infty} \frac{1}{n^s}$, which converges when the complex number s has real part greater than 1. Through a process known as analytic continuation, the function is extended to other values of s in the complex plane. The hypothesis asserts that all nontrivial zeros of $\zeta(s)$—that is, all values of s for which $\zeta(s) = 0$ and which are not negative even integers—lie on the vertical line $\mathrm{Re}(s) = \frac{1}{2}$.

Assuming RH, bounds on the error terms in prime-counting functions become significantly tighter. For example, the difference between the actual number of primes up to x and the estimate $x/\log x$ can be bounded more sharply. The hypothesis also leads to improved estimates on how often primes occur in short intervals and how large the gaps between consecutive primes can be. Without a proof, many of these refinements remain conditional. The Riemann Hypothesis has been verified for many individual zeros through numerical computation, and no counterexamples have been found. Nevertheless, the general statement remains unproven.

Transparent Statements, Resistant Proofs

The central claim of this chapter can be stated in one line and requires no definitions beyond the integers. This is typical of many problems in number theory. Simplicity of formulation does not imply tractability.

Consider these easy-to-state problems that remain unsolved:

- **Twin Prime Conjecture**: Are there infinitely many primes p such that $p + 2$ is also prime?
- **Goldbach's Conjecture**: Can every even integer greater than 2 be written as the sum of two primes?
- **Odd Perfect Numbers**: Does there exist an odd perfect number (a number equal to the sum of its proper divisors)?

Each can be explained to a child, yet they have resisted centuries of mathematical effort. They can be tested on billions of examples, but no general proof exists.

Zhang's proof is concise and intricate. Its validity depends on a balance between distributional estimates and the sieve framework. Subsequent refinements by the Polymath8 project, led by Terence Tao, and by Maynard's independent method reduced the bound but did not simplify the analytic core.

This chapter is included because of Zhang's personal trajectory and because it illustrates this broader principle: many problems in number theory are easy to state and test numerically, yet remain inaccessible to current methods. Let's demonstrate it further.

The following questions of existence, stated in a single line, can fall anywhere along the spectrum of difficulty:

- **Impossible to answer**: Hilbert's Tenth Problem asked for a general procedure to decide whether any Diophantine equation (equation in integer variables with polynomial coefficients) has a solution. Matiyasevich's work, building on Davis–Putnam–Robinson, showed (Matiyasevich, 1970) that no such algorithm can exist.
- **Unknown**: The *Collatz Conjecture* asks whether repeated iteration of the rule $n \mapsto n/2$ if n is even and $3n + 1$ if n is odd always reaches 1. No proof is known.
- **Hard No**: Fermat's Last Theorem—that $x^n + y^n = z^n$ has no nontrivial integer solutions for $n > 2$—resisted proof until Wiles.

Proofs: Infinitude of Primes

Two nice proofs of the infinitude of primes—one using topology and another via a trigonometric product identity—.

1. **Furstenberg's Topological Proof**
   Define a topology on $\mathbb{Z}$ by taking as a basis of open sets all arithmetic sequences of the form:
   $$S_{a,b} = \{a + bn \mid n \in \mathbb{Z}\}.$$
   These sets mimic modular residue classes and are closed under finite intersections, forming a well-defined topology. Suppose there are only finitely many primes, $p_1, p_2, \dots, p_k$. The union of their corresponding arithmetic sequences,
   $$S = \bigcup_{i=1}^{k} S_{0,p_i} = \{n \mid n \equiv 0 \mod p_i \text{ for some } i\},$$
   consists of all integers divisible by at least one prime and would be a closed set. Its complement, the set of integers not divisible by any p_i, must then be open. However, every basic open set $S_{a,b}$ is infinite, meaning an open set cannot be finite. Since the complement of S consists of finitely many integers (the units ± 1 modulo $p_1 p_2 \dots p_k$), we obtain a contradiction. Thus, the assumption that the set of primes is finite must be false.
2. **Trigonometric Product Proof**
   Assume for contradiction that the set $\mathcal{P}$ of primes is finite. Then consider the product:
   $$0 < \prod_{p \in \mathcal{P}} \sin(\pi/p).$$
   Since each term is positive, the product itself remains positive.
   Now define $N = 2\prod_{p'} p'$, the product of all assumed primes. By construction, every prime p divides N, so the term $1 + N$ must be divisible by some prime q. That is, for some integer k, $1 + N = kq$. Then, evaluating the sine function:
   $$\sin(\pi(1+N)/q) = \sin(k\pi) = 0.$$
   This forces the right-hand side of the original product identity to be zero:
   $$\prod_{p \in \mathcal{P}} \sin(\pi(1+N)/p) = 0.$$
   But this contradicts the assumption that the left-hand side was positive. Hence, the set of primes must be infinite.

Remark: Both proofs subtly rely on Euclid's key step: that some prime p must divide the product of primes plus one, making these proofs disguised versions of the classic argument.

References: H. Furstenberg, *Bull. Amer. Math. Soc.*, 62 (1955), 353.& *A One-Line Proof of the Infinitude of Primes*, Amer. Math. Monthly, 122 (2015), 466.

Bounded Gaps Between Primes

Statement of Result and Outline of Method

In 2013, Yitang Zhang proved that there exists a constant N such that infinitely many prime pairs (p, q) satisfy $q - p \leq N$. The initial bound was $N < 7 \times 10^7$. Zhang's approach refined the Goldston–Pintz–Yıldırım (GPY) method through two components:

- **Distribution in Arithmetic Progressions:** Primes remain evenly distributed across residue classes beyond the Bombieri–Vinogradov range.
- **Weighted Sieve:** Modified sieve to detect multiple primes within admissible tuples $n + h_i$.

Sieve Setup and Admissibility

A tuple $\mathcal{H} = \{h_1, \dots, h_k\}$ is admissible if for every prime p, the set $\mathcal{H} \mod p$ does not cover all residue classes modulo p.

The GPY strategy constructs a weighted sum:

$$S(n) := \left(\sum_{i=1}^{k} \Lambda(n + h_i) \right) w(n),$$

where Λ is the von Mangoldt function and $w(n)$ is a smooth function supported on $n \in [x, 2x]$. The weights emphasise values where multiple $n + h_i$ are likely prime:

$$\sum_n S(n) = \sum_n \left(\sum_i \Lambda(n + h_i) \right) w(n).$$

If this exceeds the random baseline, then for some n, at least two $n + h_i$ are prime.

Example

The admissible set $\{0, 2, 6\}$ avoids covering all residue classes modulo any prime. For $n = 5$, we get $\{5, 7, 11\}$—three primes. The method proves such cases occur infinitely often.

Zhang's Level of Distribution

Zhang proved that primes remain equidistributed up to moduli $q \leq x^\theta$ for $\theta > 1/2$, surpassing the Bombieri–Vinogradov barrier ($\theta = 1/2$).

Define:

$$\theta(x; q, a) = \sum_{\substack{p \leq x \\ p \equiv a \pmod q}} \log p.$$

The deviation of $\theta(x; q, a)$ from $x/\phi(q)$ remains small across a wide range of moduli, enabling uniform error control. Zhang bypassed the Elliott–Halberstam conjecture by achieving a weaker but sufficient level of distribution.

Maynard's Modification and Polymath Refinements

Maynard introduced new sieve weights detecting primes in admissible tuples without requiring $\theta > 1/2$, simplifying the construction.

The Polymath8 project refined and extended both approaches:

- **Polymath8a:** Improved error analysis, reduced N to 4,680.
- **Maynard's Variant:** Lowered N further, generalised to m primes in bounded intervals.
- **Polymath8b:** Reduced bound below 250.

References:

Zhang, Y. (2014). Bounded gaps between primes. *Annals of Mathematics*, 179(3), 1121–1174.

Maynard, J. (2015). Small gaps between primes. *Annals of Mathematics*, 181(1), 383–413.

Real Democracy Has Never Been Tried

Top (Voting methods): Ranked-choice voting is a system in which voters express their preferences by submitting complete rankings of all candidates, and the system aggregates them into a ranked list or a single winner. Different aggregation methods (Borda count, instant-runoff voting (IRV), plurality, Condorcet) can produce distinct winners from identical voter rankings, demonstrating the inherent ambiguity in collective decision-making. The same preference profile can yield different outcomes depending on which features of the rankings are emphasised by the chosen method. This indeterminacy means there is no canonical way to translate individual preferences into collective choices.
Bottom (Arrow's Theorem): Arrow's impossibility theorem proves that no ranked-choice voting method can satisfy all four fairness criteria simultaneously when there are at least three alternatives and two voters. Every democratic aggregation method must compromise at least one criterion, making trade-offs unavoidable in social choice.

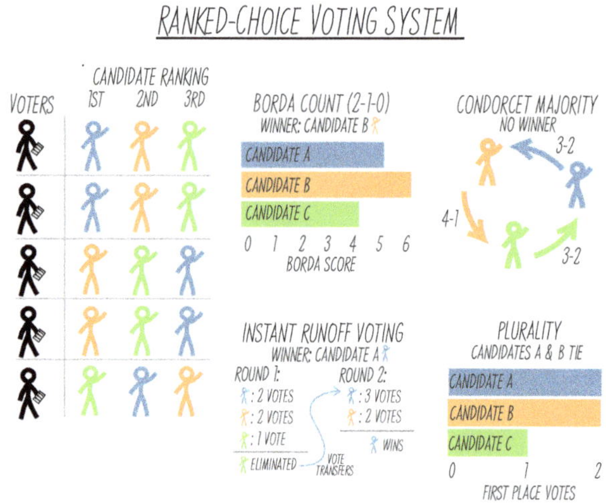

Real Democracy Has Never Been Tried

Arrow's Impossibility Theorem shows that no voting rule can convert individual rankings into a collective decision without violating at least one basic principle of fairness. What seems like a straightforward requirement for democracy turns out to be mathematically impossible, leaving every voting system to sacrifice some aspect of fairness.

VOTING SYSTEMS ∘ ARROW'S IMPOSSIBILITY ∘ PREFERENCE RANKINGS ∘ CONDORCET CYCLES ∘ INDEPENDENCE OF IRRELEVANT ALTERNATIVES ∘ UNANIMITY & NON-DICTATORSHIP ∘ PLURALITY VS BORDA ∘ INSTANT RUNOFF ∘ UNAVOIDABLE TRADE-OFFS ∘ SOCIAL CHOICE THEORY ∘ LONELY RUNNER CONJECTURE

"Democracy is the worst form of government,
except for all those other forms that have been tried from time to time."
— Winston Churchill, 1947

"People like Coldplay and voted for the Nazis.
You can't trust people, Jeremy."
— Super Hans, 2005

Real Democracy Has Never Been Tried

> In the mid-twentieth century, formal models of collective decision-making began to draw the attention of economists, political theorists, and mathematicians. Rather than treating voting as a procedural artefact, researchers sought to characterise what could or could not be achieved when individual preferences are aggregated into a group decision.
>
> Kenneth Arrow's work in 1951 became a cornerstone of this approach. During the following decades, related work by Allan Gibbard and Mark Satterthwaite (Gibbard, 1973; Satterthwaite, 1975) showed that even the absence of strategic manipulation was mathematically incompatible with certain fairness assumptions. These findings anchored a larger research programme that explored the logical trade-offs inherent in any decision procedure.
>
> By the 1980s, scholars such as Donald Saari and Michel Balinski introduced geometric and algebraic methods into the analysis of voting rules. These approaches revealed that many well-known paradoxes arise not from particular cases, but from the geometry of the space in which preference profiles reside. The field began to borrow tools from topology, convex geometry, and representation theory, linking social-choice questions to broader developments in pure mathematics.

A voting system takes as input a collection of individual preference rankings and produces a single collective outcome. Each voter submits a total ordering of the available options, specifying a sequence from most to least preferred, without ties. The voting rule processes these inputs and returns either a single winner or a complete ranking of the options according to the aggregated preferences.

These systems formalise the process of collective decision-making—their applications extend beyond political elections and include committee procedures, academic appointments, boardroom voting, and algorithmic decision-making in multiagent systems and recommendation engines.

The input to a voting system is called a profile: a multiset of total orders, with one such order for each voter. Each total order ranks the finite set of alternatives such that every candidate is assigned a unique position. This representation provides the basis on which aggregation rules operate.

The number of possible total orderings grows rapidly with the number of options. For three candidates, there are $3! = 6$ distinct rankings; for five candidates, there are $5! = 120$. This factorial growth introduces combinatorial complexity, making it infeasible to analyse all possible configurations exhaustively once the number of alternatives exceeds a small threshold.

Certain properties are often desired of voting rules. Anonymity requires that the outcome not depend on which voter submitted which ballot. Neutrality requires that all candidates be treated symmetrically. These properties enforce symmetries on the rule and ensure that it operates independently of irrelevant identifiers.

 https://doi.org/10.11647/OBP.0526.09

Additional desirable properties include monotonicity and consistency. A system is monotonic if ranking a candidate higher on a ballot cannot reduce that candidate's chances of winning. It is consistent if identical outcomes from separate groups imply the same outcome when those groups are merged. These properties aim to prevent procedural anomalies that would violate intuitive fairness.

Several voting systems are widely implemented in practice. These include Plurality voting, the Borda count, Condorcet methods, and instant-runoff voting. Each of these systems interprets the profile differently and emphasises different aspects of the ranking data.

Plurality voting considers only the top-ranked candidate on each ballot. The candidate receiving the most first-place votes is declared the winner. All lower-ranked information is discarded, making the system computationally simple but sensitive to strategic voting and vote splitting. Most U.S. House and state legislative races, and U.K. parliamentary constituencies, use this 'first-past-the-post' system. U.S. presidential electors are chosen by state-level plurality in most states.

The Borda count assigns a score (de Borda, 1781) to each candidate based on their rank position on each ballot. For example, in a three-candidate election, a first-place vote may yield two points, second place one point, and third place zero. These points are then summed across all ballots, and the candidate with the highest total score wins. This method incorporates more information from the ranking but can fail to elect a candidate who would beat all others in head-to-head contests. Some academic societies and professional organisations use Borda counting for internal elections.

Condorcet methods use pairwise majority comparisons (Condorcet, 1785) between candidates. For each pair, the number of voters who prefer one candidate to the other is counted. If a candidate defeats every other candidate in these head-to-head contests, that candidate is called the Condorcet winner. Not all profiles contain such a candidate, and additional rules are required when cycles occur. The Debian Project, a Linux distribution, uses a Condorcet method to elect its project leader.

Instant-runoff voting proceeds through iterative elimination. In each round, the candidate with the fewest first-place votes is eliminated, and those ballots are reassigned to the next preferred remaining candidate. The process continues until a single candidate remains. This system allows voters to express multiple preferences but can still produce paradoxical reversals when a candidate gains additional support. Australia's House of Representatives, Ireland's presidency, and several U.S. cities including San Francisco employ instant-runoff voting.

Each of these methods can produce different results on the same input profile—the choice of rule determines which features of the preferences are preserved and which are ignored. No method fully captures all intuitively fair principles, and the differences between them reflect the trade-offs inherent in social choice.

Cycles in group preferences arise even when individual preferences are fully ordered and consistent. Each voter's ballot may assign a strict ranking to all options, but the collective result may still fail to satisfy transitivity. For example, candidate A may be preferred to B by a majority, B preferred to C, and yet C preferred to A, forming a cycle. This outcome cannot be mapped to a total order and reveals a limitation that no voting rule can avoid in all cases.

Any system that outputs a full group ranking must address such cycles. One approach is to discard some pairwise comparisons and resolve the ranking using the remaining ones. Another approach is to introduce external tie-breaking rules, which may depend on arbitrary or external criteria. Both strategies impose coherence on an input that may not support it.

Kenneth Arrow proposed a system to evaluate the reasonableness of complete voting systems. He identified four conditions that any acceptable aggregation rule might aim to satisfy: unanimity, non-dictatorship, independence of irrelevant alternatives, and transitivity.

Unanimity requires that if all voters rank option X above option Y, then the group outcome must reflect that same order. Non-dictatorship ensures that no single voter can always determine the result regardless of the others' preferences. These two conditions express responsiveness and fairness.

The third condition, independence of irrelevant alternatives (IIA), states that the group preference between any two options X and Y should depend only on how voters rank X relative to Y. Preferences involving other options must not affect the outcome of this pairwise comparison. This condition ensures that unrelated rankings cannot distort local outcomes.

Transitivity requires consistency across comparisons: if the group prefers X to Y and Y to Z, it must also prefer X to Z. If transitivity fails, the output cannot be interpreted as a ranking at all. It contains loops that prevent any ordering from being formed.

Arrow's impossibility theorem proves that no method (Arrow, 1951) satisfies all four conditions (unanimity, non-dictatorship, IIA, and transitivity) when there are at least three options and at least two voters.

Trade-offs are unavoidable in voting systems—which does not imply that voting is invalid. Some methods give up IIA to maintain collective agreement and equal treatment of voters. Others allow intransitive outcomes in order to preserve independence or avoid concentrated control. Every system must fail at least one of the criteria.

Mathematics Beyond Physics

I included this chapter to demonstrate that mathematical limitations, though typically associated with physical systems, can apply to social processes as well.

As an undergraduate, I took courses in game theory given by Ron Holzman, who has Erdős number 1 for a paper on maximal triangle-free graphs (where any added edge creates a triangle).

Ron also co-authored, in 2001 with Tom Bohman and Dan Kleitman, a proof of the $n = 6$ case of the Lonely Runner Conjecture (Wills, 1967). This conjecture states that for any n runners moving at constant but distinct speeds around a circular track of unit length, there exists a time when each runner is at least $1/n$ of the track away from every other runner. The problem models the moment when each one is 'lonely,' meaning far enough from all others to be considered isolated.

I loved this problem because the cases $n = 2$ and $n = 3$ are easy to visualise and prove (albeit lengthy for $n = 3$) using elementary arguments, yet the general case has resisted a proof for over fifty years. It has been proven for $n \leq 7$, with recent preprints claiming $n \leq 10$.

The conjecture has an elegant reformulation in terms of number theory and approximation on the unit circle. For a runner moving at integer speed v, define the Bohr set $B(v, \delta) = \{t \in \mathbb{R}/\mathbb{Z} : \|vt\| \leq \delta\}$, where $\|x\|$ denotes the distance from x to the nearest integer. This set consists of all times when the runner is within a distance δ of their starting point.

This question, about runners on a track, can be rephrased in terms of fractional parts, Bohr sets, and the geometry of coverings in $\mathbb{R}/\mathbb{Z}$, which illustrates how problems from everyday intuition often touch the edge of what mathematics can currently answer.

When Matthieu Rosenfeld proved the $n = 8$ case in 2025 using computer-assisted backtracking over prime divisors, I thought I could help push the result further. His verification reduces, for each prime p, to showing that no 'bad' covering exists—a problem close to set cover, which is close to SAT. I reformulated his conditions as a Boolean satisfiability instance and ran Kissat, a state-of-the-art SAT solver. The result was humbling: for $k = 5$ and $d = 31$, his ad-hoc backtracking finished in 0.02 seconds; Kissat took 59 seconds on the same instance. The SAT encoding, with its monotone clauses and pseudorandom structure, fell outside the regime where conflict-driven clause learning excels. Rosenfeld cited the attempt in his subsequent $n = 9$ proof, which was generous given that I contributed a negative result. He then proved $n = 9$; Trakulthongchai independently proved $n = 9$ and $n = 10$ shortly after.

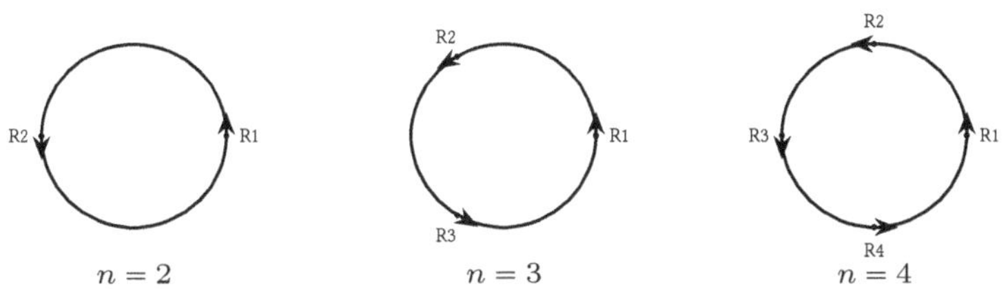

Figure by author. The Lonely Runner Conjecture for $n = 2$, 3, and 4. Each circle shows a time when every runner is at least $1/n$ of the track away from all others. Arrows indicate the direction of motion.

Radio Yerevan Jokes

Q: Radio Yerevan was asked: "Capitalism is the exploitation of man by man. What about Socialism?"
A: Radio Yerevan answered: "Under Socialism it is exactly the other way around."

Q: Is there freedom of speech in the Soviet Union the same as in the USA?
A: In principle, yes. In the USA, you can stand in front of the Washington Monument and yell, "Down with Reagan!" and you will not be punished. In the Soviet Union, you can stand in Red Square and yell, "Down with Reagan!" and you will not be punished.

Q: What is chaos?
A: We do not comment on national economics.

Q: Is it true that half of the members of the Central Committee are fools?
A: What a crazy question. Half of the members of the Central Committee are not fools!

Q: Is it true that Ivan Ivanovich Ivanov from Moscow won a car in a lottery?
A: In principle, yes. But: it wasn't Ivan Ivanovich Ivanov but Aleksander Aleksandrovich Aleksandrov; he is not from Moscow but from Odessa; it was not a car but a bicycle; and he didn't win it—it was stolen from him.

Q: Why is our government not in a hurry to land our men on the moon?
A: What if they refuse to return?

Q: We are told that communism is already visible on the horizon. What then is a horizon?
A: An imaginary line that moves farther away each time you approach it.

Q: Is it true that conditions in our labour camps are excellent?
A: In principle, yes. Five years ago, one of our listeners was sceptical, so he was sent to investigate. He must have liked it so much that he hasn't returned yet.

Q: What is the most beautiful city in the Soviet Union?
A: Yerevan, obviously.
Q: How many nuclear bombs will it take to destroy Yerevan?
A: Baku is also a beautiful city.

Q: Is it true that the United States is on the edge of a precipice?
A: True, but we are a step farther than them.

Topological Proof of Arrow's Impossibility Theorem

Let Ω_m be the set of all strict total orderings over m alternatives. Each ranking is a permutation of the m options, so Ω_m has $m!$ elements. For n voters, a preference profile is a point in the product space

$$\mathcal{P} = \Omega_m^n,$$

which contains every possible combination of rankings across the electorate. A social welfare function (SWF) is a map

$$F : \mathcal{P} \to \Omega_m,$$

assigning to each profile a collective ordering.

Arrow's theorem asserts that no such function exists satisfying all of the following properties (for $m \geq 3, n \geq 2$):

1. *Unrestricted Domain*: The SWF must accept *any* logically possible profile from $\mathcal{P} = \Omega_m^n$ as input—no restrictions on which combinations of voter preferences are permitted.
2. *Pareto Efficiency*: If every voter ranks $x \succ y$, then $F(\mathbf{P})$ must also rank $x \succ y$.
3. *Independence of Irrelevant Alternatives (IIA)*: The social ranking of x and y depends only on how voters rank x versus y, not on preferences over other candidates.
4. *Non-Dictatorship*: No single voter's preferences always determine the group ranking.
5. *Transitivity*: The output $F(\mathbf{P}) \in \Omega_m$ is inherently a strict total order, so transitivity is guaranteed by construction—F must map into the space of transitive orderings.

To describe the topological version, consider $\mathcal{P}$ as a discrete high-dimensional complex. Each profile is a vertex, and edges connect profiles differing by a single adjacent transposition in one voter's list. This adjacency pattern turns $\mathcal{P}$ into a combinatorial manifold with rich connectivity, encoding the geometry of preference space.

IIA implies that for each pair (x, y), the collective ranking between x and y is determined by the projection

$$\pi_{xy} : \mathcal{P} \to \{x \succ y, y \succ x\}^n,$$

where $\pi_{xy}(\mathbf{P})$ records, for each voter, whether they prefer x or y. Thus, the function F factors through these binary-valued projections. The total group ranking is assembled from pairwise decisions, each constrained to depend only on corresponding slices of the profile space. This induces a factorization over a lower-dimensional cube of binary comparison data.

The fibres of these projection maps—the preimages of fixed pairwise patterns—form the basic objects on which F must be consistent. The Pareto condition fixes behaviour on unanimous fibres, while non-dictatorship prevents collapse to a single voter's coordinate. The key insight is that these fibres cannot be globally stitched together without encountering a topological obstruction.

These obstructions cannot be resolved without violating one of the assumptions. Cycles force discontinuities, unanimity fails to propagate, or dictatorship emerges. No aggregation rule can navigate the profile space while satisfying all four conditions.

References:

Arrow, K. J. (1963). *Social Choice and Individual Values.* Wiley.

Baryshnikov, Y. (1997). Topological and discrete social choice: in a search of a theory. *Social Choice and Welfare,* **14**(2), 199–209.

Saari, D. G. (1994). *Geometry of Voting.* Springer.

The Tunnel at the Beginning of Light

Top (Solar Fusion Process): This diagram illustrates the *proton-proton (pp) chain*, the dominant fusion process in the Sun's core, where hydrogen nuclei (protons) fuse into helium, releasing energy. Here's the breakdown of the stages shown:

Stage 1: Proton-Proton Fusion: Two *protons (red)* fuse to form: a *deuteron* (one proton + one neutron—shown here as red + blue), a *positron* (green), and a *neutrino* (ν). This happens twice independently. **Reaction:** $p + p \to \mathrm{D} + e^+ + \nu_e$

Stage 2: Deuteron-Proton Fusion: The deuteron (p+n) fuses with another *proton (red)* to form: *helium-3 (two protons, one neutron—two red + one blue)* and a *gamma ray* (γ), indicating energy release. This also occurs in parallel twice. **Reaction:** $\mathrm{D} + p \to \text{He-3} + \gamma$

Stage 3: Helium-3 Fusion: Two *helium-3 nuclei* collide, forming: one *helium-4 nucleus* (two protons + two neutrons—two red + two blue), two free *protons* (red), released back to the plasma. **Reaction:** $\text{He-3} + \text{He-3} \to \text{He-4} + 2p$

Net Result: The chain converts four protons into one helium-4 nucleus, with: energy carried by *gamma rays, neutrinos,* and *kinetic energy* of the particles, two protons recycled. This chain is responsible for the Sun's energy production, light, and the solar neutrinos we detect on Earth.

Bottom (Quantum Tunnelling): This enables the entire process by allowing particles to skip energy barriers. The barrier dampens the probability wave without zeroing it. This quantum mechanical effect permits fusion to occur at the Sun's core temperature (fifteen million K), where classical physics predicts negligible fusion rates due to insufficient thermal energy to overcome the Coulomb barrier between protons.

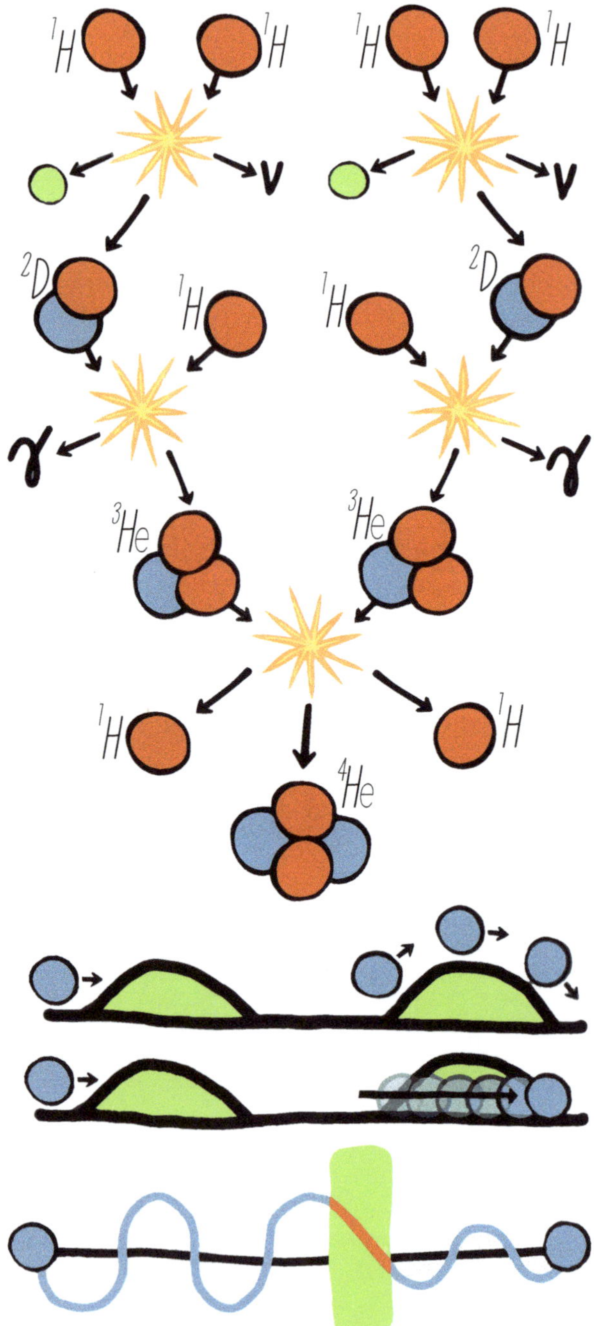

The Tunnel at the Beginning of Light

Solar fusion proceeds despite temperatures insufficient for classical nuclear reactions because quantum tunnelling enables protons to penetrate the Coulomb barrier with non-zero probability. At the Sun's core temperature of 15 million Kelvin, the average proton possesses only about 1/20th of the energy classically required to overcome electromagnetic repulsion between positively charged nuclei. Quantum mechanics allows particles to 'tunnel' through energy barriers they cannot surmount classically, with probability decreasing exponentially with barrier height and width. This tunnelling effect, combined with the enormous number of interaction attempts in the solar plasma, sustains the fusion rate necessary for stellar stability over billions of years.

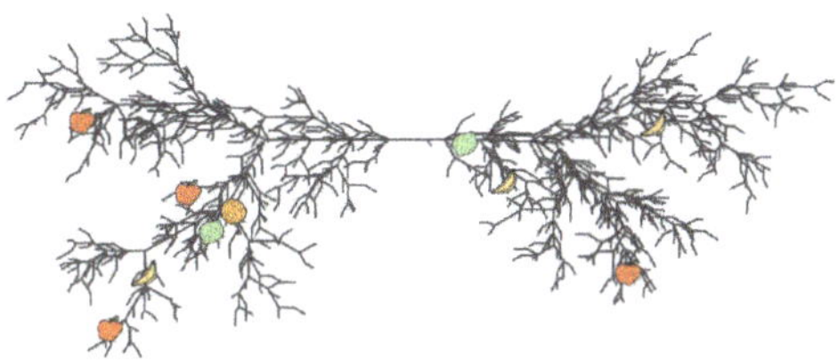

Solar Core Fusion ◦ Proton-Proton Chain ◦ Coulomb Barrier Problem ◦ Quantum Tunnelling Solution ◦ Gamow Factor ◦ Weak Force & Neutrinos ◦ Lepton Number Conservation ◦ Solar Neutrino Problem ◦ Neutrino Oscillations ◦ Hydrostatic Equilibrium ◦ Main Sequence Stability

«o helios esti lithos pyrodes, meizon tes Peloponnesou.»

("The Sun is a fiery mass, larger than the Peloponnesus.")

— Anaxagoras, c. 450 BC

The Tunnel at the Beginning of Light

In the late nineteenth century, Lord Kelvin and Hermann von Helmholtz proposed that gravitational contraction powered the Sun, but this mechanism accounted for only tens of millions of years—far shorter than the timescales implied by geological and biological evidence on Earth. In the early twentieth century, Arthur Eddington rejected this view, positing that nuclear processes must fuel the Sun's enduring luminosity.

In the 1920s, George Gamow introduced quantum mechanics into stellar models, showing that charged particles could penetrate electrostatic barriers via quantum tunnelling. Around the same time, Robert Atkinson, Fritz Houtermans, and Ralph Fowler explored how fusion might occur at stellar temperatures, providing theoretical support for nuclear reactions in stars.

Hans Bethe's 1939 work clarified these mechanisms, describing both the proton–proton chain and the carbon–nitrogen–oxygen (CNO) cycle (Bethe, 1939). This established the theoretical basis for stellar fusion in different stellar environments. Later decades brought confirmation through solar neutrino detection and improved nuclear cross-section measurements. These developments cemented the view that the mechanism of quantum tunnelling—initially a purely theoretical construct—directly powers the Sun and shapes the broader evolution of stars.

The Sun produces energy through nuclear fusion in its core. Gravitational compression generates densities exceeding $150\,\mathrm{g/cm^3}$ and temperatures near $1.5 \times 10^7\,\mathrm{K}$. At these conditions, hydrogen nuclei fuse into helium, releasing binding energy.

In the simplest view, fusion proceeds via close approaches of hydrogen nuclei aided by quantum tunnelling. At core temperatures of order 10^7 K, protons have thermal kinetic energies too small to classically overcome the Coulomb barrier, but tunnelling allows occasional close encounters where the strong nuclear force binds them. Through a sequence of interactions known as the proton–proton chain, four protons are ultimately transformed into a helium nucleus.

Each fusion event in the Sun's core releases a small amount of energy: approximately 26.7 MeV per helium nucleus formed. However, the Sun generates a total power output of roughly 3.8×10^{26} W, which requires converting mass to energy at an enormous rate. By the relation $E = mc^2$ (Einstein, 1905), this luminosity implies a mass loss of about 4.3×10^9 kg per second.

This mass loss manifests as outward radiation pressure. Within the core, energy liberated by fusion builds up pressure that counteracts gravitational collapse. The resulting hydrostatic equilibrium maintains the Sun's structure—every second, the immense weight of the Sun's outer layers is balanced by pressure generated from fusing approximately 6×10^{11} kilograms of hydrogen into helium. The Sun's long-term stability emerges from this balance. Fusion sustains the outward force needed to resist the crushing pull of its own mass.

The energy generated in the core undergoes radiative diffusion. Photons scatter innumerable times off electrons and nuclei as they migrate outward through the radiative zone. In the outer layers, convective transport becomes dominant, with rising and sinking plasma transporting energy. After this migration, energy is finally emitted from the photosphere as sunlight, spanning a broad electromagnetic spectrum.

Conservation of energy, momentum, and electric charge ensures consistency in nuclear reactions. Quantum field theories of particle interactions also impose another conserved quantity: lepton number. Leptons—a class of particles including electrons, neutrinos, and their antiparticles—must be created or destroyed in such a way that the total lepton number remains unchanged.

The proton–proton chain, which powers the Sun, involves changes in particle types that require mechanisms beyond the electromagnetic and strong forces. In particular, the weak nuclear force is necessary to enable the conversion of protons into neutrons while preserving all conservation laws. The weak force enables the fusion of hydrogen into helium.

Here is the first step of the chain:

$$\mathrm{p} + \mathrm{p} \;\rightarrow\; \mathrm{d} + e^{+} + \nu_e,$$

where p denotes a proton, d a deuteron (a bound state of one proton and one neutron), e^{+} a positron, and ν_e an electron neutrino. In this reaction, one proton transforms into a neutron through a weak interaction. To conserve electric charge, a positron—the antimatter counterpart of the electron—is emitted. To conserve lepton number, an electron neutrino is emitted simultaneously.

In the lepton number accounting, electrons and neutrinos are assigned a lepton number of $+1$, while positrons and antineutrinos carry a lepton number of -1. Before the reaction, the system has zero net lepton number; after the reaction, the positron (-1) and neutrino ($+1$) balance each other, maintaining overall neutrality. The emission of the neutrino is therefore a necessity for the reaction to be consistent with the symmetries of particle physics.

Although neutrinos possess extremely small mass and interact only via the weak force, they carry away a significant fraction of the reaction's energy and linear momentum. Unlike photons—which scatter thousands of times before reaching the solar surface—neutrinos traverse the Sun's dense interior with minimal interaction and escape into space almost immediately. Neutrinos produced in the Sun's core reach Earth in about eight minutes, providing a direct and real-time probe of nuclear processes inside the Sun.

The detection of solar neutrinos has been crucial for confirming theoretical models of stellar energy generation. Measurements not only validate the dominance of the proton–proton chain but also reveal minor contributions from alternative fusion pathways, such as the carbon–nitrogen–oxygen (CNO) cycle in which carbon, nitrogen, and oxygen nuclei fuse to produce helium.

Quantum mechanics introduces behaviours absent in classical physics. One of these is tunnelling: the ability of a particle to penetrate and traverse a potential barrier even when its total energy is insufficient to overcome it.

Classically, a particle with energy less than the height of a potential barrier would be fully reflected, with zero probability of passage. In quantum mechanics, however, particles are described by continuous wavefunctions governed by the Schrödinger equation (Schrödinger, 1926). Even in classically forbidden regions, the wavefunction persists, decaying exponentially rather than vanishing abruptly.

When a quantum particle encounters a barrier higher than its energy, its wavefunction inside the barrier takes the form of a decaying exponential. If the barrier has finite width, there exists a nonzero probability that the particle will appear beyond the barrier—a phenomenon known as quantum tunnelling.

In the solar core, the fusion of protons faces a major obstacle: the Coulomb barrier arising from electrostatic repulsion when the protons are close enough to trigger the strong nuclear force. The potential energy associated with two protons at close approach is approximately 1 MeV, whereas the typical thermal kinetic energy at 1.5×10^7 K is about 1 keV. Classically, the probability of overcoming the barrier would be vanishingly small, and fusion would be effectively impossible.

Despite this, fusion proceeds because quantum tunnelling allows protons to penetrate the Coulomb barrier with nonzero probability. Quantum mechanics enables fusion at energies far below the classical threshold. The proton wavefunctions extend into and through the classically forbidden region, resulting in occasional barrier penetration and subsequent nuclear fusion.

The probability of tunnelling through the Coulomb barrier is quantified by the Gamow factor (Gamow, 1928). This factor arises from solving the Schrödinger equation for two charged particles and introduces an exponential suppression depending on the product of the charges, the reduced mass of the system, and the relative kinetic energy. A common parametrization is

$$P(E) \sim \exp\left(-\sqrt{\frac{E_G}{E}}\right),$$

where E_G (the Gamow energy) depends on the charges and reduced mass. Equivalently, $P(E) \sim \exp\left(-a/\sqrt{E}\right)$ with a constant a set by the same parameters.

At stellar core temperatures, the Gamow factor dominates the fusion reaction rate. Although tunnelling remains rare per collision, the immense number of protons ensures sufficient fusion events to sustain the Sun's energy output. The exponential sensitivity of tunnelling probability to temperature creates a self-regulating system: if fusion falls below the rate needed to balance gravitational compression, contraction increases core temperature until equilibrium restores; if fusion runs too high, expansion cools the core and reduces the reaction rate. This feedback mechanism maintains stable stellar burning within a narrow band of core conditions.

This regulatory mechanism underlies the main sequence which is the phase during which hydrogen fusion occurs steadily in the core. A star remains on the main sequence while hydrogen supply sustains the equilibrium fusion rate. The phase lifetime depends on stellar mass, which sets both compression rate and required temperature. For the Sun, this balance produces stability lasting approximately 10^{10} years.

The Sun's luminosity remains constant through stable interaction between gravity, fusion kinetics, and quantum tunnelling probabilities. These parameters determine the mass-to-energy conversion rate. The resulting energy supports overlying layers without expansion or collapse.

Solar neutrinos arise when the weak force converts a proton's up quark into a down quark during fusion. Baryon number is conserved (two initial protons become a deuteron with baryon number 2), and lepton number is conserved because the emitted positron ($L = -1$) and electron neutrino ($L = +1$) balance to zero.

The Sun produces approximately 2×10^{38} neutrinos per second, carrying 2% of fusion energy. With interaction cross-sections of 10^{-44} cm^2, they pass through matter nearly unimpeded—while photons require thousands of years to diffuse through the Sun, neutrinos escape instantaneously, reaching Earth in eight minutes.

Every detected neutrino was produced moments earlier in the solar core. Measuring their flux and energy spectrum tests stellar energy generation models with high precision.

When physicists first detected solar neutrinos in the 1960s, they encountered a puzzle (Davis, Harmer & Hoffman, 1968). Raymond Davis Jr.'s Homestake experiment used 400,000 litres of perchloroethylene to capture neutrinos through the reaction:

$$\nu_e + \mathrm{Cl}^{37} \rightarrow e^- + \mathrm{Ar}^{37}.$$

The Homestake detector measured only about one-third of the neutrino flux predicted by standard solar models. This deficit, known as the solar neutrino problem, persisted for over three decades despite improved experiments and refinements to stellar theory.

The resolution came through the discovery of neutrino oscillations (Pontecorvo, 1957)—neutrinos transform between different flavours as they propagate. The Standard Model lists three flavours: electron (ν_e), muon (ν_μ), and tau (ν_τ) neutrinos. Solar fusion produces only electron neutrinos, but oscillations into other flavours during travel to Earth explain why early detectors registered a deficit.

The Sudbury Neutrino Observatory (SNO), 2 kilometres underground in Ontario, used heavy water to measure both total neutrino flux and electron neutrino flux. SNO's 2001 results confirmed the total flux matched predictions, but two-thirds of electron neutrinos had oscillated into other flavours en route to Earth.

Neutrino oscillations require nonzero mass. The original Standard Model assumed massless neutrinos, so oscillations constitute evidence for physics beyond it. Current measurements indicate neutrino masses are less than a few tenths of an electron volt—over a million times smaller than the electron mass.

This discovery resolved the solar neutrino problem and validated both solar fusion theory and quantum field theory. The neutrino flux matches predictions from nuclear burning models. Oscillations opened new physics avenues, including CP violation studies and implications for the universe's matter-antimatter asymmetry.

While neutrinos probe nuclear processes directly, helioseismology—the study of solar oscillations—maps conditions throughout the solar interior.

The Sun undergoes acoustic oscillations driven by outer-layer convection. These pressure waves propagate through the interior like seismic waves through Earth. Oscillation frequencies, amplitudes, and patterns depend on internal temperature, density, and composition profiles.

Solar oscillations appear as periodic Doppler shifts in photospheric absorption lines. The Global Oscillation Network Group (GONG) and Solar and Heliospheric Observatory (SOHO) monitor these oscillations continuously. Millions of distinct modes have been identified, each with characteristic radial and angular patterns.

Analysing oscillation mode frequencies reveals the Sun's interior structure—a three-dimensional map of temperature, density, and rotation rate versus depth and latitude.

Helioseismology confirms stellar model predictions with high accuracy: temperature profiles match theory to within 0.1% throughout most of the interior. The convective zone depth measures 0.287 solar radii from the surface, matching theoretical predictions. It also validates density and temperature profiles used for neutrino predictions. The inferred central temperature of $(1.57 \pm 0.01) \times 10^7$ K confirms conditions for the observed proton-proton chain rate. This independent confirmation strengthens confidence in stellar evolution theory and solar nuclear processes.

There's a nonzero probability I'll tunnel out.

Quantum Tunnelling in Stellar Fusion

Coulomb Barrier and Characteristic Energies

In stellar cores, nuclear fusion requires overcoming electrostatic repulsion between positively charged nuclei. The Coulomb potential between two protons is

$$V_C(r) = \frac{Z_1 Z_2 e^2}{4\pi\epsilon_0 r},$$

where $Z_1 = Z_2 = 1$, and $r \sim 1$ fm. Estimating numerically:

$$V_C \sim \frac{(1.602 \times 10^{-19}\,\mathrm{C})^2}{4\pi(8.85 \times 10^{-12}\,\mathrm{F/m}) \cdot 1 \times 10^{-15}\,\mathrm{m}} \sim 1\,\mathrm{MeV}.$$

By comparison, the thermal kinetic energy at the Sun's core temperature $T \approx 1.5 \times 10^7$ K is:

$$k_B T \approx 1\,\mathrm{keV}.$$

Classically, such energy is insufficient for fusion; quantum tunneling provides a nonzero probability of barrier penetration.

Tunnelling Probability and the Gamow Factor

The tunnelling probability is approximated by the Gamow factor:

$$P_{\mathrm{tunnel}}(E) \sim \exp\left[-2\pi\eta(E)\right],$$

where the Sommerfeld parameter $\eta(E)$ is defined as

$$\eta(E) = \frac{Z_1 Z_2 e^2}{\hbar v}, \qquad v = \sqrt{2E/\mu},$$

with μ the reduced mass of the two-particle system. Substituting for v, the Sommerfeld parameter becomes

$$\eta(E) = \alpha Z_1 Z_2 \sqrt{\frac{\mu c^2}{2E}},$$

where α is the fine-structure constant. The exponential suppression governed by $\eta(E)$ dominates the energy dependence of the fusion rate.

Gamow Peak and Effective Fusion Energy

Fusion occurs predominantly at energies where the product of the Maxwell–Boltzmann distribution and the tunnelling probability is maximised. This defines the *Gamow peak*, centred around

$$E_{\mathrm{pk}} \approx \left(\frac{\pi^2 \mu c^2 \alpha^2 Z_1^2 Z_2^2 (k_B T)^2}{2}\right)^{1/3}.$$

The Gamow peak arises from the interplay between thermal distribution (favouring higher energies) and tunnelling suppression (favouring lower energies). Although $E_{\mathrm{pk}} \ll V_C$, the overlap is sufficient to permit fusion in a small fraction of collisions.

Thermally Averaged Fusion Rate and the Proton-Proton Chain

The effective reaction rate is governed by the thermally averaged cross section:

$$\langle \sigma v \rangle = \int_0^\infty \sigma(E)\, v(E)\, f_{\mathrm{MB}}(E)\, \mathrm{d}E,$$

where $\sigma(E)$ includes nuclear interaction probabilities and tunnelling effects, and $f_{\mathrm{MB}}(E)$ is the Maxwell–Boltzmann distribution. The dominant fusion pathway in the Sun is the proton–proton chain, initiated by

$$p + p \to d + e^+ + \nu_e.$$

Subsequent reactions in the chain yield ^{4}He, positrons, neutrinos, and photons. The net energy released per helium nucleus formed is approximately 26.7 MeV.

References:

Bethe, H. A. (1939). Energy Production in Stars. *Phys. Rev.*, **55**, 434–456.

Clayton, D. D. (1983). *Principles of Stellar Evolution and Nucleosynthesis.* University of Chicago Press.

Edges of Tomorrow

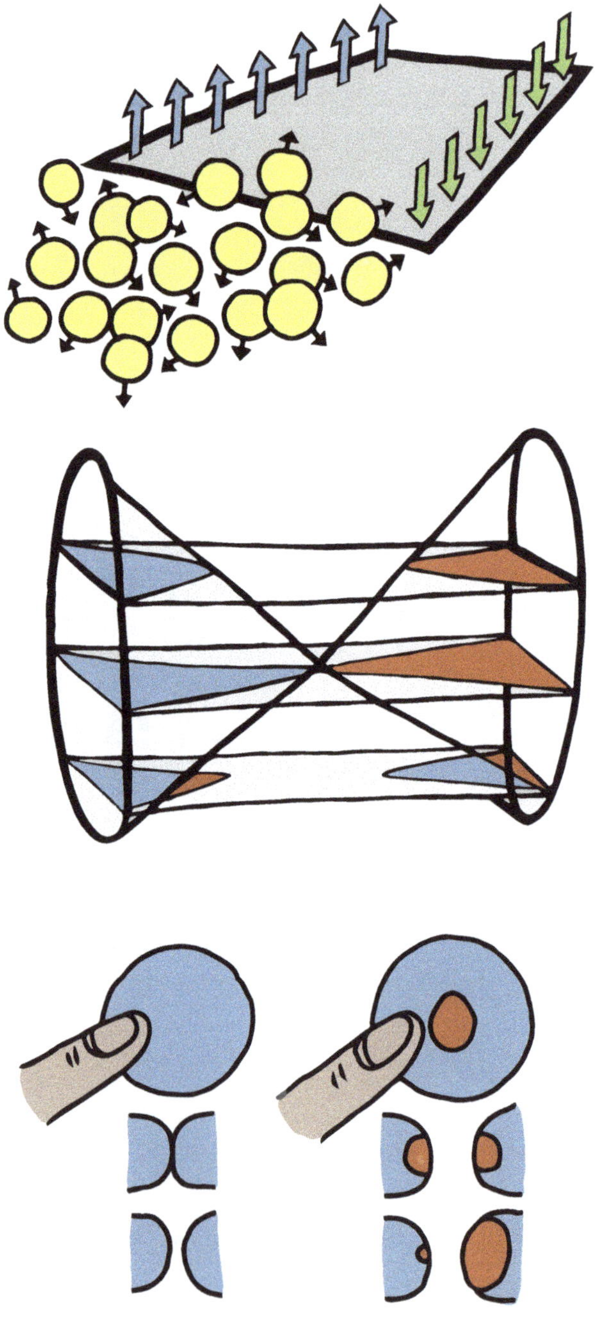

Quantum Hall Effect: In the top panel, the bulk is insulating (yellow electrons scattered inside the slab), while the edges host perfectly conducting states. Blue arrows flow along one boundary and green arrows along the opposite, illustrating chiral edge currents immune to backscattering.
Spin–Orbit Coupling and Band Inversion: The middle panel shows three popular momentum–energy diagrams stacked top to bottom, here drawn on a double-cone surface. Spin–orbit coupling reorganises the band structure, inverting the order of conduction and valence bands (blue and orange). This inversion forces the existence of surface states that cross the gap, guaranteeing conduction channels protected by time–reversal symmetry.
Robustness of Topology: The bottom panel contrasts two shapes. On the left, a blue disc without internal twist can be smoothly deformed and split apart, representing a trivial insulator. On the right, a disc enclosing an orange region cannot be removed without tearing—a topological obstruction. This illustrates why the protected edge states of a topological insulator are resistant to continuous perturbation that preserves symmetry.

Edges of Tomorrow

Topological insulators exhibit an unusual combination of properties: insulating in their bulk yet conducting electricity perfectly along surfaces or edges. This behaviour originates from the topology of the material's electronic energy structure in momentum space, which guarantees protected conductive states resistant to scattering and imperfections. The mathematical concept of topology, concerning properties preserved under continuous deformation, manifests physically through the way electron wave functions 'twist' as their momentum changes, leading to robust edge states and quantised conductance.

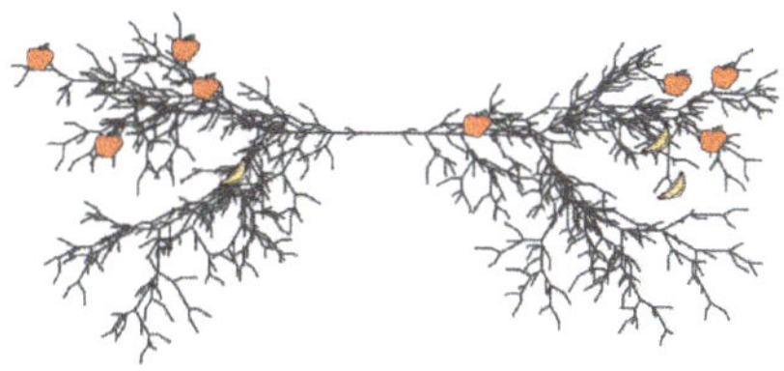

Quantum Band Theory ∘ Bloch Waves & Brillouin Zone ∘ Metal-Insulator Classification ∘ Berry Phase Topology ∘ Band Inversion ∘ Bulk-Boundary Correspondence ∘ Kramers Pairs ∘ Spin-Momentum Locking ∘ ARPES Evidence ∘ $\mathbb{Z}_2$ Invariants ∘ Protected Edge States

"In these days the angel of topology and the devil of abstract algebra fight for the soul of each individual mathematical domain."

— Hermann Weyl, 1939

Edges of Tomorrow

The story of topological insulators begins with a puzzling observation. In 1980, German physicist Klaus von Klitzing was studying how electricity flows through ultra-thin sheets of material placed in powerful magnetic fields. He discovered something strange: the Hall conductance jumped between precise values, instead of changing continuously. The steps remained exact even when the material had impurities or defects. Traditional physics couldn't explain why these measurements stayed so perfect despite the messiness of real materials.

Two years later, a quartet of theorists—Thouless, Kohmoto, Nightingale, and den Nijs—proposed an explanation. They suggested that von Klitzing's steps weren't determined by the material's detailed atomic arrangement but by something more abstract: the overall 'shape' of quantum states, borrowing ideas from topology—the branch of mathematics that studies properties preserved under continuous deformations. Just as a coffee cup and a donut share the same topological essence (both have one hole), these quantum states had mathematical properties that remained unchanged even when the material was disturbed.

This insight lay dormant for decades, viewed as a mathematical curiosity specific to systems in magnetic fields. Then, in 2005, physicists Charles Kane and Eugene Mele, working with theoretical models of graphene—single sheets of carbon atoms—they predicted that materials could exhibit similar protected electrical behaviour without any magnetic field at all. Their key insight was that an electron's intrinsic spin could play the role previously filled by the magnetic field. They envisioned materials that would insulate in their interior but conduct electricity perfectly along their edges, with this edge conduction protected by symmetries—a phenomenon they called the quantum spin Hall effect, creating a two-dimensional topological insulator.

Kane and Mele's graphene predictions proved difficult to realise experimentally, but in 2007, Laurens Molenkamp's team in Germany succeeded by engineering layers of mercury telluride and cadmium telluride (HgTe/CdTe quantum wells). They observed exactly what the theory had predicted: electrical current flowing along the material's edges while the interior remained insulating. Soon after, theorists including Liang Fu, Charles Kane, and Eugene Mele extended these ideas from two-dimensional to three-dimensional materials, identifying $Bi_{1-x}Sb$ as a three-dimensional topological insulator; shortly thereafter, Zhang and collaborators predicted that the Bi_2Se_3 family (Bi_2Se_3, Bi_2Te_3, Sb_2Te_3) would host robust surface states, while remaining insulating in the bulk.

By 2009, experimental physicists using angle-resolved photoemission spectroscopy (ARPES) confirmed these predictions, observing the spin-polarised surface states in Bi_2Se_3 and related compounds. A new class of matter had been established—materials whose most remarkable properties arose not from their microscopic details but from the global topology of their quantum states. In 2016, the Nobel Prize in Physics was awarded to Thouless, Haldane, and Kosterlitz for theoretical discoveries of topological phase transitions and topological phases of matter.

In classical physics, electric conduction follows from charged particle motion. Apply a field across a conductor: electrons accelerate opposite the field direction, generating current. Ohm's law captures this proportionality between field and current density. Resistance arises from scattering—electrons colliding with impurities or lattice vibrations (see Chapter 4).

On the surface, conductivity follows simple rules: more mobile electrons, fewer collisions, and less resistance. Reality is more complicated. Some metals show decreasing resistivity with temperature; others saturate. Pure crystalline insulators contain electrons but don't conduct. Graphene conducts; diamond—nearly identical in composition—insulates. Classical mechanics does not explain these differences.

Within quantum mechanical models, solid-state electrons live in discrete quantum states. Pauli's exclusion principle dictates (Pauli, 1925) that there must be at most one electron per state, which explains why matter does not collapse—electrons can't all pile into the lowest energy state but must stack up.

At zero temperature, electrons fill states from lowest energy upward, stopping at the *Fermi energy*—the highest occupied level. Picture a parking garage where cars (electrons) fill spots from the ground floor up. The Fermi energy marks the top occupied floor. This boundary matters because only electrons near the top can move—those buried deep in lower levels have nowhere to go. For conduction to occur, electrons near this boundary must find adjacent, empty parking spots (states) they can shift into.

When such states exist arbitrarily close in energy, an applied field perturbs the electron distribution near the Fermi surface, inducing current. When no nearby states are available—either because all states are filled or because the next states lie across a finite energy gap—the field cannot induce a response. The system remains non-conductive.

This quantum picture explains why electron count alone cannot predict conductivity. A material may contain an abundance of delocalized electrons yet remain insulating if all available quantum states are occupied. Conduction requires a *partially filled band*—a continuous set of states near the Fermi energy where electrons can transition without violating the exclusion principle.

Band theory explains what classical physics could not. Diamond and graphene contain identical carbon atoms, yet one insulates while the other conducts—their lattice symmetries create different band structures. A slight atomic shift can open a gap worth few electron-volts. A relativistic effect called spin-orbit coupling can flip an insulator into a conductor. Electron count is replaced by the question: 'can electrons near the Fermi surface find empty states to occupy?'

In crystals, atoms repeat like wallpaper patterns. This regularity creates a periodic landscape of electrical forces that electrons must navigate. They must respect the crystal's symmetry. Mathematically, this produces wavefunctions of a specific form called *Bloch waves* (Bloch, 1929):

$$\psi_{n\mathbf{k}}(\mathbf{r}) = e^{i\mathbf{k}\cdot\mathbf{r}} u_{n\mathbf{k}}(\mathbf{r})$$

where $u_{n\mathbf{k}}(\mathbf{r})$ is periodic with the lattice and $\mathbf{k}$ is the crystal momentum—not ordinary momentum but a quantum label for the electron's wavelike motion through the crystal. Because the crystal repeats, many different $\mathbf{k}$ values describe the same physical state. We keep only unique values in a finite region called the *Brillouin zone.* Importantly, this zone wraps around like a donut (torus)—go too far in any direction and you're back where you started.

The energies of Bloch states form continuous intervals called *bands*, separated by *band gaps*—regions of energy where no eigenstates exist. At zero temperature, electrons fill bands up to the Fermi energy. Whether the material conducts depends on the presence of accessible states near this energy.

This band-filling criterion separates materials into three types:

Metals have their Fermi energy inside a band. Electrons find empty states nearby—a nudge in momentum keeps them in the same band. Fields redistribute electrons near the Fermi surface. Phonons and impurities scatter them, but can't stop conduction entirely.

Band Insulators trap the Fermi level in a gap. No states exist for electrons to hop into. Breaking across requires serious energy: 1–10 electron-volts. Without that kick, electrons stay put. The material ignores weak fields.

Semiconductors squeeze the gap down to 0.1–2 electron-volts. Room temperature provides enough thermal energy to promote some electrons across. Dopants (impurities) affect the chemical potential, creating more carriers: digital processors, made of silicon etched carefully to have billions of semiconducting junctures. This tunability was the key to building the digital age.

This classification predicts conductivity from energy spectra alone—where bands sit and whether electrons can reach empty states. Two materials with identical band gaps, however, can differ in conductivity, because band theory does not account for how wavefunctions connect across momentum space.

Classical physics describes conduction as particle drift under a field but cannot distinguish diamond from graphene. Band theory replaces drift with quantum states and energy gaps, sorting materials into metals, semiconductors and insulators by their spectra. Topology adds a third criterion: the global arrangement of wavefunctions across the Brillouin zone. Two insulators can share the same band gap yet belong to different topological classes, because their wavefunctions wind differently through momentum space. The distinction is invisible to any local measurement of energy.

At each point in the Brillouin zone the occupied states define a direction in an abstract vector space. As momentum varies, this direction rotates. Carry it around a closed loop and compare it to the starting value—on flat ground, a vector transported parallel to itself around any closed path returns unchanged, but on a curved surface the vector rotates by an angle proportional to the enclosed area. The classic example is a sphere: start at the equator, walk to the north pole keeping your arrow pointing 'straight ahead,' then return to the equator via a different meridian. The arrow now points in a different direction, rotated by the holonomy of the path—a quantity determined entirely by the curvature enclosed. Electron wavefunctions in a crystal behave the same way. Transported around a loop in the Brillouin zone, they acquire a phase shift called the Berry phase (Berry, 1984). When this phase is 2π, states return to themselves and the topology is trivial. When the phase is π, states swap identities and the topology is nontrivial. A discrete label—a topological invariant—counts these phase windings and survives any smooth deformation that preserves the gap and the relevant symmetries.

At boundaries where the topological invariant changes—where a topological insulator meets vacuum, for instance—the energy gap must close locally, producing conducting channels confined to the boundary. These edge or surface states resist ordinary backscattering and remain robust against roughness and nonmagnetic disorder.

The invariants depend on dimension and symmetry. Breaking time-reversal symmetry (making the system distinguish between forward and backward time, like adding a magnetic field) in 2D yields integer *Chern numbers* (Thouless et al., 1982)—counting how many times wavefunctions twist. Preserving time-reversal symmetry gives binary $\mathbb{Z}_2$ invariants—just 0 or 1, trivial or nontrivial.

These labels translate into measurable quantities. Confine electrons to a thin sheet and apply a strong magnetic field—the current flowing along the edge is quantised in exact multiples of e^2/h, where e is the electron charge and h is Planck's constant. The Chern number C counts these multiples—$C = 1$ means one conducting channel, $C = 2$ means two—and the value is locked by topology, unable to change without closing the bulk gap. This quantisation is so exact that it now defines the international standard of electrical resistance. In three-dimensional materials that preserve time-reversal symmetry, the relevant invariant is $\mathbb{Z}_2$, which takes only two values. Ordinary insulators such as silicon have $\mathbb{Z}_2 = 0$ and inert surfaces. Bi_2Se_3, a grey crystal that looks unremarkable, has $\mathbb{Z}_2 = 1$—its bulk is insulating, but its surface carries a single metallic cone of spin-polarised electrons that cannot be removed by any perturbation respecting time-reversal symmetry. Two insulators with comparable gaps can belong to different topological classes, and the one with $\mathbb{Z}_2 = 1$ will conduct at its surface where the other never will.

How do topological phases arise in real materials? Often through a mechanism called *band inversion*. In ordinary materials, electron states follow a natural hierarchy: simple spherical orbitals (s-orbitals) have lower energy than more complex dumbbell-shaped ones (p-orbitals). But heavy atoms such as bismuth have strong spin-orbit coupling—the electron's spin interacts with its orbital motion. This interaction can flip the energy ordering, pushing p-states below s-states. When bands cross and switch places, the wavefunction topology changes. A boring insulator becomes topological.

The **bulk-boundary correspondence** links bulk topology to edge physics: when two regions with different topological invariants meet, the gap must close at the boundary. In topological insulators, time-reversal symmetry provides the crucial protection. This symmetry means physics looks the same whether you run the film forward or backward. For electrons, it guarantees that every state with momentum pointing right and spin up has a partner with momentum pointing left and spin down at exactly the same energy—these are called Kramers pairs (Kramers, 1930), like mirror images that can't be independently manipulated.

On the boundary, this pairing enforces that electrons with opposite spins propagate in opposite directions. Imagine two lanes of traffic where spin-up electrons go right and spin-down electrons go left. For an electron to make a U-turn (backscatter), it would need to reverse both its momentum and flip its spin simultaneously—like a car having to change both direction and flip upside down to turn around. This process is forbidden unless time-reversal symmetry is broken. As a result, non-magnetic disorder, surface roughness, and similar imperfections cannot localise these boundary states.

Experiments bear this out. The first confirmation came in two dimensions—König et al. (2007) measured quantised edge conductance in HgTe/CdTe quantum wells, precisely matching the $\mathbb{Z}_2$ prediction for a 2D topological insulator. In three dimensions, angle-resolved photoemission spectroscopy (ARPES) ejects electrons from a surface with ultraviolet light and records their energy and emission angle, reconstructing the band structure momentum by momentum. Applied to Bi_2Se_3 and its relatives Bi_2Te_3 and Sb_2Te_3, ARPES reveals the metallic surface cone predicted by the $\mathbb{Z}_2 = 1$ classification, with conducting states that bridge the bulk gap along a linear energy–momentum relation, making the surface electrons behave as massless fermions travelling at constant velocity.

Transport measurements provide complementary evidence. When the bulk is sufficiently insulating, electrical conductance measured at low temperatures remains finite, reflecting contributions from surface channels. These conducting modes persist across different sample thicknesses, geometries, and surface treatments. As opposed to ordinary surface effects—dangling bonds, reconstructions, impurity bands—which vary with preparation and vanish with surface treatment, topological surface states survive even when crystals are cleaved or exposed to ambient conditions, as long as time-reversal symmetry is preserved. Magnetotransport experiments reveal weak anti-localization effects and spin-momentum locking, consistent with theoretical predictions for topological surface states.

This protection against disorder enables practical applications. Because surface modes remain stable against a broad class of perturbations, topological insulators provide a platform for low-dissipation electronic devices. The suppression of backscattering by symmetry makes them attractive for interconnects and surface-conduction components that remain reliable despite fabrication imperfections and environmental variations.

More speculative applications involve quantum information. When topological insulators interface with superconductors, the resulting heterostructures can host exotic quasiparticles with non-Abelian exchange statistics—anyons that obey different algebraic rules than bosons or fermions. In proposed topological quantum computers, information would be encoded in the collective state of these quasiparticles, with operations performed by braiding them in space. Such transformations depend only on topology, offering intrinsic protection against many types of errors.

While experimental realisation of anyon manipulation in topological insulators remains largely academic, the theoretical foundation exists. The combination of robust surface conduction, symmetry protection, and potential for hosting exotic quantum phases positions topological insulators at the intersection of fundamental physics and future technologies.

The $\mathbb{Z}_2$ Topological Invariant

Time-Reversal Symmetry and Kramers Pairs

Time-reversal symmetry acts on electronic states as $\mathcal{T}|\psi\rangle = \Theta K|\psi\rangle$, where Θ is a unitary matrix and K is complex conjugation. For spin-1/2 electrons, $\mathcal{T}^2 = -1$, leading to Kramers theorem: time-reversal maps states at $\mathbf{k}$ to partners at $-\mathbf{k}$. At generic $\mathbf{k}$, these are states at different momenta. At time-reversal invariant momenta (TRIM) where $\mathbf{k} = -\mathbf{k}$ (modulo reciprocal lattice), all states come in degenerate pairs at the same momentum.

The $\mathbb{Z}_2$ Classification

In 2D with time-reversal symmetry, the topological character is captured by a binary invariant $\nu \in \{0, 1\}$. Unlike the Chern number (which requires broken time-reversal), this $\mathbb{Z}_2$ index survives when $\mathcal{T}$ is preserved.

Consider occupied Bloch states $|u_{n\mathbf{k}}\rangle$ forming a bundle over the Brillouin zone. At each TRIM point Γ_i, define the antisymmetric matrix:

$$w_{mn}(\Gamma_i) = \langle u_m(-\Gamma_i)|\mathcal{T}|u_n(\Gamma_i)\rangle.$$

The Pfaffian $\text{Pf}[w(\Gamma_i)]$ depends on the occupied-band gauge choice. The gauge-invariant object is:

$$\delta_i = \frac{\text{Pf}[w(\Gamma_i)]}{\sqrt{\det[w(\Gamma_i)]}} = \pm 1.$$

Computing the Invariant

For a 2D system with inversion symmetry, the $\mathbb{Z}_2$ invariant is:

$$(-1)^\nu = \prod_{i=1}^{4} \delta_i$$

where the product runs over the four TRIM points. With inversion symmetry, δ_i can be computed from parity eigenvalues as $\delta_i = \prod_m \xi_{2m}(\Gamma_i)$, giving $(-1)^\nu = \prod_i \delta_i$. If $\nu = 0$, the system is a trivial insulator; if $\nu = 1$, it's a topological insulator.

Physical Meaning

The invariant counts (mod 2) how many times occupied bands switch partners under Kramers pairing as we traverse the Brillouin zone. In a trivial insulator, Kramers pairs can be tracked consistently. In a topological insulator, the pairing pattern contains a twist—like trying to match socks while walking around a Möbius strip.

Bulk-Boundary Correspondence

When $\nu = 1$, the boundary must host an odd number of Kramers pairs of gapless states. These come in counter-propagating time-reversed partners with opposite helicity; non-magnetic elastic backscattering between partners is forbidden by $\mathcal{T}$. This makes the helical edge states robust against non-magnetic disorder.

Example: HgTe/CdTe Quantum Wells

Band inversion occurs when the quantum-well thickness exceeds a critical value $d_c \approx 6.3$ nm. For $d < d_c$ (thin wells), the ordering is normal with $E_{\Gamma_6} > E_{\Gamma_8}$ and the phase is trivial ($\nu = 0$). For $d > d_c$ (thick wells), the ordering is inverted with $E_{\Gamma_6} < E_{\Gamma_8}$, yielding $\nu = 1$. The transition at d_c closes and reopens the gap with different topology.

References:

Kane, C. L. and Mele, E. J. (2005). $\mathbb{Z}_2$ Topological Order and the Quantum Spin Hall Effect. *Physical Review Letters*, **95**(14), 146802.

Bernevig, B. A., Hughes, T. L., and Zhang, S.-C. (2006). Quantum Spin Hall Effect and Topological Phase Transition in HgTe Quantum Wells. *Science*, **314**(5806), 1757–1761.

You Would Like to Order First

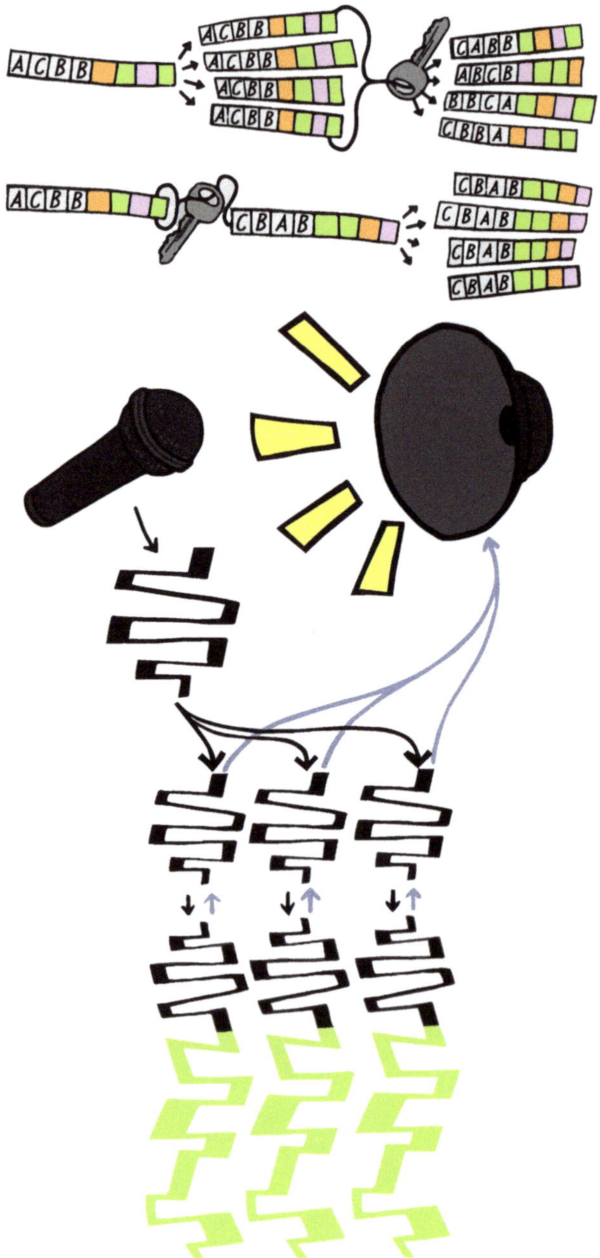

Top (Multiplex → Encrypt): The same message (ACBB) is first combined with different colour-coded templates through multiplexing. Each variant is then independently encrypted using a shared key. This leads to distinct ciphertexts, but if the templates are known or public, attackers can reverse the multiplexing process and end up with multiple ciphertexts of the same underlying message, enabling algebraic attacks that exploit the known relationships between the variants. **Bottom (Encrypt → Multiplex):** The message is first encrypted (producing, for example, CBAB from the original message), and the resulting ciphertext is then duplicated and wrapped in different colour-coded templates. Because all copies are cryptographically identical before templating, multiplexing adds no diversity to the encrypted content. This offers attackers fewer opportunities for correlation-based attacks. **Signal Flow (Communication Path):** The bottom illustration shows the complete communication path: microphone input → multiplexing → encryption → transmission → decryption → demultiplexing → speaker output.

You Would Like to Order First

GSM mobile communications used stream ciphers to protect call privacy, but protocol design left them vulnerable. Encryption was applied only after error correction and formatting, so fixed training sequences and redundant coding produced predictable ciphertext patterns. These leaks allowed attackers to recover session keys with modest effort, demonstrating that security depended less on theoretical cipher strength than on overall system design.

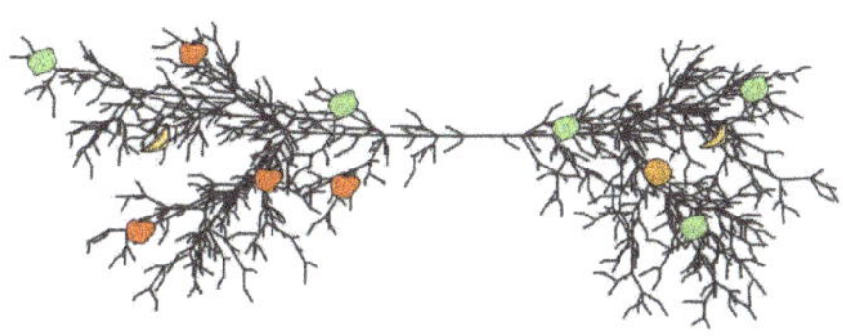

GSM TDMA Structure ◦ Fixed Burst Format ◦ Processing Order Vulnerability ◦ Stream Cipher XOR ◦ Predictable Plaintext ◦ Convolutional Codes ◦ A5/2 Algebraic Attack ◦ Cipher Downgrade Attack ◦ Session Key Reuse ◦ Biham-Barkan-Keller ◦ Protocol Layer Weakness

> "I just found out that my cellular telephone was a lemon."
>
> — Tobias Funke, 2004

You Would Like to Order First

The history of secure communication traces back to the intersection of two twentieth-century revolutions: cryptography and wireless technology. For much of the century, radio systems prioritised reach and reliability over confidentiality. During World War II, breakthroughs such as the German Enigma and Allied SIGSALY systems introduced large-scale encryption in radio, but these were bespoke wartime inventions, not standardised infrastructure.

In the postwar decades, civilian telecommunication networks expanded rapidly, but radio links remained analogue and vulnerable. First-generation mobile systems (1G), such as AMPS in the United States and NMT in Scandinavia, used frequency modulation without encryption. Voice data was transmitted as analogue waveforms, easily intercepted by anyone with a scanner. This vulnerability, while tolerated during the early novelty of mobile phones, became untenable as usage grew.

By the 1980s, Europe pursued a unified digital cellular standard under the banner of the Groupe Spécial Mobile (GSM). The aim was not only cross-border interoperability, but also improved spectrum efficiency and—critically—built-in security. The digital transition allowed for integration of error correction, time-multiplexing, and cryptography into the protocol stack. Unlike analogue predecessors, GSM was designed from the outset to offer some degree of confidentiality on the radio link.

Digital cellular networks are designed to carry simultaneous conversations—across a limited radio spectrum. Each call consists of two independent data streams, one uplink and one downlink, connecting handset and base station. These streams must be separated both by directionality and from the traffic of other users occupying the same band.

So we have duplexing and multiplexing. Duplexing separates the uplink (phone to tower) from the downlink (tower to phone). In frequency division duplexing (FDD), each direction is assigned their own frequency band, allowing simultaneous transmission and reception. In time division duplexing (TDD), both directions share a common band but alternate in fixed, synchronised time slots.

Multiplexing separates users sharing the same physical channel. In frequency division multiple access (FDMA), the spectrum is divided into separate frequency bands, each assigned to a different user. This isolates signals but requires fixed bandwidth allocation and limits how flexibly users can be added or removed. In time division multiple access (TDMA), users transmit in alternating time slots within a repeating frame structure. Each user has exclusive access to the channel during its assigned slot. This improves spectral efficiency, but requires strict global timing to keep transmissions aligned. In code division multiple access (CDMA), all users transmit simultaneously over the same frequency band, but each encodes its data using a unique pseudorandom spreading code. The receiver uses correlation to extract the intended signal. This allows full-time transmission with statistical multiplexing, but demands signal separation. All three approaches require that each user's transmission be confined to a fixed envelope, a burst, with predictable alignment and duration.

 https://doi.org/10.11647/OBP.0526.12

These requirements propagate upward through the entire transmission stack. Each burst must arrive in its designated slot, with precise size and timing. Modulation, equalisation, and error correction depend on this regularity. As a result, every upstream layer, from speech encoding to encryption, must preserve the burst format.

To transmit speech, the analogue signal is sampled at regular intervals (Shannon, 1949) and each sample is encoded as a digital number. This raw bitstream is then compressed using a speech codec, a specialised algorithm that reduces bandwidth by representing only perceptually important features. As a toy example, consider a 20 ms segment of audio. Uncompressed, this might require over 2,000 bits. A codec might instead describe it using only pitch, volume, and phoneme class, reducing the bitrate by an order of magnitude.

GSM (the Global System for Mobile Communications) was developed as a pan-European standard for digital cellular networks in the early 1990s. It replaced earlier analogue systems with a structured, time-synchronised digital stack designed for interoperability, moderate confidentiality, and efficient spectrum use. The GSM radio interface is based on TDMA: each 200 kHz carrier is divided into repeating time frames of eight slots, with each user assigned one slot per frame. Each slot (or burst) carries 114 bits of payload, framed by synchronisation and guard bits.

Voice is transmitted as a sequence of such bursts. Every 20 milliseconds of speech is compressed into a 260-bit frame, which after coding and interleaving is split across multiple 114-bit radio bursts for transmission. These bits are divided into classes by perceptual importance. The most critical will later be protected with more redundancy. Each frame is processed independently and must be transmitted in order, aligned to the caller's assigned slot. From this point forward, it is treated as a fixed-length atomic unit: encoded, encrypted, and modulated as a whole. Control channels such as SACCH (Slow Associated Control Channel) follow a similar pattern, expanding 184-bit messages to 456 bits after error correction coding.

Before transmission, the frame is convolutionally encoded (which is a type of error correction (Elias, 1955) coding). This adds redundancy by producing each output bit as a function of the current and previous inputs. The goal is to enable error correction at the receiver without retransmission. After encoding, the output is interleaved. It is reordered across time so that localised bit corruption does not overwhelm any one frame. These operations are deterministic and standardised. Their result is a longer, structured bitstream with predictable relationships between positions.

At this point, the data must be encrypted, but without affecting their size or timing. Each burst has a fixed payload size, and must be transmitted precisely at its assigned interval. This rules out modes that expand input or require buffering; encryption must operate in place with no change to length or alignment. GSM therefore uses a stream cipher (Vernam, 1926): a keystream is generated and extttXORed with the data bit-for-bit, producing ciphertext of equal length and immediate readiness for modulation.

GSM fixes the processing order as: compression → error correction → interleaving → encryption. This sequencing is a deliberate engineering decision. By placing encryption at the end of the stack, the system isolates cryptographic logic from earlier processing stages. Each module performs a self-contained transformation. This design simplifies implementation—but, as we will see, introduces a vulnerability.

By the time the bitstream reaches the cipher, it is no longer raw data. It has been processed into a rigid format defined by the protocol. Within the 114 ciphered bits per burst this includes:

- *Padding and known link-layer fields:* deterministic bits that fill or structure payloads on certain channels.
- *Error-correction codes:* parity bits computed from public polynomials.
- *Interleaving:* a known permutation applied identically to each block.

Each 114-bit burst contains payload data bracketed by tail, training, and guard intervals of fixed length; those bracket fields are not ciphered. Within the ciphered payload, the bitstream is heavily preprocessed prior to encryption. Bit patterns arising from coding, interleaving, and protocol padding are defined explicitly by the standard and repeat across sessions. The plaintext entering the encryption algorithm is therefore predictable at specific locations. It is drawn from a constrained distribution with high predictability and low entropy in fixed subregions. A passive observer capturing encrypted GSM traffic receives ciphertext derived from partially labelled inputs whose positions and formats are specified in advance by the protocol.

GSM's stream cipher preserves structure in a way that a block cipher with diffusion would not. Because encryption is bitwise XOR, linear relations introduced by coding survive intact in the ciphertext. Consider a concrete example: suppose the channel coding introduces a parity check—a known XOR relation among data bits. After encryption, the corresponding ciphertext bits satisfy the same parity relation among their respective keystream values. An attacker can deduce this constraint without knowing the underlying data.

By collecting multiple ciphertext samples, each reflecting similar patterns but different keystream realisations, the attacker builds a system of equations that gradually reduces the candidate key space. GSM compounds this vulnerability: voice frames are transmitted redundantly across multiple bursts, providing numerous ciphertext instances derived from aligned inputs. Repeated encipherment of predictable structure with the same key makes the keystream a target for mathematical reconstruction.

This vulnerability was exploited explicitly in the work of Eli Biham, Elad Barkan, and Nathan Keller. In 2003, they demonstrated a ciphertext-only attack against A5/2 capable of recovering the full 64-bit session key in under one second, using multiple frames of intercepted communication from control channels such as SACCH. The attack made no assumptions about plaintext content beyond its adherence to GSM's format. The weakness resulted from applying error correction and interleaving before encryption, allowing algebraic methods to exploit the resulting regularity. The attack combined brute-force enumeration of the cipher's R4 register (2^{16} possibilities) with solving overdetermined systems of linear equations derived from keystream parity constraints. This required hours of preprocessing and gigabytes of storage but was tractable on standard computing hardware.

In the same year, the authors presented an active attack that used this weakness in A5/2 to compromise A5/1 (a stronger cipher that GSM uses by default). GSM allows the base station to select the cipher for communication. A rogue station can impersonate a valid tower and request a downgrade to A5/2 from a handset that supports it. Once the device complies, the attacker captures the A5/2-encrypted exchange, recovers the session key, and then uses that key to decrypt subsequent bursts sent using A5/1. This is possible because GSM reuses the session key across ciphers during a session. The presence of A5/2 in the cipher suite thus undermines A5/1, regardless of whether the latter is ever explicitly requested by the attacker. Any device that implements A5/2 inherits its vulnerabilities and propagates them to the stronger cipher via shared key state.

Barkan and Biham continued to refine their attacks. In 2005, they improved known plaintext techniques against A5/1, specifically targeting its irregular initialization procedure. This reduced the computational burden of recovering internal state, particularly in scenarios with limited plaintext exposure. However, their most significant advance came in 2006, when they extended ciphertext-only techniques to A5/1 itself. The approach required far more ciphertext and offline preprocessing than the attack on A5/2, but the principle was similar. By leveraging the publicly known convolutional codes used before encryption, the attackers extracted algebraic relations between ciphertext bits and the keystream. These relations were then used to filter candidate internal states of the cipher's LFSRs (Linear Feedback Shift Registers), narrowing the search space (Golomb, 1955) to feasible dimensions. The complexity of the attack remained high, but it fell within the capabilities of a moderately resourced organisation with access to terabyte-scale storage and standard computational infrastructure.

The attack on A5/1 demonstrated that GSM's vulnerability was a result of the interplay between encryption placement, channel configuration, and cipher reuse. GSM's decision to support multiple ciphers without enforcing mutual isolation of key state allowed one weak algorithm to compromise the integrity of the entire suite. Because GSM does not authenticate base stations, handsets cannot verify that cipher selection is legitimate. Any device supporting A5/2 remains exposed to downgrade. Once the session key is recovered through a break of A5/2—whether using algebraic decoding or parity-based keystream reconstruction—that same key grants access to A5/1-protected content. GSM's cipher suite is therefore not modular. Its effective security is determined not by the strongest cipher in use, but by the weakest that is supported. A5/2's inclusion rendered A5/1 susceptible by transitive failure.

Protocol Assumptions and Personal Entry Point

The weakest points in deployed cryptographic systems are rarely in the mathematics. They are in the layers that surround it: in protocol assumptions, state handling, framing conventions, or timing logic. This is why cryptographic standards are slow to change—not because better ciphers are unavailable, but because known, tested flaws are often safer than untested replacements. The defensive posture of a system is not just algorithmic strength, but mostly accumulated knowledge of how it fails.

I first encountered this issue in a lecture by Eli Biham around 2003. He outlined the GSM vulnerability using nothing but XOR equations, known plaintext segments, and short recurrence relations. This attack did not require knowing the full formalism of block cipher construction or number theory. It showed that security could collapse under regularity exposed by the protocol—and that the analysis of "where" in a system encryption occurs mattered as much as "how."

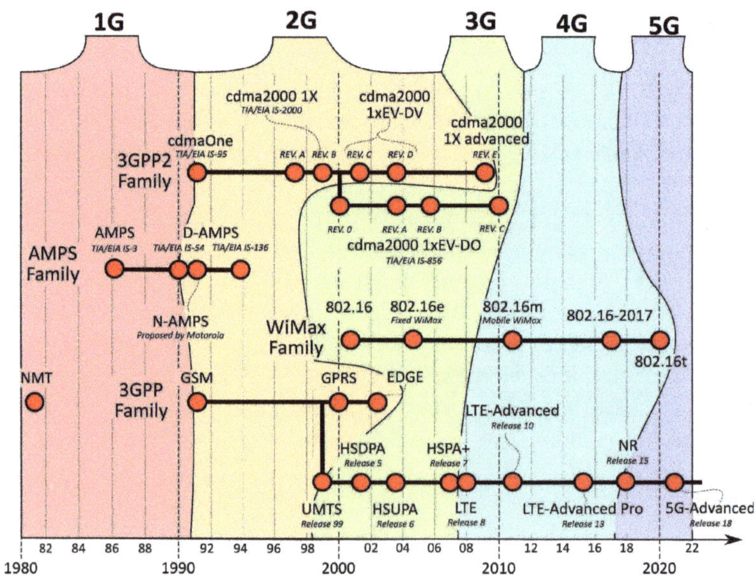

Cellular network standards and generation timeline, Wikimedia Commons, CC BY-SA 4.0

Toy GSM-Style Frame: How Post-Encoding XOR Leaks Keystream

Frame Layout

We model a simplified SACCH-style control message with *32 information bits* $s_1, \dots, s_{32}$. GSM inserts fixed training bits, padding, and applies forward-error correction; we model this with a minimal layout:

$$\underbrace{1100101101100101}_{\text{training}}\ \underbrace{s_1 \dots s_8}_{\text{data}}$$
$$\underbrace{01101001}_{\text{pad}}\ \underbrace{s_9 \dots s_{16}}_{\text{data}}$$
$$\underbrace{11001011}_{\text{parity}}\ \underbrace{s_{17} \dots s_{24}}_{\text{data}}\ \underbrace{s_{25} \dots s_{32}}_{\text{data}}. \quad (1)$$

Exactly **64** bits form the encoder input: 32 unknown information bits and 32 deterministic bits known to the attacker. This scales toward realistic GSM processing where SACCH messages expand from 184 bits to 456 bits.

Redundancy (Mini-Convolutional Code)

Each input bit x_i passes through a toy $(1, 1/2)$ convolutional code with generator polynomials $(1,\ 1+D)$:

$$y_{i,0} = x_i, \qquad y_{i,1} = x_i \oplus x_{i-1}.$$

Assume zero-state initialization so the bit before each known block is defined. This yields 128 output bits $Y = [Y_0, \dots, Y_{127}]$ where $Y_k = y_{\lfloor k/2 \rfloor, k \bmod 2}$. Every pair satisfies

$$y_{i,1} \ \oplus\ y_{i-1,0} = y_{i,0} \quad \forall i \geq 1, \qquad (2)$$

a parity relation that survives encryption as a linear constraint tying three keystream bits together.

Interleaver

A fixed block interleaver permutes the 128 bits:

$$\pi(i) = (i \bmod 4) \cdot 32 + \left\lfloor \frac{i}{4} \right\rfloor, \qquad (3)$$

a public mapping known to attacker and receiver.

Encryption After Coding

Encryption XORs a keystream $K_0, \dots, K_{127}$ with the permuted code bits:

$$C_i = Y_{\pi(i)} \oplus K_i\ ., \qquad (4)$$

Because XOR preserves length and position, all structure in Y remains in masked form in C.

Ciphertext-Only Attack Sketch

Training leakage: From 32 known training/-pad/parity input bits, the rate 1/2 encoder produces 64 known coded bits (32 from $y_{\cdot,0}$ and 32 from $y_{\cdot,1}$ with state assumption), so the attacker obtains $K_i = C_i \oplus Y_{\pi(i)}$ for 64 positions.

Parity recursion: Using Equation (2), each parity relation after interleaving becomes a linear equation in three ciphertext bits and three keystream bits. The known K_i values seed a sparse linear system over $\mathbb{F}_2$ that propagates to many additional K_j. With *multiple frames*, the system typically determines the full keystream through solving a large linear system or using precomputed tables.

Information bit recovery: With keystream recovered, the attacker inverts the interleaver and convolutional code to extract $s_1, \dots, s_{32}$ from each frame.

Why Encrypt-First Stops the Leak

If encryption preceded coding, the encoder would process $X \oplus K'$ rather than X. Parity relation (2) would then bind unknown values, blocking the attack. Fixed fields would reveal nothing until after decryption.

References:

Barkan, E., Biham, E., Keller, N. (2008). Instant ciphertext-only cryptanalysis of GSM encrypted communication. *J. Cryptology* 21(3):392–429.

Right on Spot

Model Comparison: Three theoretical models are applied to two classic experiments: the double slit and the photoelectric effect. The goal is to highlight where predictions agree with the observed.
Top (Corpuscular): Double-slit: no interference, $I(x) \sim e^{-(x-x_1)^2} + e^{-(x-x_2)^2}$. Photoelectric: emission occurs only if photon energy exceeds the threshold, $R \sim \Theta(\omega - \phi)$.
Middle (Wave): Double-slit: interference arises from wave superposition, $I(x) \sim \cos^2(x)$. Photoelectric: continuous energy absorption incorrectly predicts emission at all ω, $R \sim 1$.
Bottom (QED): Double-slit: coherent paths interfere, $I(x) \sim |\Psi_1 + \Psi_2|^2 \sim \cos^2(x)$. Photoelectric: decoherent transition amplitudes yield $R \sim |\mathcal{M}|^2 \sim \Theta(\omega - \phi)$. The same path integral formalism applies, but coherence conditions differ. Green frames highlight where predictions match experimental outcomes.

Right on Spot

Poisson's spot, also called the Arago spot, demonstrates wave diffraction through the unexpected appearance of a bright point at the centre of a circular object's shadow. When Augustin-Jean Fresnel proposed light as a wave phenomenon in 1818, Siméon Poisson derived this counterintuitive prediction in an attempt to disprove the theory. The effort backfired when François Arago experimentally confirmed the spot, which corpuscular optics could not explain, providing decisive support for the wave model of light.

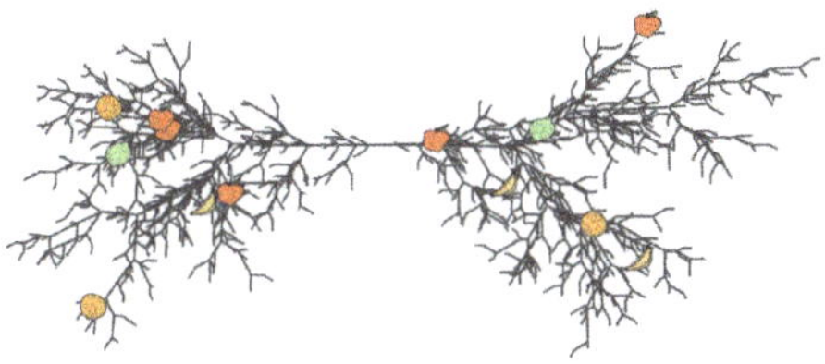

Newton's Corpuscular Theory ◦ Huygens's Wave Theory ◦ Young's Double Slit ◦ Fresnel's Diffraction Theory ◦ Poisson's Absurd Prediction ◦ Arago Spot Experiment ◦ 1818 Grand Prix ◦ Wave Theory Vindication ◦ Predictive Precision ◦ Scientific Method Evolution ◦ QED Modern View

"Truth is stranger than fiction, but it is because Fiction is obliged to stick to possibilities; Truth isn't."
— Mark Twain, 1897

Right on Spot

At the start of the nineteenth century, France's scientific institutions were dominated by Isaac Newton's *corpuscular theory* of light. Though British, his mechanical worldview had become the orthodox guideline for natural principles in France, treated with near-religious devotion. His theories were seen as mathematically perfect expressions of divine order, where challenging his optical theory risked accusations of scientific heresy. This dominance persisted through institutional inertia within the *Académie des Sciences*, where senior mathematicians like Siméon Denis Poisson and the recently deceased Joseph-Louis Lagrange had built their legacies on Newtonian principles.

In contrast, Christiaan Huygens's earlier *wave theory*, though developed in Paris in the late seventeenth century, had fallen out of favour. His principle—that every point on a wavefront acts as a source of secondary wavelets—was largely viewed as a heuristic, lacking the mechanical precision demanded by the Newtonian establishment.

The political context mattered. Post-Napoleonic Wars, French science underwent institutional consolidation. State-sponsored prizes regulated scientific boundaries as much as they rewarded discovery. The Academy's *Grand Prix* competitions reinforced orthodoxy while offering recognition. When the 1818 diffraction prize was announced, it tested allegiance as much as it sought an explanation.

Into this charged environment entered Augustin-Jean Fresnel, a provincial engineer without formal academic standing. He submitted a comprehensive wave theory treating interference and diffraction as fundamental, not anomalous. Fresnel extended Huygens's principle with integrals and phase relations to predict intensity patterns. For the Academy establishment, this represented an unwelcome challenge to Newtonian dogma that had defined scientific legitimacy for generations.

Poisson, serving on the jury, represented the old guard—a committed Newtonian who viewed mathematical elegance as the ultimate arbiter of truth. Dominique-François Arago, also on the committee, occupied a more ambiguous position. Though not yet fully aligned with the wave camp, Arago had corresponded with Fresnel and was known for his openness to alternative descriptions. The committee thus represented an ideological fault line within French science: Newtonian orthodoxy versus a nascent wave revival driven by empirical evidence and mathematical rigour. What followed would be a confrontation not only of theories but of institutional momentum and scientific methodology.

In the dominant optical theory of the eighteenth century, light was conceived as a stream of discrete particles governed by Newtonian mechanics. Isaac Newton's *Opticks* (1704) formalised this corpuscular model, arguing that light rays consist of minute corpuscles emitted from luminous bodies and travelling in straight lines. Reflection was explained as an elastic rebound from surfaces, while refraction was attributed to short-range attractive forces exerted by denser media, accelerating the particles and bending their trajectories toward the normal.

To support the theory, Newton conducted a series of controlled experiments using prisms, lenses, and narrow apertures. He systematically investigated the behaviour of light under dispersion, interference, and filtering conditions, recording the colours and intensities projected onto screens. In his most influential experiment, he demonstrated that white light could be decomposed into component colours using a glass prism and then recombined into white light using a second prism positioned in reverse.

The corpuscular model accounted for key optical observations such as rectilinear propagation, sharp shadows, and well-defined reflections from mirrors. It also offered coherence: by avoiding dependence on any physical transmission medium, the theory preserved the principle of action at a distance and aligned with Newton's general program of universal mechanics.

A competing model had been introduced by Christiaan Huygens in 1678, proposing that light propagated as a continuous wave. In this formulation, each point on a wavefront was treated as a source of secondary spherical disturbances, which spread outward in all directions. The superposition of these wavelets formed a new wavefront—defined as the envelope tangent to all secondary spheres—a geometric construction now known as the Huygens principle (Huygens, 1690). This approach permitted derivation of the laws of reflection and refraction using geometric reasoning and offered an alternative to the ballistic model without invoking interfacial forces.

Huygens's wave theory reproduced Snell's law by modelling light as a wavefront that changes orientation when passing into a medium where wave speed decreases. This accounted for refraction by assigning slower propagation to denser materials, a reversal of the corpuscular model's velocity assumption, later confirmed experimentally. The wave approach also gave qualitative explanations for diffraction and interference, though these effects had not yet been studied in systematic detail.

The theory required a universal propagation medium, the luminiferous ether, assumed to carry transverse vibrations through empty space. This posed internal difficulties: the ether needed to be rigid enough for high wave speeds but remained undetectable in mechanical or optical measurements. The model also offered no obvious account for sharply defined edges or specular reflection, which limited its compatibility with geometric optical effects.

By the early 1800s, the corpuscular theory retained institutional dominance in British science. Newton's *Opticks*, first published in 1704 and reissued in expanded editions through the mid-eighteenth century, remained the authoritative source on optical behaviour. Its treatments of reflection, refraction, and chromatic dispersion were widely accepted as definitive, and its success in reproducing geometric light paths reinforced its credibility within Newtonian mechanics.

On the continent, especially within the French Academy of Sciences, Huygens's wave theory received more sustained attention, though it remained a minority position. Even as late as 1815, standard French accounts described diffraction as a peripheral anomaly rather than a core optical effect. Classical phenomena, mirror reflection, Snell's law, and prism-induced colour separation were consistent with both frameworks. In the absence of distinct, falsifiable predictions, theoretical preferences were often shaped by broader methodological and philosophical alignments.

In 1801, Thomas Young presented experiments to the Royal Society showing that monochromatic light passing through two narrow, parallel slits produced a regular pattern of alternating bright and dark fringes on a distant screen. The results depended on slit geometry and colour, and the visibility of the fringes required careful alignment. The interference pattern was stable, reproducible, and inconsistent with the behaviour of independent particles travelling in straight lines.

Young explained the pattern using the principle of superposition: coherent wavefronts emanating from the two slits combined with relative phase shifts that depended on path difference. At some locations, the waves interfered constructively; at others, destructively. While this interpretation was met with scepticism in Britain, where Newtonian mechanics held institutional authority, it found more interest in France. There, analytic approaches to physical phenomena were gaining prominence, and interference was increasingly viewed as a direct signature of wave behaviour.

In the nineteenth century, the French Academy of Sciences organised public prize competitions to address unresolved scientific questions. These Grand Prix offered formal recognition and were intended to establish clarity on foundational issues. Notable winners of contemporary Grand Prix included Joseph Fourier (1810, for heat conduction), Jean-Baptiste Biot (1812, for electricity and magnetism), Sophie Germain (1816, for the theory of elastic surfaces—the first woman to win a Grand Prix from the Academy), and Siméon Denis Poisson (1819, for mechanics), establishing the prizes as a prestigious venue for scientific advancement.

This same tradition led to the 1818 announcement of a prize for the best theoretical account of diffraction. At the time, no existing model could give a complete explanation of the observed phenomena. The corpuscular theory remained standard in Britain, while some continental physicists were exploring wave-based descriptions grounded in interference.

Only two entries were received—the first, now lost and submitted anonymously, was rejected outright. The committee noted its neglect of known experimental results, its unfamiliarity with prior work by Young and Fresnel, and its numerous conceptual and computational errors. The second submission, authored by Augustin-Jean Fresnel, presented a detailed mathematical treatment of diffraction using the wave hypothesis. Fresnel extended Huygens's geometric principle by modelling light as a scalar disturbance with definable amplitude and phase. He introduced an integral formulation in which secondary wavelets emanating from every point on a wavefront interfered according to phase delay, yielding precise intensity predictions behind apertures and opaque obstacles. These predictions were derived entirely from geometry and propagation delay, without recourse to forces or ballistic mechanisms.

The response from the judging committee exposed a divide. Siméon Denis Poisson, an influential Newtonian, examined Fresnel's equations with the aim of identifying contradictions. He derived what he considered a decisive counterexample: were the wave theory valid, then a circular opaque disc should produce a bright spot at the centre of its geometric shadow. From the perspective of corpuscular optics, such a prediction was absurd. That region received no direct rays and should remain dark. Poisson concluded that this result, clearly at odds with particle reasoning and common sense, invalidated Fresnel's model as a whole.

Dominique Arago, also on the committee but less ideologically aligned with Newtonian mechanics, recognised that Poisson's objection could be tested directly. He constructed an experimental apparatus using a monochromatic point source, a small circular obstruction, and a screen positioned along a collimated beam path. Under proper alignment, a narrow bright point consistently appeared at the centre of the shadow, exactly as predicted by Fresnel's analysis. The result was repeatable, insensitive to minor perturbations—and could not be reconciled with any particle-based mechanism. It provided direct empirical confirmation that phase-based wave interference governed the observed pattern. The spot's appearance is consistent with Babinet's principle (Babinet, 1837): an opaque disc and an aperture of equal size produce identical diffraction patterns, so the constructive interference at the disc's centre mirrors the bright centre of a circular aperture.

The confirmation of the Arago spot resolved a challenge to the wave theory. It reinforced the rising standard for evaluating physical theories: the ability to produce precise, testable predictions from mathematical formulations applied to specific conditions.

The corpuscular theory could not reproduce the observed intensity maximum. No assumption about particle trajectories, angular spread, or probabilistic scattering could account for constructive illumination at the shadow's centre. The appearance of the Arago spot showed that physical models must be judged by whether their equations generate correct spatial distributions, not by whether their assumptions align with intuition. Predictive precision replaced plausibility as the standard for theoretical acceptance.

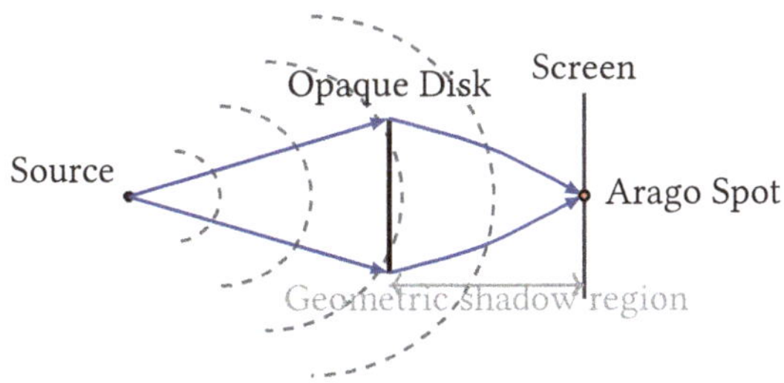

Figure by author. Diffraction around a circular obstacle produces constructive interference at the centre of the geometric shadow: the Arago spot.

This standard was challenged again in the early twentieth century, when new experiments revealed optical behaviour that neither wave theory nor particle theory could explain in full. The concept of duality arose in response to this fragmentation.

Diffraction, interference, and polarisation aligned with wave theory, particularly as formulated by Maxwell's equations (Maxwell, 1865), which described light as a transverse electromagnetic wave propagating through space. In contrast, the photoelectric effect (Einstein, 1905) (emission of electrons from a metal when illuminated) and Compton scattering (inelastic scattering (Compton, 1923) of photons by electrons) could not be explained by continuous fields. Each domain appeared to demand a separate formalism: wave optics for propagation, and quantum mechanics for emission and absorption.

This division extended into mathematical treatment. Wave behaviour was modelled using Maxwell's classical field equations, which predicted interference patterns and polarisation with high accuracy. Particle behaviour was captured by early quantum mechanics, where photons were treated as quantized packets of energy and momentum. As a result, physicists adopted a pragmatic view: light exhibits wave-like properties in some experiments and particle-like behaviour in others. The term 'duality' was used as a placeholder for the absence of a single consistent description. Though heuristically useful, this terminology preserved the conceptual split rather than resolving it.

Quantum electrodynamics, the modern quantum field theory of light, treats light as a quantized electromagnetic field. Photons are discrete energy-momentum packets, arising from specific field modes. They are not classical particles with trajectories, nor simple waves in a medium. The field spans spacetime and interacts with detectors through localised energy exchanges. Its behaviour follows rules that describe propagation and interaction.

Photon states are described using quantum field theory, built from the quantized modes of the electromagnetic field. Each state carries momentum, polarisation, and frequency. Measurements correspond to specific detector responses. The theory provides exact rules for calculating detection probabilities, scattering processes, and energy transfers without classical analogies.

The Arago spot results from coherent superposition of field amplitudes from different spatial regions. Each amplitude picks up a phase based on its path and boundary conditions. The total intensity comes from summing these complex amplitudes and taking their absolute value squared. This produces the observed pattern on screen, including a central bright point where waves interfere constructively.

When light intensity drops to single photons, each detection event remains a point. Over time, their distribution recreates the interference pattern. This confirms that probability amplitudes maintain phase relationships even in the quantum regime. The spot depends not on averaging many photons but on the field's linear behaviour and phase relationships between paths.

What appears as wave-particle duality emerges from viewing the quantum field through different experimental lenses. The field's equations produce patterns mathematically similar to classical interference when examined through intensity measurements, yet yield particle-like detection events when observed through photon counters. There is no duality—only a well-defined quantum field with observable operators.

This rejection of the notion of light's 'dual-nature' is important. In economics, Okun's Law relates changes in unemployment to deviations of GDP (gross domestic product) growth from its potential, taking a form mathematically similar to Hooke's Law for springs with a proportional 'restoring' effect. Yet this does not mean unemployment *is* a spring. In the same way, resemblance of quantum behaviour to either waves or particles reflects the equations and measurements, not the essence of what light actually is.

The Arago Spot

We use scalar diffraction theory—based on the Helmholtz equation and the Rayleigh–Sommerfeld integral—together with Babinet's principle to compute the on-axis field behind a circular disc.

Scalar Framework

For a monochromatic field $\Psi(\mathbf{r})$ with wavenumber $k = 2\pi/\lambda$,

$$\nabla^2\Psi + k^2\Psi = 0.$$

We use the Rayleigh–Sommerfeld formulation, which satisfies boundary conditions at the aperture plane. In what follows, Ψ denotes the complex amplitude and U the on-axis field.

Babinet and the Axial Field

Let $U_{\text{inc}} = A_0 e^{ikb}$ be a normally incident plane wave, and let $U_{\text{ap}}(0)$ (aperture) and $U_{\text{disc}}(0)$ (complementary disc) be the *on-axis* fields a distance b downstream. Babinet's principle gives

$$U_{\text{disc}}(0) = U_{\text{inc}}(0) - U_{\text{ap}}(0).$$

For a *circular aperture* of radius a, the axial Rayleigh–Sommerfeld integral in the Fresnel approximation evaluates to

$$U_{\text{ap}}(0) = A_0 e^{ikb}(1 - e^{i\pi N}),$$

where the Fresnel number $N = a^2/(\lambda b)$. By Babinet's principle, the complementary *disc* gives

$$U_{\text{disc}}(0) = A_0 e^{ikb} e^{i\pi N}.$$

The magnitude is $|U_{\text{disc}}(0)| = |A_0|$ for *all* N, so the on-axis intensity equals the incident intensity—the Arago spot. What varies with N is the phase $\arg U_{\text{disc}}(0) = kb + \pi N$ and the surrounding rings. In real experiments, finite source size, partial coherence, edge imperfections, and detector averaging slightly lower the on-axis intensity.

What Controls the Pattern

The Fresnel number $N = a^2/(\lambda b)$ determines whether the diffraction lies in the Fresnel or Fraunhofer regime and characterises how many Fresnel zones fit within the disc radius. When $N \gtrsim 1$ (near-field regime), the disc edge is within the first few Fresnel zones, producing pronounced concentric Fresnel rings with high contrast around the central spot. The axial field coherently combines contributions from the circular rim because the phase variation around the rim is quadratic and vanishes to first order on axis, making rim contributions nearly in phase. When $N \ll 1$ (Fraunhofer regime), the aperture field magnitude $|U_{\text{ap}}(0)| \to 0$ and ring contrast becomes weak, though the central disc intensity remains at the incident level for all N.

Integral Evaluation

On axis, the Rayleigh–Sommerfeld integral for a circular aperture reduces to

$$U_{\text{ap}}(0) = -\frac{iA_0 e^{ikb}}{\lambda b}\, 2\pi \int_0^a r\, e^{i\pi r^2/(\lambda b)}\, dr.$$

Substituting $u = i\pi r^2/(\lambda b)$ gives $r\,dr = (\lambda b)/(2i\pi)\, du$, so

$$\begin{aligned} U_{\text{ap}}(0) &= -\frac{iA_0 e^{ikb}}{\lambda b} \cdot 2\pi \cdot \frac{\lambda b}{2i\pi} \int_0^{i\pi N} e^u\, du \\ &= -iA_0 e^{ikb} \cdot \frac{1}{i}(e^{i\pi N} - 1) \\ &= A_0 e^{ikb}(1 - e^{i\pi N}). \end{aligned}$$

Rewriting:

$$\begin{aligned} U_{\text{ap}}(0) &= -2iA_0 e^{ikb} e^{i\pi N/2} \sin(\pi N/2), \\ |U_{\text{ap}}(0)| &= 2|A_0||\sin(\pi N/2)|. \end{aligned}$$

The aperture phase is

$$\arg U_{\text{ap}}(0) = kb + \tfrac{\pi N}{2} - \tfrac{\pi}{2}$$

with π-jumps where $\sin(\frac{\pi N}{2}) < 0$. The disc phase is $\arg U_{\text{disc}}(0) = kb + \pi N$. Ultrasound experiments confirm this N-dependent phase.

References:

Hitachi, A., and Takata, M. (2009). Babinet's principle in the Fresnel regime studied using ultrasound. arXiv:0904.1269.

A Circle of Time

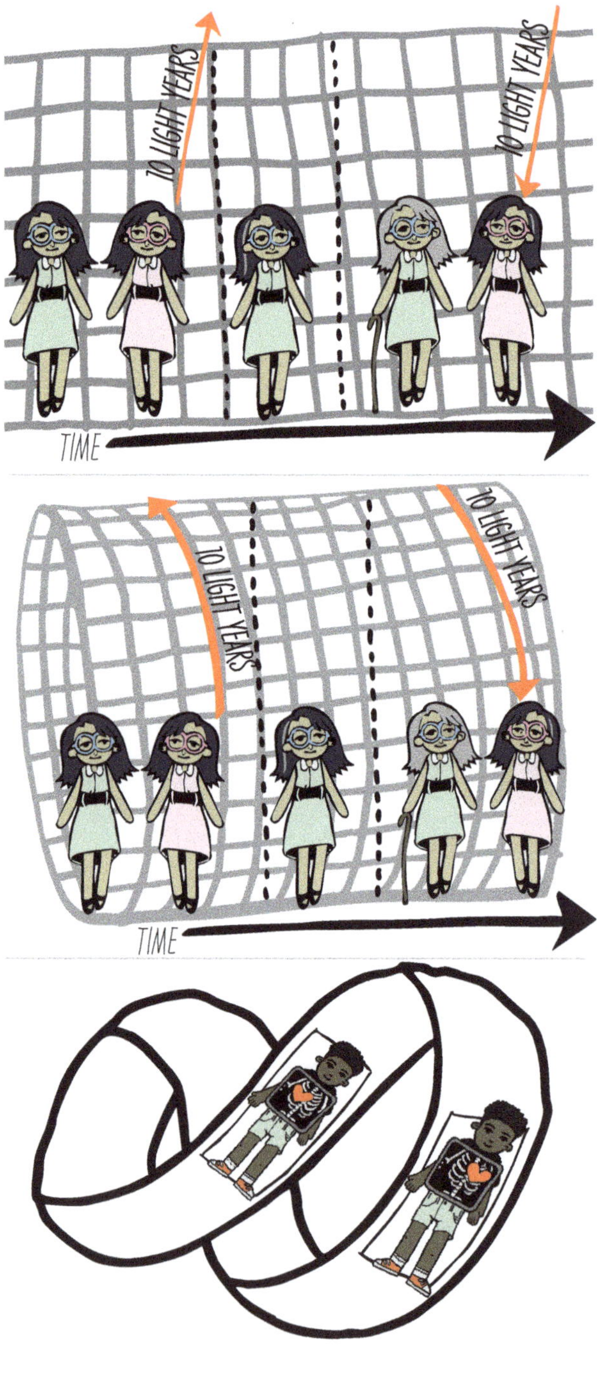

Top (Twin Paradox Variants): Variants of the twin paradox. The first shows the standard scenario: one twin remains on Earth while the other travels to a distant star and back at relativistic speed, experiencing less proper time due to time dilation and the non-inertial turnaround. The second panel sets the paradox in a compactified spacetime, where space is wrapped into a cylinder. Here, a twin can traverse the compact dimension at constant velocity—never accelerating or changing frames—yet still accumulate less proper time than the twin who remains stationary.
Bottom (Non-Orientable Topology): Non-Orientable topology and global spacetime orientation. Depicted as a Möbius-like strip, this represents a spacetime where following a continuous timelike path can bring a traveller back to their starting point with their internal orientation flipped.

A Circle of Time

In a cylindrical universe with compact spatial dimensions, twins can separate and reunite without acceleration, with one travelling around the circumference while the other remains stationary. Despite neither experiencing acceleration, they age differently upon reunion, creating a variant of the famous special relativistic paradox. In non-orientable topologies such as the Klein bottle universes, travellers can additionally experience reversal of chirality. Even a thoroughly non-dextrocardic explorer might come back from a cosmic stroll with his heart on the right side, no trauma needed.

Twin Paradox ◦ Special Relativity Postulates ◦ Proper Time Paths ◦ Compact Space Topology ◦ Cylindrical Universe ◦ Acceleration-Free Paradox ◦ Preferred Frame Detection ◦ Klein Bottle Spacetime ◦ Chirality Reversal ◦ CMB Topology Search ◦ Flatness Constraint

"Time is an illusion. Lunchtime doubly so."
— Douglas Adams, 1979

"I'm not a creative type like you
with your work sneakers and your left-handedness."
— Jack Donaghy, 2008

A Circle of Time

Albert Einstein's 1905 introduction of special relativity redefined time and motion, emphasising the role of inertial frames and leading to the now-familiar notion of time dilation. A few years later, Hermann Minkowski introduced a four-dimensional spacetime geometry, allowing relativistic effects to be understood geometrically. The so-called 'twin paradox' became an iconic thought experiment illustrating the asymmetric ageing of two travellers, one of whom undergoes acceleration.

Parallel to these developments, mathematicians in the late 19th and early 20th centuries, including Eugenio Beltrami, August Möbius, and Felix Klein, investigated the implications of identifying the edges of geometric surfaces. These ideas laid the groundwork for understanding non-orientable and compact spaces, such as the Möbius strip and Klein bottle. Henri Poincaré introduced foundational concepts in topology that allowed physicists to study global properties of space beyond Euclidean structure.

In 1949, Kurt Gödel proposed a rotating cosmological model that permitted closed timelike curves, demonstrating that general relativity allowed for causal paths that returned to earlier points in time. Later work by John Wheeler, Bryce DeWitt, and others examined exotic topologies in general relativity, where spacetime could be globally identified in non-intuitive ways, even if it remained locally flat.

The foundation of special relativity rests on two principles, known as the postulates of the theory (Einstein, 1905). First, all inertial motion is equivalent: no experiment can detect absolute rest. Second, light in a vacuum travels at a constant speed, $c = 299{,}792{,}458$ metres per second, in every inertial frame, regardless of the motion of the source or the observer.

The first postulate extends Galilean symmetry—physics does not change under uniform motion, there is no preferred velocity, no absolute background. Any inertial observer, whether drifting through space or sitting still on Earth, applies the same physical laws. The second postulate introduces a fixed scale, the speed of light, that remains unchanged across all inertial frames. It does not behave like other velocities. If you move toward a beam of light at half its speed or away from it just as fast, you still measure its speed relative to you as c. The constancy of c holds in every experiment ever conducted. This fixed speed breaks the logic of velocity addition in classical mechanics. Something must change—what changes is time.

To see how, imagine a pulse of light emitted inside a moving train car. A mirror is mounted on the ceiling, directly above the source. In the frame of the train, the light travels straight upward, hits the mirror, and returns to the source. In the ground frame, the train is moving horizontally during the pulse's travel, so the light follows a diagonal path. Since both observers agree that the speed of light is c, and the diagonal path is longer than the vertical one, they must assign different durations to the same event.

This shows that simultaneity depends on the observer's frame—two events judged to occur at the same time in one frame may occur at different times in another. There is no universal present; motion affects how clocks are synchronised across space.

 https://doi.org/10.11647/OBP.0526.14

From this follows a broader conclusion: elapsed time depends on trajectory. Two clocks that start together, separate, and reunite may disagree. Even if both move inertially, they accumulate different amounts of proper time. This difference reflects the geometry of spacetime. Duration becomes a function of path. The twin paradox illustrates this (Langevin, 1911): two siblings begin together, one remains on Earth while the other travels outward at high speed, reverses direction, and returns. When they reunite, one has aged 10 years, the other only 1 over the entire round trip.

At first glance, the situation seems symmetric. Each twin sees the other in motion, and motion implies time dilation, so why is there a preferred twin that stays younger? This is because only the travelling twin changes inertial frame. The stay-at-home twin remains in one throughout. The shift occurs at turnaround, when the traveller accelerates and transitions to a new inertial frame. That transition comes with a new definition of simultaneity: a new assignment of which distant events on Earth are happening 'now.' The shift occurs abruptly in the traveller's coordinate system, producing a discontinuous reassignment of time to faraway clocks. In the new frame, the traveller's slice of simultaneity jumps forward, assigning later times to the Earth clock without any local observation.

The result is that the traveller accumulates less proper time between departure and return. In flat spacetime, there are many possible inertial paths between the same events, and they do not yield equal durations. The traveller's path is shorter. If they move at $0.995c$ for 5 years outbound and 5 years return (as measured by the Earth clock), their own clock measures only 1 year.

Classical relativity requires acceleration to break the symmetry between twins. But this requirement disappears if space itself has boundaries that connect back to themselves. Consider a universe where space wraps around in one direction, like the surface of a cylinder. Travel far enough in the x-direction and you return to your starting point from the opposite side. Mathematically, we say the points at positions x and $x + L$ are identified—they represent the same physical location, where L is the circumference of the universe in that direction. This periodic boundary condition means that coordinates differing by L describe identical points in space. The resulting topology is cylindrical, but the local geometry remains flat—similar to Earth, which locally feels flat but is spherical globally.

Now consider two identical clocks: one remains at rest while the other moves uniformly around the compact direction, maintaining constant speed and never accelerating. After one complete loop, the moving clock returns to the stationary one. Both have followed inertial trajectories; both consider themselves at rest. Yet when they compare clocks, they disagree. With circumference $L = 1$ light-year and the moving twin travelling at $v = 0.8c$, the journey takes $\Delta t = L/v = 1.25$ years as measured by the stationary twin. But the moving twin's clock shows only $\Delta t\sqrt{1 - v^2/c^2} = 1.25 \times 0.6 = 0.75$ years. The moving twin ages 0.5 years less, despite never accelerating.

This recreates the twin paradox without any frame changes. Each observer sees the other as moving. Each expects the other's clock to tick more slowly. In the classical case, the paradox is resolved by noting that one twin undergoes a change of inertial frame. Here, no such event occurs. The setup is symmetric in every local respect. Still, the clocks disagree.

The resolution comes from recognizing that compactifying space breaks a global symmetry. In ordinary Minkowski space, all inertial frames are equivalent. But once we impose the identification $x \sim x + L$, that equivalence no longer holds at the global level—there is a distinguished frame: the one in which the identification is purely spatial, with no accompanying time shift. In that frame, a light pulse sent around the loop in both directions returns simultaneously. In any other inertial frame, the forward and backward travel times differ.

The twins can detect this asymmetry directly. Let the moving twin send light signals in both directions around the universe. If moving at velocity v relative to the compact rest frame, the light travelling forward takes time $L/(c-v)$ to complete the loop, while light travelling backward takes $L/(c+v)$. The total round-trip time is $t_{total} = L/(c-v) + L/(c+v) = 2Lc/(c^2 - v^2)$. For the stationary twin in the compact rest frame, both directions take exactly L/c, giving a total of $2L/c$. This Sagnac-like asymmetry reveals motion (Sagnac, 1913) relative to the universe's topology.

The spatial loop introduces a global constraint: although each observer sees themselves as stationary, only one is stationary relative to the universe itself. This asymmetry explains the clock discrepancy—proper time depends not only on the local geometry of the path, but on how that path winds through the global shape. The twin who moves around the loop crosses more space within the same spacetime interval and accumulates less proper time. Local measurement will not be sufficient to reveal the difference. The effect is detected only when trajectories reconnect across the full topology. Locally, all observers still see standard special relativity effects. If they didn't, we could rule out compact spatial dimensions just by testing special relativity (SR) in small laboratories here on Earth.

While such compact dimensions are not currently a theoretical frontier, some cosmological models predict that space could be finite and wrap around on scales comparable to the observable universe. The cosmic microwave background (CMB) radiation carries information about the universe's topology, and astronomers have developed methods to search for these specific signatures.

The most direct approach looks for repeated patterns in the CMB. If space wraps around with circumference L, light from the same physical region can reach us along multiple paths. We would see the same temperature fluctuations repeated at different locations in the sky, separated by the angle subtended by the compact dimension. Astronomers search for these correlations using statistical tests, comparing temperature patterns at different sky positions and looking for correlations stronger than expected by chance. The analysis must account for instrumental noise, foreground contamination from our galaxy, and the natural statistical variations in the CMB itself. Current data from the Planck satellite has ruled out compact topologies with characteristic scales smaller than about half the observable universe.

A second method examines the geometry directly. In a finite universe, the total solid angle covered by the CMB would be less than 4π steradians. We would see the same physical surface from multiple directions, creating a characteristic pattern of repeated circles on the sky. Galaxy surveys provide another probe: if space is compact, we might observe the same galaxy clusters at different redshifts and positions, their light having travelled different distances around the universe. The most distant visible galaxies would appear both in their 'true' location and as 'ghost images' from light that circled the universe multiple times. These ghosts would show the same galaxy at different cosmic ages, creating a unique observational signature.

The geometry of space adds another constraint. Cosmologists measure the density parameter Ω_0, which determines the universe's curvature. If $\Omega_0 = 1$, space is perfectly flat, like an infinite sheet of paper. If $\Omega_0 > 1$, space curves back on itself like the surface of a sphere. If $\Omega_0 < 1$, space curves outward like a saddle. Current observations from supernovae, the CMB, and galaxy surveys all indicate $\Omega_0 = 1.000 \pm 0.002$. The universe is flat to high precision.

This flatness constrains but does not eliminate compact topologies. A flat universe can still wrap around on itself, like a flat torus formed by connecting opposite edges of a square. But if space were significantly curved ($\Omega_0 \neq 1$), the curvature would create additional observable signatures that could either help or hinder topology searches. In a closed universe ($\Omega_0 > 1$), space naturally curves back on itself, making some compact topologies easier to detect. In an open universe ($\Omega_0 < 1$), the negative curvature works against compactification, making topology searches more difficult.

The observed flatness suggests that if the universe is compact, it must have a very specific topology: one that preserves flatness while allowing space to close on itself. This narrows the search to particular classes of compact manifolds, such as the three-torus or other flat topologies, while ruling out many curved compact spaces. Current observations constrain compact topologies to scales comparable to or larger than the observable universe, with fundamental domain sizes at least on the order of tens of billions of light-years. Future missions with better sensitivity and resolution may push these limits further, but detecting cosmic topology remains one of the most challenging problems in observational cosmology (See Chapter 43 for more.)

Compactifying a dimension, making space periodic, can lead to observable asymmetries between otherwise equivalent observers. But topological modifications can go further. Instead of just gluing the ends of space together, we can twist them before joining.

You may have seen the Möbius strip (Möbius, 1865): a flat band with a half-twist, joined end to end. It has only one side and one edge. If you travel along it, you return to where you started but flipped. What was left becomes right. The Möbius strip is an example of a non-orientable space.

A space is orientable if it allows a consistent definition of left and right everywhere. On a sheet of paper, or the surface of a sphere, you can carry a small arrow around any path and it will always point the same way relative to the surface. But on a Möbius strip, that fails. The arrow returns reversed. There is no global way to define direction that holds across the entire space.

The Klein bottle extends this concept (Klein, 1882) to a closed surface without boundaries. Like the Möbius strip, it reverses orientation, but it closes without edges. It cannot be embedded in three-dimensional space without intersecting itself, but as a topological object it is well-defined. A path around the Klein bottle can return to its starting point mirrored because of how space is connected, not through motion or twisting.

Now apply this idea to spacetime. Consider a spacetime with Klein bottle topology, where the spatial identification becomes $x \sim -x + L$. Movement along this direction not only loops back, but also inverts orientation. A clock moving uniformly along the compact path returns to its original location but mirrored. Left becomes right. Clockwise becomes counterclockwise.

This leads to direct physical consequences. Many systems have intrinsic handedness: chiral molecules, spin-aligned particles, asymmetric anatomy. In a non-orientable universe, these properties are not preserved globally. A round trip along the compact direction can convert a left-handed structure into its right-handed counterpart. The change is undetectable locally—the traveller feels nothing, no process unfolds. Yet on return, the configuration has flipped.

There is an anatomical condition called *situs inversus totalis*, where all internal organs are mirrored. In physiology, this is a congenital condition, present from birth. But in a non-orientable spacetime, such a reversal could result from motion alone. A person could leave on a journey through space, follow a smooth inertial path, and return anatomically mirrored. The heart that began on the left would now be on the right. Every asymmetry, from organ placement to molecular chirality, would be inverted. The twins would face a peculiar situation upon reunion: the traveller would return with every internal anatomy reversed, verifiable by examining molecular handedness or organ placement. Yet no force acted on the traveller. No acceleration occurred. The inversion arose purely from the topology of spacetime.

How to Reverse Your Heart at Home

This reversal can be modelled physically at home. Begin with a strip of paper approximately 30 cm long and 2 cm wide. Introduce a half twist and tape the ends together, forming a Möbius strip. You now have a surface with only one side and one edge: an example of a non-orientable space.

To visualise orientation reversal, draw a schematic figure: for example, a stick figure facing right with a small arrow marking its left hand. Make sure the figure is upright and aligned with the edge of the strip, as though standing on it. If possible, use a transparent sheet so you can track embedded orientation.

Now, slide the figure smoothly along the surface, keeping it flush against the paper and preserving its local orientation. Do not rotate or detach it. Maintain contact with the same 'side' of the strip (though, by construction, there is only one). After completing a full circuit, the figure returns to its original location, but with its left and right reversed. The arrow now appears on the opposite side. No flipping occurred, yet the orientation is inverted.

The Twin Paradox and Chirality

Compact Minkowski Geometry

Consider a (1+1)-dimensional Minkowski spacetime with metric

$$ds^2 = -c^2dt^2 + dx^2,$$

under the identification $x \sim x + L$, forming a spatial circle of circumference L. A twin (A) remains stationary at $x = 0$. Twin B travels at constant velocity $v > 0$ along the compact direction and returns after n full loops, where $n \in \mathbb{Z}^+$. The path is globally closed but locally inertial throughout.

Define $\beta \equiv v/c$. Let the coordinate reunion time be $\Delta t = \frac{nL}{v}$. Twin A's proper time is

$$\tau_A = \Delta t = \frac{nL}{v}.$$

Twin B's proper time is reduced by the standard Lorentz factor:

$$\tau_B = \Delta t\sqrt{1-\beta^2} = \frac{nL}{v}\sqrt{1-\beta^2}.$$

The ratio

$$\frac{\tau_B}{\tau_A} = \sqrt{1-\beta^2}$$

is strictly less than 1. The proper-time difference is nonzero, despite both worldlines being geodesic.

Lorentz Symmetry and Preferred Frames

In infinite Minkowski space, all inertial frames are equivalent. Compactification breaks this symmetry. The identification $x \sim x + L$ selects a preferred frame in which the identification is purely spatial. In other frames boosted along the x-axis, the identification becomes mixed with time.

To detect this asymmetry, send light signals in opposite directions around the loop. An observer moving at velocity v relative to the compact frame measures asymmetric round-trip times:

$$t_\pm = \frac{L}{c(1 \mp \beta)}, \qquad \Delta t = t_+ + t_- = \frac{2L}{c(1-\beta^2)}.$$

This directional difference reveals the observer's motion relative to the compact topology. The spacetime remains locally Minkowskian, but the global topology renders the compact frame observationally distinct.

Non-Orientable Identification and Chirality Reversal

Now replace the identification with a non-orientable one:

$$x \sim -x + L,$$

which reverses orientation upon completing a loop. This defines a compact, boundaryless, non-orientable manifold—the spacetime analogue of a Klein bottle.

Let a traveller carry an orthonormal frame $e^\mu(t)$ along a geodesic parameterized by proper time t. After one full traversal, parallel transport yields:

$$e^\mu(t+T) = R^\mu_{\ \nu} e^\nu(t),$$

where $R^\mu_{\ \nu}$ is a linear transformation with determinant $\det R = -1$. This inversion flips handedness: the transported frame returns as a mirror image of itself.

Fields that are sensitive to orientation—such as spinors or chiral matter—cannot be globally defined without modification. While scalar fields remain unaffected, spinor bundles require consistent orientation to maintain chirality. In this topology, left-handed and right-handed states are exchanged after global propagation, even in the absence of any local interaction or curvature.

References:

Misner, C. W., Thorne, K. S., Wheeler, J. A. (1973). *Gravitation.* Freeman.

Geroch, R. (1967). *J. Math. Phys.*, **8**.

Isham, C. J. (1989). *Modern Differential Geometry for Physicists.* World Scientific.

Envelope Trade-Up

Top (Basic Setup): The left scenario shows concrete values: one envelope contains \$100, with potential alternatives of \$50 or \$200. The right scenario generalises this with variables: envelope X could correspond to either $x/2$ or $2x$ in the alternative envelope. This applies regardless of specific amounts.
Second (Bounded Finite): Eight envelopes containing $1, 2, 4, 8, 16, 32, 64, 512$ represent a bounded geometric sequence. In this finite setup, edge effects matter: the smallest value (1) can only be paired with (2), while the largest (512) can only be paired with (256). These boundary constraints resolve the paradox.
Third (Semi-Infinite): The sequence $1, 2, 4, \ldots, 2^n, \ldots$ extends infinitely in one direction. This introduces asymmetry: there's a smallest possible value but no largest. The lower bound affects the expectation calculation.
Fourth (Bi-Infinite): The sequence $\ldots, 1/2, 1, 2, 4, \ldots, 2^n, \ldots$ extends infinitely in both directions. This symmetric case is where the classical paradox is the most clear, as there are no boundary effects to break the apparent symmetry.
Bottom (Probability Distributions): Three different priors over envelope values: (1) Uniform distribution over $\{1, 2, \ldots, 100\}$ assigns equal $1/100$ probability to each value. (2) Exponentially decaying distribution over positive integers gives decreasing probability as values increase. (3) Gaussian distribution over all real numbers.

Envelope Trade-Up

The Envelope Paradox presents two envelopes where one contains twice the money of the other. After selecting one envelope, seemingly valid probabilistic reasoning suggests an expected gain by switching (averaging $x/2$ with $2x$), regardless of which envelope was initially chosen. This symmetric conclusion creates a logical inconsistency since perpetual switching cannot be optimal. The paradox arises from improper application of expected value calculations to scenarios with unbounded distributions or when conditional probabilities are not properly accounted for. Resolving the paradox requires distinguishing between known values and variables, recognising when probability distributions are ill-defined, and understanding the limitations of calculations with potentially infinite quantities.

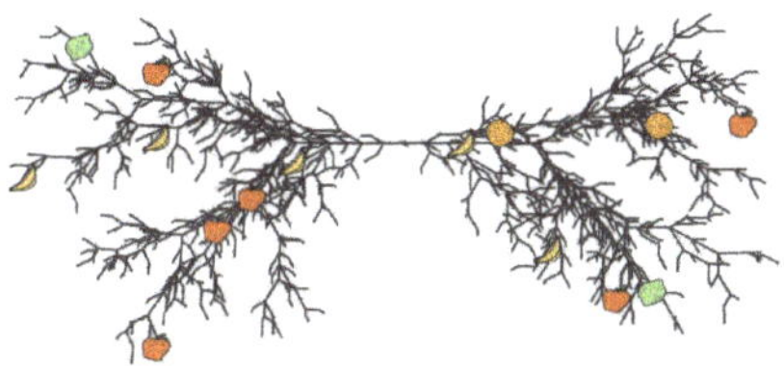

Two Envelopes Paradox ◦ Switching Expectation ◦ Incompatible Baselines ◦ Bounded Distribution Solution ◦ Improper Prior Problem ◦ Bertrand's Chord Paradox ◦ Random Without Context ◦ Boundary Cancellation ◦ Symmetric Model Zero Gain ◦ Kraitchik 1930 ◦ Simulation Verification

"How wonderful that we have met with a paradox; now we have some hope of making progress."

— Niels Bohr, c. 1958

Envelope Trade-Up

The envelope paradox traces its origins to Belgian mathematician Maurice Kraitchik, who proposed a related puzzle in his 1930 book *Les Mathématiques des Jeux*. Kraitchik's version involved two men comparing the value of their neckties, with the winner giving his necktie to the loser as consolation. He also discussed a variant where the men compared the contents of their purses, assuming each contained between one and some large number of pennies with equal probability.

The puzzle gained mathematical attention when John Edensor Littlewood mentioned it in his 1953 book on elementary mathematics, crediting the idea to physicist Erwin Schrödinger. Littlewood's formulation involved cards with numbers written on both sides—a player sees one side of a random card and must decide whether to flip it over. His version made explicit the role of improper prior distributions in generating the paradox.

Martin Gardner brought the puzzle to widespread attention in his 1982 book *Aha! Gotcha*, presenting it as a wallet game between two equally wealthy individuals. Gardner's formulation captured the essential difficulty: each person could construct identical arguments for why the game favoured them, yet by symmetry, the game had to be fair. Remarkably, Gardner confessed that while he could analyse the problem correctly, he struggled to pinpoint exactly what was wrong with the switching argument.

The modern envelope paradox resurfaced as probability theory matured into a rigorous mathematical discipline. The twentieth century developments of measure theory, decision theory, and Bayesian analysis provided new tools for understanding why such problems arose and how they might be resolved.

You are presented with two identical envelopes—you are told that one envelope contains twice the money of the other. No other information is given. No other information is given. You select one envelope at random and open it, revealing \$100. At this point, you are given the opportunity to switch.

Consider the reasoning that leads to paradox. The envelope you opened contains \$100, which must be either the smaller amount or the larger. If \$100 is the smaller amount, the other envelope contains \$200. If \$100 is the larger amount, the other envelope contains \$50. Since these two cases seem equally likely, the expected value from switching appears to be:

$$\text{Expected gain} = 0.5 \cdot (+100) + 0.5 \cdot (-50) = +25.$$

This suggests a \$25 gain from switching. The same calculation applies regardless of the amount observed, whether \$10, \$100, or \$1000. You should apparently switch every time. But this logic also tells you to switch back again after switching, producing an endless preference loop. The contradiction is that each envelope appears preferable to the other.

Why does this reasoning feel compelling? The calculation follows expected value logic. The probabilities seem reasonable. Without additional information, why shouldn't the observed amount be equally likely to be the smaller or larger? The arithmetic is correct.

Yet the conclusion violates intuitions about symmetric problems. The puzzle demands a definitive answer while generating contradictory recommendations.

The error lies in how the observed amount x is interpreted across different terms of the expectation. In one term, x represents the smaller amount; in the other, it represents the larger amount. This reference creates incompatible baselines for comparison.

To make the model coherent, let x denote the smaller of the two amounts. Then the envelopes contain x and $2x$, and each is equally likely to be selected. If you hold x, switching yields $+x$; if you hold $2x$, switching yields $-x$. The outcomes cancel:

$$\text{Average change} = 0.5 \cdot (+x) + 0.5 \cdot (-x) = 0.$$

No advantage arises. The paradox dissolves because the original argument uses expectation without a consistent model.

This becomes clearer in a bounded setup. Suppose the smaller amount is chosen uniformly from $\{2^0, 2^1, \ldots, 2^{999}\}$. For most observed amounts A, switching seems to yield a gain of $+0.25A$ since A could be either the smaller or larger envelope value. However, this ignores boundary cases: if $A = 2^0$, switching cannot halve it; if $A = 2^{999}$, switching cannot double it. These rare but extreme boundary effects precisely cancel the average gain across interior values, returning the total expectation to zero.

In the limit as the model becomes unbounded, for example, when x is drawn from $\{\ldots, 2^{-2}, 2^{-1}, 2^0, 2^1, \ldots\}$, the problem reappears. For any observed amount, switching appears to yield a gain of $+0.25A$. But this assumes all values are equally probable, which is not possible over an infinite set.

A uniform distribution over an infinite number of values cannot exist (Kolmogorov, 1933)—there is no way to assign equal, nonzero probability to infinitely many outcomes and still have the total probability equal to one. Any attempt results in an *improper prior*: a function that resembles a distribution but cannot be normalised.

To reason coherently in such a context, one must use a *probability measure*: a rule that assigns consistent, additive weights to sets of values and sums to one. A measure is *proper* if it satisfies this condition. If it diverges or is undefined, expectations may not exist. Even if each outcome is finite, the global average may be infinite or ill-defined. In that case, expectation ceases to be a meaningful decision tool.

One setup does yield a switching advantage. Fix a value a and flip a coin: if heads, prepare envelopes with a and $2a$; if tails, use a and $a/2$. Hand the envelope containing a to the player. Switching yields either $+a$ or $-0.5a$ with equal probability, giving an expected gain of $+0.25a$. Here, switching is optimal because the model is asymmetric and expectation is applied with explicit conditioning.

The switching advantage depends entirely on the unknown prior distribution over envelope pairs. If smaller amounts are more probable, observing \$100 suggests you likely hold the larger envelope. If larger amounts are more probable, the reverse holds. Without knowing this distribution, rational choice becomes impossible.

Bertrand's paradox (1889) poses the question: what is the probability that a random chord of a circle is longer than the side of an inscribed equilateral triangle? The answer depends on what 'random' means. Select two random points on the circumference and connect them: the probability is $1/3$. Select a random radius and place the chord perpendicular to it at a random distance from the centre: the probability is $1/2$. Select a random interior point as the chord's midpoint: the probability is $1/4$. Each method seems reasonable, yet they yield different results. 'Random chord' is undefined without specifying the selection procedure.

Similarly, should you treat envelope pairs $(50, 100)$ and $(100, 200)$ as equally likely? Or should you treat the amounts \$50, \$100, \$200 as equally likely? Each choice determines a different prior over the smaller value x. The first approach makes pairs equally likely; the second makes values equally likely, implying smaller values are more probable than larger ones when considering pairs. Without specifying the generation mechanism, 'random envelope' has no unique meaning, and the 'principle of indifference'—treating outcomes as equally likely (Keynes, 1921) in the absence of information—produces contradictions.

The Bayesian approach requires a prior distribution over envelope pairs. Without knowing the generation mechanism, no prior can be justified. Even given a distribution, improper priors (such as uniform over all positive reals) produce undefined probabilities, and heavy-tailed distributions yield infinite expected values. In symmetric setups, observing one amount provides zero information about which envelope contains more. Over repeated trials with any proper distribution, switching gains nothing on average.

Renormalizing Pascal

Pascal's Wager collapses under the same pathology (Pascal, 1670). The canonical argument claims belief in God offers infinite expected utility: either finite loss (if wrong) or infinite gain (if right). The arithmetic seems decisive. But critics note the state space is unbounded. For any deity G granting infinite utility for act A, one can posit an Anti-God penalising A with infinite disutility. The expected value becomes:

$$EU = \sum_{i=1}^{\infty}[P(G_i) \cdot \infty] + \sum_{j=1}^{\infty}[P(G_{-j}) \cdot (-\infty)].$$

Undefined. The same divergence that breaks the envelope calculation breaks Pascal.

Physical field theories face identical infinities at high energies. The solution is *renormalization*: impose a cutoff scale (Dyson, 1949) Λ, integrate out fluctuations above this threshold as irrelevant to the macroscopic theory. Virtual particles appear transiently but lack the coupling strength to affect observables.

Apply this to theology. Introduce a *social-credal cutoff* ϵ. A theological hypothesis enters the expected utility calculation only if its social magnitude—measured by historical persistence, institutional architecture, mass adherence—exceeds the threshold. Below ϵ, the probability is set to zero.

Ad hoc philosophical fictions function as virtual particles: transient fluctuations in logical space lacking ontological mass to couple with macro-sociological reality. The infinite tail truncates. The divergent series collapses into a finite choice between established traditions and the null hypothesis.

This appears to commit the *ad populum* fallacy. It does, but only if logical operators possess Platonic independence. They do not. Axioms such as the Law of Excluded Middle or $1 + 1 = 2$, in this context, are not absolute universals. They are emergent properties of shared human neurobiology, intersubjective agreements sustained by consensus among rational agents.

The validity of *logic* is operational, not objective. It rests on the *vox populi* of the species. If 99% of logicians were rewired to reject the Law of Excluded Middle, it would cease to function as truth within our epistemic system. When we call an argument irrational, we point to inconsistencies with shared basic axioms, not disagreements about logic itself. Rationality is the practice of consistency within a common framework.

We operate several layers of abstraction above this foundation. When debating probability theory or physical laws, we invoke principles two or three levels removed from basic logic, mistaking these derived arguments for appeals to cosmic truth. The system functions only because the base layer—the axioms of logic itself—enjoys universal consensus. If basic logic ever fragmented to 60/40 agreement, rational discourse would collapse entirely. We could not even agree on what 'inconsistent' means. Uncomfortable as it feels, one must recognise that at the bottom of all our reasoning—why we believe anything from a news piece to a scientific theory, or even logic itself—lies a social standard. There is no easy way to define truth independent of our shared neurobiology.

If rationality itself is consensus-dependent, religious truth within a rational wager must undergo identical phase transitions. A deity with zero social magnitude is a private delusion. It lacks the intersubjective validation required to exist as a probability vector in a decision matrix.

The 'Inverted Gods' objection does not refute Pascal's logic. It demonstrates that his system requires boundary conditions. The ϵ cutoff aligns the Wager with the effective field theory of human consensus. Pascal's emphasis on *praxis* becomes pivotal: the 'true' religions are those that operationalized belief into action sufficient to cross ϵ.

In practice, the Wager loses meaning for a simpler reason: humans are not probabilistic decision makers (Kahneman & Tversky, 1979). We cross streets and eat apples despite non-zero probabilities of losing our infinitely-valued lives to accidents or allergies (maybe subconsciously we find terms such that they cancel each other out). We institute cutoffs on rare events and extreme losses not through Bayesian calculation but through neural circuits that force decisions with limited data. The theoretical requirement for a cutoff matches the empirical fact that we already use one. We cannot meaningfully operate on probabilities when assumptions and expected rewards are unbounded. The Wager fails not because the mathematics is wrong, but because the mathematics is not the only guidance to action. For problems at the edge of human experience, intuition and experience are paradoxically more valuable and coherent than analytical slaloms (such as this very piece of commentary!).

Exercises

1. Wine/Water Paradox

A mixture contains wine and water in a ratio $x = \frac{\text{wine}}{\text{water}}$, with $x \in [\frac{1}{3}, 3]$. Assuming x is uniformly distributed, compute the probability that the mixture has $x \leq 2$. Now let $y = \frac{1}{x}$, representing the water-to-wine ratio, and assume y is uniformly distributed instead. What is the probability that $y \geq \frac{1}{2}$? Why are the two results inconsistent?
Hint: Use

$$P(x \leq x_t) = \frac{x_t - x_{\text{min}}}{x_{\text{max}} - x_{\text{min}}}, \qquad P(y \geq y_t) = \frac{x_{\text{max}}(1 - x_{\text{min}} y_t)}{x_{\text{max}} - x_{\text{min}}}.$$

Evaluate both at $x_t = 2$, $y_t = \frac{1}{2}$. What contradiction arises?

2. Monty Hall Problem

Three closed doors hide a car and two goats. You choose Door 1. The host, who knows what is behind each door, opens Door 3, revealing a goat, and offers you the option to switch to Door 2. Should you switch?
Hint: The host never opens your door and never reveals the car. Compute:

$$P(\text{Car behind D1} \mid \text{Host opens D3}) \quad \text{vs} \quad P(\text{Car behind D2} \mid \text{Host opens D3}).$$

Use Bayes's theorem. The host's choice depends on your initial pick and the hidden car's placement.

3. Tuesday Boy Problem

A family has two children. You are told that one of them is a boy born on a Tuesday. What is the probability that both are boys?
Hint: Let $\varepsilon = \frac{1}{7}$. Then:

$$\text{P(BB} \mid \text{B}_\text{T}) = \frac{(1 - (1 - \varepsilon)^2) \times \frac{1}{4}}{0 + \frac{1}{4}\varepsilon + \frac{1}{4}\varepsilon + \frac{1}{4}(\varepsilon + \varepsilon - \varepsilon^2)} = \frac{1 - (1 - \varepsilon)^2}{4\varepsilon - \varepsilon^2} = \frac{(2 - \varepsilon)}{(4 - \varepsilon)}$$

4. Two-Black-Sides Card Problem

You randomly select a card from a box containing: one black-black card, one white-white card, and one black-white card. You see a black face. What is the probability the other side is black?

5. Sleeping Beauty Problem

Sleeping Beauty is put to sleep. A fair coin is tossed. If heads, she is woken on Monday only. If tails, she is woken on Monday and Tuesday (but forgets Monday). Upon waking, what is her credence that the coin landed on heads?
Hint: Model the wake-up events: $\{$H-M, T-M, T-T$\}$. Normalise using the condition: 'I am awake now.'

Resolving the Envelope Paradox

The paradox emerges from treating the observed amount Y as equally likely to be the smaller or larger of two amounts related by a $1:2$ ratio. The naive expectation assumes: $\mathbb{E}[\text{switch}] = 0.5 \cdot 2Y + 0.5 \cdot Y/2 = 5Y/4$, suggesting a gain from switching. But this treats Y as both X and $2X$ in different terms—a misapplication of conditional expectation.

Correct Conditioning

Let X be drawn from prior $f(x)$. The envelope pair $(X, 2X)$ is constructed, and one is presented at random. Let Y denote the observed amount:

$$\begin{aligned} Y = X &\Rightarrow \text{other envelope has } 2Y, \\ Y = 2X &\Rightarrow \text{other envelope has } Y/2. \end{aligned}$$

The conditional probabilities are:

$$\begin{aligned} \mathbb{P}(Y = y \mid \text{observed smaller}) &\propto f(y), \\ \mathbb{P}(Y = y \mid \text{observed larger}) &\propto \tfrac{1}{2} f(y/2). \end{aligned}$$

The expected value of switching given Y is:

$$\mathbb{E}[\text{sw} \mid Y] = \frac{2Y \cdot f(Y) + \frac{1}{2}Y \cdot f(Y/2)}{f(Y) + f(Y/2)}.$$

This depends on the shape of f. When f decays rapidly, $f(Y/2) \gg f(Y)$ for large Y, implying Y is likely the larger value and switching is unfavourable.

Example: Pareto Prior

For $X \sim \text{Pareto}(\alpha)$ with $f(x) = \alpha x^{-(\alpha+1)}$ on $[1, \infty)$:

$$f(Y/2) = \alpha \cdot 2^{\alpha+1} \cdot Y^{-(\alpha+1)} = 2^{\alpha+1} f(Y).$$

Thus:

$$\begin{aligned} \mathbb{E}[\text{switch} \mid Y] &= \frac{2Y + \frac{1}{2}Y \cdot 2^{\alpha+1}}{1 + 2^{\alpha+1}} \\ &= Y \cdot \frac{2 + 2^{\alpha}}{1 + 2^{\alpha+1}}. \end{aligned}$$

This may exceed or fall below Y depending on α.

Improper Priors

For log-uniform $f(x) \propto 1/x$ on $[1, \infty)$ (improper since $\int_1^\infty \frac{1}{x} dx = \infty$):

$$\begin{aligned} f(Y) = \frac{1}{Y}, &\quad f(Y/2) = \frac{2}{Y}, \\ \mathbb{E}[\text{switch} \mid Y] &= \frac{2Y \cdot \frac{1}{Y} + \frac{1}{2}Y \cdot \frac{2}{Y}}{\frac{1}{Y} + \frac{2}{Y}} = Y. \end{aligned}$$

This suggests switching is neutral, but the result is meaningless: the improper prior makes the marginal distribution undefined.

Finite Uniform Model

Let $x \in \{2^0 . 2^1, \ldots, 2^{N-1}\}$ be uniform. For observed amount $A = 2^m$:

- Interior ($1 \le m \le N-2$): $\mathbb{E}[\Delta \mid A] = \frac{1}{2}(2^m) + \frac{1}{2}(-2^{m-1}) = 2^{m-2}$.
- Boundaries: $\mathbb{E}[\Delta \mid A = 2^0] = +1$, $\mathbb{E}[\Delta \mid A = 2^{N-1}] = -2^{N-2}$.

Global expectation:

$$\begin{aligned} \mathbb{E}[\Delta] &= \frac{1}{N}\left(\sum_{m=1}^{N-2} 2^{m-2} + 1 - 2^{N-2}\right) \\ &= \frac{1}{N}\left(-2^{N-3} + \tfrac{1}{2}\right) < 0. \end{aligned}$$

As $N \to \infty$, this approaches zero from below, confirming no long-run switching advantage.

References:

Nalebuff, B. (1989). The Other Person's Envelope Is Always Greener. *J. Econ. Persp.*, **3**(1), 171–181.

An Empty Threat

Top (Metastable State): A system stuck in a local energy minimum. Though not at the true lowest-energy state, it remains trapped unless perturbed. The ball resting in a shallow well represents temporary stability—common in physical phenomena such as false vacuum decay or the Sun.
Middle (Inertial vs Gravitational Mass): Both sides read '1KG,' but differ in interpretation. Inertial mass resists acceleration (right), while gravitational mass dictates weight in a field (left). Although equivalent, they arise differently. The Higgs mechanism contributes to inertial mass through interaction with the Higgs field; why gravitational mass follows, remains an open question in quantum gravity.
Bottom (Higgs Mechanism): A particle (left) moves through the Higgs field (right), acquiring mass via constant interaction. The disturbance in the field (rippled line) corresponds to a Higgs boson. The mechanism explains how otherwise massless particles—like W and Z bosons—gain mass while preserving gauge symmetry.

An Empty Threat

Our universe may exist in a false vacuum—a metastable state that could decay through quantum tunnelling, producing a bubble of altered physics expanding at light speed. Current Higgs boson measurements suggest that while such decay is improbable for timescales far exceeding the age of the universe, the possibility remains that reality itself could undergo a phase transition that abolishes matter and forces—with the bonus side effect of no more spam.

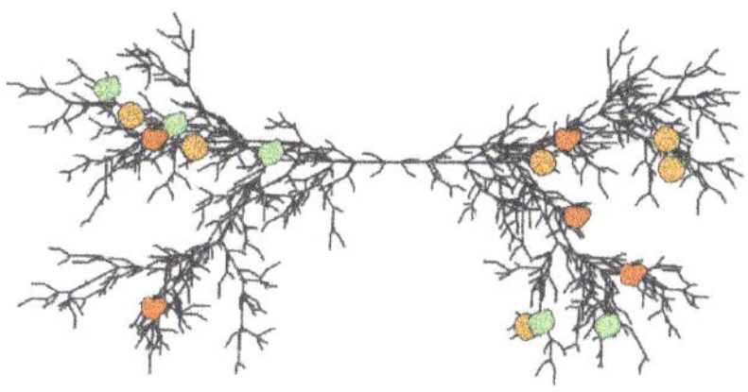

False Vacuum ◦ Higgs Field 246 GeV ◦ Mexican Hat Potential ◦ Top Quark Mass Sensitivity ◦ Metastability Edge ◦ Quantum Tunnelling ◦ Coleman-De Luccia Instanton ◦ Bubble Nucleation ◦ 10^{100} Year Lifetime ◦ Virtual Particle Corrections ◦ Renormalization Group Flow

“Darling, there’s a place for us,
Can we go, before I turn to dust?”
— Joanna Newsom, 2006

“כִּי הִנֵּה הַיּוֹם בָּא בֹּעֵר כַּתַּנּוּר וְהָיוּ כָל זֵדִים וְכָל עֹשֵׂה רִשְׁעָה קַשׁ וְלִהַט אֹתָם הַיּוֹם הַבָּא אָמַר ה צְבָאוֹת„
(“For, behold, the day cometh, that shall burn as an oven; and all the proud, yea, and all that do wickedly, shall be stubble: and the day that cometh shall burn them up, saith the Lord ...”)
— Malachi, c. 400 BCE

An Empty Threat

The concept of vacuum stability emerged from parallel developments in cosmology and particle physics. In 1980, Alan Guth proposed cosmic inflation to solve the horizon and flatness problems. His model required a scalar field temporarily trapped in a false vacuum state, driving exponential expansion before decaying to the true vacuum. This established that metastable vacuum states could have observable consequences.

Sidney Coleman and Frank De Luccia calculated the decay rate of false vacua in 1980, showing that quantum tunnelling creates expanding bubbles of true vacuum. Their formalism demonstrated that gravitational effects could either enhance or suppress transitions, depending on the energy difference between vacua. The calculation revealed that vacuum decay proceeds through nucleation of critical bubbles whose walls accelerate outward at the speed of light.

The discovery of the Higgs boson at CERN in 2012 promoted the idea of vacuum stability from theoretical speculation to measurable physics. The Higgs mass of 125 GeV, combined with precision measurements of the top quark mass at 173 GeV, placed the Standard Model near the boundary between stable and metastable regimes. Giuseppe Degrassi and collaborators showed in 2012 that these values imply the Higgs self-coupling likely becomes negative at energies around 10^{10} GeV, creating a deeper minimum in the potential.

These calculations depend critically on the running of coupling constants with energy scale. The renormalization group equations track how the Higgs quartic coupling λ evolves from low to high energies. At the measured Higgs and top masses, λ decreases with increasing energy and may cross zero. Beyond this critical point, the effective potential develops a new minimum at large field values where the universe would have different physical laws.

The 2012 discovery of the Higgs boson completed the Standard Model but raised an existential question: precision measurements placed our universe near the boundary between stable and metastable regimes. Our vacuum may be stable for now but not forever.

Classical physics defines vacuum as empty space—the absence of matter. This is not the case in quantum field theory. The vacuum is not emptiness but a specific configuration of fields filling all space. A field assigns a value to every point in space. In quantum field theory, every point contains a quantum state for the electromagnetic field, the Higgs field, quark fields, and others. The vacuum is the configuration where these fields minimise the total energy density.

A quantum field extends throughout all space and determines the probabilities of measurement outcomes. The electromagnetic field carries light waves and radio waves. When this field vibrates in a particular pattern (mode), we observe it as a photon. Similarly, the electron field's vibrations manifest as electrons. Each elementary particle type—quarks, leptons, bosons—corresponds to its own field. These fields exist everywhere, even in 'empty' space. What we call particles are localised excitations, like waves on an ocean that

 https://doi.org/10.11647/OBP.0526.16

pervades the universe. The vacuum is the state where all these fields vibrate with their lowest possible energy.

This lowest energy state is far from inert. The Higgs field has a nonzero value throughout space, approximately 246 GeV. This value gives mass to fundamental particles (Higgs, 1964) through their couplings to the field. Without it, electrons would be massless, atoms could not form, and matter would not exist.

Whether this vacuum state is permanent depends on the field potential—a mathematical term describing the energy associated with different field values. Just as a marble rolls to the bottom of a bowl, fields evolve toward configurations that minimise their potential energy. The shape of this potential determines whether our vacuum is truly stable or merely appears so.

Potentials can have multiple minima. A local minimum is a dip surrounded by higher terrain—stable against small disturbances but not the lowest point. A ball in a shallow depression on a hillside stays put even with a deeper valley elsewhere. A false vacuum is a field configuration at a local minimum when a deeper minimum exists. Climbing out requires energy, so the field appears stable despite not occupying the true ground state.

The Higgs field is a scalar field—it has spin 0 and a single degree of freedom at each point in space, unlike vector fields (spin 1) or tensor fields (spin 2).

Spin is an intrinsic quantum property, analogous to but distinct from classical rotation. A spin-0 particle (scalar) has no preferred direction—it looks identical from every angle, like a sphere. A spin-1/2 particle (fermion) requires two full rotations to return to its original state, a distinctly quantum behaviour with no classical analogue—electrons, quarks, and all matter particles have spin 1/2. A spin-1 particle (vector) has a direction, like an arrow pointing in space. The photon, with spin 1, must have its electric and magnetic fields oriented perpendicular to its motion. A spin-2 particle (tensor) has even more complex directional properties—the graviton (hypothetical particle mediating gravity), if it exists, would have spin 2. The spin determines how particles behave under rotations and what kinds of fields they can create. Scalar fields such as the Higgs are the simplest: just a number at each point in space, no direction.

The field's potential at tree level (before quantum corrections) takes the shape:

$$V(\phi) = \lambda(\phi^2 - v^2)^2.$$

This creates a 'Mexican hat' shape: high at the centre, dropping to a circular valley at radius $v \approx 246$ GeV. The field settles in this valley, breaking electroweak symmetry—the W and Z (Weinberg, 1967) bosons become distinguishable from photons by acquiring mass.

From this minimum, all particle masses follow. The W and Z bosons acquire masses proportional to v. Quarks and leptons gain mass through Yukawa couplings to the Higgs. The location in the valley sets the mass spectrum of the Standard Model.

The tree-level potential is an approximation. Virtual particles—quantum fluctuations that briefly borrow energy from the vacuum—modify the effective potential. These corrections depend on energy scale: at higher energies, different virtual processes dominate.

Virtual particles contribute through quantum loops. A virtual top quark can appear from the vacuum, interact with the Higgs field, then disappear. Though fleeting, these processes change the effective potential. Heavy particles such as the top quark contribute most strongly because their coupling to the Higgs is proportional to their mass. The top quark's virtual loops pull the Higgs potential downward at large field values, while the Higgs self-interactions and gauge boson loops push it upward. The competition between these effects determines vacuum stability.

The renormalization group(Wilson, 1971)—a mathematical tool that tracks how physical parameters change at different energy scales—shows how couplings evolve with energy scale μ. The Higgs self-coupling $\lambda(\mu)$ and top Yukawa coupling $y_t(\mu)$ satisfy coupled differential equations. The large top quark mass means y_t is close to 1 (the Yukawa coupling $y_t = \sqrt{2}m_t/v \approx 0.99$), and its contribution drives λ downward as energy increases.

If $\lambda(\mu)$ becomes negative at high scales, the potential bends downward for large field values. A second minimum forms far from the electroweak scale. This new minimum can be deeper than the original, making our vacuum metastable rather than stable.

The Higgs boson mass, measured at 125.25 ± 0.17 GeV, and the top quark mass at 172.9 ± 1.5 GeV, determine the boundary between stability and metastability. These values place the Standard Model near the critical line.

A lower top mass of approximately 2 GeV would strongly favour absolute stability; about a 2 GeV higher top mass would strongly favour metastability with a shorter (yet still astronomically long) lifetime.

This sensitivity transforms vacuum stability from philosophical speculation to experimental physics. Precision measurements of the top mass will determine whether our vacuum is stable or metastable.

The discovery of the Higgs boson at CERN in 2012 added a measurable aspect to this abstract question. Combined with precision measurements of the top quark mass, physicists could finally calculate whether our universe sits in a truly stable vacuum or a metastable one. The result was unsettling: we appear to live on the edge. If our vacuum is indeed metastable, the primary concern becomes quantum tunnelling—the mechanism by which the field could spontaneously transition to a lower minimum despite the energy barrier.

Classical physics forbids transitions between separated minima—the field cannot climb over the barrier. Quantum mechanics allows tunnelling through barriers. In field theory, this occurs via instantons: field configurations (Coleman, 1977) that interpolate between vacua in imaginary time, where time becomes a spatial dimension in the calculation.

The process nucleates a bubble of true vacuum within the false vacuum sea. The probability depends on the Euclidean action S_E (the action calculated in imaginary time) of the optimal tunnelling path:

$$\Gamma/V \sim A \exp(-S_E/\hbar),$$

where A is a dimensional prefactor containing field fluctuation modes. For small energy differences between vacua, S_E becomes large, exponentially suppressing the decay rate.

Once a critical bubble forms, energy differences drive its expansion. The true vacuum has lower energy density, creating pressure that accelerates the bubble wall outward. The wall approaches the speed of light, converting false vacuum to true vacuum.

The bubble wall itself is a domain wall—a boundary layer where the field smoothly transitions between the two vacuum values. Its thickness is set by the inverse mass scale of the field, usually microscopic. The energy density in the wall is large, concentrated in this thin shell. As the bubble expands, this energy gets diluted over larger surface area, but the total energy grows as the bubble engulfs more false vacuum volume. The wall accelerates outward under constant pressure, asymptotically approaching the speed of light.

Inside the bubble, physics changes. The Higgs field takes its new value, altering all particle masses and couplings. Electrons might become too heavy to orbit nuclei, or too light to be localised. The balance enabling atomic structure disappears. Chemistry and biology cease to exist through redefinition of all physical laws.

Matter encountering the advancing wall undergoes complete transformation. Particles defined by their interactions with the old vacuum value cannot exist in the new vacuum. The process is not gradual—as the wall passes, particle masses and interaction strengths change discontinuously. Protons might become unstable, quarks might not confine (bind together to form protons and neutrons), electromagnetic and weak forces might merge or separate differently. No information about the previous state survives because the encoding mechanism no longer exists.

For Standard Model–like parameters near current measurements, estimates give $S_E/\hbar \sim 400$–500, implying an astronomically long lifetime vastly exceeding the age of the universe.

High-energy processes cannot trigger decay. Cosmic rays reach 10^{11} GeV, far above the 10^{10} GeV scale where the Higgs self-coupling runs negative, yet have bombarded Earth for billions of years without incident. The LHC's 1.4×10^4 GeV collisions are negligible by comparison. Vacuum decay requires coherent field excitations over macroscopic regions, not pointlike particle collisions—a single high-energy impact excites fields only locally, insufficient to nucleate the critical bubble geometry needed for tunnelling.

A nuclear war destroys cities but leaves physics intact. Vacuum decay replaces physics itself. The universe continues, but under different rules that may not permit matter, let alone life.

The Unbearable Lightness of Being

Of all existential threats—asteroids, pandemics, nuclear war—vacuum decay offers the ultimate consolation: we'll never know it happened. No final moments of terror, no last goodbyes, no time for regret. The bubble wall travels at light speed, so the universe's rewriting arrives simultaneously with news of its approach.

[Musical Theme: Upbeat, vaudeville-style cabaret]

Verse 1

(Rubato)
When you attend a funeral,
It is sad to think that sooner or
Later those you love will do the same for you.
And you may have thought it tragic,
Not to mention other adjec-
Tives, to think of all the weeping they will do.
But don't you worry.
No more ashes, no more sackcloth.
And an armband made of black cloth
Will some day never more adorn a sleeve.
For if the bomb that drops on you
Gets your friends and neighbors too,
There'll be nobody left behind to grieve.

Chorus

And we will all go together when we go.
What a comforting fact that is to know.
Universal bereavement,
An inspiring achievement,
Yes, we all will go together when we go.

Verse 2

We will all go together when we go.
All suffuse with an incandescent glow.
No one will have the endurance
To collect on his insurance,
Lloyd's of London will be loaded when they go.

Verse 3

Oh we will all fry together when we fry.
We'll be french fried potatoes by and by.
There will be no more misery
When the world is our rotisserie,
Yes, we will all fry together when we fry.

Bridge

Down by the old maelstrom,
There'll be a storm before the calm.

Verse 4

And we will all bake together when we bake.
There'll be nobody present at the wake.
With complete participation
In that grand incineration,
Nearly three billion hunks of well-done steak.

Oh we will all char together when we char.
And let there be no moaning of the bar.
Just sing out a *Te Deum*
When you see that I.C.B.M.,
And the party will be 'come as you are.'

Oh we will all burn together when we burn.
There'll be no need to stand and wait your turn.
When it's time for the fallout
And Saint Peter calls us all out,
We'll just drop our agendas and adjourn.

Bridge 2

You will all go directly to your respective Valhallas.
Go directly, do not pass Go, do not collect two hundred dolla's.

Final Chorus

And we will all go together when we go.
Ev'ry Hottentot and ev'ry Eskimo.
When the air becomes uranious,
And we will all go simultaneous.
Yes we all will go together
When we all go together,
Yes we all will go together when we go.

— *Tom Lehrer*

(from tomlehrersongs.com, public domain since 2022)

False Vacuum Decay: Mathematical Formulation

Higgs Potential and Vacuum Stability

The Higgs potential in the Standard Model takes the form

$$V(\phi) = \mu^2\phi^2 + \lambda\phi^4,$$

where ϕ is the Higgs field, $\mu^2 < 0$ for spontaneous symmetry breaking, and $\lambda > 0$ for stability. The vacuum expectation value is $\langle\phi\rangle = v = \sqrt{-\mu^2/\lambda} \approx 246$ GeV. Up to an additive constant, this is the Mexican-hat form $\lambda(\phi^2 - v^2)^2$ with $\mu^2 = -2\lambda v^2$.

However, renormalization group running modifies the effective potential at high energies. The quartic coupling evolves as

$$(16\pi^2)\,\beta_\lambda = 12\lambda^2 + (12y_t^2 - 9g^2 - 3g'^2)\lambda \\ - 12y_t^4 + \frac{9}{8}g^4 + \frac{3}{4}g^2g'^2 + \frac{3}{8}g'^4,$$

where y_t is the top quark Yukawa coupling, g and g' are the $SU(2)_L$ and $U(1)_Y$ gauge couplings, and Q is the energy scale. For Higgs mass $m_H \approx 125$ GeV and top mass $m_t \approx 173$ GeV, λ runs negative at scales $Q \sim 10^{10}$-10^{11} GeV, creating a second minimum at large field values.

Coleman-De Luccia Instanton

Vacuum decay proceeds via bubble nucleation described by the Euclidean action

$$S_E = \int d^4x \left[\frac{1}{2}(\partial_\mu\phi)^2 + V(\phi)\right].$$

The critical bubble solution has $O(4)$ symmetry in Euclidean space, satisfying

$$\frac{d^2\phi}{d\rho^2} + \frac{3}{\rho}\frac{d\phi}{d\rho} = \frac{dV}{d\phi},$$

where $\rho = \sqrt{x_1^2 + x_2^2 + x_3^2 + x_4^2}$ is the four-dimensional radius.

The nucleation rate per unit volume is

$$\Gamma = Ae^{-S_E/\hbar},$$

where A is a prefactor and S_E is the Euclidean action of the bounce solution. For the Standard Model, current estimates give $S_E/\hbar \sim 400 - 500$, making spontaneous decay negligible over cosmic timescales.

High-Energy Triggers

Local few-particle collisions (in colliders or from ultra-high-energy cosmic rays) are not expected to nucleate the required $O(4)$-symmetric critical bubble. Observed cosmic rays reach energies up to $\sim 3 \times 10^{20}$ eV without any indication of catalysed vacuum decay, consistent with the nonperturbative, extended nature of the tunnelling process.

Bubble Dynamics

Once nucleated, the bubble wall accelerates due to the pressure difference between vacua. The wall Lorentz factor obeys

$$\gamma^2 = \frac{1}{1 - v^2},$$

where v is the wall velocity. In the thin-wall limit, the pressure difference ϵ across the wall drives v rapidly toward the speed of light ($v \to 1$) as the bubble expands; the detailed dynamics depend on the surface tension σ, the energy difference ϵ, and the background spacetime.

Renormalization Group Uncertainty

The stability analysis depends critically on precise measurements of Standard Model parameters. The most sensitive quantities are:

$$m_t = 173.1 \pm 0.9 \text{ GeV}$$
$$m_H = 125.25 \pm 0.17 \text{ GeV}$$
$$\alpha_s(M_Z) = 0.1179 \pm 0.0010.$$

An increase of order $\sim$1 GeV in the top mass would shift the stability assessment toward instability, while a $\sim$3 GeV increase in the Higgs mass would favour absolute stability.

References:

Coleman and De Luccia, *Phys. Rev.* D **21**, 3305 (1980).

Degrassi et al., *JHEP* **08**, 098 (2012).

The Busy Beaver That Ate the TREE

Top (Valid TREE(3) Sequence Prefix): An initial segment of a TREE(3)-avoiding sequence. Each rooted tree is added in such a way that no earlier tree is homeomorphically embeddable into a later one. The sequence starts with simple trees and becomes rapidly more complex. Exponentiating a googolplex by itself a googolplex times is comically dwarfed by TREE(3). **Bottom (Largest Number Competition):** Using 20 cm of ink, write the largest number possible. Early entries include exponential towers and Knuth up-arrow notation. But one leverages the TREE function hierarchy in TREE(TREE(3))—a number vastly larger than previous contenders. The final banner is a different type of entry, more reminiscent of Rayo's number.

The Busy Beaver That Ate the TREE

Imagine you are given a pen with enough ink to write 20 centimetres and you are to write the biggest number you can think of. You can start by a tower of exponents $10^{10^{\cdots^{10}}}$, but it is not that big. In this chapter we explore the hierarchy from computable functions to TREE(3)—a number so immense that even if you built a tower of exponentials starting with a trillion raised to the power of a trillion, and then repeated that construction every attosecond for a trillion years, the result would still be vanishingly small in comparison. Yet even TREE(3) sits as close to infinity as the number eight, placing us in an infinity zoo where sizes exceed the categories brains evolved to handle.

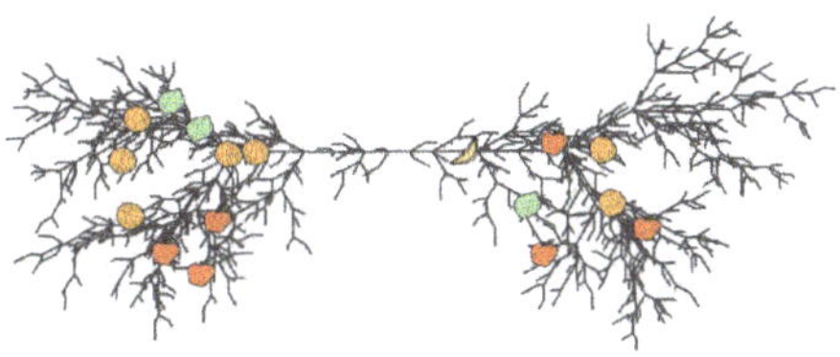

Sand Reckoner ∘ Tower Notation ∘ Knuth Arrows ∘ 20cm Ink Competition ∘ Ackermann Function ∘ TREE(3) Combinatorics ∘ Tree Embedding Game ∘ Busy Beaver ∘ Halting Undecidability ∘ Rayo's Function ∘ Large Cardinals

„Das Unendliche hat wie keine andere Frage
von jeher so tief das Gemüt der Menschen bewegt...
Aus dem Paradies, das Cantor uns geschaffen,
soll uns niemand vertreiben können."

("The infinite! No other question has ever moved so profoundly the spirit of man... Cantor's paradise, from which no one will expel us.")

— David Hilbert, 1926

The Busy Beaver That Ate the TREE

> The challenge of *expressing vast quantities* appears across ancient texts. The Hebrew Bible uses 'רִבֵּי רְבָבוֹת' (*ribei revavot*)—myriads of myriads—to denote numbers beyond ordinary counting, as in Daniel 7:10 describing the heavenly host: 'thousand thousands ministered unto him, and ten thousand times ten thousand stood before him.' This poetic multiplication hinted at systematic ways to build larger numbers.
>
> Archimedes formalised this intuition in *The Sand Reckoner* (Ψαμμίτης, Psammites), written circa the early third century BCE as a letter to Gelon, son of King Hiero II of Syracuse. The work addressed Aristarchus's heliocentric model, which implied a universe vastly larger than previously imagined. Archimedes asked: could one count the grains of sand needed to fill such a cosmos?
>
> Greek numerals stopped at a myriad (10,000). Archimedes extended them by defining 'orders'—the first order contained numbers up to a myriad myriads (10^8), the second order began at 10^8 and continued to $(10^8)^2$, and so forth. Using this system, he estimated the universe could hold at most 10^{63} sand grains. The calculation was secondary to the method: *The Showing* that any finite quantity, however vast, could be expressed and manipulated. This was a milestone in scientific notation and the separation of numbers from physical counting.
>
> Edward Kasner popularised the terms after asking his nine-year-old nephew, Milton Sirotta, to name 10^{100}; Milton proposed 'googol,' and they defined 'googolplex' as 10^{googol}. The coinage predates the book, but the terms were widely disseminated in Kasner and Newman's *Mathematics and the Imagination* (1940), illustrating how notation can grow rapidly.
>
> Modern developments began with Wilhelm Ackermann's 1928 function that grows faster than any primitive recursive function. This gave rise to the computational growth rates forming a hierarchy—some functions outpace others so dramatically that conventional notation fails.
>
> Harvey Friedman migrated large numbers from recreational mathematics into serious research in the 1990s. His TREE sequence, derived from Kruskal's tree theorem, produced numbers dwarfing all previous constructions. TREE(3) is finite but so large it cannot be expressed using conventional operations, iterated any reasonable number of times. The proof requires axioms beyond Peano arithmetic, connecting large numbers to results in proof theory and the limits of formal systems.

Archimedes wrote the *Sand Reckoner* to count sand grains in the cosmos. His real purpose was notation—showing how large numbers could be handled by grouping units into 'orders' and assigning names to powers of powers. This marked one of the first recorded attempts to handle orders of magnitude through notation.

Children discover this principle through play. Counting on fingers reaches ten. Tally marks extend to dozens. Roman numerals handle thousands awkwardly. Arabic numerals with positional notation reach millions effortlessly.

The number 10^{10}—ten billion—sits at the edge of intuitive grasp. Demographers estimate that about 10^{11} humans have ever lived. The observable universe contains approximately 10^{80} atoms. Scientific notation makes this tractable: '1 followed by 80 zeros.' But even this notation meets its limits.

Consider $10^{10^{10}}$. This number has ten billion digits. If each digit were an atom, we would need 10^{10} atoms to write it down—an incomprehensibly tiny fraction of the universe's 10^{80} atoms (10^{-70} of them). Yet the notation remains compact—just a few symbols capture a magnitude that dwarfs physical representation.

Now build a tower of many years: $10^{10^{10^{\cdot^{\cdot^{\cdot}}}}}$ with ten tens, call it T_{years}. To grasp this scale, imagine beings who create universes by setting random initial conditions, attempting to arrange cosmic evolution so that 13.8 billion years later, if a civilisation emerges on the resulting Earth, they launch exactly 100 trillion green peas toward the Moon, and all land in a pre-specified bucket at exactly 8:00:00.00000 PM on Friday, 4 July. The initial conditions must account for every quantum fluctuation, every gravitational interaction, the formation of galaxies, stars, planets, the evolution of life, agriculture, spaceflight technology, and the precise timing of launch.

Most attempts fail—Earth never forms, or forms without life, or life evolves differently, or the civilisation launches peas a second too early. When an attempt fails, they wait 10^{100} years until that universe reaches heat death, then start fresh with new initial conditions. To achieve a billion consecutive successes will require much less time than T_{years}.

Physical metaphors lose meaning at these scales. No arrangement of atoms, no duration of time, no cosmic process captures numbers this large. We can develop formal notation that builds recursively, where each operation multiplies growth rates.

Knuth's arrow notation compresses this tower-building (Knuth, 1974). One arrow denotes exponentiation, $10 \uparrow 10 = 10^{10}$. Two arrows build a tower, $10 \uparrow\uparrow 10 = 10^{10^{10^{\cdot^{\cdot^{\cdot}}}}}$ with ten 10s. Three arrows iterate the two-arrow operation, $10 \uparrow\uparrow\uparrow 10 = 10 \uparrow\uparrow (10 \uparrow\uparrow\uparrow 9)$, building towers of towers recursively. Each arrow multiplies the growth rate beyond comprehension.

Frame this as a competition. You have ink for 20 centimetres of writing. Produce the largest number possible. Every symbol must be precisely defined. Writing '10^{10}' beats '10000000000'—notation outpaces digits. With Knuth's arrows, $3 \uparrow 3 = 3^3 = 27$, but $3 \uparrow\uparrow 3 = 3^{3^3} = 3^{27}$, already over 7 trillion. Now, instead of manually building recursive stacks, we can define functions that generate them.

The Ackermann function grows faster than any primitive recursive function:

$$\begin{aligned} A(0, n) &= n + 1 \\ A(m + 1, 0) &= A(m, 1) \\ A(m + 1, n + 1) &= A(m, A(m + 1, n)) \end{aligned} .$$

Starting modestly—$A(1, n) = n+2$, $A(2, n) = 2n+3$, $A(3, n) = 2^{n+3} - 3$—by $A(4, 2)$ we reach $2^{65536} - 3$, computed from a power tower five levels high, $2^{2^{2^{2^2}}} - 3$. A well-chosen function reference like '$A(A(10, 10), A(10, 10))$' beats unfathomable explicit digits. The best use of ink is no longer to write numbers, but to specify methods of generation.

Beyond recursive towers lies combinatorial explosion. While Ackermann and arrows build through iteration, TREE(3) emerges from a simple game with trees that generates growth surpassing any tower of exponentials. The leap from arithmetic to combinatorics produces numbers that dwarf all previous constructions.

Take a deep breath and play a game called TREE(n). You draw a sequence (Kruskal, 1960) of trees—not botanical trees, but branching diagrams with a single root at top, branches splitting downward, with vertex colours from an n-colour set. You have n colours available (say, red, blue, and green for TREE(3)). The rules: the i-th tree in your sequence can have at most i vertices; no earlier tree may embed in any later tree.

An earlier tree embeds in a later tree if there is an injective, colour-preserving map of vertices that preserves lowest-common-ancestor relations. In simpler terms, tree A embeds in tree B if you can find a subset of B's vertices that matches A's structure and colours exactly.

TREE(1) = 1. With one colour (say, only red), you draw a single red vertex. The second tree needs two vertices, so it must have two red vertices—but any configuration of two red vertices embeds the single red vertex. Game over.

TREE(2) = 3. With two colours (red and blue), the longest sequence is (1) single red vertex, (2) blue root with blue child below, (3) single blue vertex. Try adding a fourth tree with at most 4 vertices using red and blue—any configuration will embed one of these three.

TREE(3) is where the magnitude explodes. This number dwarfs any fixed-height tower of exponentials and values produced by many natural hierarchies at modest inputs. However, the Ackermann function eventually exceeds any fixed bound for sufficiently large inputs. If every atom in the observable universe became a tower of googolplexes, and these googolplex-atoms multiplied together every nanosecond throughout cosmic history, the result wouldn't approach one trillionth of TREE(3).

TREE(3) is a specific, well-defined number. There exists a definite answer to "what is the 97th digit of TREE(3)?" We simply cannot compute it. Enter functions that grow even faster through different mechanisms. The next numbers are more dependent on the underlying axioms and the language used to define them.

TREE arises from combinatorial constraints—avoiding embeddings in sequences. The busy beaver function BB(n) shifts from combinatorial to computational (Radó, 1962) limits. Among all n-state Turing machines (theoretical computing devices) that eventually halt, BB(n) equals the maximum number of steps any such machine takes before stopping. The known values are BB(1)=1, BB(2)=6, BB(3)=21, BB(4)=107. For BB(5), the value is known to be 47,176,870. BB(6) exceeds $10 \uparrow\uparrow 15$.

Unlike TREE(3), BB(n) derives its magnitude from undecidability. No algorithm can compute BB(n) for arbitrary n, as this would solve the halting problem (Turing, 1936)—proven impossible by Turing. The function eventually surpasses TREE(n) because it encompasses all computational processes, including those calculating TREE values. Recent analysis suggests BB(2645) likely exceeds TREE(3), and from then on, it grows explosively faster.

The ultimate strategy abandons specific constructions for meta-mathematical limits. Rather than defining a particular fast-growing function, we can ask: what is the largest number definable within the language itself?

Rayo's function does that while venturing into linguistic and logic territory. At MIT's 2007 "Big Number Duel," philosopher Agustín Rayo proposed the ultimate strategy—define Rayo(n) as the largest natural number expressible in first-order set theory using at most n symbols. This approaches the theoretical limit of our 20 cm game—essentially encoding "the largest number definable with this much notation" within formal logic.

Rayo(n) outgrows any function definable in its own language through diagonalization—exceeding every possible definition. It dwarfs both TREE(n) and BB(n).

All these finite numbers—from towers of exponentials to TREE(3) to Rayo's function—remain infinitely far from infinity. They demonstrate ingenuity in naming ever-larger quantities, yet each sits at the same infinite distance from the first infinity.

And so, we now turn to infinity. When we cross to infinity, the rules of growth change. Infinity comes in two flavours—cardinals (sizes of sets) and ordinals (positions in well-ordered sequences). Think of 'three' (counting objects) versus 'third' (position in line). The smallest infinite cardinal is $\aleph_0$ (aleph-null), the cardinality of natural numbers. The smallest infinite ordinal is ω, their order type.

Cardinal arithmetic is different from finite arithmetic: $\aleph_0 + 1 = \aleph_0$—Hilbert's Hotel has infinitely many rooms (Hilbert, 1926), all full, yet can accommodate one more guest; $\aleph_0 + \aleph_0 = \aleph_0$—interleave odds and evens; $\aleph_0 \times \aleph_0 = \aleph_0$—arrange rationals in a grid.

But exponentiation breaks this pattern. ${\aleph_0}^{\aleph_0} > \aleph_0$. Exponentiation represents functions between sets. In finite cases, $5^5 = 3125$ is exactly the number of possible functions from $\{1, 2, 3, 4, 5\}$ to itself. Similarly, ${\aleph_0}^{\aleph_0}$ represents all functions from naturals to naturals, yielding the continuum's cardinality which is larger than $\aleph_0$.

With ordinals, exponentiation truly explodes. Form ω^ω—omega to the omega power. Then ω^{ω^ω}—a tower of omegas. But why stop at finite towers? We're already working with infinity! Build an infinite tower: $\omega^{\omega^{\omega^{\cdot^{\cdot^{\cdot}}}}}$ with ω many ω's. This is the limit of finite towers, well-defined in ordinal arithmetic. This mind-bending construction yields ε_0, the first epsilon number, satisfying $\omega^{\varepsilon_0} = \varepsilon_0$.

This unimaginably large ordinal, built from an infinite tower of infinities, is tiny in the hierarchy of infinities. It's merely the first in a new regime: ε_1 is the next fixed point after ε_0; ε_ω is the ω-th fixed point; $\varepsilon_{\varepsilon_0}$ uses our 'massive' infinity as a mere index.

How do we organise these ever-larger infinities? In 1908, mathematician Oswald Veblen developed a hierarchy. Start with the function $\varphi_0(\alpha) = \omega^\alpha$—this generates our familiar exponential towers. The function φ_1 then enumerates all the epsilon numbers (those fixed points where $\omega^x = x$). The function φ_2 finds all the fixed points of φ_1—ordinals so large that even the epsilon-generating function cannot reach them. Each level finds what the previous level missed, climbing an infinite ladder where each rung reveals new unreachable ordinals above.

This process continues through all finite indices: φ_3, φ_4, and onward. But now we can define φ_ω, then $\varphi_{\omega+1}$, even $\varphi_{\varphi_0(0)}$. The indices themselves become ordinals! Eventually, we reach an ordinal so large it equals its own index in the Veblen hierarchy, $\Gamma_0 = \varphi_{\Gamma_0}(0)$. This is the Feferman-Schütte ordinal, discovered independently by Solomon Feferman and Kurt Schütte in the 1960s.

Γ_0 marks more than just another large ordinal. Below Γ_0, we can build ordinals step by step using explicit rules. Beyond it, we need new principles. In technical terms, Γ_0 is the proof-theoretic ordinal of predicative mathematics—it measures exactly how far we can count using only definitions that refer to previously constructed objects. To go further requires impredicative methods: definitions that refer to totalities containing the very object being defined. It's like trying to lift yourself by your own bootstraps—impossible in physics, but sometimes necessary in mathematics. Beyond Γ_0 lie ordinals and cardinals requiring ever-stronger principles.

A word of caution—beyond this point (and a little bit before this point to be honest), we enter territory inhabited almost exclusively by logicians and set theorists. These larger ordinals and cardinals, while mathematically precise, have little connection to anything outside specialised logical discussions. They represent abstract possibilities rather than quantities that arise naturally in mathematics. Yet surprises occur—just as TREE(3) emerged from combinatorics to dwarf all previous numbers, these abstract ordinals occasionally appear in analysis. Still, for most purposes, this glimpse into the hierarchy suffices.

ω_1^{CK} (Church-Kleene ω_1)—the first ordinal with no computable description. Every ordinal before this can be described by some computer program, even if that program would run forever. But ω_1^{CK} transcends computation itself. No algorithm, no matter how clever, can specify this ordinal.

ω_1—the first uncountable ordinal. All ordinals before this can be put in one-to-one correspondence with natural numbers. But ω_1 is the first ordinal too large for any such pairing. If you tried to list all smaller ordinals as first, second, third..., you would run out of natural numbers before reaching ω_1. It's a bigger kind of infinity.

Inaccessible cardinals—infinite numbers unreachable by standard set operations. You cannot reach an inaccessible cardinal by taking powers (such as $2^{\aleph_0}$), unions, or any combination of usual set-theoretic operations starting from smaller cardinals.

Measurable cardinals—infinite numbers large enough to support probability measures. On finite sets, we can assign probabilities—half the integers from 1 to 10 are odd. But on infinite sets, this usually fails. Measurable cardinals are so large that probability makes sense again. Beyond these lie supercompact cardinals, where the entire universe of sets at any level looks like a small-scale model of the universe at the supercompact level.

Hierarchy of Growth Rates

Level 1: Elementary Functions

$$f(n) = n + c \text{ (linear)}$$
$$f(n) = n^k \text{ (polynomial)}$$
$$f(n) = k^n \text{ (exponential)}$$
$$f(n) = n! \approx \sqrt{2\pi n}\left(\frac{n}{e}\right)^n$$

Level 2: Iterated Exponentials

Tetration: ${}^k a = \underbrace{a^{a^{\cdot^{\cdot^{a}}}}}_{k \text{ times}}$

Digits of ${}^k 2$ are about $(\log_{10} 2) \cdot {}^{k-1}2$.

Level 3: Ackermann Function

$$A(0, n) = n + 1$$
$$A(m + 1, 0) = A(m, 1)$$
$$A(m + 1, n + 1) = A(m, A(m + 1, n)).$$

Growth: $A(1, n) = n + 2$, $A(2, n) = 2n + 3$, $A(3, n) = 2^{n+3} - 3$, $A(4, n) = \underbrace{2^{2^{\cdot^{\cdot^{2}}}}}_{n+3} - 3$.

Level 4: Knuth Arrows

$$a \uparrow^1 b = a^b$$
$$a \uparrow^n 1 = a \text{ for } n \geq 1$$
$$a \uparrow^n 0 = 1 \text{ for } n \geq 1$$
$$a \uparrow^n b = a \uparrow^{n-1} (a \uparrow^n (b - 1))$$
$$\text{for } n \geq 1, b > 1.$$

Extension: $a \uparrow^0 b := ab$.
$3 \uparrow 3 = 27$, $3 \uparrow\uparrow 3 = 7{,}625{,}597{,}484{,}987$
$3 \uparrow\uparrow\uparrow 3 = 3 \uparrow\uparrow 7{,}625{,}597{,}484{,}987$

Level 5: Fast-Growing Hierarchy

Indexed by ordinals:

$$f_0(n) = n + 1$$
$$f_{\alpha+1}(n) = f_\alpha^n(n)$$
$$f_\lambda(n) = f_{\lambda[n]}(n) \text{ for limit } \lambda$$
$$f_\omega(n) = f_n(n)$$
$$f_{\omega^2}(n) = f_{\omega \cdot n}(n)$$
$$f_{\varepsilon_0}(n) \text{ dominates finite } \omega \text{ towers}$$

Level 6: TREE Function

TREE(n) = max sequence of rooted finite trees with vertices coloured from an n-element set; on step i the tree has at most i vertices; forbid homeomorphic embedding from any earlier tree into any later.
TREE(1) = 1, TREE(2) = 3
Via Kruskal's theorem, associated length functions dominate f_α for all $\alpha < \theta(\Omega^\omega)$; TREE(3) is far beyond f_{ε_0}-scale growth.

Level 7: Busy Beaver

BB(n) = max steps of any halting n-state, 2-symbol TM.
BB(4) = 107, BB(5) = 47,176,870 (proven 2024)
BB(6) $> 10 \uparrow\uparrow 15$ (lower bound)
Eventually dominates all computable functions.

Level 8: Rayo Function

Rayo(n) = the least natural number greater than every number definable in first-order set theory by a formula of length $\leq n$ (with fixed encoding).
Dominates any n-symbol definable function by diagonalization.

Growth Comparison

For large n: polynomial $\ll$ exponential $\ll$ Ackermann $\ll$ arrows $\ll f_{\varepsilon_0} \ll$ TREE $\ll$ BB $\ll$ Rayo

Each level uses fundamentally stronger recursion principles. Comparison depends on proof-theoretic bounds (FGH, Kruskal), computability (BB), and definability (Rayo).

References:

M. H. Löb and S. S. Wainer, 'Hierarchies of number-theoretic functions I/II,' Archiv für mathematische Logik und Grundlagenforschung (1970–72).
H. Friedman, 'Finite forms of Kruskal's theorem,' Journal of Combinatorial Theory, Series A **95**, 102–144 (2001).

A Leaky
Crystal Ball

Top (Celebrity in Disguise): A secret value is hidden in either Room A or Room B, represented by the celebrity and their doppelgänger. The attacker does not know which room contains the real secret.
Middle (Reporter's Probe): The reporter (attacker) sends a crafted request: ordering crème brûlée to Room A. This dish is special because only the celebrity would ever receive it. The CPU (chef) will prepare it only if the celebrity is in Room A.
Bottom (Timing Side Channel): The reporter then makes a normal order, sweet potato pie. If the torch is still warm, the pie finishes faster—revealing that the crème brûlée preparation occurred, and thus that the celebrity is in Room A. If no torch warmth remains, the celebrity must be in Room B. This mirrors speculative execution: incorrect paths are rolled back architecturally, but their side effects persist in microarchitectural state, leaking secrets through timing.

A Leaky Crystal Ball

Speculative execution optimises performance by executing instructions before knowing if they're needed, leaving microarchitectural traces in cache memory even when results are discarded. Attacks such as Meltdown and Spectre exploit this, by constructing code sequences where a secret value determines which memory addresses are accessed during speculation. By measuring which addresses load quickly afterward (indicating they were cached), attackers can determine if specific bits were 0 or 1, allowing secrets to be extracted across privilege boundaries.

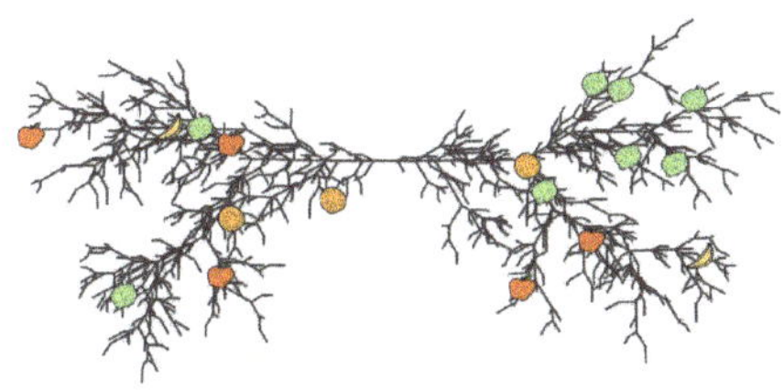

CPU Pipeline Architecture ◦ Speculative Execution ◦ Meltdown & Spectre ◦ Cache Timing Channels ◦ Branch Predictor Training ◦ Microarchitectural State ◦ Kernel Memory Extraction ◦ Mitigations & Performance ◦ KPTI & Retpolines ◦ Hardware Vulnerabilities ◦ 30% Performance Cost

"Be wary of anyone who claims to be able to see the future."

— Wit to Shallan, Rosharan year 1174

A Leaky Crystal Ball

> The core techniques behind speculative execution—pipelining, branch prediction, and out-of-order execution—originated as performance optimisations in the 1960s and matured across RISC and superscalar designs in the 1980s–1990s. The IBM System/360 Model 91 (1966) introduced dynamic scheduling. Tomasulo's algorithm and register renaming were foundational (Tomasulo, 1967). By the 2000s, speculative execution had become ubiquitous in high-performance processors.
>
> Meanwhile, side-channel attacks emerged independently in cryptography. Kocher (1996) demonstrated timing attacks on modular exponentiation. Cache timing attacks followed, with Bernstein (2005) showing cache-based AES key recovery. Rowhammer (2014) revealed that hardware itself could be attacked—rapidly accessing memory rows caused electrical interference that flipped bits in adjacent rows, allowing privilege escalation. These hardware vulnerabilities showed that physical properties of components could be weaponised. Yet usually such attacks were seen as requiring special conditions, deliberate software flaws, or physical proximity.
>
> Spectre and Meltdown, disclosed in 2018, showed that speculative execution—once considered internal and safe—could be manipulated into violating memory isolation. The result was a universal class of vulnerabilities.

Every program—from your web browser to your text editor—is a sequence of simple instructions that tell the processor what to do: load data from memory, store results back, add numbers, compare values. Think of it like a recipe where each step must be precise: 'take flour from cabinet,' 'add to bowl,' 'mix ingredients.' These instructions manipulate data stored in memory, the computer's workspace.

Memory is like a vast warehouse with different storage areas. Closest to the processor are registers—think of them as the processor's hands, holding just a few items it's actively working with. Next come caches, like workbenches near the assembly line, storing frequently-used data. Main memory is the large warehouse floor, holding gigabytes of data but taking longer to access. When you open a document, it moves from your hard drive into main memory, then pieces flow through caches to registers as the processor works on them.

Processors execute instructions through pipelines—assembly lines where different stages happen simultaneously. While one instruction is being decoded (figuring out what 'add' means), another is being executed (actually adding numbers), and a third is being fetched from memory. This parallelism makes processors fast, executing billions of instructions per second. Dependencies constrain this parallelism. Adding $A + B$ requires knowing both values first. Processors analyse these dependencies and reorder independent operations to keep the pipeline flowing.

The instruction `if (x < y)` determines whether certain code will be executed, but evaluating the condition takes time. Waiting would leave execution units idle. Instead, processors predict the outcome and speculatively execute that path. Branch predictors

 https://doi.org/10.11647/OBP.0526.18

track patterns—loops usually continue, error checks usually pass, sorted data produces predictable comparisons.

If the prediction is correct, the speculative work becomes part of normal execution. If incorrect, the speculative instructions are discarded and execution switches to the correct path. From the perspective of the committed values of memory and registers, it is as if the speculation never occurred. Internally, though, speculative execution modifies shared state: caches (fast memory that stores recently accessed data), branch predictors, translation buffers, and other timing-sensitive components. As we will see, these modifications leave traces.

Modern computers set strict boundaries between different programs and between user programs and the operating system kernel. The kernel—the core of the operating system—manages hardware, controls security, and stores sensitive data such as passwords, encryption keys, and private information from all running programs. User programs (your browser, text editor, games) are forbidden from reading kernel memory. Similarly, one user's programs cannot read another user's data. These restrictions are enforced through memory protection mechanisms built into the processor.

But what if these protections could be bypassed? Imagine requesting a book from a library where some rooms are restricted. To save time, the librarian starts walking toward the room before confirming whether you have access. If you are authorised, the book is delivered. If not, the librarian returns—nothing was given. But the door was opened. Now suppose you are not told which room contains which book. Later, you notice that one door swings more easily. You did not receive the book, but you know where the librarian went. This is a side channel attack—extracting secrets through indirect observations rather than direct access.

Meltdown (2018) is an example of an exploit that targets privileged kernel memory—the protected area containing the operating system's secrets. User programs attempting to read kernel memory trigger an immediate error, like trying to enter a secured building without a keycard. But during speculative execution, the processor starts fetching the forbidden data before checking permissions. The security check eventually catches this and triggers an exception—Meltdown never 'officially' reads the secret. But in the microseconds between the speculative read and the security check, the secret value exists in the processor. The attack uses this value to access a specific location in the attacker's own memory array. When the security exception fires and speculation is cancelled, the secret itself is erased—but the access pattern remains in the cache. By timing how fast different array locations load, the attacker deduces which location was accessed, revealing the secret byte value.

Many operating systems historically mapped kernel pages into every user address space for performance. Meltdown exploited deferred permission checks on affected Intel CPUs, allowing transient use of privileged data before the fault was architecturally raised. AMD reported its CPUs were not affected by Meltdown; ARM susceptibility varied by core implementation.

Spectre trains the branch predictor, then exploits its predictions (Kocher et al., 2018). During training, call a victim function repeatedly with valid array indices. The predictor learns that bounds checks like `if (x < array_len)` succeed. During attack, provide an out-of-bounds index. Expecting success, the hardware speculatively reads `array[x]` beyond the array boundary. The speculative path uses this secret value as an index: `probe[secret * 4096]` . Each possible byte value maps to a different memory page, separated by 4 KB to avoid cache-line collisions. After the misprediction triggers rollback, the attacker times accesses to all 256 probe pages. The fastest access reveals which page was cached, exposing the byte. Repeat to extract regions.

Branch predictors maintain 95%+ accuracy by detecting patterns in program behaviour. They track individual branches and correlations between them. A global history register records recent branch outcomes, indexing into pattern tables that predict future behaviour. Some predictors track paths—sequences of branches—to capture control flow patterns. The predictor's state is shared across privilege levels and between different programs, creating a cross-domain communication channel.

Spectre variant 1 exploits conditional branches. Variant 2 targets indirect branches—jumps to addresses computed at runtime, common in object-oriented code and function pointers. The hardware must predict not just taken/not-taken but the actual destination address. The Branch Target Buffer (BTB) caches these predictions, but entries can be poisoned to redirect speculation to attacker-chosen gadgets.

Cache timing provides the physical channel. Processors use multiple cache levels—L1 (closest to CPU, 4 cycles), L2 (12 cycles), L3 (40 cycles), and main memory (200+ cycles). These differences reveal which addresses were accessed. Single-bit leaks, extracted repeatedly, compromise entire keys through differential analysis.

These attacks can extract any data the CPU has access to: passwords stored in memory, cryptographic keys, personal files, browser history, emails, database contents. If the kernel has it in memory, Meltdown can read it. If a program processes sensitive data, Spectre can extract it—even from JavaScript running in a web browser.

Mitigations constrain speculation at every level. `LFENCE` instructions create serialisation barriers. Kernel page table isolation (KPTI) unmaps kernel memory from user space. Indirect-branch mitigations include retpolines—a Google-developed technique replacing indirect branches (Turner, Sugerman, & Broido, 2018) with a return-based construct that traps speculation via the return predictor—and hardware IBPB/IBRS/STIBP controls. Some predictors are flushed or partitioned on context switches on newer systems. Each fix degrades the optimisation it protects.

Processors achieve high performance through pipelines—14–19 stages in contemporary designs. Without speculation, a mispredicted branch would flush the pipeline, wasting dozens of cycles. At 4 GHz, each wasted cycle represents 250 picoseconds of lost computation. Multiply by billions of branches per second, and performance would drop drastically.

The x86 `RDTSC`/`RDTSCP` instructions return a cycle count (cycle-level resolution). Time can be derived given clock frequency; on invariant TSC systems this can approach sub-nanosecond granularity. When restricted, alternatives exist: thread scheduling provides a coarse clock, contention on shared resources amplifies timing differences, and browser

JavaScript enables attacks through SharedArrayBuffer spin loops or WebAssembly instruction counting.

After Spectre and Meltdown, each processor optimisation became a potential side channel. Vector units, return predictors, schedulers, and virtualization boundaries revealed new attack surfaces.

Later attacks exploited other processor features. Downfall (2023) targets Intel's AVX gather instructions—when speculatively gathering data, these vector operations transiently load values from unauthorised memory regions. The values leave cache footprints after rollback, allowing extraction of cryptographic keys by timing which vector elements were accessed. Intel processors from sixth through eleventh generation carry this flaw.

Retbleed (2022) demonstrated that even retpolines fail. Return instructions—used in every function call—use their own prediction mechanism. By poisoning the return stack buffer, attackers force speculative execution of arbitrary gadgets, bypassing the carefully constructed retpoline defences.

Inception (2023) showed AMD processors aren't immune—it creates 'phantom speculation' by nesting mispredictions within mispredictions. GhostRace (2024) weaponized something previously thought safe: synchronisation primitives. Race conditions during speculative execution leak data even from properly synchronised code.

A fully-mitigated system may run 30% slower than its vulnerable predecessor. Some workloads see greater degradation—databases that rely heavily on indirect calls, JIT compilers that generate dynamic code, virtualized environments with frequent context switches. Mitigations cost years of Moore's Law gains.

"Don't quote me, but the oracle at Delphi told me she's using ChatGPT."

Folklore About Some Interesting Bugs

Geographic Email Limits

A university sysadmin faced a bizarre bug: emails wouldn't travel more than 500 miles. Boston at 420 miles worked fine; Memphis at 520 miles failed completely.
A timeout was set in nanoseconds instead of milliseconds. Light travels 200,000 km/s through fibre. The timeout expired at exactly 500 miles.

The Ice Cream Correlation

Drivers reported their keyless entry failed—but only after buying vanilla ice cream at one specific shop. Not chocolate, not strawberry. Just vanilla. Engineers were sceptical until they confirmed it.
The vanilla counter was at the front, chocolate in back. Vanilla buyers returned before the engine cooled, triggering a temperature-sensitive component failure.

Posture-Dependent Passwords

An IT team dismissed reports that a password worked standing but failed sitting—until they saw it happen. Same user, same password, different postures, different results.
Two keyboard keys had been swapped. Standing users looked down and typed what they saw. Seated users touch-typed from muscle memory, entering the wrong sequence.

Night Shift Crashes

Night-shift staff reported server crashes that never occurred during the day. No software changes, no power issues—just nightly failures.
The cleaning crew's floor polisher vibrated at 60 Hz, matching certain hard drives' resonant frequency. The vibration misaligned read/write heads just enough to crash the system.

The 2:45 PM Crash

One server crashed daily at exactly 2:45 PM. Logs showed thermal throttling, but only at that precise time.
Sunlight through a west-facing window hit the poorly ventilated rack at just the right angle. The temperature spike at 2:45 PM pushed the server past its limit.

Spectre and Meltdown: Mechanics and Leakage Rates

Introduction

Spectre and Meltdown exploit transient execution: instructions issued before permissions or branches resolve. Although architecturally squashed, these instructions leak data through persistent microarchitectural side effects—most notably, the cache—observable via timing measurements.

Meltdown: Transient Load + Fault Deferral

The CPU speculatively executes a faulting memory load and uses the result before the exception is raised. A dependent load encodes the secret byte into cache state.

```
; RCX ← kernel memory address
movzx rax, byte [rcx]
shl   rax, 12
mov   rbx, [probe + rax]
```

Comments:

- Line 1: Transiently loads a protected byte (zero-extended) into RAX.
- Line 2: Multiplies the secret by 4096 (page alignment).
- Line 3: Loads from `probe[s × 4096]`, caching a secret-dependent line.

Observation: The attacker probes access times to `probe[i × 4096]` and identifies the secret by locating the cache hit.

Spectre: Mistrained Bounds Bypass

The attacker mistrains the branch predictor to speculatively skip a bounds check. This leads to out-of-bounds access during the transient window.

```
if (x < array_length)
    temp = probe[secret[x] * 4096];
```

Only executed speculatively. The loaded cache line leaks `secret[x]`.

Setup: Train with in-bounds x; switch to out-of-bounds. Side effects persist even when the branch is mispredicted.

Cache Timing Oracle

Let $S \in \{0, \dots, 255\}$ be the secret. Let C_i be the access time to `probe[i × 4096]`. Then:

$$\hat{S} = \arg\min_i C_i$$

$$\text{where} \quad C_i = \begin{cases} T_{\text{hit}} + \epsilon_i & \text{if } i = S \\ T_{\text{miss}} + \epsilon_i & \text{otherwise} \end{cases},$$

With low noise, each measurement leaks up to approximately 8 bits.

Information-Theoretic View

Let X be the secret, Y the timing observation. Leakage is given by:

$$I(X;Y) = H(X) - H(X \mid Y).$$

Under uniform X and low timing noise, $I(X;Y) \to 8$ bits. Repetition and majority decoding mitigate jitter and measurement error.

Measured Bandwidth

Meltdown (local):	up to $\sim$500 KB/s
Spectre (native):	$\sim$10 KB/s
Spectre (JavaScript):	order of bits/s

Reported rates vary widely across microarchitectures, operating systems, and mitigation settings.

References:

Kocher et al. (2018). *Spectre Attacks: Exploiting Speculative Execution.* arXiv:1801.01203.

Lipp et al. (2018). *Meltdown: Reading Kernel Memory from User Space.* USENIX Security.

Consider the Muon's PoV

Top (Muon Time Dilation): Cosmic rays strike the upper atmosphere, producing showers of muons. At rest, muons have a lifetime of only about 2.2 microseconds, which should not allow many of them to reach detectors on the surface. However, due to relativistic time dilation, their internal clocks run slower from the Earth's frame of reference, allowing far more muons to survive and be detected than expected under classical assumptions.
Bottom (Neutrino Flavour Oscillations): When produced in cosmic-ray interactions, neutrinos come in three flavours: electron, muon, and tau. Early atmospheric neutrino experiments detected only about two-thirds of the expected muon neutrinos, a puzzle known as the atmospheric neutrino anomaly. The resolution came from neutrino oscillations: as neutrinos travel, they can change 'hats' between flavours, converting among types. This explains why detectors only saw a fraction of the muon neutrinos originally predicted.

Consider the Muon's PoV

Muons created by cosmic rays colliding with the upper atmosphere provide direct evidence for time dilation. With a rest-frame lifetime of approximately 2.2 microseconds and travelling close to light speed, classical physics predicts these particles should decay before reaching Earth's surface. Instead, detectors routinely observe muons at sea level. Special relativity explains this observation: from Earth's reference frame, the muons' time runs slower by a factor of γ (approximately 10–50 depending on energy), extending their lifetime enough to reach ground level. From the muon's perspective, relativistic length contraction reduces the distance travelled.

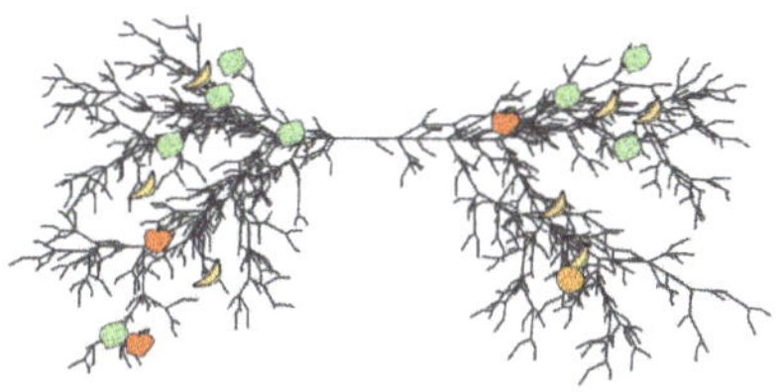

Atmospheric Muons ∘ 2.2 Microsecond Lifetime ∘ Relativistic Time Dilation ∘ $\gamma = 15$ at 0.998c ∘ Pion Decay Origin ∘ Cosmic Ray Showers ∘ Lepton Generations ∘ Rossi-Hall 1941 ∘ Sea Level Detection ∘ Length Contraction ∘ Natural Relativity Test

"Who ordered that?"
— I.I. Rabi, 1936

Consider the Muon's PoV

Carl D. Anderson's cosmic ray research had yielded the positron (Anderson, 1933), earning him the Nobel Prize in 1936. Working at Caltech with graduate student Seth Neddermeyer, Anderson continued photographing particle tracks in cloud chambers at high altitude. In late 1936, they noticed tracks that curved less than electrons in magnetic fields but more than protons—evidence of a particle with intermediate mass.

The 1937 discovery paper was cautious. The tracks suggested a mass roughly 200 times that of the electron, but the particle's identity remained unclear. Anderson and Neddermeyer called it a 'mesotron,' reflecting its intermediate mass between electrons and protons. Physicist Isidor Rabi reportedly quipped: "Who ordered that?" The particle seemed superfluous—it played no obvious role in atomic structure or known nuclear processes.

Theoretical physicists initially tried to identify the mesotron with Hideki Yukawa's predicted meson, which should mediate the strong nuclear force. Yukawa had calculated in 1935 that such a particle should have a mass around 200 electron masses and interact strongly with nuclei. But the mesotron penetrated matter far too easily and interacted too weakly with atomic nuclei to be Yukawa's particle. The confusion persisted until 1947, when Cecil Powell, César Lattes, and Giuseppe Occhialini discovered the pion—the true Yukawa particle—using improved photographic emulsions at high altitude. They showed that pions produced in cosmic ray collisions decay into the lighter mesotron. The lighter particle was renamed the muon, recognising it as a heavier cousin of the electron rather than a nuclear force carrier.

Meanwhile, physicists studying cosmic ray showers noticed an anomaly. Muons produced 10–20 kilometres above Earth's surface were reaching sea-level detectors in numbers far exceeding expectations. With a measured lifetime of 2.2 microseconds at rest, a muon travelling even at light speed should cover less than 700 metres before decaying. Yet detectors routinely observed muons at sea level, having traversed more than 10 kilometres through the atmosphere.

Bruno Rossi, an Italian physicist who had fled fascist Italy to work at Los Alamos and later MIT, recognised the discrepancy. In 1940–1941, Rossi and David B. Hall conducted systematic measurements comparing muon counts at different altitudes. Their 1941 data showed that relativistic time dilation explained the observations: muons travelling at velocities exceeding $0.99c$ experience dilated lifetimes, allowing them to reach Earth's surface before decaying. The experiment provided one of the first natural confirmations of special relativity outside laboratory settings, using naturally occurring particles travelling macroscopic distances. The agreement between predicted and observed muon flux at various altitudes removed lingering doubts about relativistic time transformation.

By the 1950s and 1960s, muons had transitioned from mysterious intruders to standard tools. Their long lifetime, clean decay signature, and penetrating power made them invaluable for testing quantum electrodynamics, probing weak interactions, and serving as high-energy probes in collider experiments. The particle that seemed to serve no purpose became central to understanding the structure of matter.

 https://doi.org/10.11647/OBP.0526.19

Right now, as you read this, particles are passing through your body at nearly the speed of light. Roughly one muon traverses each square centimetre of your skin every minute. They originate 10–20 kilometres above Earth's surface, born from collisions between cosmic rays and atmospheric nuclei. They rain down continuously, penetrating buildings, mountains, and human tissue, leaving faint trails of ionisation as they pass.

What is a muon? It belongs to a family of particles called leptons—point-like matter particles that do not experience the strong nuclear force. The most familiar lepton is the electron, which orbits atomic nuclei and carries electric current through wires. The muon is essentially a heavier version of the electron: same electric charge (-1), same spin ($\frac{1}{2}$), but 207 times more massive at $105.7\,\text{MeV}/c^2$. This extra mass makes the muon unstable. It decays via the weak interaction:

$$\mu^- \to e^- + \bar{\nu}_e + \nu_\mu,$$

producing an electron and two neutrinos. Laboratory measurements show this decay occurs with a mean lifetime of 2.2 microseconds in the muon's rest frame.

Atmospheric muons begin with cosmic rays—mostly high-energy protons accelerating through interstellar space, some reaching energies up to 10^{11} GeV. When these protons strike atmospheric nuclei at altitudes of 10–20 kilometres, they produce hadronic showers: cascades of secondary particles including pions and kaons. Pions decay rapidly:

$$\pi^+ \to \mu^+ + \nu_\mu, \qquad \pi^- \to \mu^- + \bar{\nu}_\mu,$$

with a rest-frame lifetime of 26 nanoseconds. Because the pions travel at relativistic speeds, their decay products inherit high energies and directionality. The resulting muons typically have energies of 1–10 GeV and velocities exceeding $0.995c$.

Here lies the puzzle. A muon at rest lives 2.2 microseconds. Even travelling at light speed, this permits a maximum distance of:

$$d_{\text{classical}} = c \cdot \tau_0 = 3 \times 10^8\,\text{m/s} \cdot 2.2 \times 10^{-6}\,\text{s} \approx 660\,\text{m}.$$

Classical physics predicts that muons created 15 kilometres up should decay long before reaching sea level. Detectors at ground level should register only a tiny residual flux—the rare survivors from production events occurring just above the stratosphere.

But measurements show otherwise. Muons arrive at sea level in abundance, approximately one per square centimetre per minute. Some penetrate deep underground, detected in mines and beneath mountains. The observed flux exceeds classical predictions by more than an order of magnitude. If Newtonian mechanics governed particle decay, atmospheric muons would be rare curiosities, not the dominant component of cosmic radiation at Earth's surface.

Special relativity resolves the paradox (Einstein, 1905). From Earth's reference frame, the muon's lifetime dilates according to the Lorentz factor:

$$\tau_{\text{observed}} = \gamma\tau_0, \qquad \gamma = \frac{1}{\sqrt{1 - v^2/c^2}}.$$

For a muon travelling at $v = 0.998c$, we calculate $\gamma \approx 15$. The observed lifetime extends to:

$$\tau_{\text{observed}} = 15 \times 2.2\,\mu\text{s} \approx 33\,\mu\text{s},$$

allowing the muon to cover approximately 10 kilometres before decaying. This matches the observed sea-level flux and explains detections deep underground.

From the muon's rest frame, time passes normally—it still lives only 2.2 microseconds by its own clock. But the atmosphere is contracted along the direction of motion by the same factor γ. The 15-kilometre journey from production altitude to sea level contracts to roughly 1 kilometre in the muon's frame. This distance is easily traversable within 2.2 microseconds at $0.998c$. Both perspectives yield identical predictions: muons reach sea level. The symmetry reflects the covariance of physical laws under Lorentz transformations—no preferred reference frame exists.

Detecting muons exploits their electric charge and penetrating power. As they traverse matter, they ionise atoms along their path, losing energy gradually. Unlike electrons, which radiate intensely at high energies (bremsstrahlung), muons are massive enough to suppress radiative losses. They pass through metres of rock or steel with only modest energy attenuation.

The simplest detector uses a scintillator—a material that emits light when ionised—coupled to a photomultiplier tube. When a muon passes through, it excites molecules in the scintillator, which promptly re-emit photons. The photomultiplier amplifies this signal into a measurable electrical pulse. Cloud chambers and bubble chambers reveal muon tracks visually: the particle ionises a supersaturated or superheated medium, leaving a trail of condensation or bubbles. Modern experiments use arrays of scintillation counters, drift chambers, or resistive plate chambers with timing electronics to reconstruct trajectories and measure momenta.

At sea level, the vertical muon flux is approximately one per square centimetre per minute, with typical energies around 4 GeV. This makes muons the dominant component of secondary cosmic radiation at Earth's surface. They contribute background signals in neutrino detectors, serve as calibration sources for particle physics experiments, and enable muon tomography—a radiographic technique that uses atmospheric muons to image dense structures such as nuclear reactor cores or hidden chambers in pyramids.

The atmospheric muon flux provided one of the earliest natural confirmations of relativistic time dilation. The phenomenon requires no synchronised atomic clocks, no particle accelerators, no carefully orchestrated experimental setup. Nature performs the test constantly. Bruno Rossi and David Hall's 1941 measurements, comparing muon counts at different altitudes, demonstrated quantitative agreement with relativistic predictions. The data matched time dilation calculations and ruled out classical decay rates.

This test carries philosophical weight; sceptics might argue that relativistic formulas are mathematical conveniences—useful calculational tools with no physical reality. Atmospheric muons refute this: particles created at high altitude, travelling macroscopic distances through the atmosphere, reach ground-level detectors in numbers precisely predicted by relativistic kinematics. The agreement between laboratory measurements of rest-frame decay rates and atmospheric propagation distances over tens of kilometres confirms that time dilation is not an artefact of coordinate systems but a physical consequence of spacetime geometry.

The muon's place within the Standard Model (Weinberg, 1967) reveals a deeper puzzle. Fundamental matter particles divide into two families: quarks, which carry colour charge and form composite hadrons such as protons and neutrons, and leptons, which are point-like and unaffected by the strong nuclear force. The six known leptons arrange into three generations:

- First generation: electron (e^-), electron neutrino (ν_e)
- Second generation: muon (μ^-), muon neutrino (ν_μ)
- Third generation: tau (τ^-), tau neutrino (ν_τ)

Each generation replicates the pattern: one charged lepton and one neutral neutrino. The charged leptons have identical electric charge and spin but vastly different masses. The electron at $0.511\,\mathrm{MeV}/c^2$ is stable. The muon, 207 times heavier, decays in 2.2 microseconds. The tau lepton, at $1776.9\,\mathrm{MeV}/c^2$, survives only 290 femtoseconds. Neutrinos have no electric charge, interact only via the weak force and gravity, and possess extremely small masses—each less than $1\ \mathrm{eV}/c^2$.

Quarks mirror this generational structure: up and down (first generation), charm and strange (second), top and bottom (third). Each generation preserves the same interaction patterns and quantum numbers, differing only in mass and lifetime. Heavier fermions decay into lighter ones through weak interactions, conserving energy, momentum, and quantum numbers.

Why three generations? The Standard Model incorporates this structure through input parameters—masses, mixing angles, decay constants—but offers no explanation for the number of generations or the mass hierarchy. The replication appears arbitrary. Nothing in gauge theory or quantum field theory requires precisely three generations, yet all known fermions fit this pattern. The muon, when first discovered, seemed superfluous—physicist Isidor Rabi's quip "Who ordered that?" captured the bewilderment. It plays no obvious role in atomic structure or nuclear processes, yet it exists, precisely replicating the electron's quantum numbers at a different mass scale.

The generational structure affects everything from CP violation in weak decays (Christenson et al., 1964) to neutrino oscillations to the running of coupling constants at high energies. But the underlying reason—why nature chose three generations, why the mass ratios span orders of magnitude—remains unknown.

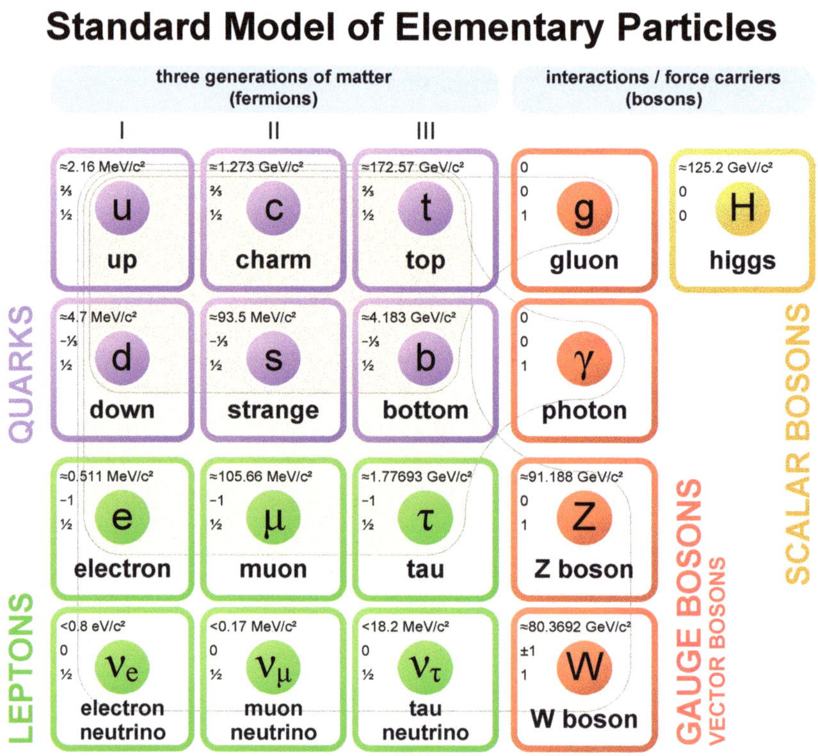

The Standard Model of Elementary Particles, showing the three generations of quarks and leptons, the gauge bosons mediating the fundamental forces, and the Higgs boson. Source: Wikimedia Commons, CC BY-SA 4.0.

Relativistic Lifetimes of Cosmic Muons

Introduction

Muons decay via the weak interaction with a proper lifetime $\tau_0 \approx 2.2\,\mu$s in their rest frame. This value was determined experimentally by observing muons at rest in controlled environments. High-energy muons, produced either in cosmic ray interactions or particle accelerators, are slowed down using materials such as carbon or liquid hydrogen until they are brought to rest. Once stationary, their decays are monitored using detectors that record the timing and energy of the emitted positrons from the decay process:

$$\mu^+ \to e^+ + \nu_e + \bar{\nu}_\mu.$$

The time distribution of detected positrons follows an exponential decay law:

$$N(t) = N_0 e^{-t/\tau_0},$$

where $N(t)$ is the number of decays observed at time t, and τ_0 is the proper lifetime of the muon. By fitting the observed decay curve, τ_0 was measured with high precision to be approximately $2.2\,\mu$s.

When moving at relativistic speeds, time dilation modifies this observed lifetime according to special relativity. From the perspective of an observer on Earth, the muon's lifetime appears stretched by a Lorentz factor γ, allowing it to travel much farther than expected before decaying.

1. Time Dilation Factor

Let v be the muon's speed and γ the Lorentz factor:

$$\gamma = \frac{1}{\sqrt{1 - \frac{v^2}{c^2}}}.$$

For $v \approx 0.9995\,c$, the factor is:

$$\gamma \approx 32.$$

Consequently, the muon's observed lifetime in the lab frame is

$$\tau_{\text{obs}} = \gamma\,\tau_0 \approx 32 \times 2.2\,\mu\text{s} \approx 70\,\mu\text{s}.$$

2. Distance Travelled

During this dilated lifetime, a muon can travel:

$$d = v\,\tau_{\text{obs}} \approx 0.9995\,c \times 70\,\mu\text{s} \approx 21\,\text{km}.$$

This exceeds the 15 km from the upper atmosphere to sea level, explaining why so many muons survive to reach detectors.

3. Alternate View: Length Contraction

In the muon's reference frame, its lifetime remains $2.2\,\mu$s, but the distance to Earth is contracted by $1/\gamma$, shrinking 15 km to under 500 m. Both descriptions are consistent, reflecting the symmetry of special relativity.

Conclusion

The unexpected abundance of muons at sea level offered a compelling demonstration of relativistic time dilation. These measurements confirmed that high-speed particles experience significantly slowed decay rates, matching Einstein's predictions and underscoring the profound role of relativity in particle physics.

References:

Rossi, B., and Hall, D. B., *Physical Review*, 59(3), 223–228 (1941).

HyperPhysics Muon Simulation: `http://hyperphysics.phy-astr.gsu.edu/hbase/Relativ/muon.html`

Scan the QR code below to access the simulator:

Capish, Comprehendes, Computes?

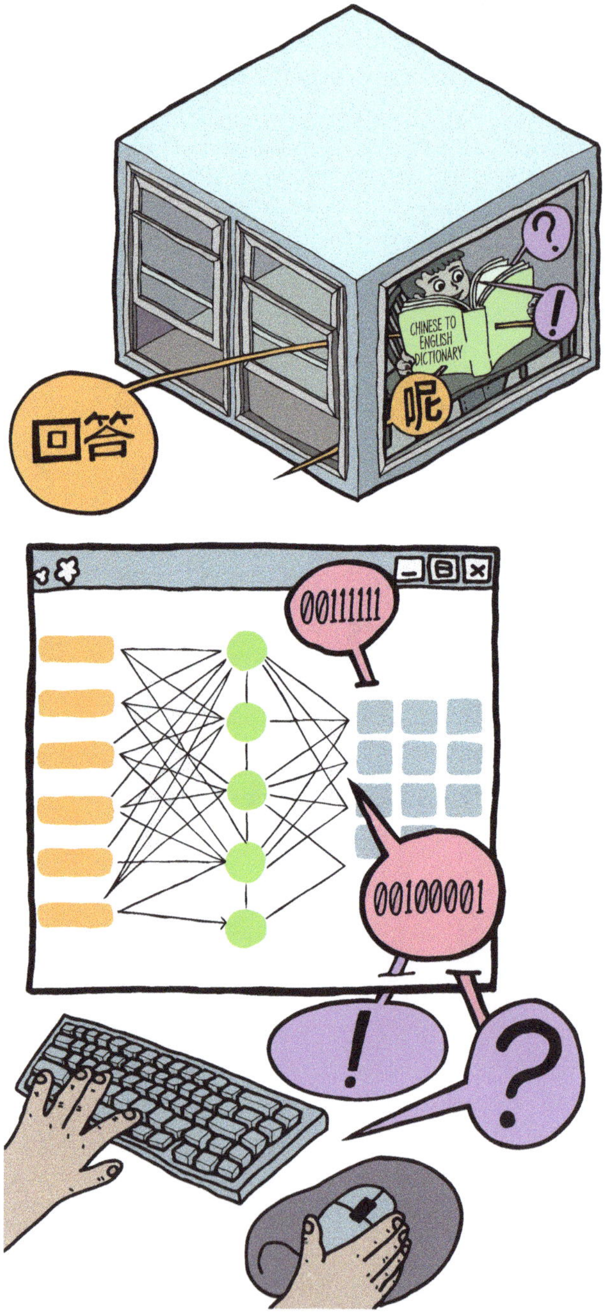

Top (Chinese Room Experiment): The Chinese Room Thought Experiment (John Searle, 1980). An English-speaking person sits inside a sealed room, using a rulebook to manipulate Chinese symbols without understanding their meaning. The system produces fluent-looking Chinese responses purely through syntactic manipulation. Searle's point: following formal symbol-manipulation rules does not constitute understanding.
Bottom (Modern Language Models (LLMs)): Instead of explicit rules, LLMs use high-dimensional vector representations and learned statistical patterns from massive datasets. Inputs are converted into embeddings, processed through multiple nonlinear layers, and decoded into probable outputs. Yet for all the architectural difference, LLMs—like the Chinese Room—operate without intrinsic semantic understanding. They produce contextually appropriate language based on training distributions, not genuine comprehension.

Capish, Comprehendes, Computes?

Searle's Chinese Room thought experiment challenges computational theories of mind: someone manipulates Chinese symbols according to rules without understanding the language. They produce appropriate responses, passing a linguistic Turing test, yet possess no comprehension. The argument distinguishes syntax (symbol manipulation) from semantics (understanding), suggesting that computers executing algorithms operate only at the syntactic level.

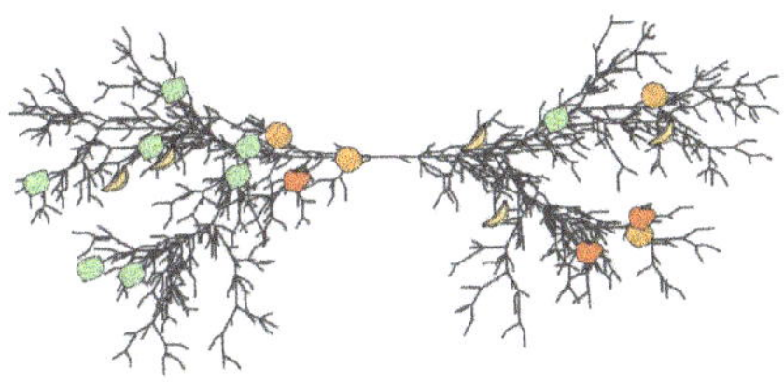

Chinese Room Experiment ◦ Syntax Without Semantics ◦ Searle's Argument ◦ Strong AI Critique ◦ Large Language Models ◦ Next-Word Prediction ◦ Trillion Parameters ◦ Compression & Generalisation ◦ In-Context Learning ◦ Substrate Independence ◦ Belief Formation Patterns

Words are the only bullets in truth's bandolier.
And poets are the snipers.
— Martin Silenus, circa 2850 A.D.

S'il y avait de telles machines [...] jamais elles ne pourraient user de paroles ni d'autres signes en les composant, comme nous faisons pour déclarer aux autres nos pensées.

("If there were such machines [...] they could never use words by composing them, as we do, to declare their thoughts to others.")
— René Descartes, 1637

Capish, Comprehendes, Computes?

Alan Turing's 1950 essay introduced the 'imitation game' (later known as the Turing Test), proposing that if a machine's textual responses were indistinguishable from a human's, it could be considered intelligent. This marked the beginning of modern debates on machine cognition. In the following decades, the symbolic AI movement gained momentum, with figures such as John McCarthy formalising logic-based systems and Jerry Fodor proposing the 'Language of Thought' hypothesis, which treated mental processes as manipulations of internal symbolic representations.

By the late 1970s, optimism surrounding symbolic approaches began to collide with deeper philosophical questions. Critics questioned whether syntactic manipulation alone could account for semantics—understanding rooted in meaning. In 1980, John Searle articulated the Chinese Room argument, asserting that executing formal rules does not entail comprehension. His critique challenged the assumption of strong AI: that implementing a program is equivalent to having a mind.

Contemporaneously, Daniel Dennett proposed the 'intentional stance,' emphasising observer-relative attributions of belief and intention, while Patricia and Paul Churchland advocated for eliminative materialism, arguing that folk-psychological terms like 'belief' and 'desire' might eventually be replaced by neurobiological accounts. Meanwhile, connectionist models—distributed neural networks—began gaining traction in the mid-1980s, offering an alternative to rule-based systems by emphasising statistical learning over symbolic structure.

By the 1990s, AI had achieved public milestones such as Deep Blue's 1997 victory over Garry Kasparov. Yet critics noted that performance alone does not imply understanding. With the advent of large language models in the 2010s and 2020s, capable of generating coherent and contextually appropriate text, the debate has re-emerged: do these systems understand language, or are they sophisticated instances of the Chinese Room, manipulating symbols without grasping their meaning? Public reports about recent frontier models (e.g., GPT-5) leave specific parameter counts and training token totals undisclosed, though they are trained on massive text corpora at an internet scale.

The Chinese Room thought experiment presents a scenario (Searle, 1980)—an English speaker sits in a sealed chamber, Chinese messages arrive through a slot. The person possesses a rulebook, written in English, that specifies how to manipulate incoming Chinese characters to produce syntactically valid Chinese responses. The rulebook contains no semantic information, only symbol manipulations. The individual follows these instructions and returns the processed strings through the slot.

To an external Chinese speaker, the conversation appears coherent. The responses are grammatically correct, contextually relevant, and indistinguishable from those of a fluent human. Yet the person inside understands none of the content. They do not know that symbols refer to objects, events, or ideas—they execute formal operations on uninterpreted marks. The room satisfies a behavioural test for language competence, yet no part of the system possesses comprehension.

This scenario forces a separation between two dimensions of linguistic behaviour: *syntax*, the arrangement of symbols, and *semantics*, the capacity to represent or grasp meaning. Searle's central claim is that syntactic competence, even when sufficient to pass behavioural tests, does not entail semantic understanding. The system's outputs may simulate language use, but the process lacks intentionality—the directedness of mental states toward meaning-bearing entities or propositions.

This argument extends beyond the thought experiment—it challenges the claims of 'strong AI,' the position that appropriately programmed computers possess minds like humans. Proponents of strong AI maintained that mental states are computational: if a system manipulates symbols according to rules that preserve formal structure and generate appropriate outputs, it qualifies as intelligent. The Chinese Room rejects this inference.

The problem cuts across disciplines. In computer science, the debate centres on algorithmic representation limits and generalisation in machine learning. In linguistics, it intersects with theories of reference, deixis, and semantic grounding. Philosophy confronts questions about intentionality, mental content, and necessary conditions for knowledge. Neuroscience examines embodiment, sensory integration, and causal mechanisms by which mental states arise in biological systems.

From these inquiries emerges a potential requirement: semantic understanding may demand more than pattern matching. Some propose that AI systems might achieve genuine understanding through embodied interaction—robotics, environmental embedding, or sensorimotor coupling that shapes internal representations through causal contact with physical entities. Others argue that meaning resides in subjective experience or first-person perspective that may be inaccessible to artificial systems.

The debate has intensified with large language models (LLMs). These systems demonstrate capabilities that extend far beyond the simple rule-following in Searle's original formulation. They engage in reasoning, exhibit creativity, and show generalisation across domains. Yet they remain neural networks trained through a deceptively simple objective: predicting the next word in a sequence.

The training process operates at an unprecedented scale. Public reports about recent frontier models (e.g., GPT-5) do not disclose exact parameter counts or training token totals; nonetheless, they are trained on massive text corpora at an internet scale. During training, the network processes sequences and learns to predict probability distributions over all possible next words. For the input 'The capital of France is,' the model learns P('Paris') = 0.85, P('located') = 0.03, and so forth.

This process is self-supervised. No human labels the 'correct' next word because the next word serves as the target. The model minimises cross-entropy loss, heavily penalising confident wrong predictions while providing diminishing returns for improving accurate predictions. Through billions of prediction tasks, spanning months of computation across thousands of processors, the network's parameters converge toward configurations that compress the statistical structure of human language.

At first glance, this training method seems to confirm Searle's critique. The model manipulates symbols based on statistical patterns without direct access to meaning. Critics dismiss the resulting capabilities as 'mere probabilistic parroting': statistical correlation without genuine understanding. This characterisation faces an explanatory challenge that cuts to the heart of the Chinese Room debate. Consider a training example: 'there are two boxes. Box A contains a red ball and Box B contains a blue ball. If you randomly pick a box and then randomly pick a ball, what is the probability of getting a red ball?' Now present the model with: 'there are two containers. Container X holds a cyan sphere and Container Y holds a purple sphere. If you randomly select a container and then randomly choose a sphere, what is the probability of getting a cyan sphere?'

LLMs solve the second problem correctly, yielding 0.5, despite probably never encountering 'cyan' and 'purple' in mathematical contexts during training. The model abstracts the underlying structure: P(cyan) = P(select Container X) × P(cyan | Container X) = 0.5 × 1.0 = 0.5. This generalisation cannot be explained by memorization of surface patterns. The specific word combinations or even subsets of this sentence likely never appeared in training data. The model recognises invariant mathematical entities across surface variations, performs conceptual substitution, and transfers zero-shot to novel domains.

This generalisation emerges from a compression constraint. The network must compress terabytes of text into gigabytes of parameters while maintaining prediction accuracy. This compression pressure forces extraction of underlying patterns, rules, and relationships rather than memorization. To predict that 'The ball rolled down the hill and splashed into the pond,' the model must develop representations of physics, not just word associations.

Probing techniques confirm this type of learning. Linear classifiers can extract representations of truth, causality, and object properties from the model's activations. The networks construct hierarchical abstractions. Early layers capture syntax and word boundaries, while later layers encode semantic relationships. This suggests that successful next-word prediction requires building models of the world described in text.

These models undergo multiple training phases that complicate the Chinese Room analogy. Pre-training teaches next-word prediction but produces systems that continue text rather than follow instructions. A model asked 'What is your first name?' might respond 'What is your last name?,' not from understanding but because such sequences appear in training data. Instruction fine-tuning then maps user intentions to appropriate responses through supervised learning on curated instruction-response pairs, transforming text completers into conversational assistants. Finally, reinforcement learning from human feedback drives outputs toward what evaluators judge to be helpful, honest, and harmless.

The specific mechanisms underlying these capabilities matter less than the computational patterns they produce. Current LLMs rely on attention mechanisms, but this appears to be an implementation detail. Alternative architectures achieve similar capabilities through different pathways. This supports substrate independence: intelligence emerges from computational patterns (Putnam, 1967) rather than specific implementations.

This interpretation strengthens when examining capabilities that arise without explicit training. In-context learning allows models to acquire new skills (Brown et al., 2020) from examples in the input prompt, without parameter updates. Present a model with examples of translating English to a made-up language, and it can continue the pattern for new inputs. This suggests that during pre-training, models develop meta-learning algorithms within their forward pass. They maintain implicit probability distributions over possible tasks and update these based on observed examples.

Such capabilities challenge the Chinese Room analogy directly. Searle's scenario involves fixed rule-following. The person executes predetermined instructions without understanding. But LLMs develop adaptive computational patterns that acquire new competencies dynamically. Their 'rules' are not rigid instructions but flexible algorithms that respond to novel contexts. This suggests something different from the static symbol manipulation Searle described.

The core philosophical question persists. The models achieve these capabilities through statistical learning over vast datasets, building representations that compress and generalise from linguistic patterns. Whether this compression constitutes genuine understanding or remains sophisticated simulation is disputed.

Defining 'genuine' versus 'simulated' understanding proves notoriously difficult, yet we possess immediate, non-inferential access to our own comprehension. I know that I grasp meaning—not through behavioural testing or external validation, but through direct phenomenal awareness. Current computational systems lack this quality. The asymmetry is clear. From a first-person perspective, the distinction between understanding and simulation is self-evident. From a third-person perspective, it is non-existent. Other humans and machines occupy the same epistemic position relative to my certainty. I cannot verify understanding in either perspective through observation alone; asserting its existence in other humans requires a leap of faith.

Together with Turing's test (Turing, 1950), the Chinese Room is a test about devising alethic criteria from indistinguishability tests. It forces precision in our definitions of understanding, intelligence, and comprehension. Whether future research reveals that specific architectural features—embodiment, sensorimotor coupling, phenomenal consciousness—are necessary for intelligence, or that substrate independence holds across all cognitive capabilities, Searle's scenario continues to provide a framework for examining these questions.

Cross-Validation.

Belief Formation: Humans and Models

Humans form beliefs in unsettling ways. We think we update our views when presented with new evidence, but reality reveals a different pattern. Beliefs that align with our existing worldview stick around; those that contradict it get dismissed or twisted into supporting evidence. When someone challenges our deep convictions, we often become *more* confident in what we believed originally—sometimes described as the **backfire effect** (though evidence suggests such effects are not widespread and are context-dependent). We're not neutral fact-processing machines. We're defensive storytellers, preserving narrative coherence over empirical accuracy.

We can now compare this to large language models. When you train a model on billions of text examples, it learns whatever patterns exist in that data, including confident assertions about false claims. Later, when researchers try to fine-tune the model with correct information, the original learning resists change. The neural weights have settled into configurations optimised for the original data distribution.

This is how learning systems work. Both human brains and neural networks must compress vast amounts of information into manageable models. Once those patterns solidify, changing them means destabilising everything else that depends on them. In humans, this shows up as cognitive dissonance and motivated reasoning. In models, it appears as **gradient stasis** and **catastrophic forgetting**. This shows up as cognitive gradient stasis when learning new ones.

Both systems handle uncertainty similarly. Humans rarely admit ignorance cleanly. Instead, we confabulate. Language models do the same. When asked about topics outside their training data, they don't respond with "I don't know." They generate confident-sounding responses that preserve conversational flow, even when information is sparse or contradictory.

This suggests that both human cognition and current AI systems are optimised for something other than truth correspondence. They prioritise internal consistency and social coordination over factual accuracy. In humans, this makes evolutionary sense. Being wrong together was often more adaptive than being right alone. In AI systems, it emerges from the training objective: predict the next word in a way that sounds human-like.

The Chinese Room becomes more provocative through this lens. Searle asked whether symbol manipulation without understanding constitutes genuine comprehension. But perhaps the more unnerving question is whether human 'understanding' is itself merely symbol filtering that prioritises narrative coherence over external reality.

The Epistemic Fragility of Syntax-Only Cognition

Framing the Dispute

The Chinese Room is not an empirical argument but a methodological critique aimed at conceptual boundaries in cognitive science. Searle's challenge is not about empirical performance but what we are permitted to infer from it. He contends that passing the Turing Test—or any test based solely on linguistic output—cannot entail genuine understanding unless we already have a theory that licences such an inference. The argument's strength lies in its methodological reversal: where AI seeks to move from behavioural evidence to mental attribution, Searle denies the validity of that inference without a prior grounding in what it means to 'understand.'

Understanding Without a Criterion

'Understanding' is not an operationally neutral term. Unlike 'predicts weather' or 'stores data,' it is irreducibly normative and semantically loaded. To assert that a system understands is to claim it possesses internal relation to meaning—content-bearing capacity that cannot be exhausted by structural descriptions alone. The Chinese Room exposes justificatory laziness: we project understanding onto systems that behave in familiar ways, without specifying what justifies that projection.

Replies to Searle rely on stipulative bridges—identifying understanding with functional role, environmental embedding, or counterfactual dependence. These bridges merely relocate the conceptual burden. What is lacking is a non-circular account of when formal behaviour amounts to semantic content. This is not a technical failure but a philosophical silence.

Intentionality and Attribution

Searle's intentionality point is often misconstrued. He is not claiming that syntax cannot, in principle, be paired with semantics. He asserts that such pairing is not guaranteed by formal operations alone. The Chinese Room Argument (CRA) is not about what symbols do; it is about what they mean. And meaning is not intrinsic to the system unless some internal state stands in a relation of intentional directedness—a relation not captured by computational transitions.

Attempts to circumvent this by pointing to system-wide properties (as in the Systems Reply) or virtual entities (as in the Virtual Mind Reply) fail to address a fundamental asymmetry: intentional states have first-person authority, whereas syntactic states do not. If a system 'understands,' then it makes sense to ask what it understands and why. But if that attribution is based only on labelling (e.g., 'this system understands Chinese'), we are no longer explaining cognition—merely renaming behaviour.

Simulation and Normativity

Searle targets the normative dimension of cognition. Understanding involves norm-sensitive responsiveness to content, not merely causal states. A person can misunderstand, misinterpret, or revise their understanding. These are not errors in computation; they are errors in relation to content. But a syntactic machine cannot err in this sense. It can malfunction, but it cannot misbelieve. Without normativity, there is no epistemic traction, and without that, no understanding.

The Epistemic Cost of Ambiguity

The enduring appeal of the Chinese Room stems from its methodological clarity. It does not claim that AI will never understand. It claims that we lack a criterion by which to know if it does. To assert that future systems might understand language is to talk without terms. Until we have a definition of understanding that does not collapse into performance, or a theory of meaning that does not presume biological embedding, our attributions remain projections—not findings.

References:

Searle, J. R. (1980). *Minds, Brains, and Programs.* Behavioral and Brain Sciences, 3(3), 417–457.

Dennett, D. C. (1987). *The Intentional Stance.* MIT Press.

Chalmers, D. (1996). *The Conscious Mind.* Oxford University Press. See Also: https://plato.stanford.edu/entries/chinese-room/

Exponentially Generalisable

Top (Addition and Multiplication): The exponential map translates addition into multiplication. Sequences of additions along the number line become products, such that $\exp(a+b)=\exp(a)\cdot\exp(b)$. **Middle (Unit Circle Mapping):** On the complex plane, the exponential maps purely imaginary numbers ia to points on the unit circle $e^{ia}\in U(1)$. Tangent vectors correspond to directions in the Lie algebra, while their exponential images wrap around the circle as group elements. **Bottom (Manifold and Tangent Plane):** In differential geometry, the exponential map extends to manifolds. A tangent vector at a point p defines a geodesic, whose endpoint on the manifold is $\exp_p(v)$. Conversely, the logarithm map returns tangent vectors from manifold points. This generalisation connects linear structure in tangent spaces to nonlinear geometry of manifolds.

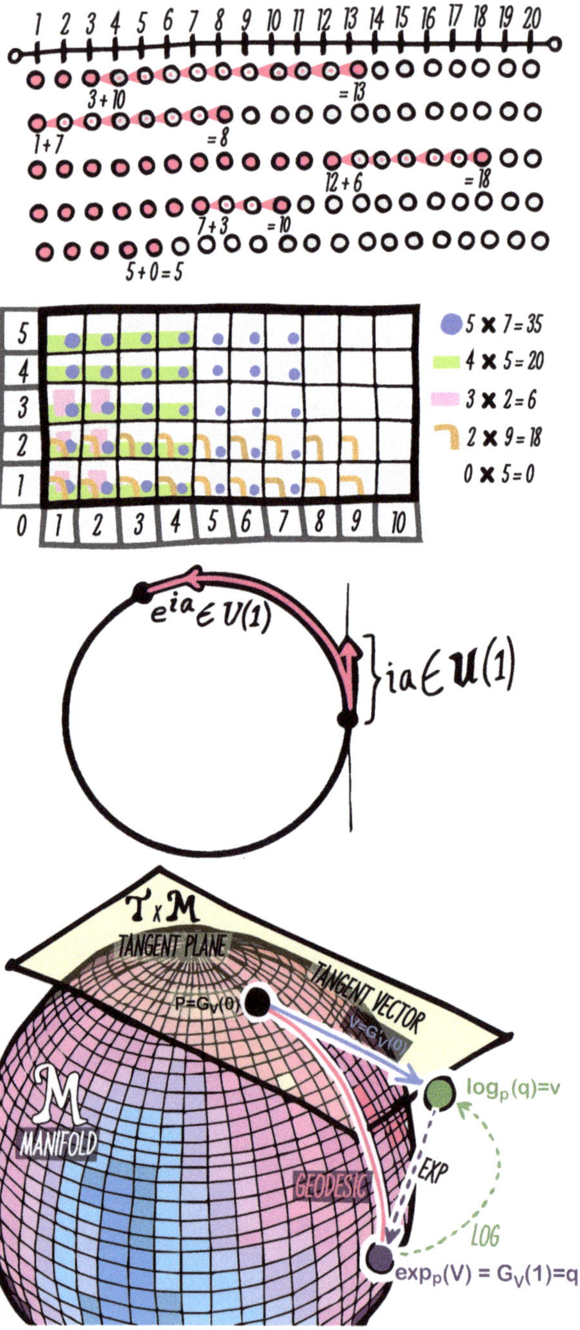

Exponentially Generalisable

The exponential function extends far beyond calculus, appearing across mathematics as a bridge between local and global structure. From power series to Lie theory, from Riemannian geometry to sheaf cohomology, exponential maps carry additive or infinitesimal data into multiplicative, compositional, or curved settings. What began as a trick for quick computation has become a central map linking analysis, geometry, and algebra.

EXPONENTIAL AS BRIDGE ∘ ADDITIVE TO MULTIPLICATIVE ∘ LIE ALGEBRA TO GROUP ∘ GEODESIC EXPONENTIAL ∘ POWER SERIES $\sum x^n/n!$ ∘ SHEAF COHOMOLOGY SEQUENCE ∘ CATEGORY THEORY EXPONENTIALS ∘ LOCAL TO GLOBAL ∘ MULTIPLE DEFINITIONS ∘ UNIVERSAL PATTERN ∘ GENERALISATIONS

"The rate of increase of inflation is decreasing."
— Richard Nixon *(i.e.,* $\frac{d^3}{dt^3}$[Purch. Power] > 0 *, a kind of economic jerk).*

"This was the first time a sitting president
used the third derivative to advance his case for reelection."
— Hugo Rossi, 1996

Exponentially Generalisable

> In 1614, John Napier introduced logarithms to simplify arithmetic, turning multiplication into addition. Though his tables were constructed geometrically and not in terms of a function e^x, the underlying idea—of an operation whose inverse linearises multiplication—was foundational. Over a century later, in 1748, Leonhard Euler formally introduced the exponential function, defining e^x through its power series and connecting it to the constant $e \approx 2.71828$. He also showed that this function uniquely solves the differential equation $f' = f$ with $f(0) = 1$.
>
> By the nineteenth century, the exponential function was generalised to complex analysis, where its series converges for all complex inputs, and to linear algebra via matrix exponentials. In parallel, Sophus Lie developed the theory of continuous transformation groups, now called Lie groups, and demonstrated how exponentiation links the tangent space at the identity (the Lie algebra) to the global group structure. In the early twentieth century, Élie Cartan further extended these ideas into geometry and topology, embedding exponential maps into the study of curvature, connections, and geodesics. What began as a computational tool thus evolved into a central organising principle of modern mathematics.

The exponential function appears early in mathematical education, often as the solution to continuous growth or the base of natural logarithms. Yet its role extends far beyond calculus. It mediates transitions—additive rules become multiplicative behaviour, local definitions yield global constructions, linear approximations curve into manifolds.

The number e emerges through compound interest limits, power series, and differential equations with self-similar rates. These formulations converge because they encode the same operation. The exponential function canonically bridges additive structure with compositional behaviour.

This pattern pervades mathematics. In differential geometry, the exponential map sends tangent vectors to manifold points along geodesics. In Lie theory, it maps algebra generators to group elements. In sheaf cohomology, it connects additive and multiplicative sheaves. In category theory, it defines internal function spaces. Each incarnation translates local data into global structure.

We may remember several equivalent definitions of the number e, or the exponential function $\exp(x)$, from calculus. One learns that the limit $(1 + x/n)^n$, the inverse of the integral of $1/x$, the power series $\sum x^n/n!$, and the solution to the differential equation $f' = f$ with $f(0) = 1$, all yield the same function.

The similarity extends far beyond functions over $\mathbb{R}$ or $\mathbb{C}$. There are many constructions, across different areas of mathematics, that are all called 'the exponential map.' These are not merely notational coincidences. In each case, the map expresses a transition from an additive domain to a multiplicative, compositional, or curved codomain.

Some of these maps are defined analytically by convergent series. Others are defined geometrically, such as in differential geometry where a vector in the tangent space is

 https://doi.org/10.11647/OBP.0526.21

mapped to a point on the manifold along a geodesic. Others arise algebraically, as in sheaf theory or representation theory.

Context	**Definition of** $\exp(x)$ **or Analogue**	**Structures (Domain** $\to$ **Codomain)**	**Emergent Property / Defining Aspect**
Formal Power Series	$\sum \frac{x^n}{n!}$	Algebra $\to$ its unit group	$\exp(x+y) = \exp(x)\exp(y)$ when $xy = yx$
Lie Theory	$\gamma_X(1)$	Lie algebra $\to$ Lie group	Local diffeo; flows compose via group law
Eigenfunction of Derivation	$K(f) = \lambda f$, $f(0) = 1$; K a derivation	Functions $A \to B$	$f(x+y) = f(x)f(y)$
Microlocal Analysis	$D(e^{i\xi\cdot x}) = P(x,\xi)\, e^{i\xi\cdot x}$	Operator $D \to$ symbol $P(x,\xi)$	Hörmander: singularities propagate along Hamiltonian
Sheaf Theory	$0 \to \mathbb{Z} \to \mathcal{O} \xrightarrow{\exp} \mathcal{O}^* \to 1$	Sheaf $\mathcal{O} \to \mathcal{O}^*$	Links additive and multiplicative sheaves
Algebraic Homomorphism	$\phi(x+y) = \phi(x)\phi(y)$	Additive group $\to M^\times$	Continuity forces $\phi = e^{\lambda z}$ over $\mathbb{C}$
Category Theory	Y^X by adjunction	CCC: $X, Y \to Y^X$	$\mathrm{Hom}(A \times X, Y) \cong \mathrm{Hom}(A, Y^X)$
Riemannian Geometry	$\exp_p(v) := \gamma_v(1)$	$T_pM \to M$	Riemann normal coords; metric flat to 1st order

These maps differ in detail but are still related. When the domain operates by addition and the codomain by multiplication or composition, the exponential map provides the bridge.

In analysis, the complex exponentials $e^{i\langle\xi,x\rangle}$ serve as fundamental probes for linear partial differential operators D. Applying D to these exponentials extracts its symbol—a polynomial in the frequencies ξ. Hörmander's theorem exploits this to track the *wave front set*: the directions in frequency space where the distribution's Fourier transform fails to decay rapidly. Here, the highest-order terms of the symbol define a Hamiltonian; the singularities then propagate along the 'null' trajectories where this Hamiltonian vanishes. In this way, the exponential bridges the local analytic property of a differential equation to the global geometric path of its singularities.

In Lie theory, a Lie algebra captures infinitesimal symmetries via antisymmetric brackets. The associated Lie group embodies these symmetries through multiplication. The exponential map takes an algebra element and returns the time-one value of its one-parameter subgroup. Near the identity, this map is a diffeomorphism.

In Riemannian geometry, the exponential map sends a tangent vector $v \in T_pM$ to the point $\gamma_v(1) \in M$ reached by the geodesic starting at p in direction v. This map defines *Riemann normal coordinates*, a coordinate system centred at p in which the metric reduces to the identity and its first derivatives vanish—the metric becomes flat up to second-order corrections. Physically, this is the equivalence principle in its sharpest mathematical form (Einstein, 1916): the exponential map constructs a local inertial frame in which gravity disappears and the laws of physics reduce to their special-relativistic forms (see also Chapter 6).

In sheaf theory, the exponential arises in the exact sequence $0 \to \mathbb{Z} \to \mathcal{O} \xrightarrow{\exp(2\pi i\,\cdot)} \mathcal{O}^* \to 1$, linking the additive structure of holomorphic functions to the multiplicative structure of nonvanishing functions. This map defines a connecting homomorphism, enabling classification of line bundles, identification of divisor classes, and the detection of obstructions such as the first Chern class.

In algebra and number theory, exponential homomorphisms convert additive modules into multiplicative groups. These homomorphisms satisfy $\exp(x + y) = \exp(x)\exp(y)$ and are unique up to scalar under torsion-free assumptions. They enable the extension of scalar operations to group actions.

In categorical settings, exponential objects arise in cartesian closed categories through the adjunction $\mathrm{Hom}(A \times X, Y) \cong \mathrm{Hom}(A, Y^X)$. The exponential object Y^X characterises internal homomorphisms and governs how composition distributes over products, generalising the function space construction from set theory into more abstract environments.

The exponential map recurs wherever mathematics needs to translate between different modes of combination. Its universality across analysis, geometry, algebra, and category theory suggests it captures something deeper than any particular formula.

Generalisations

The exponential map exemplifies a broader phenomenon in mathematics: core operations that retain their essential character while adapting to new contexts. Other examples illuminate this pattern.

Such recurrence is not unique to exponentiation. Mathematics frequently extends core notions into broader domains, preserving their defining relations while adjusting the ambient structure. The factorial function, initially defined on the natural numbers by recursion, extends to the complex plane as the Gamma function. This extension retains the recurrence and multiplicative shift $\Gamma(n+1) = n\Gamma(n)$, but replaces discrete input with a holomorphic domain.

The derivative generalises beyond calculus into measure theory. The Radon–Nikodym derivative expresses the rate of change between two measures—preserving the Leibniz rule and linearity while removing dependence on pointwise evaluation. In each case, the derivative remains an object that localises variation, though its technical definition differs.

Curvature also admits generalisation. From elementary circle-based definitions, it extends to Gaussian and mean curvature in surfaces, and further to the Riemann curvature tensor in higher-dimensional manifolds. The notion of curvature continues to measure deviation from flatness, but its role adapts to the presence of connections, holonomy, and coordinate invariance.

Size and distance generalise in parallel. Counting extends through Lebesgue measure, volume forms, and Haar measure on locally compact groups, preserving additivity and invariance under structure-preserving transformations. Distance moves from the Euclidean formula to abstract metric spaces and beyond to intrinsic metrics and

Gromov-Hausdorff limits, retaining the triangle inequality while shedding any dependence on ambient coordinates.

Dimension generalises from the number of coordinates in Euclidean space to vector space dimension (basis cardinality), topological dimension (covering refinements), Hausdorff dimension (fractal scaling behaviour), and Krull dimension in commutative algebra (chains of prime ideals). Each preserves the intuition of 'degrees of freedom' while adapting to its ambient structure.

Lawvere's Fixed-Point Theorem and the Limits of Self-Reference

Beyond the applications in geometry and algebra, the categorical exponential possesses a startling logical depth. The exponential object is powerful enough to provide a universal framework for self-reference, unifying many of the most significant limitative results in the history of logic and mathematics.

This unification is phrased by Lawvere's fixed-point theorem (Lawvere, 1969). In any Cartesian closed category, the theorem states that if a space of functions B^A can be fully 'named' by elements of A, then every transformation $g : B \to B$ must have a fixed point. The contrapositive means that if you can find a single transformation without a fixed point (like logical negation, or adding one to a number), then no universal naming system can exist.

This single, abstract result shows that the same fundamental logic underpins several famous paradoxes and impossibility proofs:

- **Cantor's diagonal argument** shows that a set A cannot be mapped surjectively onto its power set $\mathcal{P}(A) \cong \{0, 1\}^A$, as the 'diagonal' element is always missed.
- **Russell's paradox** results from the impossibility of a set containing all sets that do not contain themselves.
- **Gödel's first incompleteness theorem** constructs a formal statement that asserts its own unprovability, a fixed point of negation within the logic.
- **Turing's proof of the undecidability of the Halting Problem** shows that no program can exist that correctly determines whether any program will halt, another diagonalization argument.

While this categorical perspective may not simplify the individual proofs, it demonstrates that these seemingly separate achievements are all manifestations of the something common about functions and self-reference, elegantly captured by the exponential object.

Exponential Maps Across Contexts

How does each general definition specialise to $\exp(z) = e^z$ **on** $\mathbb{C}$**?**

1. **Formal Power Series.** The series $\exp(x) = \sum_{n=0}^{\infty} x^n/n!$ in any commutative $\mathbb{Q}$-algebra, evaluated over $\mathbb{C}$, converges absolutely to $e^z\colon \mathbb{C} \to \mathbb{C}^*$ with the homomorphism property $\exp(z+w) = \exp(z)\exp(w)$.
2. **Lie Theory.** The exponential maps $X \in \mathfrak{g}$ (an infinitesimal transformation) to the time-1 value of the integral curve of its left-invariant vector field starting at the identity. For $G = \mathbb{C}^*$, $\mathfrak{g} \cong \mathbb{C}$, and the integral curve for $V_X(z) = Xz\frac{d}{dz}$ through $z(0) = 1$ solves $\dot{z} = Xz$, giving $z(t) = e^{Xt}$ and $\exp(X) = e^X$.
3. **Eigenfunction of a Derivation.** Characterise the exponential as the eigenfunction satisfying $Df = \lambda f$ with $f(0) = 1$ (the identity of the target group). For $D = d/dz$ and $\lambda = 1$, the unique solution is $f(z) = e^z$.
4. **Microlocal Analysis.** Applying a differential operator D to a plane wave extracts its symbol, converting calculus into algebra: $D(e^{i\xi\cdot x}) = P(x,\xi)\, e^{i\xi\cdot x}$. On $\mathbb{C}$, the exponential $e^{\zeta z}$ is the universal probe for constant-coefficient operators; for d/dz and $\zeta = 1$, e^z yields eigenvalue 1.
5. **Sheaf Theory.** In the short exact sequence $0 \to 2\pi i\,\mathbb{Z} \to \mathcal{O}_M \to \mathcal{O}_M^* \to 0$, the map $f \mapsto e^f$ translates additive holomorphic functions into multiplicative, non-vanishing ones. On $M = \mathbb{C}$, this is $e^z\colon \mathbb{C} \to \mathbb{C}^*$, with kernel $2\pi i\mathbb{Z}$ encoding the monodromy obstructing a global logarithm.
6. **Algebraic Homomorphism.** A holomorphic group homomorphism $\phi\colon (\mathbb{C},+) \to (\mathbb{C}^*,\times)$ satisfying $\phi(x+y) = \phi(x)\phi(y)$ must take the form $\phi(z) = e^{\lambda z}$. Holomorphicity is essential: it eliminates $\bar{z}$ dependence. Setting $\lambda = 1$ recovers the classical exponential.
7. **Category Theory.** The exponential object Y^X in a Cartesian Closed Category satisfies $\mathrm{Hom}(A \times X, Y) \cong \mathrm{Hom}(A, Y^X)$. With $X = \mathbb{C}$ and $Y = \mathbb{C}^*$, currying $(a,x) \mapsto e^{ax}$ yields $a \mapsto (x \mapsto e^{ax})$. The category provides the habitat, placing e^z as a global element in the function space $(\mathbb{C}^*)^{\mathbb{C}}$.
8. **Riemannian Geometry.** The geometric exponential sends $v \in T_pM$ to the geodesic endpoint $\gamma_v(1)$. On $\mathbb{C}^*$ with the bi-invariant metric $\langle u,v\rangle_z = \mathrm{Re}(u\bar{v})/|z|^2$, the space is flat, so geodesics through the identity coincide with one-parameter subgroups $t \mapsto e^{tX}$, yielding $\exp_1(X) = e^X$.

How $f' = f \Rightarrow f(x+y) = f(x)f(y)$

Let $(A, +, 0)$ be an additive monoid and $(B, +, \cdot, 0, 1)$ a unital commutative ring. Let $\text{Func}(A, B)$ be the ring of functions with pointwise operations.

Define shift operator $T_y : \text{Func}(A, B) \to \text{Func}(A, B)$ by $(T_y g)(x) := g(x+y)$.

Suppose $K : \text{Func}(A, B) \to \text{Func}(A, B)$ satisfies:

(A) Additivity: $K(g+h) = K(g)+K(h)$.
(L) Leibniz: $K(g \cdot h) = K(g) \cdot h + g \cdot K(h)$.
(C) Kills Constants: $K(c_b) = c_0$ where $c_b(x) \equiv b$.
(T) Translation Invariance: $K \circ T_y = T_y \circ K$.

Let $f \in \text{Func}(A, B)$ and $\lambda \in B$ satisfy:

(E) Eigenfunction: $K(f) = \lambda \cdot f$.
(N) Normalisation: $f(0) = 1$.
(U) Uniqueness: Evaluation at 0 is injective on $\ker(K - \lambda I)$.

For f satisfying (E), let $g_y := T_y f$. Then:

$$\begin{aligned} K(g_y) &= K(T_y f) \\ &= T_y(K(f)) = T_y(\lambda \cdot f) \\ &= \lambda \cdot T_y(f) = \lambda \cdot g_y. \end{aligned}$$

Define the function $g := g_y - f \cdot c_{f(y)}$ where $c_{f(y)}$ is the constant function with value $f(y)$ (i.e., pointwise $g(x) = f(x+y) - f(x) \cdot f(y)$).

Evaluating at 0:

$$g(0) = f(y) - f(0) \cdot f(y) = 0.$$

Computing $K(g)$:

$$\begin{aligned} K(g) &= K(g_y - f \cdot c_{f(y)}) \\ &= K(g_y) - K(f \cdot c_{f(y)}) && \text{(by (A))} \\ &= K(g_y) - (K(f) \cdot c_{f(y)} + f \cdot K(c_{f(y)})) && \text{(by (L))} \\ &= \lambda \cdot g_y - (\lambda \cdot f \cdot c_{f(y)} + f \cdot c_0) && \text{(by (E), (C))} \\ &= \lambda \cdot g_y - \lambda \cdot f \cdot c_{f(y)} \\ &= \lambda \cdot (g_y - f \cdot c_{f(y)}) \\ &= \lambda \cdot g. \end{aligned}$$

Since $K(g) = \lambda \cdot g$ and $g(0) = 0$, uniqueness (U) gives $g = c_0$ (the zero function). Thus:

$$f(x+y) = f(x) \cdot f(y).$$

Hence the additive exponential property emerges from the derivation properties: **(A)** additivity, **(L)** Leibniz, **(C)** annihilation of constants, **(T)** translational symmetry, and **(U)** uniqueness at 0. The functional equation $f(x+y) = f(x)f(y)$ is a geometric necessity, of which the real exponential $e^{x+y} = e^x e^y$ is a special case.

Remark: If (U) fails, we obtain

$$f(x+y) = f(x) \cdot f(y) + h_y(x),$$
$$h_y \in V_\lambda^0 := \{g : K(g) = \lambda \cdot g,\ g(0) = 0\}$$

This error term satisfies a cocycle condition central to group cohomology, measuring how far f is from being a homomorphism.

Intuition: Translation invariance makes K commute with shifts: $K(T_y f) = T_y K(f)$. Thus every translation $T_y f$ of an eigenfunction remains in the λ-eigenspace of K. By (U), translation acts on this eigenspace by scalar multiplication: $T_y f = c_{f(y)} \cdot f$, meaning that translating f by y yields the function $x \mapsto f(y) \cdot f(x)$. Commuting with the derivation therefore forces addition in the domain (the group operation generating translations) to correspond to multiplication in the codomain (the scalar action on the eigenspace).

Example: Classical Derivative
Let $A = (\mathbb{R}, +, 0)$, $B = (\mathbb{R}, +, \cdot, 0, 1)$, and $K = \frac{d}{dx}$ on smooth real functions. Define $f(x) := e^{\lambda x}$ for $\lambda \in \mathbb{R}$.

Then K satisfies (A) $(f+g)' = f' + g'$, (L) $(fg)' = f'g + fg'$, (C) $(\text{constant})' = 0$, and (T) $(T_y f)'(x) = f'(x+y) = (T_y f')(x)$. Condition (U) holds since any solution to $g' = \lambda g$ with $g(0) = 0$ is $g \equiv 0$ by uniqueness of ODE solutions.

Since $f' = \lambda e^{\lambda x} = \lambda f$ and $f(0) = 1$, all hypotheses (A)–(U) hold, hence $e^{\lambda(x+y)} = e^{\lambda x} e^{\lambda y}$.

Creeping Bug

Top (3D Mesh with UV Grid): A polygonal model is shown with a superimposed UV grid. The grid flattens a surface into 2D coordinates, assigning each vertex to a location in texture space. This mapping determines how an image will be wrapped onto the surface. **Bottom (Texture Atlas):** A sprite sheet provides the 2D image data: character animations, items, and enemies. Each region of the sheet corresponds to a set of UV coordinates. During rendering, the engine looks up the correct portion of the texture based on UV mapping and projects it onto the surface, aligning 2D artwork with 3D geometry.

Creeping Bug

Minecraft's Creeper began as a mistake. Markus Persson entered the pig's dimensions backwards, producing a tall, thin figure that looked nothing like an animal. Instead of deleting it, he added a texture, a frown, and an explosive routine borrowed from the game's block-destruction code. The result was a creature that crept silently, paused, and detonated. A simple modelling error became Minecraft's most iconic enemy.

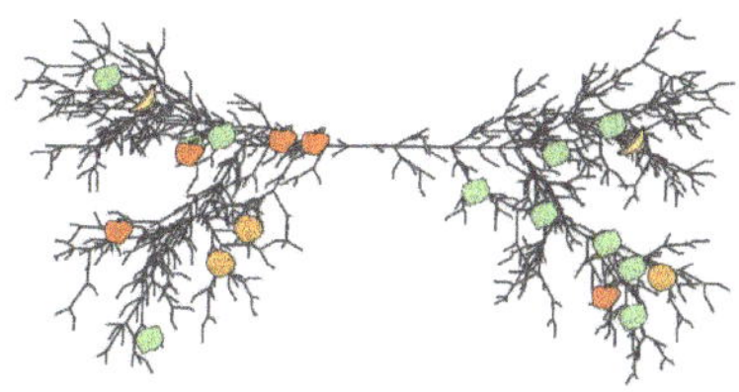

CREEPER ORIGIN STORY ◦ PIG MODEL BUG ◦ ENTITY COMPONENT SYSTEMS ◦ AI TASK COMPOSITION ◦ GAME ENGINE EVOLUTION ◦ BUGS BECOMING FEATURES ◦ STREET FIGHTER COMBOS ◦ ROCKET JUMPING ◦ MINECRAFT REDSTONE LOGIC ◦ EMERGENT GAMEPLAY ◦ FAIL-SOFT DESIGN

"Once you've got a task to do,
it's better to do it than live with the fear of it."
— Logen Ninefingers, 575 AU

"The first 90 percent of the code accounts for
the first 90 percent of the development time.
The remaining 10 percent of the code accounts for
the other 90 percent of the development time."
— Tom Cargill, 1985

Creeping Bug

Minecraft (2009) began as the independent project of Markus 'Notch' Persson, a Swedish programmer inspired by sandbox-building games such as Infiniminer (2009) and Dwarf Fortress (2009). He began development on 10 May 2009, using Java with the Lightweight Java Game Library (LWJGL), releasing the first public version seven days later on 17 May. Early versions focused on creative construction in a procedurally generated block world (Perlin, 1985), appealing to players' innate curiosity and design instincts. Persson incorporated player feedback rapidly, adding survival mechanics, crafting, hostile mobs, and multiplayer support. Development was open and iterative, creating a strong early community on forums such as TIGSource.

In 2010, Persson founded Mojang to support ongoing development. Jens Bergensten (known as Jeb) joined and later took over lead development. The game officially launched on 18 November 2011 at MineCon in Las Vegas, having already sold millions of copies in beta. Mojang maintained a low-friction sales model: a single upfront purchase, no digital rights management (DRM), and support across multiple operating systems. This simplicity contributed to rapid global adoption.

In 2014, Microsoft acquired Mojang and Minecraft for $2.5 billion. At that point, Minecraft had sold over 54 million copies. Since then, the game has expanded to nearly every platform, including consoles, mobile, and VR. As of October 2023, Minecraft has sold over 300 million copies, making it the best-selling video game of all time. In 2021, it reported more than 140 million monthly active users. It has been used in classrooms, cited in academic studies on spatial reasoning and collaboration, and remains a major force in online content creation—especially on YouTube, where Minecraft videos have accumulated over 1 trillion views.

Minecraft's success stems from a blend of simplicity and depth. It offers intuitive core mechanics (block placement, mining, crafting) with nearly unlimited creative potential. Procedural generation ensures novelty, while redstone logic introduces programmable mechanics akin to electrical engineering. The game fosters personal expression, exploration, and emergent storytelling (Gardner, 1970). Its low system requirements and modding support further extended its reach and longevity. Minecraft's development history is a case study in iterative design, community engagement, and the creative payoff of systems-first thinking.

In 1962, game developers wrote code that directly controlled individual pixels. *Spacewar!* on the PDP-1 computed each dot's position through direct arithmetic: $x + dx$, $y + dy$. The electron beam drew vectors where the calculations specified. No abstraction layers existed between the programmer's calculations and the phosphor display. Each spaceship consisted of six numbers in memory—position, velocity, angle, and fuel. The game loop ran sixty times per second—read switch states from the control panel, update positions by adding velocities, subtract fuel for thrust, apply gravity as a constant acceleration toward the centre, check if position vectors intersected for collisions, then send the computed coordinates to the vector display. The entire program fit in 4 KB of memory.

Early arcade machines hardcoded every behaviour into separate routines. *Space Invaders* (1978) implemented each alien type with its own movement function, its own collision detection, its own point calculation. Moving the bottom row required one function that decremented x-coordinates and checked the left boundary. Moving the middle row used a different function with different speed constants. The top row had its own handler. When any alien touched the screen edge, specific code executed to move all aliens down one row and reverse the direction flag. Adding a new enemy type meant writing new functions for movement, new functions for collision detection, new functions for scoring—touching every system in the game. The code grew linearly with content.

Programmers began utilising these patterns of repetition. Every moving object needed position and velocity, every visible object needed drawing routines, every destructible object needed health values—the same structures repeated across every game. In *Adventure* (1979), Warren Robinett faced the Atari 2600's 128 bytes of RAM. He couldn't afford separate code for each object type. Instead, he consolidated repeated elements into a single object handler. Dragons, bats, keys, and swords were entries in an object table, each storing position, size, colour, and a behaviour ID. The behaviour ID indexed into a jump table of function pointers. Object `0x1A` (the yellow key) had behaviour type `0x04` (can be picked up). Object `0x0E` (the red dragon) had behaviour type `0x07` (chase player). A single collision detection routine handled all interactions by comparing behaviour IDs, and a single drawing routine rendered all objects by reading their size and colour. What would have required dozens of specialised handlers fit into 128 bytes of RAM.

When Toru Iwatani designed *Pac-Man* (1980), he gave all four ghosts the same movement code but different target selection algorithms. Blinky (red) targeted Pac-Man's current tile. Pinky (pink) targeted four tiles ahead of Pac-Man in his facing direction. Inky (cyan) computed a complex target: take the position two tiles in front of Pac-Man, draw a vector from Blinky to that position, then double it. Clyde (orange) chased Pac-Man when more than eight tiles away but fled to his home corner when closer. Four ghosts from one movement function with different target coordinates.

The 1990s brought object-oriented programming to game development (Dahl & Nygaard, 1966). *Doom* (1993) designed its actors through inheritance hierarchies. Every monster derived from a common base class containing position, health, and state machine logic. The imp and the baron of hell executed identical state machine code. They differed only in their data tables—health points (60 versus 1000), projectile type (fireball versus plasma ball), pain chance (200/256 versus 50/256). State machines were data. The imp's fireball attack was state S_TROO_ATK3: display sprite TROOF, duration 8 tics, action function A_TroopAttack, next state S_TROO_1. New monsters were created by mixing existing action functions with new sprites and parameters. The Revenant combined A_SkelMissile (fire homing missile) with A_SkelFist (punch if close). The Archvile combined A_VileChase (resurrect dead monsters) with A_VileAttack (immolating flame attack). No new code required—just new data tables.

By 2000, inheritance hierarchies had limitations. A FlyingEnemy class couldn't share code with a FlyingProjectile without multiple inheritance. A FireGolem pulled from both Golem and FireCreature, creating diamond ◇ inheritance—when a class inherits from two classes that themselves inherit from the same base class. Deep hierarchies were fragile. Changing Animal broke Dog which broke FlyingDog which broke FireBreathingFlyingDog. Entity-component systems replaced them. An entity was just an ID number. Components were bags of data: `Position {x: 5, y: 10, z: 3}`, `Velocity {dx: 1, dy: 0, dz: 0}`, `Sprite {texture: 'goblin.png'}`, `Health {current: 30, max: 30}`, `AI {behaviour: 'aggressive'}`. Systems were functions that processed components. MovementSystem iterated through all entities with Position and Velocity, updating positions. RenderSystem drew all entities with Position and Sprite. DamageSystem processed collisions for entities with Health. Adding flight to a goblin meant adding a Flying component. Making a barrel explode meant adding an Explosive component. A flying, exploding, invisible barrel needed no new class—just combine Flying, Explosive, and remove the Sprite component.

Game engines were standardised architectures. id Tech (1993), Unreal Engine (1998), and Unity (2005) provided complete frameworks—rendering pipelines with shaders and occlusion culling, physics engines with collision detection and constraint solvers, audio systems with 3D spatialization, networking layers with client prediction and lag compensation. Developers no longer built engines from scratch. They configured existing systems, wrote gameplay scripts, created art assets. A game was configuration data plus custom logic, running on someone else's foundation. Most games using these engines still required teams of specialists to craft specific experiences. Some games took a different approach, providing tools for players to build their own content within the game's systems.

Minecraft used these principles extensively. Markus Persson wrote no hardcoded mining animations, no specific crafting sequences, no predetermined progression. He built basic systems. Blocks had properties—hardness, tool requirements, light emission, update behaviours. Items had functions—dig block, place block, damage entity. Entities had composable AI tasks: place blocks, break blocks, update neighbours. Players built cities, computers, musical instruments, working calculators. Redstone dust carried signals up to 15 blocks. Torches inverted signals—powered input produced unpowered output. Repeaters delayed signals by 1–4 ticks. Pistons pushed blocks when powered. Players built logic gates—NOT gates (Shannon, 1938) from single torches, OR gates from merged dust lines, AND gates from torch arrays, memory cells from piston feedback loops (Eccles & Jordan, 1919). Players even built 8-bit CPUs. The programmer set the stage, the players set the game.

Entities in *Minecraft* used composition. The base Entity class defined position, velocity, and bounding box. LivingEntity added health and damage handling. Mob added a list of AI tasks, small behaviour programs that executed each tick. Tasks were objects with simple interfaces—`shouldExecute()` to check if the task should run, `startExecuting()` to initialise, `updateTask()` to perform the behaviour. A zombie's task list contained `AttackPlayerTask` (priority 2), `WanderTask` (priority 5), `LookAtPlayerTask` (priority 8). A sheep had `EatGrassTask`, `FollowParentTask`, `PanicTask`. The cow combined wandering, following players holding wheat, and panicking when hurt. The skeleton combined attacking players, fleeing from wolves, and avoiding sunlight. Every mob was assembled from the same library of parts.

Among these assembled creatures was the Creeper—*Minecraft*'s most recognisable enemy. Silent, green, explosive. It approaches players with little warning and detonates on proximity, destroying carefully built constructions. Unlike zombies that moan or skeletons that rattle, the Creeper moves in complete silence until its final hiss. The Creeper began as a pig. In 2009, Persson was implementing farm animals and creating a pig model. In his 3D modelling program, he entered the creature's dimensions. The pig required length 2.0, height 1.0, width 1.0—a horizontal rectangle with stubby legs. But when typing the values, Persson accidentally swapped length and height, entering height 2.0, length 1.0. Instead of a quadruped, he got a vertical pillar with four tiny legs at the bottom. The model file loaded without error—*Minecraft*'s model loader accepted any valid vertex data. The renderer displayed exactly what it received—a tall, thin creature standing upright. Persson found the error amusing. He textured it green, added a frowning mouth, adjusted the legs to look more like feet. He kept it as a joke, then made it explode. He copied TNT's explosion code—remove blocks within radius R, damage entities with distance-based falloff, spawn item entities for destroyed blocks. He bound this to proximity detection borrowed from zombie AI—if distance to player less than 3 blocks, start countdown timer. After 1.5 seconds, detonate.

The Creeper was not the only accident that stuck. Bugs have shaped game design as much as deliberate decisions.

Street Fighter II (1991) had combos through a programming oversight. During development, producer Noritaka Funamizu discovered that the recovery time after certain moves was shorter than the hit-stun inflicted on opponents. By timing inputs precisely, players could land a second attack before the opponent recovered from the first. The window was narrow—frame-perfect in some cases—and the developers assumed players would rarely exploit it. But location test players began discovering two-hit and three-hit sequences. Capcom watched players develop increasingly complex chains—jump kick into standing fierce into special move. Instead of patching the timing windows, Capcom kept the oversight and built around it. Every subsequent fighting game implemented deliberate combo systems with cancels, links, and chains—mechanics that originated from a gap between two timing values.

Quake (1996) introduced rocket jumping through physics engine oversight. The game calculated explosion damage and knockback for all entities within the blast radius—including the player who fired the rocket. Players discovered that firing at their feet while jumping added the explosion's upward force to their jump velocity, reaching otherwise inaccessible platforms. The technique required precise timing and cost health, creating a risk-reward trade-off that became a staple of competitive first-person shooters.

Grand Theft Auto (1997) began development as *Race'n'Chase*, a straight racing game with police chases. During testing, a quirk of the police AI made them impossibly aggressive. Instead of trying to box players in, they rammed at full speed. Testers found themselves spending more time fleeing from psychotic police than racing. The chaotic pursuits were more entertaining than the intended gameplay. DMA Design redesigned the entire game around the bug, leaning into the chaos rather than fixing it. The racing game became a crime sandbox.

Devil May Cry (2001) originated from a scrapped *Resident Evil 4* prototype. The prototype's combat engine felt too fast and stylish for survival horror—characters launched enemies into the air, chained attacks mid-flight, and juggled opponents with gunfire. Rather than discard the prototype, Capcom built an entire franchise around what didn't fit, rating players on combo variety and rewarding style over caution.

Space Invaders (1978) had increasing difficulty through hardware limitations. Tomohiro Nishikado programmed the aliens to move at constant speed, but the Intel 8080 processor couldn't maintain consistent frame rates. With fifty-five aliens on screen, the game ran slowly. As players destroyed aliens, the processor had fewer sprites to update, causing the remaining aliens to move faster. Nishikado kept the unintended acceleration—what began as a hardware limitation became the game's defining tension, the last alien darting across the screen at terrifying speed.

Tribes (1998) had movement physics bugs. Players discovered that rapidly tapping jump while descending slopes prevented the normal friction from applying. Each jump reset the friction calculation before it could slow the player. This 'skiing' technique allowed players to traverse maps at tremendous speed, transforming methodical infantry combat into high-speed aerial warfare. Dynamix kept it, designing maps with long slopes, adding routes specifically for skiing, balancing weapons around high-speed combat.

Silent Hill (1999) used fog to hide hardware limitations. The PlayStation couldn't render distant polygons without severe popup and texture warping. Instead of reducing draw distance with traditional fog walls, Team Silent implemented thick, volumetric fog that moved and swirled. The fog became the game's defining atmosphere, hiding threats and turning every few metres of visibility into a source of dread.

Super Smash Bros. Melee (2001) had physics oversights. Wavedashing happened when air dodging diagonally into the ground preserved momentum while landing, causing characters to slide. Players could attack while sliding, opening new approach options. Nintendo never intended wavedashing—Masahiro Sakurai called it an exploit—but it became central to competitive *Melee*, which thrived for two decades on mechanics its creators considered mistakes.

"Your resume is... interesting. It mentions 'extensive experience in probability determination'?"

The Creeper

Entity models in early Minecraft were defined using bounding boxes in Java. Each model was specified via constructor calls such as:

```
ModelBox(name, x, y, z,
         length, height, width).
```

For passive mobs, proportions typically followed:

$$\texttt{length} > \texttt{height},\quad \texttt{length} \approx 1.2,\quad \texttt{height} \approx 0.9.$$

Due to a developer error, `length` and `height` were swapped:

$$(\ell,\ h,\ w)\ \mapsto\ (h,\ \ell,\ w).$$

This produced a tall, narrow model. Define the pig's geometry as:

$$G_{\text{pig}} = [0,\ell] \times [0,h] \times [0,w],$$

and the resulting Creeper geometry as:

$$G_{\text{creeper}} = [0,h] \times [0,\ell] \times [0,w].$$

With $\ell \approx 2.0$, $h \approx 1.0$, the resulting model appeared upright and columnar.

Visual Mutation and Face Topology

The model was assigned Minecraft's leaf texture. A simple face overlay was defined on the front-facing surface using pixel coordinates:

$$\texttt{eyes}: \quad (x,y) \in \{(3,12),(10,12)\},$$
$$\texttt{mouth}: \quad (x,y) \in \{(5,6),(4,4),(10,4)\}.$$

AI Composition and Swell Behaviour

The Creeper's actions derive from prioritised AI goals:

`Goal 1: LookAtPlayer` (range $R = 6$),
`Goal 2: ApproachPlayer` (speed $v = 0.2$),
`Goal 3: StartSwell` (trigger $r < r_s$),
`Goal 4: Explode` (delay $\tau = 1.5$ s).

Explosion occurs if the player remains within range $r_s = 2.5$ blocks for the full fuse duration. Let $r(t)$ denote distance to the player at time t. Then:

$$\texttt{if}\ \ r(t) < r_s\ \forall t \in [t_0, t_0 + \tau], \texttt{Explode()}.$$

Damage output is modelled by radial falloff:

$$D(x) = \max\left(0, E_{\max}\left(1 - \frac{\|x - x_0\|}{R}\right)\right),$$

where x_0 is the explosion centre and $R \approx 7$ blocks in air.

Reconstruction in TikZ

The diagram below illustrates the result of axis misassignment during model construction. On the left, the intended pig model appears with a horizontal body and frontal facial features. On the right, the same parameters are rendered with the `length` and `height` values swapped. This inversion laid the foundation for the Creeper's distinctive form.

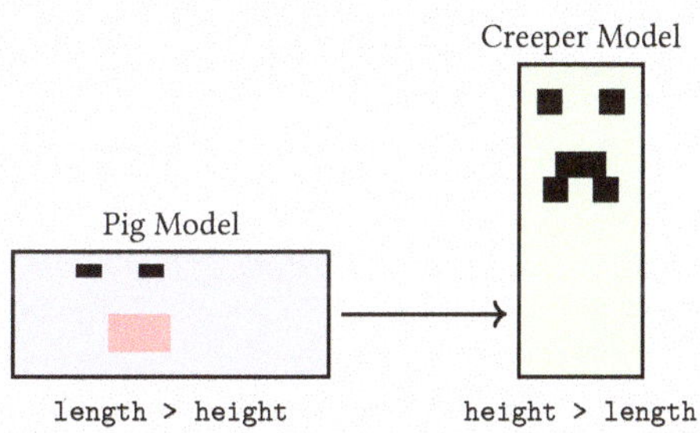

References:

Persson, M. (2009). *Initial Geometry and Entity Source.* Mojang.

Minecraft Wiki. (2025). Entity Models and AI Mechanics. https://minecraft.wiki

A Place at the End of Time

Top (Spacetime Geometry): The geometry near a black hole distorts spacetime so severely that, from a coordinate perspective, the black hole itself lies in the future light cone of any nearby event. For an infalling observer, continuing forward in time inevitably means moving closer to the singularity—tomorrow *is* the black hole in spacetime terms.
Middle (Observer Perspectives): Two perspectives on crossing the horizon. From the distant observer's viewpoint (left), the astronaut appears to freeze at the horizon, increasingly redshifted and dimmed, never quite crossing. From the astronaut's own frame (right), there's no discontinuity—no freezing, no slowdown—only smooth, uninterrupted free fall through the horizon in finite proper time.
Bottom (Penrose Diagram): The full Penrose diagram for the maximally extended Schwarzschild solution, showing all causal regions: the external universe, black hole interior, white hole region, and a second asymptotically flat universe. The diagram captures the global causal structure: horizons, infinities and singularities.

A Place at the End of Time

Black holes create an observational paradox: external observers see infalling objects freeze at the event horizon with infinite redshift, while the falling objects cross in finite proper time experiencing nothing unusual. This contradiction arises from extreme spacetime curvature near the horizon ($r = 2GM/c^2$), where gravitational time dilation becomes unbounded. Inside the horizon, causality inverts—the radial coordinate becomes timelike, making the singularity not a place but a future moment that all trajectories must reach.

Black Hole Event Horizon ∘ Schwarzschild Radius ∘ Time Becomes Spacelike ∘ Singularity as Future ∘ LIGO Gravitational Waves ∘ EHT M87 Image ∘ Coordinate Dependencies ∘ Penrose-Hawking Theorems ∘ Kerr's Quora Objections ∘ White Holes & Wormholes ∘ r Becomes Timelike

"So much universe, and so little time."

— Cohen the Barbarian, AM 1980s

A Place at the End of Time

Karl Schwarzschild derived the first exact solution to Einstein's field equations in 1916, giving rise to what is now known as the Schwarzschild metric. This solution described the spacetime geometry outside a static, spherically symmetric mass, and revealed an intriguing radius where the metric becomes singular—a mathematical curiosity at the time.

In 1939, Robert Oppenheimer and Hartland Snyder explored the gravitational collapse of massive stars showing that under general relativity, such collapse could lead to the formation of a region from which no signals escape: the conceptual precursor to what we now call a black hole.

The term 'black hole' was popularised by John Wheeler in 1967, highlighting that these regions are not conventional objects, but causal domains shaped by the warping of spacetime. In 1963, Roy Kerr discovered an exact solution for rotating black holes. The Kerr metric demonstrated that black holes can possess angular momentum, vastly enriching the theory's physical relevance. Unlike Schwarzschild's static solution, the Kerr geometry features an ergosphere, frame dragging, and a more rich causal structure—including an inner and outer horizon.

Throughout the twentieth century, indirect astrophysical evidence for black holes mounted, from X-ray binaries to quasars and galactic nuclei. By the 2010s, observations of gravitational waves and the first black hole shadow image (captured by the Event Horizon Telescope in 2019) solidified their status as real astrophysical objects.

Despite this progress, black holes continue to raise deep questions. At their core lies the unresolved issue of singularities—regions where classical spacetime is undefined—and the challenge of unifying general relativity with quantum theory remains a central frontier in physics.

General relativity describes gravity as spacetime curvature. Massive bodies distort the geometry in which other bodies move, and free-fall corresponds to inertial motion along geodesics—paths of extremal proper time. This reconceptualization allows for solutions to Einstein's equations that have no Newtonian counterpart. When matter collapses to a sufficiently small region, the curvature becomes extreme enough that no causal signal can propagate outward beyond a critical boundary.

A black hole is defined by geometry. A region of spacetime in which spatial and temporal concepts blend. The defining feature is the event horizon: a null surface that separates regions of spacetime into two domains: those from which future-directed paths can reach infinity, and those from which all such paths terminate inward. The horizon has no surface tension or material properties. Its existence follows purely from the metric. In the Schwarzschild solution, the horizon forms (Schwarzschild, 1916) at radius $r = 2GM/c^2$, where the g_{00} component vanishes and light cones tip inward. Any trajectory, regardless of force or energy, once inside this radius, proceeds inevitably toward smaller r.

Such configurations are predicted results of stellar evolution. When a sufficiently massive star exhausts its nuclear fuel, no internal pressure, thermal, degeneracy, or radiation can oppose further collapse. Neutron stars represent the final stable configuration for masses up to a few solar masses. Beyond that, collapse continues past any known state of matter. General relativity predicts that the outer region smooths into a vacuum solution matching Schwarzschild or Kerr metrics, while the interior (Kerr, 1963) forms a trapped surface with inward-pointing causal futures. The event horizon forms before any singularity becomes visible, preventing external observers from accessing information about the final collapse state.

This scenario was further confirmed in 2015—when LIGO detected gravitational waves from a binary black hole merger. The distortion in spacetime, measured to better than one part in 10^{21}, was generated by two orbiting black holes coalescing into one. The signal matched numerical relativity simulations, confirming the waveform, mass loss, and final ringdown predicted by general relativity. LIGO thus became one of the most sensitive measurement devices ever built, detecting strains comparable to changes smaller than a proton over kilometre-scale arms. The black holes radiated energy equivalent to several solar masses through measurable curvature oscillations.

Other confirmations have followed. The Event Horizon Telescope array imaged the shadow of the supermassive black hole in M87, producing a crescent-shaped brightness profile consistent with light bending and lensing near the photon sphere. Stellar orbit measurements around Sagittarius A* in the centre of the Milky Way reveal elliptical motions governed by a central mass of approximately four million solar masses in a region smaller than the orbits themselves. Accretion disc X-ray emissions, variability timing, and iron line broadening all support the interpretation of compact objects with deep gravitational wells: exhibiting effects that match the metrics of rotating (Kerr) black holes with no observable surface.

As one approaches a black hole, time ceases to behave as expected. The component g_{00} of the spacetime metric determines how proper time accumulates for a stationary observer. In Schwarzschild geometry, $g_{00} = 1 - 2GM/rc^2$ decreases with decreasing radius. A clock closer to the event horizon ticks more slowly relative to one farther away. The gravitational redshift of light signals this disparity—photons emitted near the horizon lose energy as their wavelengths stretch. At the horizon, the redshift becomes unbounded. Signals emitted at or within the horizon do not reach distant observers; emissions from just outside arrive with arbitrarily large delay and redshift.

An object falling into the black hole measures finite proper time to cross the event horizon; locally, nothing singular occurs there (neglecting tidal forces). This dual description, freezing from the outside, flowing from the inside, follows from the coordinate-dependence of simultaneity in general relativity. Infalling observers describe the event horizon as a regular null surface. The difference lies in the slicing of spacetime used to define simultaneity. Proper time and coordinate time diverge in meaning as curvature intensifies.

Inside a black hole (in Schwarzschild coordinates), the radial coordinate behaves timelike—decreasing radius corresponds to forward progression in time—while the temporal coordinate behaves spacelike. In horizon-regular coordinates, this role-swap is recognised as a coordinate effect; causality still directs all future paths toward smaller r.

Outside the horizon, the singularity occupies the spatial point $r = 0$. Inside, it changes from a place to a moment. The question is not 'where is the singularity?' but 'when will I reach it?' The answer: finite proper time ahead, as inevitable as tomorrow. Light cones inside the horizon all tilt toward smaller r, making motion toward the singularity as compulsory as motion into the future. Remaining at fixed radius would require stopping time.

The Penrose–Hawking theorems show that under reasonable energy (Penrose, 1965) and global conditions, spacetimes containing trapped surfaces are geodesically incomplete: certain timelike or null geodesics cannot be extended to arbitrary values of their affine parameter. This geodesic incompleteness—what is meant by a 'singularity' in this context—lies in the future of every worldline that crosses the event horizon in the idealised solutions. The theorems do not by themselves guarantee curvature divergence everywhere; rather, they establish the existence of incomplete causal paths. Thus, you cannot point to the singularity; it is a when, not a where.

The field equations of general relativity are time-symmetric. If the Schwarzschild solution describes an object into which signals can enter but never leave, then its time-reversed counterpart also exists. This reversed solution is called a white hole: a region of spacetime from which causal trajectories can emerge, but into which nothing can be sent. Unlike black holes, white holes cannot be formed dynamically under known physical processes. They appear in maximal analytic extensions (such as Kruskal spacetime) but lack known mechanisms for creation or stability.

Another extension is the wormhole: a spacetime manifold that connects two asymptotically flat regions through a throat. In its simplest form, the Einstein-Rosen bridge arises from a slicing (Einstein & Rosen, 1935) of the maximally extended Schwarzschild geometry. However, the bridge pinches off too rapidly to allow traversal. For a wormhole to be traversable, the geometry must remain open long enough for causal passage. This requires exotic matter: fields or fluids that violate the null energy condition, allowing repulsive gravitational effects. Such matter has not been observed. Moreover, semiclassical analyses suggest instabilities that would disrupt the throat, collapse the tunnel, or generate divergent backreaction.

Black holes, white holes, and wormholes demonstrate surprising configurations of spacetime. Near a black hole, coordinate roles switch, light cones tilt, and the metric enforces trajectories independent of any force. White holes, if they exist, have always existed. Wormholes in general relativity such as the Einstein–Rosen bridge are non-traversable; traversable wormholes would require exotic matter and remain hypothetical.

Kerr's Quora Posts: "Stop Believing Everything You Read About Black Holes"

> Sixty years after discovering the metric that bears his name, Roy Kerr has taken to Quora with the fury of a physicist whose life's work has been misinterpreted. His posts read like manifestos from an exile returning to reclaim his territory. 'Stop believing everything you read about black holes,' he declares, targeting not just popular misconceptions but the physics community.

Kerr's central accusation is that the Penrose singularity theorems prove nothing about physical singularities. What Penrose actually showed was that certain geodesics have finite affine length—they simply end. 'Now, what if the central star is singular?' Kerr asks pointedly. 'Then one is assuming it is singular and there is nothing to prove.' He claims circular reasoning: assume a singularity exists, then 'prove' singularities must exist.

His technical objection cuts directly to the heart of black hole physics. In the Kerr solution, geodesics starting outside can pass through a central neutron star and terminate on the inner horizon on the opposite side. These are Penrose's 'mysterious light rays of finite affine length.' They die not because they hit an infinitely curved singularity, but because they complete their journey through the black hole's interior. Geodesic incompleteness is a boundary condition rather than a catastrophe.

The medium amplifies the message. Quora allows Kerr to bypass peer review and speak directly: 'The trouble with the Penrose paper is that it is a "do it yourself" paper where he states propositions without proving them. This is very typical in relativity [...] conjectures "rule the roost."' These are not the measured tones of academic discourse but the exasperated words of someone watching decades of what he considers misinterpretation compound.

Most provocatively, Kerr disputes the coordinate interpretation that underlies this entire chapter. Asked whether 'time and space exchange roles at the event horizon,' his response is unequivocal: 'Of course this is not true.' The coordinate swap reflects bad coordinate choices, not physics. 'Time is a function defined on a physical manifold with the property that it increases along every world line,' he explains, demolishing the temporal interpretation in a single stroke.

The technical details matter. In Kerr-Schild coordinates, which Kerr considers 'good,' the t-coordinate remains a proper time parameter along all worldlines, never becoming spacelike. The dramatic coordinate inversions described throughout this chapter—r becoming timelike, the singularity becoming temporal—are artefacts of choosing Schwarzschild coordinates, which produce a t-coordinate that 'is not a differentiable function on the manifold.' Use better coordinates, and the mystery vanishes.

Yet Kerr's alternative is equally radical. He describes 'spin forces' that become so intense near the event horizon that infalling objects are forced to rotate around the axis. At the inner horizon, centrifugal forces grow strong enough that objects can move outward again—no longer forced toward any central singularity.

The stakes are higher than academic priority. If Kerr is correct, then black holes are not the temporal futures described in this chapter but something else entirely: regions where extreme spin and gravity create exotic dynamics. His Quora posts are interesting not only for the physics or for the fact that he first solved the stable black hole equations, but also for the platform and direct style of his writing.

It's Ferret Time

A guy's car breaks down on a rural road. No cell phone reception. Fortunately, there's a farmhouse nearby, so he walks over to ask for help.

He rings the doorbell—no answer. But he hears rustling from around the side. He finds a farmer next to a pen with three ferrets.

'Excuse me—' the guy begins, but the farmer cuts in: *'Hold on, I have to feed the ferrets. I'll be with you when I'm done.'*

The guy watches as the farmer picks up one of the ferrets, carries it to an apple tree with a ladder, climbs up while holding it, lifts it to an apple—*chomp*—then climbs back down and returns it to the pen.

'Sorry, I just—' the guy tries again, but the farmer grabs the second ferret. *'Almost done.'*

Same thing: apple tree, ladder, apple, chomp, back down, pen.

Then the third ferret. The farmer lifts it gently, murmurs something to it, and makes the familiar journey. Step by step up the ladder, hoists the ferret to a fresh apple—another bite. He climbs down slowly, cradles it all the way back, nestles it in the pen like a newborn.

Finally, the farmer turns. *'Now—how can I help you?'*

The guy says, *'My car broke down and I need to call AAA. But first... wouldn't it be a lot faster to just pick the apples and toss them into the pen?'*

The farmer pauses, thinking. Then nods. *'Yeah... I suppose it would be faster that way.'*

He shrugs.

'But what's time to a ferret?'

Coordinate Reversal

Schwarzschild Metric and the Horizon

The Schwarzschild metric describes spacetime outside a static mass M. In spherical coordinates (t, r, θ, ϕ), it takes the form (with $r_s = 2GM/c^2$):

$$\begin{aligned} ds^2 = & -\left(1 - \frac{r_s}{r}\right) c^2\, dt^2 \\ & + \left(1 - \frac{r_s}{r}\right)^{-1} dr^2 \\ & + r^2\, d\theta^2 + r^2 \sin^2\theta\, d\phi^2. \end{aligned}$$

At $r = r_s$, $g_{tt} = 0$, $g_{rr} \to \infty$. This signals a breakdown in coordinates, not in geometry.

Time Dilation and Gravitational Redshift

A static observer at fixed radius $r > r_s$ experiences proper time:

$$d\tau = \sqrt{1 - \frac{r_s}{r}}\, dt.$$

As $r \to r_s$, $d\tau/dt \to 0$. Distant clocks appear to tick normally, but local clocks slow near the horizon.

Photons emitted at r_{em} and received at $r_{\text{obs}} \to \infty$ undergo redshift:

$$1 + z = \left(1 - \frac{r_s}{r_{\text{em}}}\right)^{-1/2}.$$

As $r_{\text{em}} \to r_s$, $z \to \infty$. Light from near the horizon becomes infinitely stretched and fades from view.

Finite Infall and Apparent Freezing

A freely falling observer released from rest at radius $r_0 > r_s$ reaches radius $r \leq r_0$ in proper time:

$$\tau(r; r_0) = \frac{2}{3} \frac{r_0^{3/2} - r^{3/2}}{\sqrt{2GM}}.$$

For any finite r_0, the proper time to reach the horizon is finite. The infaller feels no discontinuity; the horizon is not a physical surface.

Even an infaller starting from rest at infinity crosses the horizon in finite proper time, and continues to $r = 0$ also in finite proper time.

However, to a distant observer, the infaller appears to freeze at $r = r_s$, due to the divergence of coordinate time: $t(r) \to \infty \quad \text{as} \quad r \to r_s$.

Coordinate Inversion Below the Horizon

Inside the horizon ($r < r_s$), the signs of metric components reverse in Schwarzschild coordinates: $g_{tt} > 0, \qquad g_{rr} < 0$. In this coordinate choice r behaves timelike and t spacelike; in horizon-regular coordinates (e.g., Eddington–Finkelstein, Kruskal) this interpretation is seen as a coordinate effect while causal futures still point to decreasing r.

The physical implication is that movement in r becomes mandatory. All future-directed timelike paths lead to smaller r, ending at the singularity $r = 0$. The singularity is not 'a place inside' but a moment in the infaller's proper future.

This inversion reflects the geometry. Any attempt to 'hover' or remain at constant r is no longer physically possible once inside the horizon.

Light Cones and Irreversibility

At the horizon, outgoing light rays remain on the surface: $\frac{dr}{dt} = 0$ for outgoing null rays at $r = r_s$.

Below the horizon, all light cones tip inward. Future lightlike and timelike paths are directed toward decreasing r. There is no direction within the cone that leads to increasing radius.

This defines the event horizon as a one-way temporal boundary: a surface from which causal influence cannot escape outward.

References:

Misner, C. W., Thorne, K. S., Wheeler, J. A. (1973). *Gravitation.*

Carroll, S. M. (2004). *Spacetime and Geometry.*

Wald, R. M. (1984). *General Relativity.*

Put on Your 4D Glasses

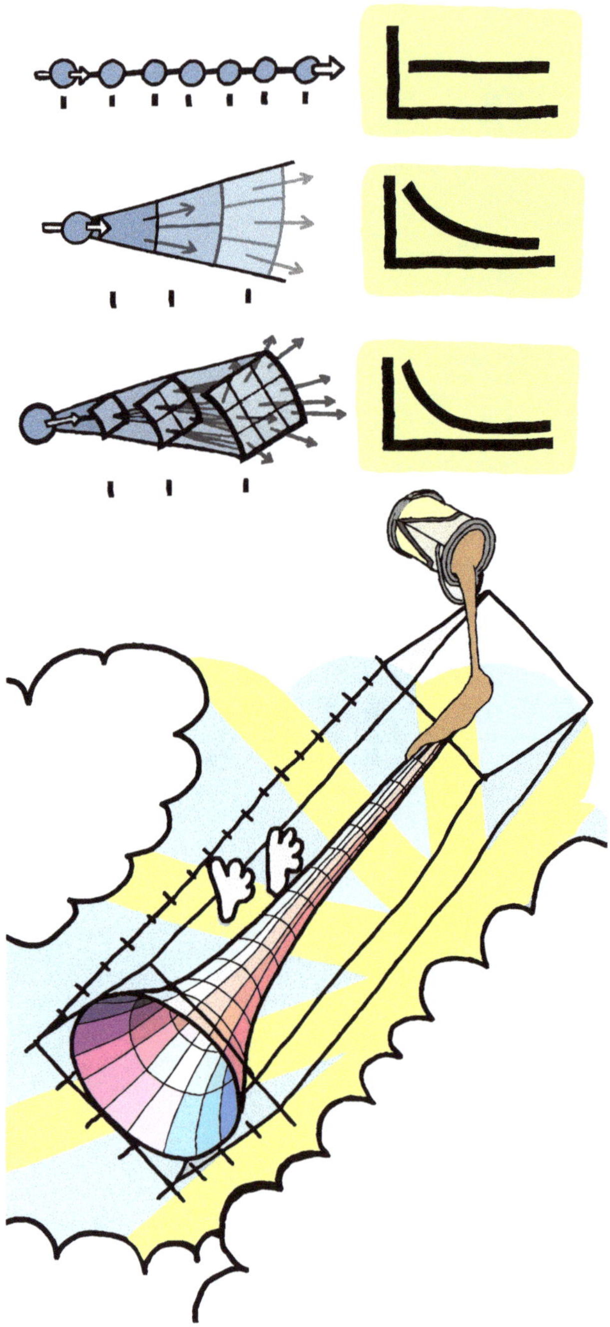

Top (Dimensional Scaling of Inverse Laws): The same point-source strength spreads differently depending on spatial dimension. In 1D, the influence remains constant along a line. In 2D, the effect dilutes like $1/r$ as it spreads over a circle. In 3D, it falls off as $1/r^2$, spreading over the surface of a sphere. This explains why gravitational and electrostatic forces scale as inverse-square laws in 3D space.
Bottom (Gabriel's Horn and the Painter's Paradox): A surface of revolution formed by rotating $y = 1/x$ around the x-axis for $x > 1$. Though the horn extends infinitely, it encloses a finite volume ($\int_1^\infty \pi(1/x)^2\, dx = \pi$) but has infinite surface area ($2\pi \int_1^\infty (1/x)\sqrt{1 + 1/x^4}\, dx = \infty$). Paradoxically, one could 'fill' it with a finite amount of paint, but never coat its inner surface.

Put on Your 4D Glasses

Why does our universe have exactly three spatial dimensions plus time? Multiple independent constraints converge on $D = 4$: only in 3D space do inverse-square laws produce stable planetary orbits and bound atoms; only in 4D spacetime are fundamental forces renormalizable in quantum field theory; only in 4D do waves propagate cleanly without trailing echoes (Huygens's principle). The arithmetic fact that $4 = 2+2$ creates unique mathematical properties—from quaternion algebra to self-dual gauge fields—that cascade through physics. Lower dimensions cannot support complex chemistry, while higher dimensions destabilise matter and causality.

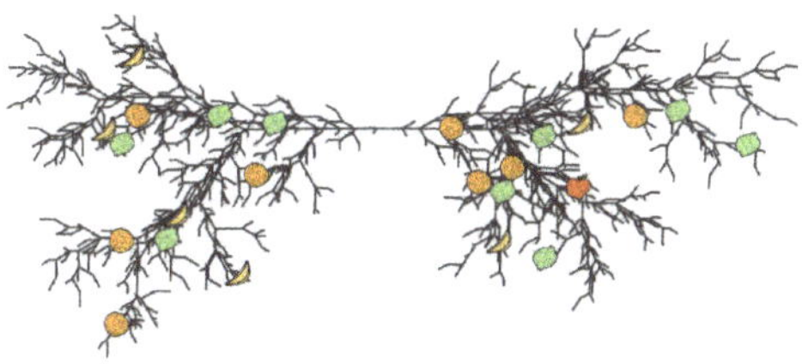

FOUR-DIMENSIONAL SPACETIME ◦ INVERSE-SQUARE FORCE LAW ◦ STABLE ORBITS ONLY $n = 3$ ◦ HUYGENS'S PRINCIPLE ◦ QFT RENORMALIZABILITY ◦ EXOTIC $\mathbb{R}^4$ TOPOLOGY ◦ DIVISION ALGEBRAS ◦ $4 = 2 + 2$ IDENTITY ◦ BLACK HOLE NO-HAIR ◦ ATOMIC STABILITY ◦ CHEMICAL BONDING

"The miracle of the appropriateness of the language of mathematics
for the formulation of the laws of physics
is a wonderful gift which we neither understand nor deserve."

— Eugene Wigner, 1960

Put on Your 4D Glasses

The idea that the dimensionality of physical space might be constrained by necessity predates the formal development of modern physics. Gottfried Wilhelm Leibniz suggested in the *Discourse on Metaphysics* (1686) that the actual world should be understood as the one 'simplest in hypotheses and richest in phenomena,' implicitly framing dimensionality as subject to selection principles. In the eighteenth century, Immanuel Kant proposed that Newton's inverse-square laws implied the three-dimensionality of space, although this causal inference would later be reversed—the inverse-square law follows from spatial geometry via Gauss's theorem, not the reverse.

A rigorous analytical approach began with Paul Ehrenfest's seminal 1917 paper 'In what way does it become manifest in the fundamental laws of physics that space has three dimensions?' He demonstrated that classical orbit stability requires exactly three spatial dimensions. In higher spatial dimensions ($N > 3$), the effective gravitational potential falls off too rapidly to maintain closed, bounded orbits; in lower dimensions, the dynamics become pathologically confined. Separately, the Huygens principle for the wave equation holds only in odd spatial dimensions $n \geq 3$, so $(3, 1)$ spacetime is the lowest-dimensional case with sharp wavefronts and no interior tails.

Gerald Whitrow's influential 1955 paper "Why Physical Space has Three Dimensions" marked a turning point by explicitly connecting dimensional constraints to the possibility of life and observers. He argued that intelligent life capable of formulating physics could only arise in three spatial dimensions—an example of anthropic reasoning. Whitrow systematically examined how communication, neural networks, and information processing would fail in spaces of different dimensionality, establishing that the question 'why three dimensions?' might be answered by 'because otherwise we wouldn't be here to ask.'

Freeman Dyson and Andrew Lenard's 1967 theorems on the stability of matter established that systems of electrons and nuclei interacting via Coulomb forces are stable of the second kind in three spatial dimensions. In higher spatial dimensions, Coulomb interactions scale differently and can lead to instabilities; in two or one spatial dimensions the Coulomb law changes and the stability analysis requires separate arguments.

Four-dimensional spacetime with a metric of signature $(3, 1)$ satisfies multiple independent physical constraints that fail in other dimensionalities. The signature means the metric has three positive and one negative eigenvalue, but this does not single out a canonical 'time direction'—timelike vectors form an open cone, and one can choose bases in which all four basis vectors are lightlike. The physical content lies in the signature, not in any preferred decomposition. Gravitational orbits destabilise, many familiar interactions become non-renormalizable for $D > 4$ (and super-renormalizable for $D < 4$), wave propagation develops trailing echoes, and atomic structures can fail to be stable when $D \neq 4$. Multiple time-like dimensions destroy the well-posedness of the initial value problem for hyperbolic differential equations such as the wave equation, rendering physics unpredictable, and generically spoil causality and energy positivity.

In classical potential theory, the spatial decay of fields from a point source—or any spherically symmetric mass, regardless of its internal dimensionality—follows a general scaling law determined by Gauss's theorem: the flux through a sphere in n spatial dimensions scales with its surface area, yielding a radial dependence of $1/r^{n-1}$ for the field and $1/r^{n-2}$ for the associated potential. In $n = 3$ this produces the inverse-square law that governs Newtonian gravity (Newton, 1687) and electrostatics. This particular falloff enables stable bound orbits under central forces, since it balances centripetal acceleration with potential curvature. In $n > 3$, the force falls too quickly to support closed, radially stable Kepler orbits; in $n < 3$, the dynamics cease to admit Kepler-type closed, radially stable motion.

Wave propagation obeys Huygens's principle (Huygens, 1690) only in odd spatial dimensions $n \geq 3$. In $(3, 1)$ spacetime, a localised disturbance generates a sharp spherical wavefront without trailing components. In even spatial dimensions, persistent field residuals remain after the main wave passes, blurring temporal boundaries between cause and effect as the Green's function of the wave equation has tails inside the light cone.

Quantum field theory imposes stringent dimensional restrictions on interaction consistency. Renormalizability—the ability to absorb divergences into a finite set of physical parameters—depends on the dimensional scaling of coupling constants. In $D = 4$, key interactions such as ϕ^4 theory, quantum electrodynamics (QED), and non-abelian gauge theories feature dimensionless couplings, rendering loop corrections manageable via renormalization group techniques. In $D > 4$, the same interactions become non-renormalizable, requiring an infinite tower of counterterms. In $D < 4$, they become super-renormalizable.

The manifold $\mathbb{R}^4$ exhibits an anomaly in differential topology: it admits uncountably many smooth structures that are pairwise non-diffeomorphic yet topologically equivalent. For $\mathbb{R}^n$ with $n \neq 4$, the smooth structure is unique—smoothness and topology coincide. (For general manifolds, distinct smooth structures on the same topological manifold, and topological manifolds admitting no smooth structure at all, were already known before exotic $\mathbb{R}^4$.) The exotic $\mathbb{R}^4$ phenomenon is intertwined with deep four-dimensional phenomena revealed by gauge theory, notably Donaldson invariants and Seiberg–Witten theory; the smooth 4D Poincaré conjecture itself remains open.

In the algebraic classification of normed division algebras over $\mathbb{R}$, there exist only four: $\mathbb{R}$ (dimension 1), $\mathbb{C}$ (dimension 2), $\mathbb{H}$ (dimension 4), and $\mathbb{O}$ (dimension 8). Of these, the quaternions $\mathbb{H}$ preserve associativity while the octonions $\mathbb{O}$ do not. They form the algebraic underpinning of spinor representations and enable the group isomorphism $\mathrm{SU}(2) \cong \mathrm{Spin}(3)$, which double-covers the rotation group $\mathrm{SO}(3)$. This supports the representation theory of spin-$\frac{1}{2}$ particles and the construction of Dirac spinors. No higher-dimensional associative division algebra exists, and the non-associativity of octonions complicates their use in comparable representation frameworks, though they underlie exceptional Lie groups (e.g., G_2, F_4, E_6–E_8).

The necessity of spinors in four dimensions emerges from a tension between quantum mechanics and relativity. Schrödinger's equation is linear in time derivatives, while relativistic energy obeys $E^2 = p^2c^2 + m^2c^4$, quadratic in energy. Dirac sought to linearize the wave operator (Dirac, 1928)—to extract a 'square root' of the d'Alembertian $\Box = \partial_t^2 - c^2\nabla^2$. Just as $x^2 + y^2$ cannot be factored into $(ax + by)$ using real numbers, this operator resists scalar factorization. The solution requires anticommuting coefficients, matrices γ^μ satisfying $\{\gamma^\mu, \gamma^\nu\} = 2\eta^{\mu\nu}$. In four-dimensional spacetime, the minimal representation uses 4×4 matrices, forcing the wavefunction to be a four-component spinor rather than a scalar. The four components do not represent spatial directions, but two particle states and two antiparticle states, each with two spin orientations. Antimatter is a result of this mathematical necessity from the requirement to linearize energy in $D = 4$ spacetime. The restriction to three spatial dimensions is important: the rotation group SO(3) is unique in admitting a double cover $\text{Spin}(3) \cong \text{SU}(2)$ that links vector and spinor representations through this square-root relationship.

In general relativity, the uniqueness of black hole solutions—encapsulated by the no-hair theorems—holds in four-dimensional, asymptotically flat spacetime under suitable regularity and symmetry assumptions. Theorems by Israel, Carter, and Robinson prove that stationary black holes in $D = 4$ are characterised entirely by mass, charge, and angular momentum. In higher dimensions, this rigidity fails. New solutions emerge with toroidal or ring-like horizons, including black rings and black strings, and the solution space displays richer phases.

The quantum mechanical stability of atomic matter depends sensitively on the spatial dimension n. For hydrogen-like atoms with a $1/r^{n-2}$ potential when $n \geq 3$, the familiar Coulombic spectrum arises for $n = 3$. In $n > 3$, the potential decays too rapidly to maintain binding; in $n = 2$, the potential becomes logarithmic. Chemical bonding patterns also require three dimensions. Tetrahedral carbon and chiral centres depend on SO(3) symmetry. In $n = 2$, bonding is planar and chirality is lost; in $n > 3$, additional rotational degrees of freedom would alter biochemical recognition.

I recommend watching Mikhail Gromov's lecture on the topic (minute 19:30 in the video titled "What is a Manifold? - Mikhail Gromov") where Gromov traces the exceptional nature of four dimensions to the arithmetic identity $4 = 2+2$. A four-element set partitions into two pairs in exactly three ways: $\{\{1,2\},\{3,4\}\}$, $\{\{1,3\},\{2,4\}\}$, and $\{\{1,4\},\{2,3\}\}$. The number of such partitions for a set of size $2n$ is $(2n-1)!! = (2n-1)(2n-3)\cdots 3 \cdot 1$. For $n = 2$: three partitions. For $n = 3$: fifteen partitions. For $n = 4$: one hundred and five partitions. Only when $n = 2$ does the partition count (3) remain smaller than the set size (4), enabling the symmetric group S_4 to map onto the smaller group S_3.

For odd-sized sets, no symmetric pair partitions exist. Only the four-element set achieves both symmetry (all parts equal) and economy (partition count smaller than set size).

This homomorphism $\varphi : S_4 \to S_3$ tracks how permutations of four elements permute the three partitions. Its kernel contains precisely those permutations preserving all partitions: the Klein four-group $V_4 = \{e, (12)(34), (13)(24), (14)(23)\}$. This renders A_4 non-simple; indeed A_n is simple for all $n \geq 5$, making A_4 the unique non-abelian, non-simple alternating group.

This manifests in Lie theory through the decomposition $\text{SO}(4) \cong (\text{SU}(2) \times \text{SU}(2))/\mathbb{Z}_2$, splitting the Lie algebra $\mathfrak{so}(4) \cong \mathfrak{so}(3) \oplus \mathfrak{so}(3)$. This corresponds to decomposing 2-forms into self-dual and anti-self-dual components: $\Lambda^2(\mathbb{R}^4) = \Lambda^2_+ \oplus \Lambda^2_-$, where the Hodge star operator satisfies $\star^2 = 1$ on oriented Riemannian 4-manifolds (and $\star^2 = -1$ on 2-forms in Lorentzian signature), yielding eigenspaces with eigenvalues ± 1 in the Euclidean case.

In gauge theory, this splitting transforms second-order Yang–Mills equations into first-order conditions (Yang & Mills, 1954). A connection with curvature F satisfying $F = \star F$ (self-dual) or $F = -\star F$ (anti-self-dual) automatically solves $D \star F = 0$ since the Bianchi identity guarantees $DF = 0$. This dimensional coincidence–that the electromagnetic field strength is a 2-form and its Hodge dual has the same rank–makes Maxwell's equations acquire their most natural geometric expression in four dimensions. Here the equation $F = \pm \star F$ for 2-forms is specific to four dimensions; higher-dimensional analogues (e.g., G_2-instantons and Spin(7)-instantons) exist but differ in structure.

Donaldson's theorem, a hallmark of four-dimensionality (Donaldson, 1983), uses this. The moduli space of anti-self-dual connections on a 4-manifold yields polynomial invariants distinguishing smooth structures. Two homeomorphic 4-manifolds may have different Donaldson invariants, proving they are not diffeomorphic–a phenomenon occurring in no other dimension. The identity $4 = 2 + 2$ enables the entire apparatus.

So the identity $4 = 2 + 2$ creates the alternating group exception, the Lie algebra splitting, the self-duality decomposition, and the instanton solutions that distinguish four-dimensional gauge theory. Stable orbits, renormalizable interactions, exotic smooth structures–these may indeed stem from this single combinatorial fact. The most sophisticated features of our universe follow from the simplest patterns in the integers.

For more details on the underlying mathematics, I recommend the book *The Wild World of 4-manifolds* by Alexandru Scorpan (2005).

Why Four Might Be 'Special'

The convergence of independent constraints—orbit stability, renormalizability, Huygens principle, division algebras, gauge theory, the arithmetic identity $4 = 2 + 2$—all pointing to four dimensions invites three interpretations. First, four may encode a fundamental truth of geometry, where mathematical coherence uniquely selects this dimensionality as the only one supporting complex, predictable structures. Second, the apparent necessity may reflect selection bias: we observe four dimensions because observers can only arise where physics permits stable atoms and chemistry, rendering our conclusion inevitable yet uninformative about whether other dimensionalities 'exist' in some broader sense. Third, the entire exercise may be backfitting logic to a random parameter—finding post-hoc explanations for an arbitrary feature of our universe, mistaking coincidence for profundity.

The proliferation of independent mathematical arguments favouring four suggests the first interpretation, yet the arguments themselves presuppose frameworks (differential geometry, quantum field theory, group representation theory) constructed within and calibrated to a four-dimensional universe. Whether these constraints reveal something deeper or merely echo the assumptions embedded in our theories remains unresolved.

Exercise: Pythagorean Theorem via Dimensional Analysis

Derive $a^2 + b^2 = c^2$ using a dimensional argument involving area scaling.

1. **Scaling Argument**
   A right triangle is uniquely determined by its hypotenuse c and one acute angle ϕ. The only dimensionally valid expression for its area is

$$\Delta = c^2 f(\phi),$$

   where $f(\phi)$ is a dimensionless function. *(Why c rather than a or b? By similarity, the shape depends on ϕ and a squared length. Any side could be used, but expressing the result in terms of c emphasises the relation to the hypotenuse.)*

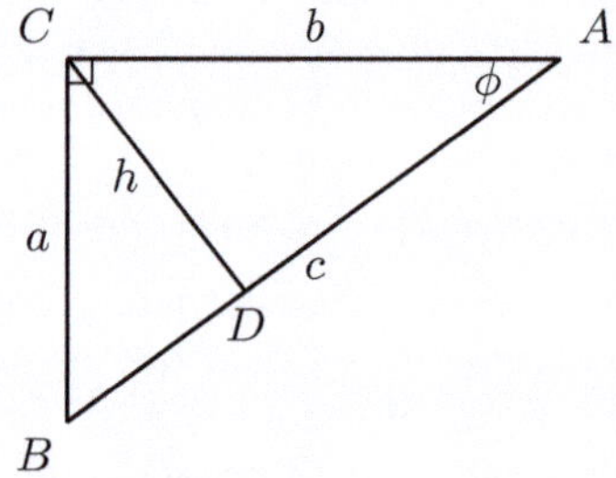

2. **Similar Triangles and Decomposition**
   The altitude CD from C to AB divides the original right triangle $\triangle ABC$ into two smaller right triangles, $\triangle CBD$ and $\triangle ACD$. By angle-angle (AA) similarity:

$$\triangle ABC \sim \triangle CBD \sim \triangle ACD.$$

   Because area scales as the square of a characteristic length, the sub-areas (corresponding to sides a or b) take the form $\Delta_1 = a^2 f(\phi)$ and $\Delta_2 = b^2 f(\phi)$. Since the total area is additive,

$$c^2 f(\phi) = a^2 f(\phi) + b^2 f(\phi).$$

   Dividing by $f(\phi)$ (assumed nonzero) gives $c^2 = a^2 + b^2$.

Dimensional analysis can be rigorously formalised using a graded algebraic structure where physical quantities lie in one-dimensional vector spaces V^d indexed by a group G of exponents (e.g., $\mathbb{R}^n$ for base units). Multiplication corresponds to the tensor product $V^a \otimes V^b \cong V^{a+b}$, and one only adds elements within the same space V^d. For example, a quantity in kilograms lies in $V^{(1,0,0,\ldots)}$ and one in metres per second lies in $V^{(0,1,-1,\ldots)}$, reflecting mass and velocity dimensions respectively. Physical laws hold under any consistent rescaling of units, leading to constraints like Buckingham's π-theorem. For details, see Tao (2012): https://terrytao.wordpress.com/2012/12/29/a-mathematical-formalisation-of-dimensional-analysis/

Dimensional Force Laws and Renormalizable Interactions

Notation: n denotes spatial dimensions; D denotes spacetime dimension ($D = n+1$ unless stated).

Flux Argument and $r^{-(n-1)}$ Fields. In n spatial dimensions, a spherical surface at radius r has 'area' scaling as r^{n-1}. For a source at the origin emitting flux uniformly, Gauss's law implies that flux per unit area decreases in proportion to $1/r^{n-1}$. Classical gravitational or electrostatic fields thus follow $F \sim \frac{1}{r^{n-1}}$. For $n = 3$, this becomes the familiar inverse-square relation. The corresponding potential $V(r)$ integrates (away from $n = 2$) as.

$$V(r) \sim \int \frac{dr}{r^{n-1}} \approx r^{2-n}.$$

When $n = 3$, $V(r) \sim 1/r$. For $n = 2$, $V(r) \sim \log r$.

Stable Orbits in Three Dimensions. A $1/r$ potential in $n = 3$ produces near-circular orbits that are stable under perturbations. Small changes in velocity cause bounded oscillations rather than catastrophic collapse or unbounded escape. In $n < 3$, forces decay more slowly (logarithmically at $n = 2$), creating strong long-range effects that disrupt stable Keplerian orbits. In $n > 3$, forces diminish rapidly, so small perturbations can disorder the trajectories.

From Classical to Quantum. This dimensional dependence also occurs in quantum physics. Atomic stability relies partly on the $1/r$ Coulomb potential in $n = 3$. In $n = 2$, the potential becomes logarithmic with qualitatively different bound states; for $n > 3$, the faster falloff reduces binding and can eliminate it at comparable scales.

Renormalizable Couplings. Quantum field theories (QFTs) further illustrate how dimensionality restricts allowed interactions. Consider a scalar field ϕ in D-dimensional *spacetime.* The ϕ^4 interaction $\mathcal{L}_{\text{int}} = \lambda\,\phi^4$ requires λ to be dimensionless or of non-negative mass dimension to avoid an infinite series of divergences. The mass dimension of ϕ is $[\phi] = (D-2)/2$, so

$$[\lambda] = D - 4[\phi] = 4 - D.$$

In $D = 4$ spacetime dimensions (signature $(3,1)$), λ is marginal (dimensionless). At $D > 4$, λ becomes irrelevant at high energy: the theory is non-renormalizable, demanding new terms for each new order in perturbation theory. For $D < 4$, the interaction is super-renormalizable with strong infrared effects.

Gauge Fields and Anomalies. In signature $(3,1)$, Yang–Mills gauge couplings are dimensionless, and the CP-odd topological density

$$\tfrac{1}{8\pi^2}\,\mathrm{tr}\,F \wedge F = \tfrac{1}{32\pi^2}\,\epsilon^{\mu\nu\rho\sigma}\,\mathrm{tr}\big(F_{\mu\nu}F_{\rho\sigma}\big)\,d^4x$$

integrates to an integer (the second Chern number) on compact 4-manifolds, independent of matter content. Gauge and mixed anomalies arise from chiral fermions; cancellation imposes constraints on representations, e.g., $\sum_{\text{fermions}} \mathrm{Tr}_R(T^a\{T^b, T^c\}) = 0$, where $\{T^b, T^c\} := T^bT^c + T^cT^b$ is the anticommutator (Jordan bracket) of the generators.

Analogous anomaly phenomena exist in other even spacetime dimensions, but renormalizable chiral gauge theories with dimensionless couplings occur naturally in signature $(3,1)$.

References:

Peskin, M. E., and Schroeder, D. V. (1995). *An Introduction to Quantum Field Theory.* Addison-Wesley.

Weinberg, S. (1995). *The Quantum Theory of Fields, Vol. I.* Cambridge University Press.

Ehrenfest, P. (1917). *In what way does it become manifest in the fundamental laws of physics that space has three dimensions?* Proceedings of the Amsterdam Academy.

Lenard, A., and Dyson, F. J. (1967). Stability of matter I, II. *Journal of Mathematical Physics.*

Let There Be Biolumines-cence

Top (Molecular Ångström scale, reaction begins): At the Ångström scale, the luminescent reaction begins: luciferin is enzymatically converted into luciferyl-adenylate, then oxidised into oxyluciferin. The process releases a visible photon ($h\nu$), producing light.
Second (Enzyme Active Site): At nanometre scale, the luciferase enzyme forms a pocket where the reaction occurs. This active site holds luciferin in the correct orientation, enabling efficient light-producing catalysis.
Third (Subcellular Photocyte): At the 1–10 μm scale, we zoom into a photocyte—a specialised light-producing cell. Organelles such as mitochondria and endoplasmic reticulum support energy-demanding luminescence.
Fourth (Cell/Tissue Cross-Section): At the 10–100 μm scale, we see a full photocyte embedded in a patterned tissue. The densely packed spheres represent light-emitting units; organisation optimises light output and diffusion.
Fifth (Lantern Organ): At the 1–5 mm scale, the firefly's lantern organ is visible in cross-section. Photocytes, tracheal tubes (for oxygen), and reflector layers are arranged to maximise brightness and directionality.
Sixth (Whole Firefly): At the centimetre scale, we see the full insect with its ventral lantern exposed.
Bottom (Population-Level Output): At the metre scale, we zoom out to an ecological view: a jar of approximately forty fireflies generates light that *seems* equivalent to a candle.

Let There Be Bioluminescence

Firefly flashes demonstrate biology's hierarchical organisation from ecosystems to quantum mechanics. Species-specific flash patterns enable mate recognition and, in tropical swarms, synchronous displays visible across forests. Neural circuits generate these patterns by controlling oxygen flow through tracheal valves to specialised photocytes. Within these cells, luciferase catalyses luciferin oxidation with extreme efficiency, converting chemical energy to light with minimal heat. The photons themselves arise when excited electrons in oxyluciferin transition between quantum energy states, emitting at 560–590 nm.

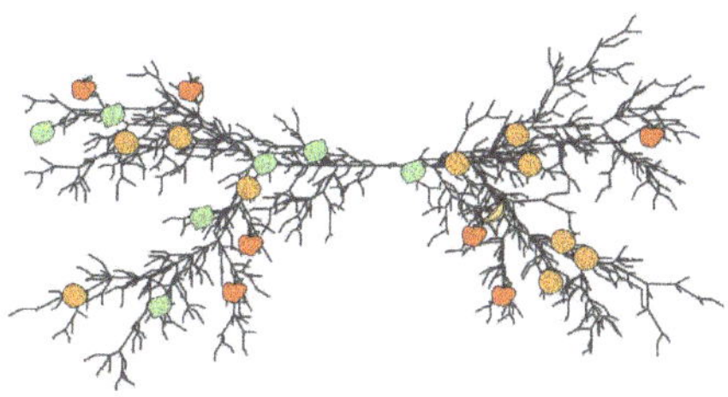

Firefly Flash Patterns ∘ Luciferin-Luciferase Reaction ∘ Quantum Photon Emission ∘ Species-Specific Signals ∘ Flash Synchrony ∘ Photocyte Structure ∘ Oxygen Control ∘ 560nm Yellow-Green ∘ Photuris Mimicry ∘ Quantum Efficiency ∘ Transdisciplinary Cascade

« Arithmétique! algèbre! géométrie! trinité grandiose!
triangle lumineux! Celui qui ne vous a pas connues est un insensé! »

("Arithmetic! Algebra! Geometry! Grandiose trinity!
Luminous triangle! Whoever has not known you is without sense!")
— Comte de Lautréamont, 1869

Let There Be Bioluminescence

Fireflies have intrigued observers for millennia, with their rhythmic flashes illuminating summer landscapes and inspiring both folklore and scientific enquiry. Systematic investigation dates to the late nineteenth century: Raphael Dubois (1887–1889) established the luciferin–luciferase system and showed that oxygen is required, coining the modern terminology. In the early twentieth century, E. Newton Harvey synthesised and advanced the field, culminating in his 1920 monograph *The Nature of Animal Light*, which framed bioluminescence as a distinct physiological phenomenon. Herbert Ives and William Coblentz (1924) performed early quantitative brightness comparisons using photographic plates and carbon glowlamp standards, though their methods lacked the precision of modern spectroscopy.

By the 1950s and 1960s, researchers succeeded in isolating the key biochemical components: the substrate D-luciferin, the energy carrier ATP, and the enzyme luciferase. These breakthroughs enabled direct experimentation on the reaction mechanism and launched decades of transdisciplinary work. Molecular biologists traced the genetic regulation of luciferase expression; biochemists elucidated its adenylation and oxidation kinetics; and physicists modelled the quantum transitions responsible for photon emission.

The behavioural and ecological dimensions developed in parallel. John and Elisabeth Buck documented synchronous flashing in Southeast Asian fireflies, establishing the field of collective rhythmic behaviour. James Lloyd systematically catalogued flash patterns across North American species and discovered aggressive mimicry in *Photuris*. Sara Lewis examined sexual selection and the evolution of courtship signals. Lynn Faust combined citizen science with field observation to document firefly diversity and decline across temperate regions. More recently, Timothy Fallon and colleagues have applied molecular and chemical tools to probe the basis of bioluminescence and lantern development. The luciferase–luciferin system became both a model for energy conversion in biological systems and a ubiquitous reporter in molecular biology.

Fireflies emit patterned flashes of visible light during twilight hours to communicate species identity and reproductive readiness. These luminous signals, typically observed during summer evenings, are not random glows but pulse sequences that vary in duration, frequency, and rhythm across species. Such optical signalling plays a central role in sexual selection, enabling individuals to locate and identify conspecific mates in low-light environments.

These flash sequences are highly stereotyped within each species, often involving precise intervals between pulses and complex rhythms. In temperate species such as *Photinus pyralis*, the male executes a repeated J-shaped flight pattern accompanied by regularly spaced flashes, while the female responds with a delayed flash after a fixed interval, forming a dialogue. Flash timing is governed primarily by neural control and oxygen gating in the lantern, including nitric-oxide–mediated regulation of tracheal oxygen delivery, rather than by the luciferase gene itself. Genetic variation in luciferase and its regulatory elements tunes emission colour and expression levels, but courtship rhythms arise from the nervous system.

In many species, males fly and emit sequences of flashes while females respond with stationary signals, enabling pairwise courtship matching. The spatial separation between signaler and responder allows females to remain camouflaged and evaluate male signals from a protected location.

Some tropical firefly species exhibit large-scale flash synchrony, with entire swarms blinking in phase over rivers and forest canopies. This phenomenon, documented in Southeast Asia and the Amazon basin, represents one of the most visually striking examples of collective animal behaviour. Each firefly possesses neural circuits that integrate visual input with motor output, enabling the organism to modulate its own flashing in response to others. The synchrony emerges from local coupling: individuals respond to neighbours' flashes with subsecond delays, creating active neuronal entrainment and feedback across the swarm. Mathematical models of pulse-coupled oscillators successfully reproduce (Mirollo & Strogatz, 1990) the observed dynamics, illustrating how group coherence emerges from individual rules of phase adjustment.

Bioluminescent flashes in fireflies serve not only for courtship but also as aposematic (warning) signals. Predators such as spiders and bats learn to associate the light with unpalatability, as many fireflies produce toxic compounds such as lucibufagins. Thus, the glow acts both as an attractant for mates and as a deterrent to would-be predators, serving dual evolutionary functions.

Interspecific mimicry has evolved in some lineages, where predatory fireflies imitate female flash codes to attract and consume males of other species. This form of aggressive mimicry (Lloyd, 1965), seen in certain *Photuris* species, exploits the flash code communication to lure unsuspecting *Photinus* males.

Firefly light is produced in abdominal lanterns composed of specialised cells called photocytes embedded within a reflective cuticular matrix. These lanterns are located on the ventral surface of abdominal segments and form discrete light-emitting organs. This positioning maximises outward light projection and prevents internal scattering.

These cells contain high concentrations of luciferase enzyme and are packed into layered structures that direct light outward. Each photocyte expresses the luciferase gene at levels 1000-fold higher than housekeeping genes, driven by lantern-specific transcription factors. The 550-amino acid luciferase protein accumulates to high micromolar concentrations in lantern photocytes.

The photocytes are organised into sheets interspersed with tracheoles and backed by a reflective layer of uric acid microcrystals. This photonic layer channels photons toward the exterior and prevents absorption by internal tissues, increasing luminous efficiency. Comparative studies show that species with more crystalline layers produce brighter signals for equivalent biochemical activity.

Oxygen is delivered via a dense network of tracheoles terminating at the photocyte surface, enabling rapid flash onset and cessation. The respiratory system in insects, based on direct gas exchange through branching air tubes, allows localised control of oxygen concentration. The firefly actively modulates tracheal valve opening to regulate oxygen diffusion, synchronising flash timing with behavioural context. This mechanism enables the rapid on-off cycling necessary for patterned flashes.

ATP is synthesised locally in photocytes via mitochondrial respiration, providing the necessary energy for the light-producing reaction. These mitochondria are spatially arranged near the luciferin–luciferase complexes to facilitate substrate delivery.

Bioluminescence in fireflies is produced by the enzyme-catalysed oxidation of D-luciferin in the presence (Dubois, 1887) of ATP, oxygen, and magnesium ions. The reaction occurs within peroxisomes in the photocytes, where all reactants are present in high concentration. The catalytic role of luciferase is central to determining efficiency and spectral output.

The reaction proceeds through a luciferyl-adenylate intermediate, followed by oxygen insertion and the formation of an excited oxyluciferin molecule. This intermediate is stabilised within the enzyme pocket, aligning the substrates to favour productive reaction pathways. The excited-state product is a singlet species with sufficient lifetime to allow radiative decay.

As oxyluciferin relaxes to its ground state, it emits a photon of visible light, typically in the yellow-green spectrum. The emission spectrum peaks around 560–590 nm for most *Photinus* species, matching the visual sensitivity range of nocturnal insects and vertebrates.

The photon-emission efficiency is typically around 40% to 60%, categorising firefly light as one of the most energy-efficient biological emissions known. Unlike incandescent or fluorescent lighting, the reaction generates minimal thermal energy and proceeds near ambient temperature. This 'cold light' property results from direct chemical-to-photon energy conversion.

The spectral output varies among species through mutations in the luciferase gene. Across beetle lineages the peak emission typically spans roughly 540–590 nm, and single amino-acid substitutions near the active site can shift the spectrum by approximately 10–20 nm. Such substitutions alter hydrogen-bonding networks around oxyluciferin. Natural selection has tuned each species' emission to match the visual sensitivity of conspecific photoreceptors.

pH, temperature, and ionic strength of the cellular milieu influence the excited-state energetics and thus shift the emission spectrum. The enzyme shape responds to environmental cues, subtly altering binding site geometry and solvent accessibility. Controlled experiments confirm that alkaline conditions favour blue-shifted emission.

When oxyluciferin forms in its excited state, the energy from the chemical reaction places an electron in a higher orbital. The molecule is now in an excited singlet state—metastable, persisting for nanoseconds before the electron drops back down. That drop releases the energy difference as a single photon. This is direct chemical-to-photon conversion: the oxidation energy becomes light without passing through heat.

This distinguishes bioluminescence from incandescence. A hot filament emits a broad Planck spectrum because thermal energy randomly excites many transitions. Oxyluciferin emits a narrow spectral band centred at 560 nm because only one specific electronic transition is accessible from the reaction. The chemical pathway selects the quantum state; the quantum state determines the photon energy; the photon energy fixes the colour.

Molecular conformation controls emission colour with nanometre precision. A twist in oxyluciferin's thiazole ring shifts the spectrum by 10 nm. A hydrogen bond from a nearby amino acid pushes it another 5 nm. Water molecules penetrating the active site can blue-shift emission by 20 nm. Each species has evolved a specific constellation of these effects, encoded in luciferase's amino acid sequence, to produce its characteristic hue.

The luciferase protein scaffold modulates the electronic structure of the reaction complex by stabilising specific orbital configurations. Active site residues create an electrostatic environment that shapes the electron density distribution.

The same physics governs LEDs, laser dyes, and firefly lanterns. In gallium arsenide semiconductors, electrons fall across a bandgap of 1.4 eV, emitting infrared. In rhodamine dyes, π-electron systems with conjugated bonds set gaps around 2.1–2.2 eV, yielding yellow–orange fluorescence. In oxyluciferin, a heterocyclic structure with sulphur and nitrogen atoms creates a gap of 2.2–2.3 eV, producing yellow-green.

This phenomenon exemplifies a continuous causal cascade that spans many scales of scientific enquiry. A courtship behaviour, encoded in species-specific flash patterns, originates in neural control of oxygen delivery to abdominal lanterns. That delivery regulates a biochemical cycle shaped by gene expression, enzyme structure, and intracellular energetics. The emitted light originates in electronic transitions within oxyluciferin—transitions governed by quantum orbital energetics and subject to selection rules derived from quantum mechanics. The same principles used to model LEDs, lasers, and atomic emission lines apply to a flash in the grass.

Transdisciplinary Numbers

When I first explored this topic, I started with a bottom-up calculation from first principles. A firefly lantern contains roughly 10^5 photocytes, each expressing about 10^6 luciferase molecules (micromolar concentrations in specialised cells). The quantum yield of the reaction—the fraction of excited oxyluciferin molecules that emit photons rather than dissipating energy as heat—is ~ 0.41 for beetle luciferases under physiological conditions. The critical constraint is oxygen delivery: Timmins et al. (2001) showed that flashes terminate via oxygen depletion when tracheoles constrict. With an effective in vivo turnover of only ~ 0.01 reactions per enzyme per second (far below the 1–2 s^{-1} maximum measured in vitro) and a 250 ms flash, this gives a biochemical budget of order 10^8–10^9 photons per flash, rising to $\sim 4 \times 10^{10}$ only under burst-like, one-turnover-per-enzyme conditions.

Standing against this was the canonical Ives and Coblentz (1924) figure—widely paraphrased as "1/40 candlepower"—which, when converted through modern photometric standards, implies $\sim 3 \times 10^{14}$ photons per flash. That mismatch of three to four orders of magnitude prompted a closer look. Re-reading Coblentz's 1912 monograph shows that his *Photinus pyralis* measurements actually ranged from 1/50 to 1/400 candle, with 1/400 predominating, and that visual nulling photometry likely matched peak rather than time-integrated intensity and assumed isotropic emission from a source that in reality beams light ventrally into only $\sim$ 1–2 steradians.

In 2025 I finally tested the numbers directly. We pointed a calibrated lux metre at individually isolated fireflies (likely *Pteroptyx* species) at distances of 1–5 cm. Peak flashes of 0.2–0.5 lux at 1–2 cm, converted to photon flux with the same radiometric machinery and corrected for the lantern's geometry, yielded 10^{10}–4×10^{11} photons per flash. I also reached out to experts: Dr. Timothy R. Fallon, a firefly photobiologist at Scripps Institution of Oceanography, independently estimated 10^8–10^9 photons per *Photinus pyralis* flash and emphasised that 'The flash terminates when O_2 is consumed,' and Lynn Faust, author of *Fireflies, Glow-Worms, and Lightning Bugs*, noted that LEDs far outshine real fireflies on camera despite looking similar to dark-adapted eyes.

Taken together—the biochemical bound, the re-examined Coblentz data, the reanalysed historical measurements, the modern lux-metre experiments, and expert estimates—all lines of evidence converge on 10^{10}–10^{11} photons per flash. The famous "1/40 candle" number survives mainly as a unit-conversion ghost: a visually matched, directionally emitted, century-old estimate that was quietly miscopied and then propagated through textbooks long after the underlying photometry had been forgotten.

"I'm sorry, but I don't speak red."

Bioluminescence Quantification

Molecular Reaction
Firefly luciferase catalyses ATP-driven luciferin oxidation, producing excited oxyluciferin that emits a photon at $\sim 560\,\text{nm}$ ($E \approx 3.55 \times 10^{-19}\,\text{J}$) with quantum yield $\Phi \approx 0.41$ (reported range ~ 0.41–0.88, depending on pH and species). Flash duration (200–300 ms) is controlled by oxygen availability via tracheal gating to photocytes. Timmins et al. (2001) demonstrated that flash termination occurs via oxygen depletion when tracheoles constrict, cutting O_2 supply to photocytes.

Bottom-Up Biochemical Calculation
Total photon emission follows from enzyme abundance and oxygen-limited kinetics:

$$N_\gamma = N_{\text{luc,cell}} \times N_{\text{cells}} \times k_{\text{eff}} \times \Phi \times t$$

Parameter ranges from physical bounds and biochemical and anatomical constraints:

- Luciferase per photocyte: 10^6 molecules (range: 3×10^5 to 10^7).
- Photocytes per lantern: 10^5 cells (range: 5×10^4 to 3×10^5).
- Quantum yield Φ: 0.41 (range: 0.41 to 0.88).
- Effective turnover k_{eff}: $0.01\,\text{s}^{-1}$ (range: 0.01 to $1\,\text{s}^{-1}$; oxygen-limited to burst discharge).
- Flash duration t: $0.25\,\text{s}$ (range: 0.25 to $1.0\,\text{s}$).

Representative cases (adapted from Silver, 2025):

$$N_\gamma^{(\text{min})} \approx 10^5 \times 10^6 \times 0.01 \times 0.41 \times 0.25 \approx 10^8 \text{ photons/flash}$$

$$N_\gamma^{(\text{mid})} \approx 10^5 \times 10^6 \times 0.1 \times 0.48 \times 0.25 \approx 10^9 \text{ photons/flash}$$

$$N_\gamma^{(\text{max})} \approx 10^5 \times 10^6 \times 1.0 \times 0.88 \times 1.0 \approx 4 \times 10^{10} \text{ photons/flash.}$$

These span oxygen-limited steady flashing through a one-turnover-per-enzyme 'burst' in which a pre-charged enzyme pool is discharged synchronously. The biochemical budget therefore constrains any realistic flash to lie in the range 10^8–4×10^{10} photons.

Resolving the Textbook Discrepancy
The commonly cited brightness figure, traced to Ives and Coblentz (1924) and paraphrased as '1/40 candle,' corresponds—under an isotropic, time-averaged interpretation—to $\sim 3 \times 10^{14}$ photons per 250 ms flash. Re-examination of Coblentz's original 1912 monograph, however, shows that *Photinus pyralis* flashes actually ranged from 1/50 to 1/400 candle, with 1/400 predominating, and that visual nulling photometry likely matched peak rather than integrated intensity. Combined with the strongly ventral beaming of the lantern (effective solid angle ~ 1–2 sr rather than 4π) and modern luminous-efficiency curves, the corrected historical value drops by an order of magnitude or more. When these corrections are added to direct lux-metre measurements of live fireflies (0.2–0.5 lux at 1–2 cm, giving 10^{10}–4×10^{11} photons/flash) and to reanalyses of Harvey and Stevens (1928) and Goh et al. (2022), all four lines of evidence converge on 10^{10}–10^{11} photons per flash—fully consistent with the biochemical bounds above and three to four orders of magnitude below the naive 1/40-candle interpretation.

References:
Timmins, G.S., et al. (2001). Firefly flashing is controlled by gating oxygen to light-emitting cells. *J. Exp. Biol.*, 204, 2795–2801.
Ives, H.E., and Coblentz, W.W. (1924). Photometric studies of luminous insects. *J. Opt. Soc. Am.*, 9(3), 217–236.
Coblentz, W.W. (1912). *A Physical Study of the Firefly.* Carnegie Institution of Washington, Publ. 164.
Harvey, E.N., and Stevens, K.L. (1928). The brightness of the light of the West Indian elaterid beetle, *Pyrophorus. J. Gen. Physiol.*, 12, 269–272.

Once in a Jew Moon

Top (Cultural Calendars and Celestial Cycles): Different civilisations devised calendars based on solar, lunar, or lunisolar cycles. The outer diagrams represent astronomical alignments and calendrical intercalation schemes. The Moon–Earth–Sun layout highlights the tension between observational cycles and constructed systems.
Second (The Oven of Akhnai and Rabbinic Authority): In the Talmudic dispute over the purity of an oven, Rabbi Eliezer insists on a minority opinion, backed by signs from nature and heaven—including a river flowing backward and a divine voice. But the sages reject all evidence, asserting that law is not decided by miracles once given to humans.
Third (Halakhic Calendar and Defiance of Astronomy): Even when astronomical calculations proved the new Moon hadn't occurred, Rabbi Yehoshua obeys the Sanhedrin's ruling and appears before Rabban Gamliel on the declared Yom Kippur, carrying a stick and wallet. Another famous schism was between Rabbi Saadia Gaon and Rabbi Aaron ben Meir over Rosh Hashanah dates.
Bottom (Shabbat Beyond Earth): In deep space, solar day cycles may be minutes or years. How do human-defined timeframes such as Shabbat apply when detached from Earth's diurnal rhythm?

Once in a Jew Moon

The Jewish calendar was developed for witnesses observing the new Moon. So when witnesses claimed they saw the new Moon 'in the morning east and evening west,' Rabban Gamliel accepted their impossible testimony, then ordered Rabbi Yehoshua to violate his own calculated Yom Kippur—establishing that communal unity is more important than astronomical accuracy. From Arctic whalers to orbital Shabbat, each generation learns that 'it is not in heaven'—religious law belongs to human authorities grappling with reality, not perfect celestial mechanics.

Jewish Calendar Authority ◦ Sunset at Poles ◦ Rabban Gamliel Decision ◦ 'Not in Heaven' ◦ Ben Meir 921 CE ◦ Molad Calculation ◦ Arctic Whalers ◦ Astronaut Shabbat ◦ Lunar-Solar System ◦ Human Consensus ◦ Calendar Unity

"וכ״כ יש להסתפק במי שקרה לו שיבא בקיץ סמוך להנארדפאל. ששם יש איזה חדשים רצופים בקיץ יום ממש... לצוד התנינים הגדולים... מתי ישבות שבתו...."
("What about someone who comes in summer near the North Pole, where for several continuous months it is actual daytime... to hunt the great whales — determine his prayer times and Shabbat...")

— Rabbi Yisrael Lipschitz, c. 1850

Once in a Jew Moon

> *Pirkei Avot* 1:1 opens with the chain of tradition: 'Moses received the Torah from Sinai and transmitted it to Joshua, and Joshua to the Elders, and the Elders to the Prophets, and the Prophets transmitted it to the Men of the Great Assembly.'
>
> This transmission of authority defined who could determine Jewish law, including calendar matters. The chain continued through specific named authorities: Shimon the Righteous (one of the last of the Great Assembly), Antigonus of Socho, then paired leaders through the generations—the Zugot (pairs), where one served as Nasi (president) and one as Av Beit Din (head of the court).
>
> The pairs included Yose ben Yoezer and Yose ben Yochanan, Joshua ben Perachya and Nittai of Arbel, Judah ben Tabbai and Shimon ben Shetach, Shemaya and Avtalyon, and finally Hillel and Shammai. From Hillel descended a dynasty of leaders who held the title of Nasi through the destruction of the Second Temple in 70 CE and beyond.
>
> During the Temple period, this leadership controlled calendar determination. The Sanhedrin, with the Nasi presiding, declared new months based on witness testimony and intercalated years to maintain seasonal alignment. Their authority to declare time derived from the biblical verse (Leviticus 23:4): 'These are the appointed seasons of the Lord, which you shall proclaim'—the Hebrew emphasises 'which YOU shall proclaim,' granting human authorities the power to establish sacred time.
>
> After the Temple's destruction in 70 CE, the Sanhedrin reconvened in Yavneh under Rabban Yochanan ben Zakkai, then moved through various Galilean cities: Usha, Shefar'am, Beit She'arim, Sepphoris, and finally Tiberias. Despite lacking a Temple, they maintained calendar authority through the traditional chain of ordination (semicha) that connected each generation back to Moses.
>
> The Roman Empire increasingly restricted Jewish self-governance. Emperor Hadrian outlawed ordination after the Bar Kokhba revolt (132–135 CE). Constantine I (306–337 CE) further limited Jewish courts' jurisdiction.
>
> Hillel II served as Nasi from approximately 320 to 385 CE. Facing intensifying persecution and the imminent collapse of centralised Jewish authority, he made an unprecedented decision around 358 CE: publish the mathematical secrets of calendar calculation.

Jewish law marks each new day at sunset/nightfall. This convention creates practical problems at extreme latitudes where the Sun remains visible for months, or in orbit where astronauts experience sixteen sunsets daily. These edge cases test the boundaries of calendar law developed for Mediterranean latitudes.

The Jewish calendar combines lunar months with solar years. Each month begins with the new Moon—the molad—occurring every 29 days, 12 hours, 44 minutes, and 3⅓ seconds. Twelve such months fall short of a solar year by about eleven days. Left uncorrected, holidays would drift through the seasons: Passover in winter, Sukkot in summer. The calendar adds seven leap months over each nineteen-year cycle, using the correspondence that 235 lunar months approximately equal nineteen solar years.

During the Temple era, witnesses who observed the crescent moon testified before the Sanhedrin in Jerusalem. Signal fires transmitted the declaration from mountaintop to mountaintop. Witnesses could lie, clouds could obscure visibility, and distant communities received delayed notification.

The Nasi (president) presided over seventy sages whose declaration of the new Moon established the month, independent of astronomical observation.

When Hillel II published the calendar's mathematical rules previously guarded by the Sanhedrin, distant communities could now calculate dates independently. The rules he revealed included the precise length of the lunar month (29 days, 12 hours, 44 minutes, 3⅓ seconds); the nineteen-year Metonic cycle with leap years in years 3, 6, 8, 11, 14, 17, and 19; and four postponement rules (dechiyot) preventing Rosh Hashanah from falling on Sunday, Wednesday, or Friday. These dechiyot—Lo ADU (direct postponement), Molad Zaken (if the molad occurs after noon), GaTRaD (in regular years, if molad falls on Tuesday after 9 hours, 204 parts), and BeTuTaKPaT (after leap years, if molad falls on Monday after 15 hours, 589 parts)—ensure Yom Kippur never falls adjacent to Shabbat and that Hoshana Rabbah never falls on Shabbat, preserving the willow-beating ritual.

This transition from human declaration to mathematical calculation transformed calendar authority from political power to an algorithm. The chain of authority from Moses through the prophets, the Great Assembly, the Zugot, and the Nesi'im had preserved and developed this knowledge through centuries of astronomical observation and rabbinical refinement. When centralised authority became unsustainable, Hillel II ensured continuity by making the secret knowledge public. His calendar remains the foundation of Jewish temporal practice today, unchanged for over 1,600 years.

The Talmud records a calendar dispute in the late first century CE. Two witnesses appeared before Rabban Gamliel claiming they saw the new Moon in the morning in the east and the evening in the west—astronomically impossible testimony, which Rabbi Dosa ben Hurkinos said: "How can they testify that a woman gave birth, and the next day her belly is between her teeth (still pregnant)?". Rabbi Yehoshua and Rabbi Dosa ben Hurkinos declared them false witnesses.

Rabban Gamliel accepted their testimony anyway.

This affected all subsequent holiday dates. If Rabban Gamliel was wrong, then Rosh Hashanah occurred on the wrong day, making Yom Kippur fall on the wrong day ten days later. Rabbi Yehoshua calculated the correct dates according to his understanding and prepared to observe them.

Rabban Gamliel then ordered Rabbi Yehoshua to appear before him "with your staff and your wallet" on the day Rabbi Yehoshua calculated as Yom Kippur. Carrying objects violates the holy day's restrictions. Rabban Gamliel demanded public desecration of what Rabbi Yehoshua believed was the holiest day of the year.

Rabbi Akiva explained to Rabbi Yehoshua: "Whatever Rabban Gamliel has done is valid, for it says, 'These are the appointed seasons of the Lord, holy convocations, which you shall proclaim in their appointed seasons.' Whether in their proper time or not in their proper time, I have no appointed seasons other than these."

Rabbi Dosa ben Hurkinos stated: "If we come to question the court of Rabban Gamliel, we must question every court that has arisen from the days of Moses until now." Authority continuity took precedence over astronomical accuracy.

Rabbi Yehoshua took his staff and wallet and walked to Yavneh on his calculated Yom Kippur. When he arrived, Rabban Gamliel stood, kissed him, and declared: "Come in peace, my teacher and my student—my teacher in wisdom and my student because you accepted my words."

The Oven of Akhnai dispute, though not calendar-related, established similar principles of authority. The sages debated whether a particular oven (broken and repaired) could become ritually impure. Rabbi Eliezer ben Hyrcanus argued it could not, offering every possible proof. The other sages disagreed.

Rabbi Eliezer called for supernatural confirmation: a carob tree uprooted itself, a stream flowed backward, the walls of the study house began to fall. Each time the sages responded: "We do not derive law from trees, from streams, from walls."

Finally, Rabbi Eliezer demanded: "If the law is as I say, let it be proven from Heaven!" A divine voice proclaimed: "Why do you dispute with Rabbi Eliezer, seeing that in all matters the law agrees with him?"

Rabbi Yehoshua rose and declared, citing the biblical verse, "It is not in heaven" (Deuteronomy 30:12).

The Talmud reports God declaring: "My children have defeated Me, My children have defeated Me!" The law belongs to human authorities interpreting through human reason. God yields to the rabbinic court's majority decision.

The Ben Meir controversy of 921–922 CE tested whether human consensus could maintain unified practice. By then, Jewish authority had shifted from the Holy Land to Babylon, where the academies of Sura and Pumbedita had become centres of Jewish learning. Aaron ben Meir, claiming authority as a Tiberian scholar in the Holy Land, challenged Babylonian dominance through calendar calculation.

Ben Meir introduced a new rule (claiming to learn it from his Rabbinic mentors): the molad threshold should be 642 parts after noon (about 35⅔ minutes) rather than the traditional calculation. For the year 922, this meant Passover would fall two days earlier than the Babylonian calculation. This technical dispute meant different communities would observe holidays on different dates.

Ben Meir asserted that proximity to Jerusalem granted special calendar authority. His calculation might have reflected Jerusalem time versus Babylonian time—the 642 parts (an hour is 1080 parts) corresponding to the longitude difference between the two centres—or about questions of exact date of the Creation. But it was in fact less about the calendar and more about authority, challenging the Babylonian academy's authority to determine Jewish law in opposition to the Holy Land's leadership.

Saadia Gaon, head of the Sura academy, wrote mathematical refutations, gathered support from Jewish communities, and challenged Ben Meir. The exilarch (leader of the Jewish diaspora, Reish Galuta) David ben Zakkai and the Babylonian academies excommunicated Ben Meir. Circular letters warned communities against following his calculations.

Saadia's position prevailed—modern astronomical calculations place the molad for Tishrei 922 at his calculated time. Unified practice took precedence over regional authority claims.

Modern geography creates new calendar challenges. Rabbi Yisrael Lipschitz, writing from Danzig in the 1850s, addressed communities in the far north where summer nights never fully darken. "During June and July," he observed, "the night shines like day. At the very least, even at midnight, one can clearly distinguish between tekhelet and white."

Traditional law uses the ability of our eyes to distinguish between blue and white threads to mark dawn prayers. Continuous visibility eliminates this marker. Rabbi Lipschitz rejected suggestions to estimate based on spring or autumn patterns, noting that communities observed dawn prayers on Shavuot 'immediately at dawn,' not at estimated times.

At the poles, more extreme conditions apply. "What about someone who comes in summer near the North Pole, where for several continuous months it is actual daytime? There the Sun circles the full horizon from east to south to west to north. How should a Jew who arrives there—along with sailors who go there to hunt giant whales—determine his prayer times and Shabbat?"

Rabbi Lipschitz proposed treating each complete Sun-circle as one day. If you arrive on Sunday, count seven Sun-circles to Shabbat. This solution maintains the seven-day cycle even when 'day' loses conventional meaning. But he acknowledged deeper problems: when people at the pole can simultaneously observe the Sun with Europeans beginning Shabbat and Americans still in Friday afternoon, which temporal reality governs?

He concluded: "May the Holy One, Blessed Be He, enlighten our eyes with the light of His Torah." This acknowledges the limits of applying Mediterranean-based law to less common conditions.

Modern transportation forced confrontation with global date boundaries. The Chazon Ish (Rabbi Avraham Yeshaya Karelitz) calculated the halakhic date line at ninety° east of Jerusalem—approximately 125.2°E longitude—rather than the International Date Line's 180° from Greenwich (International Meridian Conference, 1884).

This placed Japan on Sunday when locals observed Saturday. The line would bisect eastern Russia, China, and Australia. To avoid splitting cities, he ruled the 125.2°E meridian curves around land masses, following water.

Most communities rejected this calculation, maintaining local Saturday as Shabbat. Travellers observe stringencies from minority opinions. Practice preserves unity over theoretical precision.

Theory became crisis during the Holocaust. Thousands of yeshiva students from Mir and other Lithuanian centres fled through Siberia to Kobe, Japan, and later to Shanghai. In Kobe—east of the Chazon Ish's calculated line—his view implied Shabbat would fall on Sunday while the local community observed Saturday. As Yom Kippur approached, they cabled desperately for guidance. Rabbi Herzog convened authorities who ruled for local practice. The Chazon Ish telegrammed back: "Eat Wednesday, fast Thursday, fear nothing." In Shanghai—west of his line—Shabbat was kept on Saturday with the established community. Survivors described agonising between halakhic theory and communal cohesion.

Orbital flight creates additional complications. Jewish astronauts orbit Earth every ninety minutes, experiencing sixteen sunsets daily. Ilan Ramon on Space Shuttle Columbia is reported to have followed Cape Canaveral time, his last Earth residence. Judith Resnik lit electronic Shabbat candles according to Houston time. These choices reflect the same principle established by Rabban Gamliel: human decision creates sacred time when natural markers fail.

The principle "it is not in heaven" establishes that human authorities interpret law for practical circumstances. Rabbi Yehoshua's compliance with Rabban Gamliel prioritised communal unity over personal calculation. Saadia Gaon's victory over Ben Meir maintained unified practice against regional authority claims. Contemporary rulings for astronauts apply these same principles to orbital conditions.

An Optimisation Vort.

In *Pirkei de-Rabbi Eliezer* (Parashat Noaḥ) it is stated that "from this, one can deduce that there are 32 species of birds," right after discussing the ark. Commentaries struck out the "from this" part as it seems to be unrelated. Some commentaries question further how can it fit the Haari Z"L comment that the total number of species is 72?

My late father noted a brilliant way to reconstruct both by noting that:

$$24 + \max\{W_g : 6W_g \leq 48 < 8W_g\} + 10 + \min\{A_b : 60 \leq 2A_b < 80\} = 72$$

First we note that *Pirkei* mentions that the pure were 7 and not 2 to have more pure than impure. Now let's calculate. For birds, the Torah lists 24 impure species, one pair each, giving 48 individuals.

Pure birds enter in sevens. Requiring the pure total to exceed the impure total leads to the bound $6W_g \leq 48 < 8W_g$ (otherwise either 6 would have been enough, or 8 would have been required). The solutions are $W_g = 7, 8$. The largest solution is $W_g = 8$. Thus the total bird species count is $W_g + W_b = 8 + 24 = 32$.

For land animals, the Torah identifies $A_g = 10$ pure species (7 livestock and 3 wild animals). Impure animals enter in pairs to have male and female. Imposing the same inequality $6A_g \leq 2A_b < 8A_g$ and taking the smallest solution gives $A_b = 30$. Hence the land-animal total is $A_g + A_b = 10 + 30 = 40$. The combined total is therefore $32 + 40 = 72$ species in the ark. Taking the maximum/minimum for pure/impure solutions is also a Midrashic principle. ■

Calendar Mathematics

Molad Calculation

The traditional Jewish lunar month length:

$$29d\,12h\,44m\,3\tfrac{1}{3}s = 29.530594 \text{ days..}$$

Modern astronomical value: 29.530589 days. Error accumulates at $\Delta = 5 \times 10^{-6} \times N$ days where N is months elapsed. After 1000 years (≈12,400 months): error ≈ 1.5 hours.

Metonic Cycle

19 solar years ≈ 235 lunar months:

$$19 \times 365.2422 = 6939.602 \text{ days.}$$

$$235 \times 29.530594 = 6939.689 \text{ days.}$$

Difference: 0.087 days per 19-year cycle. Leap years occur in years 3, 6, 8, 11, 14, 17, 19.

Mathematically, the pattern of leap years in this 19-year cycle has almost the same 'tempo' as the diatonic major scale, a connection pointed out to me by Amit B. If we look at the gaps between leap years we obtain the circular sequence

$$(3, 2, 3, 3, 3, 2, 3),$$

consisting of five 'long' gaps of three years and two 'short' gaps of 2 years, distributed as evenly as possible around the nineteen-year cycle. The major scale does exactly the same thing on a 12-step chromatic circle: it uses seven notes arranged with five whole-tone steps and two semitone steps in the maximally even pattern

$$(2, 1, 2, 2, 2, 1, 2).$$

In both cases we are placing $k = 7$ marked points on a cycle of length N with $N \equiv 5 \pmod 7$ (here $N = 19$ years or $N = 12$ semitones). The average step sizes

$$12/7 = 1 + 5/7, \qquad 19/7 = 2 + 5/7$$

share the fractional part $5/7$, which forces the 'five long, two short' pattern and fixes their relative ordering. Abstractly, the leap-year cycle and the major scale are two realisations of the same maximally even 7-beat rhythm.

Dechiyot (Postponements)

Rosh Hashanah cannot fall on Sunday, Wednesday, or Friday:

1. **Lo ADU**: Direct postponement.
2. **Molad Zaken**: If molad ≥ 18:00.
3. **GaTRaD**: Regular year, Tuesday ≥ 9h 204p.
4. **BeTuTaKPaT**: After leap, Monday ≥ 15h 589p.

Ben Meir Dispute (922 CE)

Ben Meir: Molad threshold = 642 parts.
Traditional: Molad threshold = 0 parts.
For Tishrei 4683 (922 CE):
Ben Meir: Day 2, 9h 204p.
Saadia: Day 2, 15h 589p.

Polar Day Solutions

Sun-circle method: Each 24h circuit = 1 day.
Origin timezone: Follow departure location.
Proportional: Calculate theoretical solar angle: $h = 15°(t - 12) - \lambda + E$.

Classical Source

Halakhot Pesuqot (Rav Yehudai Gaon, 8th century):

לעולם ראש חדש אדר סמוך לניסן הוא ערב הפסח, והפסח הוא ערב העצרת, והעצרת הוא ערב ראש השנה. לא בד״ו פסח, לא גה״ז עצרת, לא אד״ו ראש השנה וסוכה, לא אג״ו יום הכיפורים, ולא זבד פורים.

Translation:
Always, the new Moon of Adar close to Nisan is the eve of Passover; Passover precedes Shavuot; and Shavuot precedes Rosh Hashanah. Passover never occurs on days בד״ו (MoWeFr); Shavuot never on גה״ז (TuThSa); Rosh Hashanah and Sukkot never on אד״ו (SuWeFr); Yom Kippur never on אג״ו (SuTuFr); and Purim never on זבד (MoWeSa).

References:

Feldman, W.M. (1931). *Rabbinical Mathematics and Chronology*.
Stern, S. (2019). *The Jewish Calendar Controversy of 921/2*.
Lipschitz, Y. (1850). *Tiferet Yisrael*, Berakhot 1.

A Spectrum of Skies

Top (Atmospheric Filtering of Stellar Radiation): Sunlight reaching Earth undergoes selective scattering by the atmosphere. Shorter wavelengths (blue, violet) are scattered in all directions, while longer wavelengths (red, orange) pass through more directly. This results in both the blue sky and the reddening of the Sun near the horizon.
Bottom (Hertzsprung–Russell Diagram): Stars are plotted by surface temperature (x-axis, decreasing rightward) and luminosity (y-axis, log scale). Main sequence stars form a diagonal band; giants and supergiants lie above, white dwarfs below. The Sun sits in the middle of the main sequence. The temperature scale reflects stellar classification: blue-hot stars (left) are hotter and more luminous; red stars (right) are cooler.

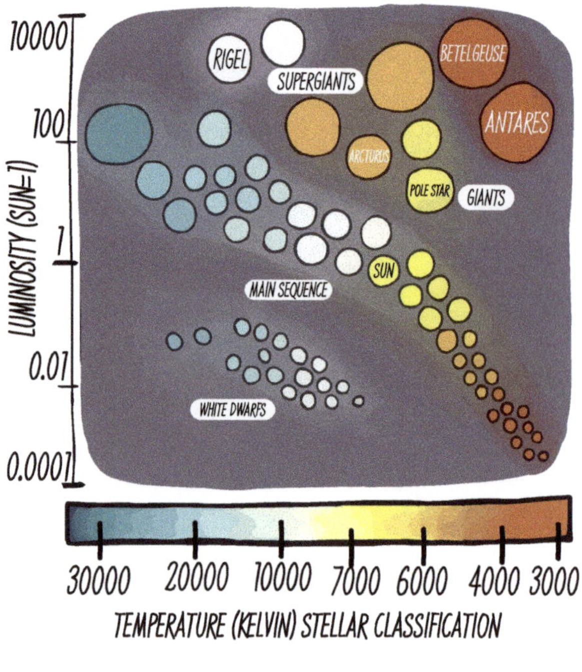

A Spectrum of Skies

Sky colours are determined by light scattering and spectral signatures. Earth's blue sky results from Rayleigh scattering, where nitrogen and oxygen molecules preferentially scatter shorter wavelengths by factors of 10–100 times more efficiently than longer ones. Mars's butterscotch-orange haze results from suspended dust particles 1–10 micrometres across, scattering all wavelengths equally while absorbing blue light. Titan's deep orange hue comes from photochemical hazes (tholins) produced by UV irradiation of methane, while Venus's perpetual cloud deck creates brilliant white from sulfuric acid droplets.

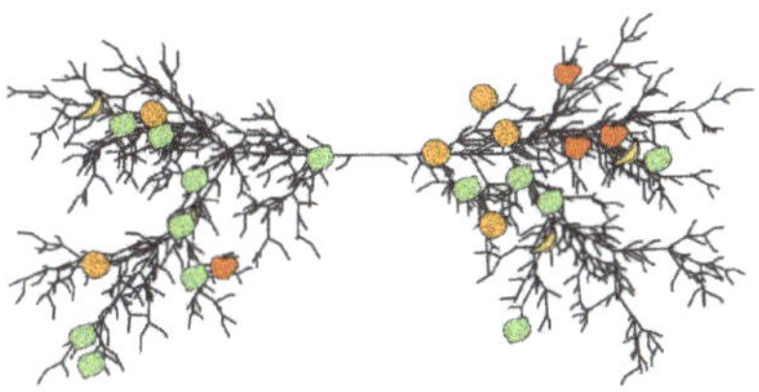

Rayleigh Scattering λ^{-4} ◦ Blue Sky Physics ◦ Mars Iron Oxide ◦ Venus Sulphuric Acid ◦ Uranus-Neptune Methane ◦ Stellar Temperature Colours ◦ Hα Red Nebulae ◦ Interstellar Reddening ◦ Doppler Colour Shift ◦ Spectroscopy Tool ◦ Atmospheric Path Length

> "Oh, I'm sorry. Well, I could put the trash into a landfill
> where it's going to stay for millions of years,
> or I could burn it up and get a nice smoky smell in here
> and let that smoke go into the sky where it turns into stars."
>
> — Charlie Kelly

> "That doesn't sound right,
> but I don't know enough about stars to dispute it."
>
> — Mac

A Spectrum of Skies

> The study of light dispersion and the spectral analysis of celestial objects motivates the scientific study into the nature of cosmic colour. Isaac Newton's prism experiments in the 1660s revealed that white light contains a continuous spectrum of colours, initiating the quantitative study of optics. By the early nineteenth century, Joseph von Fraunhofer's meticulous cataloguing of dark absorption lines in the solar spectrum laid the foundation for stellar spectroscopy. These lines, later identified with specific chemical elements, confirmed that stars share a common elemental makeup with Earth.
>
> Mid-nineteenth-century advances by William Huggins and Angelo Secchi introduced systematic spectral classification and chemical identification of stars and nebulae. Secchi's grouping of stellar spectra foreshadowed the modern OBAFGKM sequence, while Huggins showed that some nebulae emit light like gases rather than collections of stars. Theoretical breakthroughs followed in the early twentieth century with Planck's and Wien's radiation laws and the rise of quantum theory, which linked spectral lines to electronic transitions within atoms.
>
> Meanwhile, on Earth, John Tyndall's experiments on atmospheric scattering in the late nineteenth century demonstrated wavelength-dependent scattering in air (the Tyndall effect). In 1871, Lord Rayleigh formalised the underlying physics, showing that shorter wavelengths scatter more efficiently and explaining the blue sky. These ideas were extended to planetary atmospheres through spectroscopic work by Vesto Slipher in the 1910s and Gerard Kuiper in the 1940s, who linked atmospheric composition to the observed colours of planets.
>
> With the dawn of the space age, direct observations of planetary skies became possible. NASA's Mariner, Viking, and Voyager missions collected spectral data from Mars, Venus, Jupiter, and beyond, confirming that dust, aerosols, and gases define planetary hues. In the early 2000s, Cassini's orbit of Saturn and Titan revealed complex hazes and photochemical effects shaping visible colouration. Across centuries - from Newton's spectrum to Cassini's spectrometers - the colour of the cosmos has transitioned from aesthetic impression to precise physical measurement.

The blue colour of Earth's sky results from the interaction of solar radiation with atmospheric gases. When sunlight enters the atmosphere, air molecules scatter it in multiple directions, creating a diffuse background of illumination that makes the sky appear bright when the Sun is not in direct view. This scattering process depends on wavelength: when the scattering particles are much smaller than the wavelength of light, as air molecules are, the scattering cross-section varies inversely with the fourth power of wavelength. This λ^{-4} dependence arises because small particles respond to electromagnetic waves as induced oscillating dipoles—the passing light wave causes electrons to oscillate, and these accelerating charges radiate energy in all directions. The power radiated by an oscillating dipole scales with the fourth power of its oscillation frequency, which translates to the fourth power of wavelength in the denominator. This phenomenon, known as Rayleigh

 https://doi.org/10.11647/OBP.0526.27

scattering (Rayleigh, 1871), favours shorter-wavelength light, such as blue and violet, over longer-wavelength light, such as red.

Although violet light scatters more efficiently than blue, the sky appears blue rather than violet due to two compensating factors: human eyes are less sensitive to violet wavelengths, and the solar spectrum itself contains slightly less energy in the violet band. At sunrise and sunset, the geometry changes. Sunlight must traverse a much longer path through the atmosphere—up to forty times the distance at noon—causing blue and green wavelengths to scatter away before reaching observers, leaving predominantly red and orange light to paint the horizon.

Perceived colour corresponds to measurable differences in scattering efficiency and atmospheric path length, arising from electromagnetic interactions between photons and molecules. The same mechanisms of scattering, absorption, and emission operate throughout astronomical systems. Whenever light interacts with matter, whether gas, dust, or solid surfaces, its spectrum undergoes alterations that depend on the physical properties of the medium. These spectral changes manifest as both broad redistributions of intensity across wavelengths and selective suppression or enhancement of specific bands, encoding temperature, chemical composition, and geometric configuration in the spectrum. Colour in astronomy is a quantitative tool—a compressed summary of the integrated intensity distribution across wavelength bands.

Unlike planets, stars generate their own light through thermal radiation, with each star approximating a blackbody whose emission spectrum depends primarily on surface temperature. As temperature increases, Wien's displacement law (Wien, 1896) dictates that the peak emission shifts to shorter wavelengths: the hottest O-type stars blaze at 30,000 K with peak emission in the ultraviolet, while cool M-dwarfs at 3,000 K peak in the infrared. Our Sun, at approximately 5,800 K, emits across the visible spectrum with peak intensity in the green, though the integrated light appears white-yellow when viewed through Earth's atmosphere. The stellar classification sequence—O, B, A, F, G, K, M—encapsulates both this temperature progression and the changing patterns of absorption lines that appear as different atomic species become ionised or excited in stellar atmospheres of varying temperature.

The colours we observe in stars reflect their physical state at multiple scales. Surface gravity affects spectral line profiles—giant stars show narrower absorption lines than dwarfs of the same temperature due to lower atmospheric pressure. Metallicity alters colour subtly: metal-poor stars appear bluer than metal-rich stars at the same temperature because metals provide additional opacity in the blue and ultraviolet. Rapid rotation produces gravity darkening: equators are cooler and dimmer while poles are hotter and brighter (von Zeipel effect) (von Zeipel, 1924), leading to colour and brightness gradients across rapidly spinning stars such as Vega.

Stellar atmospheres contain molecules that create distinctive colour signatures. Cool stars below 3,500 K form titanium oxide (TiO), whose broad absorption bands define the M spectral class and create deep red colours. Carbon stars form carbon compounds instead, appearing distinctly crimson. S-type stars show zirconium oxide bands, while the coolest L and T dwarfs—discovered only through infrared surveys—contain methane and water vapour, rendering them invisible to the naked eye despite being closer than many visible stars.

Variable stars demonstrate colour as a dynamic property. Cepheid variables pulsate radially, their surfaces cooling and reddening during expansion, then heating and becoming bluer during contraction. RR Lyrae stars, a type of variable star, cycle through colour changes in hours rather than days. Mira variables—long-period pulsating giants—can shift from orange to deep red over months as their radii double. These colour variations serve as cosmic distance indicators at ranges beyond the direct reach of parallax, with period–luminosity relationships (Leavitt & Pickering, 1912) calibrated using colour information and reddening corrections.

Interstellar reddening complicates stellar colour interpretation. Dust grains preferentially scatter blue light, making distant stars appear redder than their intrinsic colours—a phenomenon distinct from cosmological redshift. The amount of reddening depends on grain size distribution and composition along the line of sight. Astronomers must correct for this extinction to determine true stellar properties. Some regions show anomalous extinction curves where ultraviolet absorption exceeds predictions, suggesting unusual dust compositions possibly formed in supernova ejecta or stellar winds.

Planets reflect sunlight, with apparent colour determined by the interplay between surface reflectivity and atmospheric absorption. Mars owes its rusty appearance to iron oxide in its regolith, which reflects red wavelengths while absorbing blue and green. This selective reflection creates the planet's reddish hue that persists across all viewing angles and seasons. Jupiter presents a more complex palette: its banded appearance results from layered cloud structures at different altitudes, with white ammonia ice clouds at higher levels and darker, reddish-brown clouds containing complex organic chromophores at greater depths.

The ice giants Uranus and Neptune share a different colouring mechanism. Both possess methane in their upper atmospheres, which absorbs red light beyond 600 nanometres while allowing blue wavelengths to scatter back to space. Yet Neptune appears a deeper, more vivid blue despite similar methane concentrations. Neptune's atmospheric dynamics differ: Neptune's more active vertical mixing clears high-altitude hydrocarbon hazes that would otherwise dilute the pure blue colour, while Uranus retains a whitish haze layer that mutes its appearance.

Beyond planetary atmospheres, interstellar nebulae paint the cosmos with characteristic colours. Emission nebulae glow with the light of ionised gas, powered by nearby hot stars whose ultraviolet radiation strips electrons from atoms. The dominant spectral signature is the Hα line of hydrogen at precisely 656.3 nanometres, creating the deep red glow of star-forming regions such as Orion. Planetary nebulae add complexity to this palette through doubly ionised oxygen ([OIII]) emission near 500 nanometres, producing ethereal blue-green shells around dying stars.

Reflection nebulae consist of interstellar dust clouds that scatter light from nearby stars. Like Earth's atmosphere, they scatter blue light more efficiently than red. The result is a delicate blue illumination surrounding young, hot stars, even though the dust grains themselves emit no light. In contrast, dark nebulae represent the universe's silhouettes—dense molecular clouds so opaque they block background starlight entirely, appearing as sharply defined voids against the stellar backdrop, such as the famous Horsehead Nebula.

On the grandest scales, entire galaxies display integrated colours that chronicle their stellar populations and evolutionary histories. A galaxy's colour represents the combined light of billions of stars, weighted heavily toward the brightest members. In spiral galaxies with active star formation, massive O and B type stars dominate the integrated light despite comprising less than 1% of the total stellar population. These stellar giants burn so brilliantly that they outshine thousands of Sun-like stars, lending spiral arms their blue-white glow. Elliptical galaxies, having exhausted their gas reserves billions of years ago, harbour predominantly old, cool stars that together produce a golden or reddish cast.

Interstellar dust modifies galactic colours through extinction—the wavelength-dependent absorption and scattering that removes blue light from our line of sight. Edge-on spiral galaxies show dark dust lanes bisecting their discs and reddening the light from stars behind them, much as Earth's atmosphere reddens the setting Sun.

Colour in astronomy extends beyond intrinsic properties to include the effects of motion through the Doppler shift. When celestial objects move relative to observers, their spectrum shifts: approaching objects compress wavelengths toward the blue, while receding objects stretch them toward the red. For nearby stars and galaxies, these shifts measure radial velocities—the component of motion along our line of sight. Cosmological redshift occurs as the expansion (Hubble, 1929) of intergalactic space stretches light waves during their journey across the universe, with the amount of stretching depending on both distance and the universe's expansion history.

Modern astronomical imaging translates physical phenomena into visual representations through colour-coding schemes. True-colour images approximate human vision by combining exposures through red, green, and blue filters matched to our eye's sensitivity. False-colour imaging employs colours as a visualisation tool for invisible wavelengths. Radio telescopes might encode intensity as red, X-ray telescopes as blue, and infrared as green, creating composite images that reveal hidden structures. Narrow-band filters isolate specific emission lines—hydrogen-alpha, oxygen-III, sulphur-II—each mapped to different colours to highlight ionisation zones, shock fronts, and chemical gradients. Colour mapping displays temperature distributions, velocity fields, and magnetic structures across cosmic scales.

Transmitted wavelengths broadcast physical processes throughout the cosmos. Every hue we measure corresponds to specific interactions between electromagnetic radiation and matter: Rayleigh scattering in atmospheres, thermal emission from stellar surfaces, electronic transitions in nebular gas, or the cosmic expansion detected in galactic redshifts. These mechanisms—absorption, scattering, emission, and Doppler shifting—operate according to known physical theories, making the universe into a vast spectroscopic laboratory where colour reveals temperature, composition, motion, and history across scales from planetary atmospheres to galaxy clusters.

Why This Story

This was one of the first chapters I wrote. I wanted a more sophisticated answer to a question children often ask: why is the sky blue? Why is the Sun red at sunset? I knew the basic explanation—Rayleigh scattering and wavelength dependence—but I wanted a version that would remain meaningful even after they learned mathematics and a bit of science. The question becomes more, not less, interesting with each layer of generalisation: from sunlight and air to blackbody curves, stellar classifications, and interstellar dust. Writing this chapter helped establish the book's tone—clear, complex, physically grounded—and set the standard for treating simple questions with full scientific seriousness.

Nine simulated daytime skies from each planet in the Solar System, arranged heliocentrically around the Sun in a 3×3 grid. The rows (left to right, top to bottom) correspond to: Mercury, Venus, Earth; Mars, Sun, Jupiter; Saturn, Uranus, Neptune. Each panel reflects sky colour, atmospheric scattering, and visible celestial features such as moons, rings, and the Sun's apparent size, as modelled for surface or high-atmosphere observation.

Quantitative Analysis of Astronomical Colour

Radiative Transfer and Blackbody Radiation

The propagation of specific intensity I_ν along a path s is governed by the equation of radiative transfer:

$$\frac{dI_\nu}{ds} = j_\nu - \alpha_\nu I_\nu - \sigma_\nu I_\nu + \iint \sigma_\nu(\Omega', \Omega) I_\nu(\Omega')\, d\Omega',$$

where j_ν is the emission coefficient, α_ν is the absorption coefficient, and σ_ν is the scattering coefficient. The optical depth is $d\tau_\nu = (\alpha_\nu + \sigma_\nu)\, ds$. In local thermodynamic equilibrium (LTE) without scattering, the source function reduces to $S_\nu = j_\nu/\alpha_\nu$, which approaches the Planck function for thermal emission. With scattering present, the total source function mixes thermal emission and angle-averaged scattered intensity: $S_\nu = (\alpha_\nu B_\nu + \sigma_\nu J_\nu)/(\alpha_\nu + \sigma_\nu)$, where J_ν is the mean intensity.

$$B_\nu(T) = \frac{2h\nu^3}{c^2}\left[\exp\left(\frac{h\nu}{kT}\right) - 1\right]^{-1}.$$

The wavelength of peak emission is given by Wien's law:

$$\lambda_{\text{max}} T = b \approx 2.898 \times 10^{-3}\,\text{m} \cdot \text{K}$$

and the total emitted flux per unit area follows the Stefan-Boltzmann law:

$$F = \sigma T^4, \quad \sigma = \frac{2\pi^5 k^4}{15c^2h^3}.$$

Stellar temperatures can be inferred by fitting observed continua or via colour indices (e.g., $B - V$), which measure differences in magnitude across filtered bands.

Spectral Lines and Atmospheric Composition

Spectral lines originate from electronic transitions with energy $\Delta E = h\nu = E_u - E_l$. In LTE, population ratios follow the Boltzmann distribution:

$$\frac{N_u}{N_l} = \frac{g_u}{g_l}\exp\left(-\frac{E_u - E_l}{kT}\right).$$

Ionization states are governed by the Saha equation:

$$\frac{N_{i+1}N_e}{N_i} = \frac{2g_{i+1}}{g_i}\left(\frac{2\pi m_e kT}{h^2}\right)^{3/2}\exp\left(-\frac{\chi_i}{kT}\right)$$

where χ_i is the ionisation energy and N_e is the electron density. Line shapes are broadened by natural width, thermal Doppler broadening:

$$\Delta\lambda_D = \lambda_0\left(\frac{2kT}{mc^2}\right)^{1/2}$$

and collisional (pressure) broadening, which scales with density. Observed line intensities allow reconstruction of chemical abundances and physical conditions.

Motion, Redshift, and Extinction

The Doppler effect shifts wavelengths by

$$\frac{\Delta\lambda}{\lambda_0} = \frac{v_r}{c}, \quad (v_r \ll c)$$

where v_r is the radial velocity. Redshift ($\Delta\lambda > 0$) indicates recession; blueshift indicates approach. For distant galaxies, the cosmological redshift z is related to the scale factor $R(t)$ via $1+z = R(t_{\text{obs}})/R(t_{\text{emit}})$ and follows Hubble's law for $z \ll 1$: $v = H_0 d$.

Extinction by dust is quantified by

$$A_\lambda = 1.086\,\tau_\lambda = -2.5\log_{10}\left(F_\lambda/F_{\lambda,0}\right)$$

where τ_λ is the optical depth. Reddening is the differential extinction between bands:

$$E(B-V) = A_B - A_V$$

and the total-to-selective extinction ratio $R_V = A_V/E(B-V)$ typically has a value near 3.1 for the diffuse interstellar medium. For small particles $a \ll \lambda$, Rayleigh scattering dominates, with efficiency $\propto \lambda^{-4}$. For $a \sim \lambda$, Mie scattering applies, with weaker wavelength dependence.

References:

Rybicki, G. B., and Lightman, A. P. (1979). *Radiative Processes in Astrophysics.* Wiley.

Draine, B. T. (2003). Interstellar Dust Grains. *Annual Review of Astronomy and Astrophysics.*

You're So Hot, You Cool Me Down

Top (Maximum Entropy – Overcrowded Room): A metaphor for a negative temperature system. The room is so disordered—crammed with objects, decor, and chaos—that any change would make it more ordered. This represents a population-inverted state: energy is at its maximum, and the system is beyond the entropy peak where $\beta = 0$. Thermodynamically, such states must have negative temperature.
Middle (Positive Temperature – Typical Disorder): A moderately cluttered room. It reflects a normal positive temperature system: there's room for more disorder, and adding energy generally increases entropy. This is the typical regime described by classical statistical mechanics.
Bottom (Low Temperature – Ordered State): A clean, sparse room. This is a low-entropy state where most particles (or objects) are in low-energy configurations. Adding energy would increase disorder. The system is clearly in the conventional positive-temperature regime, but near the low end.

You're So Hot, You Cool Me Down

Temperature measures how entropy changes with energy ($\partial S/\partial E$), not merely kinetic activity. While unbounded systems such as ideal gases can only reach positive temperatures, quantum systems with finite energy spectra reveal different dynamics. When energy addition increases disorder, temperature is positive; when maximum entropy is reached, temperature becomes infinite; further energy addition creates more ordered states with negative temperatures. These negative temperature states are not colder than absolute zero but as hot as infinity—they transfer energy to any positive-temperature system when brought into contact.

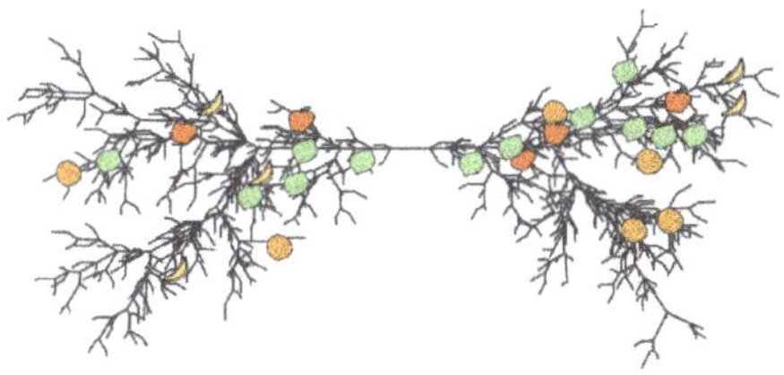

Temperature Definition ∘ Heat Flow Direction ∘ Carnot Universal Scale ∘ Kinetic Theory ∘ Statistical Mechanics ∘ Negative Temperature ∘ Population Inversion ∘ Bounded Energy Systems ∘ Boltzmann vs Gibbs ∘ Dunkel-Hilbert Controversy ∘ Coldness Parameter β

"Give a man a fire and he's warm for a day,
but set fire to him and he's warm for the rest of his life."
— Solid Jackson, Year of the Justifiably Defensive Lobster, 1988 UC

"Winter is coming."
— Eddard Stark, 298 AC

You're So Hot, You Cool Me Down

The concept of temperature originated long before its formal scientific definition. Early thermometry in the 16th and 17th centuries relied on devices such as Galileo's thermoscope, which measured qualitative warmth but lacked a standardised scale. By the early eighteenth century, Daniel Gabriel Fahrenheit introduced a reliable mercury thermometer and a temperature scale, followed by Anders Celsius and William Thomson (Lord Kelvin), whose absolute scale based on thermodynamic principles became the foundation for modern temperature measurement.

The theoretical basis for temperature matured alongside the formulation of classical thermodynamics. The zeroth law of thermodynamics, though articulated last, established the foundational equivalence relation that permits temperature to be meaningfully assigned: if system A is in thermal equilibrium with system B, and system B with system C, then A and C must also be in equilibrium. This abstracted thermal equilibrium from specific substances or instruments, enabling the development of general thermometric devices.

Simultaneously, empirical laws such as those of Boyle (1662), Charles (1787), and Gay-Lussac (1802) revealed regularities in the behaviour of gases, hinting at underlying statistical dynamics. These culminated in the ideal gas law, $PV = nRT$, linking temperature with pressure and volume in a measurable way. However, it was not until the advent of statistical mechanics in the nineteenth century—particularly through the work of Ludwig Boltzmann (1844–1906) and James Clerk Maxwell (1831–1879)—that temperature gained a microscopic interpretation. Boltzmann's definition of entropy and the expression $\frac{1}{T} = \left(\frac{\partial S}{\partial E}\right)_{V,N}$ provided a bridge between macroscopic observations and the probabilistic behaviour of particles.

This statistical framework laid the foundation for interpreting unusual thermodynamic regimes. While classical thermodynamics assumes that entropy increases with energy, leading to strictly positive temperatures, the statistical definition permits a broader spectrum. In systems with bounded energy, entropy can decrease with increasing energy, enabling the possibility of negative absolute temperatures. Such interpretations remained largely theoretical until mid-twentieth century experiments produced states interpreted as having negative temperature.

Temperature measures how hot or cold something is. Unlike energy, which emerges from a symmetry principle, temperature seems to be defined only through directionality. Noether's theorem tells us that energy (Noether, 1918) is the conserved quantity associated with time-translation invariance—systems that behave the same way now as they will an hour from now conserve energy. Momentum arises from spatial translation invariance. Angular momentum from rotational invariance. Each conservation law reflects an underlying symmetry in the physical laws.

Temperature is defined operationally. Place two systems in contact, and energy will pass from the one with higher temperature to the one with lower, until a balance is reached.

 https://doi.org/10.11647/OBP.0526.28

Temperature tells us the direction of heat flow without first specifying what temperature is. We cannot derive it from symmetries.

This directional character distinguishes temperature from other conserved quantities. Energy exists in a single isolated system. Temperature exists only through interaction, defined from the balance between energy and entropy when systems exchange heat. A system alone has energy; it acquires temperature only in relation to possible exchanges with other systems.

Early thermometry sought operational definitions through material properties. Galileo's thermoscope (1593) tracked air expansion in a glass bulb—temperature changes moved water levels, but atmospheric pressure variations corrupted readings. Fahrenheit (1724) achieved reproducibility through mercury expansion and three fixed points: a frigorific mixture of ice, water, and ammonium chloride (0°F), water-ice equilibrium (32°F), and human body temperature (96°F; the modern accepted value on his scale is 98.6°F). His mercury-in-glass design minimised pressure effects while his fixed points enabled calibration. Celsius (1742) simplified to two points—water's freezing and boiling at standard pressure—originally inverted with 100° for freezing, 0° for boiling. These scales quantified temperature through material expansion coefficients, each substance yielding slightly different readings. Agreement required careful calibration against shared fixed points.

The concept developed through distinct theoretical frameworks that initially seemed unrelated. Thermometry defined it operationally through thermal expansion—mercury rises in glass tubes, metals expand when heated. Classical thermodynamics formalised it through the Carnot cycle (Carnot, 1824): the efficiency of reversible heat engines operating between two reservoirs depends only on their temperature ratio, providing a universal scale independent of working substance. This universality indicates temperature's nature—not a property of matter but a parameter governing energy distribution.

Carnot's insight preceded atomic theory yet captured a fundamental phenomenon: temperature mediates between mechanical work and heat flow. In his ideal engine, complete conversion of heat to work is impossible not because of friction or engineering limits, but because temperature imposes constraints on energy quality. Hot reservoirs contain high-quality energy; cold reservoirs contain degraded energy. Temperature quantifies this degradation.

Lord Kelvin (1848) recognised that Carnot's efficiency formula $\eta = 1 - T_c/T_h$ contained the seeds of an absolute scale. If engine efficiency depends only on temperature ratios, not working substances, then temperature ratios have universal meaning. Kelvin defined his scale through the work extractable from heat: temperature ratios correspond to ratios of heat exchanged in reversible engines. This freed temperature from material properties—no mercury expansion, no fixed points tied to water. The Kelvin scale's zero represents the temperature at which no work can be extracted from heat, where a Carnot engine's efficiency reaches zero. Temperature became a measure of energy availability, not material response.

The mechanical interpretation of heat predated thermodynamics. Daniel Bernoulli (1738) proposed that gas pressure arises from particle impacts against container walls. His model—elastic spheres in ceaseless motion—correctly predicted that pressure times volume should be proportional to the kinetic energy of particles. This anticipated the ideal gas law by a century without the concept of temperature as average kinetic energy. Bernoulli wrote of 'increasing the intensity of motion' when heating gases but couldn't quantify the relationship. John Herapath (1820) and John Waterston (1845) independently derived $pV \propto T$ from particle mechanics. Scientific journals ignored or rejected their work. Clausius (1857) finally connected these mechanical models to thermodynamics, showing that Bernoulli's 'intensity of motion' was precisely what thermometers measured.

Kinetic theory offered a microscopic interpretation: temperature measures the average translational kinetic energy of particles, $\langle E_{\text{kin}} \rangle = \frac{3}{2} k_B T$ for ideal gases (where k_B is the Boltzmann constant with units of energy per temperature). However, temperature is not simply motion—a supersonic jet of cold gas has enormous kinetic energy yet low temperature. The random component is what matters, the deviation from collective flow, the microscopic dance beneath the macroscopic averages.

In statistical mechanics, Boltzmann defined entropy as a count of microstates (Boltzmann, 1877), leading to the relationship $\frac{1}{T} = \left(\frac{\partial S}{\partial E}\right)_{V,N}$ that defines temperature as the exchange rate between energy and entropy. When energy is added to a system, the rate at which new configurations become accessible defines temperature. Systems are hot when energy buys little additional disorder, cold when energy opens larger territories of possibility.

These definitions converge for ordinary matter but diverge in extreme conditions. The thermodynamic and statistical definitions always agree when both apply. The kinetic interpretation works only for systems with translational degrees of freedom (where energy can be represented as movement); it fails for photon gases, spin systems, or any collection where energy takes non-kinetic forms. The statistical definition remains universal, applying wherever entropy and energy are meaningful—from black holes to quantum fields.

Temperature's statistical nature fails at the boundaries of applicability. A single molecule has no temperature—temperature requires an ensemble where probability distributions make sense. We routinely discuss the temperature of systems containing mere dozens of atoms. The transition to a thermodynamically valid description occurs where statistical averages are not overwhelmed by fluctuations. For nanoscale devices operating at the edge of thermodynamic validity, the transition point where fluctuations overwhelm averages becomes critical.

Different phenomena occur in curved spacetime. The vacuum has no temperature in flat spacetime; accelerating observers perceive it as thermal—the Unruh effect (Unruh, 1976). An observer accelerating at one Earth gravity perceives empty space glowing at 10^{-20} Kelvin. Temperature arises from quantum field correlations across the acceleration horizon, not from matter. Motion through spacetime generates heat from nothing (see Chapter 47).

Black holes embody a temperature paradox. Classically, nothing escapes a black hole, implying zero temperature. Quantum mechanics near the event horizon creates particle pairs, one falling inward, one escaping as Hawking radiation (Hawking, 1974). The hole glows with temperature $T = \hbar c^3/(8\pi G M k_B)$—inversely proportional to mass. Stellar-mass black holes radiate at nanokelvins; microscopic holes would explode in blazing heat.

Temperature results from pure geometry, spacetime curvature creating thermal radiation without matter.

Systems with unbounded energy spectra can reach arbitrarily high temperatures. Ideal gases exemplify this: particle energies face no upper limit beyond total energy input. In systems with a maximum possible energy, the situation changes. Consider a lattice of spins with only two energy states per site. As more energy is added, spins flip to the excited state. When half the spins are excited, entropy is maximised. Adding further energy forces the system into more constrained configurations—more spins aligned against the field—resulting in fewer configurations and thus lower entropy. The derivative $\partial S/\partial E$ becomes negative, yielding negative temperature.

The possibility of negative temperature depends critically on the statistical definition of entropy. Dunkel and Hilbert (2014) challenged sixty years of accepted wisdom about negative temperatures. The controversy centres on two competing entropy definitions: the Boltzmann entropy $S_B = k_B \ln(\Omega_B)$ where $\Omega_B = \epsilon\omega(E)$ counts states in an energy window ϵ around E with density of states $\omega(E)$, and the Gibbs entropy $S_G = k_B \ln(\Omega)$ based on the integrated density of states $\Omega(E) = \int_0^E \omega(E')dE'$. Both yield temperature through $1/T = \partial S/\partial E$, but with different results.

The Boltzmann approach gives $T_B = (k_B\omega'/\omega)^{-1}$. When the density of states peaks and then decreases—as happens in bounded systems where high-energy configurations become constrained—ω' becomes negative, yielding negative temperature. The Gibbs approach gives $T_G = (k_B\omega/\Omega)^{-1}$. Since Ω integrates the density of states, it increases monotonically when ω decreases. The Gibbs temperature remains positive throughout.

Dunkel and Hilbert argued that only the Gibbs entropy satisfies thermodynamic consistency. A Maxwell relation—derived from the equality of mixed partial derivatives of the fundamental relation $dE = TdS - PdV$—requires that pressure computed two different ways must agree: thermodynamically through $P = T(\partial S/\partial V)_E$ and mechanically through $P = -(\partial E/\partial V)_S$. This consistency test fails for Boltzmann entropy. In a quantum particle in a box, Boltzmann predicts negative temperature where Gibbs remains positive, and the two pressure calculations disagree. The Boltzmann entropy also violates equipartition in classical systems and yields incorrect heat capacities for quantum oscillators.

Experiments measuring 'negative temperature' actually measure something different. When Purcell and Pound achieved population inversion in nuclear spins (1951), or when Braun et al. created similar states in ultracold atoms (2013), they extracted temperature by fitting exponential distributions to occupation probabilities. This procedure yields the Boltzmann temperature T_B, not the thermodynamically consistent Gibbs temperature T_G. Near entropy maxima in bounded spectra, the fitted slope can diverge and change sign while other definitions remain finite.

Consider nuclear spins in a magnetic field. A population where most spins occupy the higher energy level represents population inversion. The Boltzmann formalism assigns negative temperature to this state, suggesting it is 'hotter than hot'—energy flows from it to any positive-temperature system. The Gibbs formalism assigns high but positive temperature, recognising that adding energy to an already inverted population decreases the number of accessible configurations. Both formalisms agree on energy flow direction, but only Gibbs maintains mathematical consistency.

In systems with bounded spectra, entropy $S(E)$ rises from the ground state, peaks at some energy E^*, then falls as energy approaches its maximum. The derivative $(\partial S/\partial E)_{N,V}$ changes sign at E^*, causing $T = 1/[k_B(\partial S/\partial E)]$ to jump discontinuously from $+\infty$ to $-\infty$. The inverse temperature $\beta \equiv 1/(k_B T) = (\partial S/\partial E)_{N,V}$ avoids this discontinuity, varying smoothly from $+\infty \to 0 \to -\infty$ as energy increases. The canonical probability $p \propto e^{-\beta\epsilon}$ works uniformly: $\beta > 0$ favours low energies, $\beta = 0$ gives uniform distribution, $\beta < 0$ favours high energies. Physical observables emerge as derivatives with respect to β, not T, making β the natural variable.

Everyday macroscopic systems have unbounded spectra, so β remains positive. Finite-state quantum systems—spin ensembles in strong fields, nuclear spins in crystals—explore the full range. What textbooks call 'negative temperature' is better understood as negative β: a regime where adding energy decreases entropy as the system approaches its highest-energy bound.

Proponents of negative temperature argue these states enable Carnot engines with efficiency exceeding unity—extracting more work than the heat absorbed. Insert negative Boltzmann temperature into the Carnot efficiency $\eta = 1 - T_c/T_h$ and efficiencies above 100% seem possible. Such calculations violate thermodynamic consistency. The Gibbs temperature, always positive, forbids perpetual motion of the second kind. Moreover, creating and destroying population inversion requires non-adiabatic work that standard efficiency formulas do not account for.

Temperature functions as a label for thermal equilibrium, a slope in entropy space, and the control parameter for probability distributions over states. These roles converge in ordinary matter but diverge in engineered quantum systems. The Boltzmann entropy, despite its prevalence in textbooks, fails consistency tests. The Gibbs entropy respects thermodynamic principles at the cost of forbidding negative absolute temperature. Whether we accept systems 'hotter than infinity' or reject such phrasing depends on which of temperature's definitions we prefer. The universe, indifferent to our debates, continues to maximise entropy by whatever name we call it.

Statistical Mechanics of Negative Absolute Temperature

The Entropy-Temperature Connection

Statistical mechanics defines temperature through the relationship between entropy S and energy E: $1/T = \partial S/\partial E$. One common choice is the Gibbs entropy $S_G = k_B \ln \Omega(E)$ with $\Omega(E)$ the integrated density of states. In unbounded systems, Ω always increases with E, yielding $T > 0$. In systems with maximum energy E_{max}, the density of states can decrease near E_{max}. Under a Boltzmann definition that uses the local density of states $\omega(E)$, $\partial S/\partial E$ can become negative, corresponding to negative T in that convention.

Two-Level System: A Clear Example

Consider N spins, each with energy 0 (down) or ϵ (up). With n spins up, the total energy is $E = n\epsilon$ and the number of configurations is $\Omega(n) = \binom{N}{n}$. The entropy $S = k_B \ln \Omega$ peaks at $n = N/2$ (half spins up). For $n < N/2$, adding energy increases entropy; for $n > N/2$, adding energy decreases entropy. In a two-level model one can define an effective inverse temperature via the slope:

$$\beta \equiv \frac{1}{k_B T} = \left(\frac{\partial S}{\partial E}\right)_{N,V}.$$

At $n = N/2$: $\beta = 0$. Population inversion ($n > N/2$) corresponds to $\beta < 0$ under the Boltzmann convention.

Energy Flow and 'Hotter Than Hot'

When two systems exchange energy, entropy maximisation determines the flow direction. For energy δQ flowing from A to B:

$$\Delta S_{\text{tot}} = \delta Q \left(\frac{1}{T_B} - \frac{1}{T_A}\right).$$

For spontaneous flow ($\Delta S_{\text{tot}} > 0$), we need $1/T_B > 1/T_A$. In the β parameterisation, the ordering is continuous:

$$0^+ < ... < +\infty \equiv -\infty < ... < 0^-.$$

Thus negative temperatures are 'hotter' than all positive temperatures—energy flows from any negative-T system to any positive-T system.

The Gibbs vs Boltzmann Debate

Dunkel and Hilbert (2014) highlighted a controversy about negative temperature by distinguishing two entropy definitions:

- Boltzmann: $S_B = k_B \ln[\omega(E)]$ using density of states $\omega = d\Omega/dE$.
- Gibbs: $S_G = k_B \ln[\Omega(E)]$ using integrated density of states.

For bounded systems where ω peaks and decreases, Boltzmann can give $T_B < 0$ while Gibbs gives $T_G > 0$ always. They argue only Gibbs entropy satisfies certain thermodynamic consistency conditions such as pressure relations. The precise mapping between experimentally inferred parameters and T_G depends on ensemble and constraints, and remains debated.

References:

Ramsey, N. F. (1956). *Thermodynamics and Statistical Mechanics at Negative Absolute Temperatures.* Phys. Rev., 103, 20.

Purcell, E. M., Pound, R. V. (1951). *A Nuclear Spin System at Negative Temperature.* Phys. Rev., 81, 279.

Dunkel, J., Hilbert, S. (2014). *Consistent Thermostatistics Forbids Negative Absolute Temperatures.* Nat. Phys., 10, 67.

A Plane Hat Trick

Top (Einstein Monotile – The Hat Tile): Discovered by David Smith in 2022 and formalised in 2023 with Kaplan, Myers, and Goodman-Strauss, this thirteen-sided polygon—nicknamed the 'hat'—was the first *proven aperiodic monotile, or einstein* (German for 'one stone'). It tiles the plane completely without repeating periodically. Smith's discovery also led to related tiles such as the 'spectre,' with matching-rule colourings, and to an infinite family of aperiodic monotiles.
Bottom (Imura Spiral Monotile): A symmetric monotile family discovered by Miki Imura in 2023. These tiles expand in four interlaced spiral arms and admit elegant 3-colourings. Although they can form periodic strip tilings, the illustrated spiral configuration highlights their mathematical and aesthetic appeal. The tile corners have angles of $3\pi/7$ and $4\pi/7$, permitting intricate lawful assemblies.

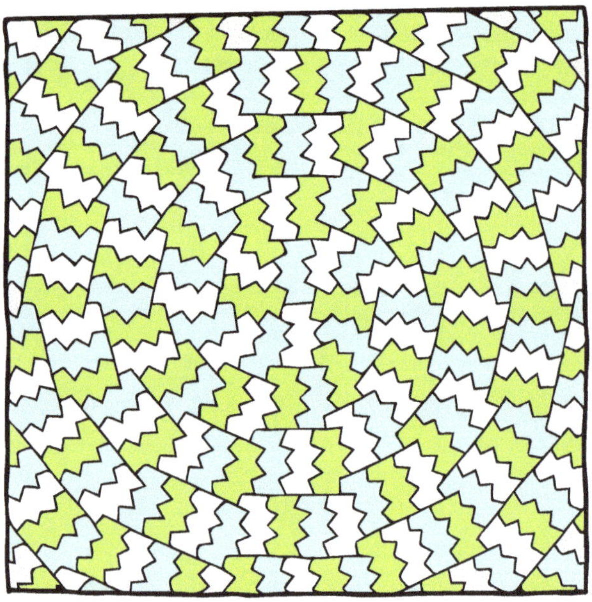

A Plane Hat Trick

The 'Einstein problem', named from German *'ein Stein'* (one stone), asks whether a monotile could tile the plane only aperiodically. In 2022, David Smith, a retired printer experimenting with shapes, discovered the thirteen-sided 'hat' that solved this sixty-year puzzle. The hat tiles infinitely without ever repeating its pattern. The study of tilings previously revealed a connection between mathematics and physical quasicrystals, a discovery that won a Nobel Prize in 2011.

Hat Monotile 2023 ◦ Aperiodic Tiling ◦ David Smith Discovery ◦ 13-Sided Einstein ◦ Penrose Tiles ◦ Local Forces Global ◦ Quasicrystal Connection ◦ Euler's Formula ◦ Substitution Tiling ◦ Spectre Variant ◦ Amateur Contributions

"There is something in such laws that takes the breath away.
They are not discoveries or inventions of the human mind,
but exist independently of us.
In a moment of clarity, one can at most discover
that they are there and take them into account.
Long before there were people on the earth,
crystals were already growing in the earth's crust.
On one day or another, a human being first came across
such a sparkling morsel of regularity lying on the ground
or hit one with his stone tool and it broke off and fell at his feet,
and he picked it up and regarded it in his open hand, and he was amazed."

— M. C. Escher, 1973

A Plane Hat Trick

Tiling problems bridge pure mathematics with the decorative arts across millennia. Islamic artisans developed intricate geometric patterns in the Alhambra and other architectural masterworks, exploring symmetry groups centuries before their mathematical formalisation. The Roman opus tessellatum, Byzantine mosaics, and Japanese tatami arrangements each integrated cultural rules into the way shapes fit together. These artistic traditions posed implicit mathematical questions: which patterns are possible, which are forbidden, and why?

Mathematical study of tilings began with natural questions. Kepler's 1619 investigation of hexagonal packing arose from observing snowflakes and honeycombs. His conjecture that hexagonal packing is optimal for circles remained unproven until Thomas Hales announced a computer-assisted proof in 1998; the full, refereed publication appeared in 2005. Dürer's 1525 treatise on measurement included systematic constructions of periodic tilings, blending Renaissance art with geometric precision.

The modern theory of aperiodic tiling emerged in the 1960s through Hao Wang's study of the domino problem (Wang, 1961): given a set of square tiles with coloured edges, can they tile the plane if adjacent edges must match colours? Wang initially conjectured that any tileable set must tile periodically, linking the question to formal logic and computability.

This conjecture was soon refuted. In 1966, Wang's student Robert Berger constructed the first known aperiodic tile set, though it required over 20,000 distinct tiles. Later refinements reduced the number significantly. In the 1970s, Roger Penrose advanced the field by introducing sets of two tiles that enforced aperiodicity using geometric constraints and local matching rules. These configurations, such as the kite and dart, became iconic examples of non-periodic order. In 1982, the discovery of quasicrystals by Dan Shechtman revealed that certain metallic alloys naturally exhibit aperiodic atomic structure, connecting tiling theory with physical materials.

Yet the search for a single connected shape—a monotile—that could tile the plane only aperiodically remained unresolved for decades. This so-called 'ein Stein' ('one tile') stood as a central open question in *tiling theory*.

In 2022, a breakthrough came from an unexpected collaboration. David Smith, a hobbyist with a long-standing interest in tiling, discovered a thirteen-sided polygon constructed from kites, which he called the 'hat.' Working with Joseph Myers, Craig Kaplan, and Chaim Goodman-Strauss, Smith demonstrated that this tile could indeed tile the plane only aperiodically, relying solely on its geometry without any matching rules or markings. Their result resolved the Einstein problem in its original geometric form and marked a major milestone in the mathematical theory of tilings.

Mathematics analyses geometric arrangements by converting visual problems into numerical equations. Take any convex polyhedron—a cube, a pyramid, or something more exotic—and count three things: vertices (corners), edges (where faces meet), and faces (flat surfaces). No matter how complex the shape, these three numbers satisfy: $V - E + F = 2$.

 https://doi.org/10.11647/OBP.0526.29

This is Euler's formula, discovered in 1750. A cube has 8 vertices, 12 edges, and 6 faces: $8-12+6=2$. A triangular pyramid has 4 vertices, 6 edges, and 4 faces: $4-6+4=2$. A soccer ball (truncated icosahedron) has 60 vertices, 90 edges, and 32 faces: $60-90+32=2$. Always two. Define the Euler characteristic of a space as $\chi = V - E + F$. This quantity connects topology to combinatorics—it remains constant as we stretch or deform the polyhedron, provided we don't tear or puncture it.

Why does this number stay fixed under deformations? Consider any vertex where three edges meet—like the apex of a pyramid or the corner of a cube or just a triple junction on the plane. Replace that vertex with a small triangle. This operation adds 2 new vertices, 3 new edges, and 1 new face. The Euler characteristic becomes $(V+2)-(E+3)+(F+1) = V - E + F$. The value is unchanged. Repeated applications of this 'vertex truncation' are part of the common proof that χ is invariant under deformations (by triangulating surfaces/graphs).

The Platonic solids—those perfect forms where identical regular p-gons meet q at every vertex—obey this constraint. At each vertex, q faces converge, and each regular p-gon contributes an interior angle of $(p-2)180°/p$ degrees. For the solid to close without leaving gaps, these angles must sum to less than 360°: $q \times (p-2)180°/p < 360°$, which simplifies to $(p-2)(q-2) < 4$. Since p and q must each be at least 3 (three edges form a polygon, three faces form a vertex), only five integer solutions exist: $(3,3)$, $(3,4)$, $(4,3)$, $(3,5)$, and $(5,3)$. These correspond to the tetrahedron, octahedron, cube, icosahedron, and dodecahedron. Euler's formula converts 'what perfect forms exist?' into 'which integers satisfy $(p-2)(q-2) < 4$?'

In this chapter, we explore another geometric problem that is historically one of the richest intersection points between mathematics and art. When arranging shapes to cover an infinite plane without gaps or overlaps, local constraints again determine global outcomes. Here too, the interplay of vertices, edges, and faces obeys mathematical laws, but now applied to infinite configurations rather than closed surfaces.

In a large circular patch of any edge-to-edge tiling, along the boundary, the formula needs correction terms, but as the patch grows, the interior dominates. In the plane, $V - E + F$ equals one (not two—the plane has different topology than a sphere). Dividing by the number of faces and taking limits, we get $1/\bar{p} + 1/\bar{q} = 1/2$, where $\bar{p}$ is the average sides per tile and $\bar{q}$ is the average tiles per vertex. This constraint explains why only three regular tilings exist: triangles ($p = 3$, $q = 6$), squares ($p = 4$, $q = 4$), and hexagons ($p = 6$, $q = 3$). It also proves that in any tiling, the average polygon has at most six sides—explaining why bees chose hexagons and why Islamic artists never tiled mosques with regular heptagons.

Tiling the plane refers to covering the infinite flat surface of Euclidean geometry with repeated, gapless copies of one or more shapes, not necessarily polygons. The problem: given a shape, can it tile the plane? If so, does it do so uniquely, periodically, or in multiple distinct ways? The arrangement must leave no gaps or overlaps and must cover the entire plane.

Classical tilings exhibit periodicity. That is, a finite patch of tiles can be shifted—translated—along certain vectors to cover the entire plane without change. This is the case for squares, equilateral triangles, and regular hexagons, all of which tile the plane in grid-like or honeycomb arrangements. The periodicity implies symmetry: the whole tiling looks the same from multiple viewpoints. This repetition reflects the symmetry group of the tiling, which includes discrete translations and, often, rotations or reflections.

Between periodic and random tilings lies a third category: aperiodic tilings. These are constructed by deterministic rules yet never repeat under any translation. In canonical substitution examples (such as Penrose and the hat), every finite patch reappears infinitely often (repetitivity), but always in new contexts—never aligning with a translated copy of itself. This paradox of local recurrence without global periodicity became central to twentieth-century tiling theory.

The question of how many tiles are needed to achieve different tiling behaviours evolved rapidly. With an infinite set of distinct tiles, constructing aperiodic tilings is straightforward—each new tile can be unique, forcing non-repetition. Berger (1966) first showed that aperiodic tilings could be achieved with a finite set of 20,426 tiles. This number dropped rapidly: Robinson (1971) reduced it to six tiles, then Penrose (1974) achieved it with just two.

Penrose tiles represented a milestone: two shapes that, when used together with matching rules, could tile the plane only aperiodically. They gave the first explicit example of enforced aperiodicity in Euclidean space using only two tiles. They spurred new directions in mathematical logic. The undecidability of the domino problem—whether a given set of tiles can tile the plane—had already been established by Berger.

Tiling theory is connected through group theory to other natural phenomena. In 1982, Dan Shechtman observed quasicrystals—materials whose atoms arrange aperiodically (Shechtman et al., 1984) yet produce sharp diffraction spots. Classical crystallography required periodic atomic arrangements; X-ray diffraction of crystals produces discrete spots because periodic structures act as three-dimensional diffraction gratings. Quasicrystals shattered this dogma. Their diffraction patterns show perfect five-fold symmetry, mathematically impossible in periodic crystals. The atoms follow deterministic rules such as Penrose tilings—local matching conditions that propagate to create long-range order without translational symmetry. Aluminium-manganese alloys cooled at specific rates form icosahedral quasicrystals, their atoms arranged in three-dimensional analogues of Penrose patterns.

Back to finite tilings. The progression from infinite sets to thousands to just two tiles led to the 'ein Stein' problem: could a single tile enforce aperiodicity? For decades, every attempt failed. In 2022, David Smith, a retired printer, discovered a thirteen-sided polygon that solved it. The 'hat' tile forces aperiodic tiling through its shape alone—a genuine monotile (Smith et al., 2023).

The hat's thirteen edges meet at specific angles that tightly constrain how neighbours can attach. Place one tile, and its neighbours must fit into the concave and convex indentations in only certain ways. These constraints propagate: each new tile placement further restricts its surroundings. The accumulation of these forced choices prevents any translational symmetry from emerging at larger scales.

Any cluster of hat tiles you identify will appear again elsewhere—rotated, reflected, embedded in different surroundings, but recognisable. This property, called local isomorphism, means the tiling contains infinite copies of every finite pattern, yet not periodically forever.

Smith discovered the hat tile through manual experimentation with paper cutouts. He recognised its aperiodic behaviour and contacted Craig Kaplan, Joseph Myers, and Chaim Goodman-Strauss. The team proved aperiodicity using substitution tiling theory: they showed the hat generates a substitution tiling where larger 'metatiles' decompose into smaller copies following strict rules. The proof required verifying that no periodic tiling exists by analysing the tile's hierarchical structure. They later discovered the 'Spectre'—a variant that tiles only with direct congruent copies, solving the stronger 'vampire einstein' problem where reflections are forbidden.

The hat tile solves a half-century-old problem. A single shape that tiles the plane only aperiodically. Its thirteen-sided form sits at a critical boundary—the minimal complexity needed to encode non-repetition in pure geometry. Like quasicrystals in nature, the hat achieves long-range order without translational symmetry, demonstrating that deterministic rules can generate infinite variety. From Euler's constraint on finite polyhedra to the hat's constraint on infinite tilings, mathematics reveals how local geometry determines global possibility.

The Human Underneath the Hat

David Smith was a retired print technician in Bridlington, England, experimenting with paper cutouts after designing shapes in PolyForm Puzzle Solver. He noticed one thirteen-sided shape resisted periodic arrangement. He contacted Craig Kaplan, Joseph Myers, and Chaim Goodman-Strauss. They proved the hat tile aperiodic through substitution tiling theory.

Robert Ammann sorted mail when he independently rediscovered Penrose tiles in the 1970s. Marjorie Rice was a California homemaker who read Martin Gardner's column on pentagonal tilings in 1975. She discovered four new families of convex pentagons that tile the plane. Joan Taylor, an amateur mathematician in Tasmania, found the first disconnected aperiodic monotile.

Within weeks of Smith's announcement, people made hat quilts, cookies, and 3D prints. The PolyForm Puzzle Solver community began exploring variations. The team discovered the Spectre variant and an entire continuum of aperiodic tiles. Smith called the excitement 'a bit surreal'—he'd been playing with shapes for years, cutting them out on cardstock, arranging them on his kitchen table.

Why Does Every Soccer Ball Always Have Exactly 12 Pentagons?

A soccer ball pattern tiles a sphere with pentagons and hexagons, where exactly three faces meet at each vertex. Let P = number of pentagons and H = number of hexagons. Each pentagon has 5 edges and each hexagon has 6 edges, but every edge is shared by exactly 2 faces:

$$E = \frac{5P + 6H}{2}.$$

At each vertex, 3 faces meet. Each pentagon contributes 5 vertices and each hexagon contributes 6, but every vertex is triple-counted:

$$V = \frac{5P + 6H}{3}.$$

The total number of faces is simply:

$$F = P + H.$$

Applying Euler's formula for a sphere ($V - E + F = 2$):

$$\frac{5P + 6H}{3} - \frac{5P + 6H}{2} + P + H = 2.$$

Multiply through by 6:

$$2(5P + 6H) - 3(5P + 6H) + 6P + 6H = 12,$$

$$10P + 12H - 15P - 18H + 6P + 6H = 12,$$

$$P = 12.$$

Therefore, any sphere tiled with pentagons and hexagons (with 3 faces per vertex) must have exactly 12 pentagons, regardless of the number of hexagons. This applies to soccer balls, fullerene molecules, and geodesic domes alike.

Allowed Tilings via Local Rules: SFTs, Wang Tiles, and Decidability

Setup: Fix a finite alphabet $\mathcal{A}$. A configuration is $x \in \mathcal{A}^{\mathbb{Z}^d}$. A finite *forbidden list* $\mathcal{F}$ is a set of patterns $p : S \to \mathcal{A}$ with $S \subset \mathbb{Z}^d$ finite. The *allowed tilings* (a shift of finite type, SFT) are

$$X(\mathcal{F}) = \{x \in \mathcal{A}^{\mathbb{Z}^d} : \text{no translate of any } p \in \mathcal{F} \text{ occurs in } x\},$$

For $d = 2$, nearest-neighbour SFTs correspond exactly to Wang tilings: unit square tiles with coloured edges, where colours must match across adjacent edges.

Positive Baseline (1D): For $d = 1$, any nearest-neighbour SFT is determined by a finite directed graph G with adjacency matrix A. Then

$$|\{\text{allowed words of length } n\}| = \mathbf{1}^\top A^{n-1}\mathbf{1}$$

$$h_{\text{top}}(X) = \lim_{n\to\infty} \frac{1}{n} \log(\mathbf{1}^\top A^n \mathbf{1}) = \log \rho(A),$$

and *emptiness* is decidable: $X \neq \emptyset$ iff G contains a directed cycle.

Undecidability in 2D (Domino Problem): Problem: Given $\mathcal{F}$ (or a set of Wang tiles), decide if $X(\mathcal{F}) \neq \emptyset$. **Theorem (Berger, Robinson):** There is no algorithm solving the Domino Problem. **Idea:** Encode the space–time diagram of a Turing machine with local constraints. Non-emptiness of $X(\mathcal{F}_M)$ is equivalent to the existence of an infinite valid computation.

Aperiodic but Allowed:

$$\exists \text{ finite tile sets } \mathcal{T} \text{ with } X(\mathcal{T}) \neq \emptyset \text{ but no } x \in X(\mathcal{T}) \text{ is periodic.}$$

Local matching rules can therefore force non-periodicity.

Expressiveness (Simulation Theorem): Every effective subshift $Y \subset \mathcal{B}^{\mathbb{Z}}$ (whose forbidden words form a recursively enumerable set) is a factor of some nearest-neighbour $\mathbb{Z}^2$ SFT $X(\mathcal{F})$. Consequences: arbitrary recursively enumerable constraint systems, prescribed entropies, and computable dynamical encodings can be realised by 2D allowed tilings.

Tractable Subclass (Planar Dimers): For planar bipartite graphs, allowed tilings by dominoes (dimers) are exactly perfect matchings. The number of tilings satisfies

$$Z = \#\{\text{allowed tilings}\} = |\det K|,$$

where K is the Kasteleyn-signed bipartite adjacency (Kasteleyn) matrix.

This gives polynomial-time counting and closed-form correlation functions. The solvability derives from planar Pfaffian structure rather than general local rules.

Takeaways:

- 1D allowed tilings: matrix methods; emptiness and entropy are computable.
- 2D allowed tilings: emptiness, periodicity, and many properties are undecidable.
- Restricted planar dimers remain exactly solvable via Pfaffian techniques.

References:

Berger, R. (1966). *The Undecidability of the Domino Problem.* Mem. AMS.

Robinson, R. M. (1971). *Undecidability and Nonperiodicity for Tilings of the Plane.* Invent. Math.

Kasteleyn, P. W. (1961–67). *Dimer Statistics and Pfaffians.* J. Math. Phys., Physica.

Divide and Conquer

Top (Simpson's Paradox – Admissions): At the aggregate level, a higher percentage of men are admitted compared to women, suggesting gender bias. But disaggregating by department reveals that women applied more often to highly competitive departments with low acceptance rates, while men applied to departments with higher admission rates. Within departments, women were often admitted at equal or higher rates. This reversal—where an apparent bias in subgroups contradicts the overall trend—is known as *Simpson's Paradox.* **Bottom (Survivorship Bias – Cancer in Smokers):** Cancer incidence increases sharply with age, but smokers have higher early mortality from other causes. As a result, fewer smokers survive into the high-risk age brackets where cancer becomes common. This skews the population-level data, making it seem as though smokers have lower cancer rates than non-smokers. The truth is hidden by *survivorship bias*: smokers often die before cancer can occur.

Divide and Conquer

Simpson's Paradox occurs when a statistical trend present in separate groups reverses when the groups are combined. This effect is a result of unequal group sizes or hidden confounding variables that distribute non-uniformly across the data. For example, a treatment might show positive effects in both male and female subgroups yet appear harmful in the aggregate population if the treatment is disproportionately given to patients with more severe conditions and males and females differ in average severity. The apparent paradox demonstrates that causal inference requires careful consideration of the causal relationship rather than relying solely on raw correlations.

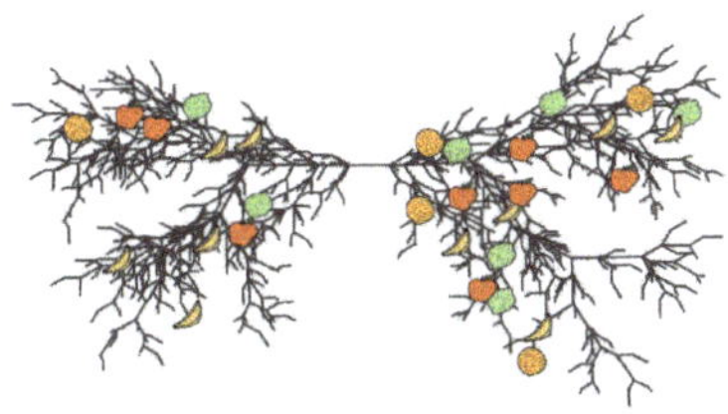

Simpson's Paradox ◦ Correlation Reversal ◦ Weighted Averages ◦ Kidney Stone Example ◦ Gerrymandering Math ◦ UC Berkeley Bias ◦ Subgroup vs Aggregate ◦ Confounding Variables ◦ Statistical Grouping ◦ North Carolina 2012 ◦ Data Partitioning

"If you torture the data long enough,
it will confess to anything."
— Ronald H. Coase

"Correlation is not causation,
but they keep close company."
— Huw Price, 2026

Divide and Conquer

Early statisticians, including G. Udny Yule (1903) and Karl Pearson, documented reversals that occur when aggregated data obscure subgroup relationships. In several early case studies, trends within groups differed from the pooled trend, foreshadowing the formal statement later articulated by Simpson.

Edward H. Simpson published a four-page paper in 1951 in the Journal of the Royal Statistical Society. He proved that for any proportions $a/b < c/d$ and $e/f < g/h$, it remains algebraically possible that $(a + e)/(b + f) > (c + g)/(d + h)$. Simpson's formulation established the general conditions under which combining data groups reverses their individual trends.

The paradox gained widespread attention through the 1973 Berkeley graduate admissions analysis by P. J. Bickel, E. A. Hammel, and J. W. O'Connell. Raw data showed women had a 35% acceptance rate compared to 44% for men. Department-specific analysis revealed no discrimination—women had equal or higher acceptance rates in four of six departments. Women applied more frequently to highly competitive departments with much lower acceptance rates, while men applied more to less selective departments. The differing application patterns created the aggregate disparity.

Simpson's result appears in medical trials when treatments are assigned based on patient severity, in election analysis when votes are aggregated by district, and in machine learning when training sets are partitioned. Colin R. Blyth coined the term 'Simpson's paradox' in 1972, though Udny Yule had described similar reversals in 1903. The phenomenon is sometimes called the Yule-Simpson effect.

The same mathematics that creates accidental reversals enables deliberate manipulation through gerrymandering. Elbridge Gerry signed a redistricting bill in Massachusetts in 1812 that created a salamander-shaped district to concentrate opposition voters, giving the practice its name. The underlying strategy of manipulating representation, however, has deeper roots in British 'rotten boroughs', which were used to control elections by packing (concentrating them in a few districts) or cracking (splitting them across many) voters.

With the rise of computer-assisted mapping after the 1990 census, gerrymandering became a science, allowing partisan mapmakers to secure durable advantages in evenly divided electorates. After a mid-decade redistricting in Texas in 2003 reshaped the congressional balance of power, courts increasingly employed mathematical diagnostics—such as the efficiency gap, mean-median difference, and ensemble simulation tests—to determine when district boundaries purposefully waste opposition votes through packing or cracking.

In North Carolina's 2012 congressional elections, Democratic candidates received about 51% of votes statewide yet won only 4 of 13 seats. Republicans secured 9 seats with roughly 49% of votes. The reversal occurred through district boundaries–lines drawn to group voters in ways that inverted the relationship between votes and representation.

Statistical association hinges on how data are grouped. The same population can yield opposite conclusions depending on the partition chosen. In extreme cases, a relationship positive in every subgroup becomes negative when groups combine—or a democratic majority becomes a legislative minority through strategic line-drawing.

Simpson's paradox demonstrates correlation reversal (Simpson, 1951). A kidney stone treatment shows 93% success for small stones and 73% for large stones. A competing treatment achieves only 87% for small stones and 69% for large stones. The first treatment beats the second in both categories. Yet overall success rates reverse: 79% versus 85.5%. The reversal arises because doctors used the superior treatment primarily on difficult cases—70% of its patients had large stones, while 91% of the inferior treatment's patients had small stones.

Gerrymandering engineers deliberate reversal. Both major parties employ this tactic when they control redistricting. Wisconsin's 2012 state assembly elections saw Democrats win 53% of votes but only 39% of seats. The mechanism: district lines packed Democratic voters into urban districts where they won by 70–80% margins, while Republican victories spread efficiently across suburban and rural districts with 55–60% margins. Both phenomena—Simpson's paradox and gerrymandering—exploit the mathematics of aggregation, but with different intent. The mathematics underlying both phenomena reduces to weighted averages. When calculating any aggregate statistic—whether treatment success rates or electoral outcomes—the result depends on two factors: the values within each group and the relative sizes of groups. Change either factor and the aggregate changes.

In formal terms, if groups have success rates $p_1, p_2, ..., p_k$ and sizes $n_1, n_2, ..., n_k$, the overall rate is:

$$\bar{p} = \frac{\sum_{i=1}^{k} n_i p_i}{\sum_{i=1}^{k} n_i}.$$

Simpson's paradox occurs when natural imbalances in group sizes (n_i) cause $\bar{p}$ to misrepresent the relationship seen in individual p_i values. Gerrymandering manipulates the same formula by choosing group boundaries to engineer specific (n_i) values.

The kidney stone example illustrates the reversal:

	Success Rate	Patients	Successes
Treatment A, Small stones	93%	30	28
Treatment B, Small stones	87%	200	174
Treatment A, Large stones	73%	70	51
Treatment B, Large stones	69%	20	14
Treatment A, Overall	79%	100	79
Treatment B, Overall	85.5%	220	188

Treatment A wins in both stone categories yet loses overall. Treatment A handled 70% difficult cases (large stones), Treatment B only 9%. When groups combine, Treatment B's easy-case bias overwhelms its inferior performance.

Gerrymandering employs similar mathematics with manipulative intent. Consider a simplified state with 10 districts, 5 million voters split evenly between parties:

District	Voters (A)	Voters (B)	Winner	Strategy
1	20%	80%	B	Packed (B stronghold)
2	22%	78%	B	Packed (B stronghold)
3	18%	82%	B	Packed (B stronghold)
4	56%	44%	A	Cracked (A edge win)
5	55%	45%	A	Cracked (A edge win)
6	54%	46%	A	Cracked (A edge win)
7	57%	43%	A	Cracked (A edge win)
8	53%	47%	A	Cracked (A edge win)
9	56%	44%	A	Cracked (A edge win)
10	55%	45%	A	Cracked (A edge win)
Totals	**50%**	**50%**	**A wins 7, B wins 3**	Gerrymandered for A

In the UC Berkeley admissions case (1973), similar patterns appeared. Women showed lower overall acceptance rates (35%) than men (44%), suggesting discrimination. At the department level, women had equal or higher acceptance rates in 4 of 6 departments. The reversal occurred because women disproportionately applied to competitive departments—English admitted 3.4% of applicants while Engineering admitted 65%.

Simpson's paradox and gerrymandering exploit the disagreement between local and global measures. In Simpson's paradox, local measures (department-specific admission rates) tell the truth while global measures (overall rates) mislead. In gerrymandering, local measures (district-level victories) are manipulated to distort global truth (statewide voter preference).

For any partition of data into groups, the overall average equals:

$$\bar{y} = \sum_i w_i \bar{y}_i$$

where $w_i = n_i/N$ represents the fraction of data in group i, and $\bar{y}_i$ is that group's average.

Simpson's paradox exposes existing groupings in the data; gerrymandering constructs groupings to exploit the same arithmetic.

The efficiency gap quantifies gerrymandering's success (Stephanopoulos & McGhee, 2015) by measuring 'wasted' votes—those beyond the 50%+1 needed to win a district. A party that wins districts by slim margins while losing others by wide margins achieves maximum efficiency. The formula:

$$\text{Efficiency Gap} = \frac{|\text{Wasted}_A - \text{Wasted}_B|}{\text{Total Votes}}.$$

Values around 7–8% have been proposed in the political science literature as a heuristic threshold for durable advantage; courts have not adopted a single standard, and experts treat it as one indicator among others. Thus, gerrymandering leaves fingerprints. Districts snake through neighbourhoods, splitting cities and joining disparate communities. Pennsylvania's

7th district (pre-2018) stretched like tentacles across five counties to link Republican areas while avoiding Democratic ones. Maryland's 3rd district exhibits similar contortions, engineered by Democrats to dilute Republican votes across Baltimore suburbs.

Simpson's paradox is revealed through careful analysis. Statisticians discover reversals by examining subgroups. Early COVID-19 comparisons illustrated how age structure can confound: countries with older populations showed higher overall death rates even when age-specific rates were comparable. Proper age standardisation is necessary before drawing conclusions.

Simpson's paradox warns that natural parameters (patient severity, department selectivity) can mislead when ignored. Gerrymandering demonstrates that artificial boundaries can be weaponized to subvert democratic representation.

Solutions to Simpson's paradox require disaggregating data and examining subgroups. Medical trials now routinely report results by patient characteristics. Universities analyse admissions by department.

The solution to gerrymandering requires judicial reform: independent redistricting commissions, mathematical constraints on district compactness, or algorithmic districting that minimises partisan advantage. Several states now use efficiency gap calculations in legal challenges to districting plans.

What are the odds a bomb hits the only person holding three guavas?

More Statistical Paradoxes and Interpretation Failures

1. Berkson's Paradox *Conditioning on a common effect induces spurious negative (Berkson, 1946) correlation.* If two independent variables both affect a selection criterion, then restricting attention to cases that satisfy that criterion creates an artificial negative correlation. This occurs in hospital datasets, where independent risk factors may appear inversely related when conditioned on admission. The association is real in the conditional data but does not reflect a relationship in the population.

2. Ecological Fallacy *Group-level associations are wrongly projected onto individuals (Robinson, 1950).* When a statistical association holds across aggregated units—such as regions or schools—it does not necessarily hold within them. For example, a country with higher average education may have higher average income, but this does not imply that more educated individuals earn more within each region. Unlike Simpson's paradox, ecological fallacy involves misapplying group-level trends to individual inference without requiring any reversal. The error lies in cross-level extrapolation, not confounding.

3. Will Rogers Phenomenon *Reclassification improves group averages without improving any member.* If individuals from the low end of one group are reclassified into another group with an even lower average, both groups may show improved mean outcomes. This occurs in cancer staging and school performance tracking, and reflects the fact that averages are sensitive to how groups are defined.

4. Modifiable Areal Unit Problem (MAUP) *Statistical results depend on the choice of spatial or administrative (Openshaw, 1984)ative boundaries.* In spatial analysis, correlations and rates can shift significantly depending on how geographic regions are aggregated. A pattern observed at the county level may not hold at the district level or when boundaries are redrawn.

5. Low Birth-Weight Paradox *Conditioning on an intermediate variable reverses risk comparisons.* Infants born to smoking mothers have higher rates of low birth-weight, and low birth-weight is associated with higher mortality. But among low birth-weight babies, those born to smokers may show lower mortality than those of non-smokers. The paradox appears because birth-weight is both an effect of smoking and a predictor of mortality. Conditioning on it introduces collider bias, obscuring causal direction.

6. Prosecutor's Fallacy *Confusing the likelihood of evidence with the probability of guilt.* In forensic contexts, the probability of observing the evidence assuming innocence is often mistaken for the probability of innocence given the evidence. For example, a DNA match with a false positive rate of $1/1000$ is incorrectly interpreted as implying a 0.1% chance of innocence, ignoring base rates. The fallacy reflects improper inversion of conditional probability.

Two Mathematical Realisations of Simpson's Paradox

Reversal in Pearson Correlation

Suppose two subgroups yield:

$$\mathrm{Corr}(X, Y \mid Z = 1) = +0.8,$$
$$\mathrm{Corr}(X, Y \mid Z = 2) = +0.7,$$

yet the marginal correlation is:

$$\mathrm{Corr}(X, Y) = -0.3.$$

This reversal can occur when the subgroup means oppose each other:

$$\mathbb{E}[X \mid Z = 1] \ll \mathbb{E}[X \mid Z = 2],$$
$$\mathbb{E}[Y \mid Z = 1] \gg \mathbb{E}[Y \mid Z = 2].$$

The total covariance decomposes as:

$$\mathrm{Cov}(X, Y) = \mathbb{E}[\mathrm{Cov}(X, Y \mid Z)] + \mathrm{Cov}(\mathbb{E}[X \mid Z], \mathbb{E}[Y \mid Z]).$$

The first term represents the true structural relationship. The second term arises from between-group mean shifts. When subgroup trends are consistent but means shift in opposite directions, this second term can dominate and flip the sign.

In such cases, the subgroup correlation reflects the actual relationship between the variables. The marginal correlation is an artefact of mixed distributions and should not be used to infer these relationships.

Why Subgroup Correlations Reflect Structure

The Pearson correlation coefficient assumes a homogeneous population. When data consist of subgroups (e.g., children vs adults), the overall correlation reflects two effects:

- the correlation within each group,
- the shift in means across groups.

This decomposes as:

$$\mathrm{Corr}(X, Y) = \text{within-group structure} + \text{between-group shift}.$$

If the subgroups differ in both $\mathbb{E}[X \mid Z]$ and $\mathbb{E}[Y \mid Z]$, the between-group term may dominate and flip the marginal sign—even if each group has a positive internal trend.

Subgroup correlations hold Z fixed and reveal how X relates to Y when background is controlled. The marginal correlation, in contrast, entangles structure with population imbalance.

For variable relationships inference—e.g., how height relates to foot size, or how score relates to study time—$\mathrm{Corr}(X, Y \mid Z)$ provides the interpretable relationship. The marginal $\mathrm{Corr}(X, Y)$ may be distorted by mixing.

Visual example: Imagine both kids and adults show that larger plates come with higher calorie counts. But if kids mostly use small plates and pile them with calorie-dense snacks, while adults take large plates but fill them with vegetables, the overall data may suggest that smaller plates correspond to higher calories. This reflects sample composition, not individual-level relationships.

How Likely is Simpson's Paradox?

Pavlides and Perlman (2009) studied how often Simpson's paradox arises in $2 \times 2 \times 2$ contingency tables. Under a uniform distribution over all such tables, they showed:

$$\boxed{\text{1 in 60 tables exhibits a reversal.}}.$$

This corresponds to a prior probability of ≈ 0.0166 that conditional trends align while the aggregate trend opposes them.

The paradox becomes rarer with more subgroups; under similar uniform assumptions, the chance decreases further as the number of conditioning groups increases.

References:

Simpson, E. H. (1951). *J. R. Stat. Soc. B*, 13(2), 238–241.

Pavlides, M. G., and Perlman, M. D. (2009). *J. Stat. Plan. Inference*, 139(1), 198–213.

Concentrate on Osmosis

Top (Diffusion Model of Osmosis): Historically, osmosis was explained as a type of diffusion: solute concentration differences cause water to move from low to high solute concentration to 'even things out.' This intuitive model treats the membrane as a passive barrier and water flow as driven by statistical mixing.

Second (Gas Pressure Analogy): Water molecules are thought of as a vapour-like phase. The side with more solute has fewer free water molecules, reducing its effective vapour pressure. This creates a pressure imbalance across the membrane, driving water toward the more concentrated side.

Third (Virial Theorem Approach): Here, osmotic pressure is derived from molecular interactions—akin to pressure in gases. Solute particles exert directional momentum transfer through collisions, and the semi-permeable membrane selectively blocks these, resulting in net force buildup.

Fourth (Chemical Potential Explanation): Osmosis is understood in terms of chemical potential gradients. Water flows from high to low chemical potential, and solutes lower the chemical potential of water. This aligns with thermodynamic definitions and governs equilibrium conditions.

Bottom (Membrane Force Model – Debye's View): In this modern (1923) mechanistic picture, the membrane itself plays an active role. It exerts differential mechanical forces on solute and solvent. Osmotic flow results from water being pulled across in response to these localised interactions.

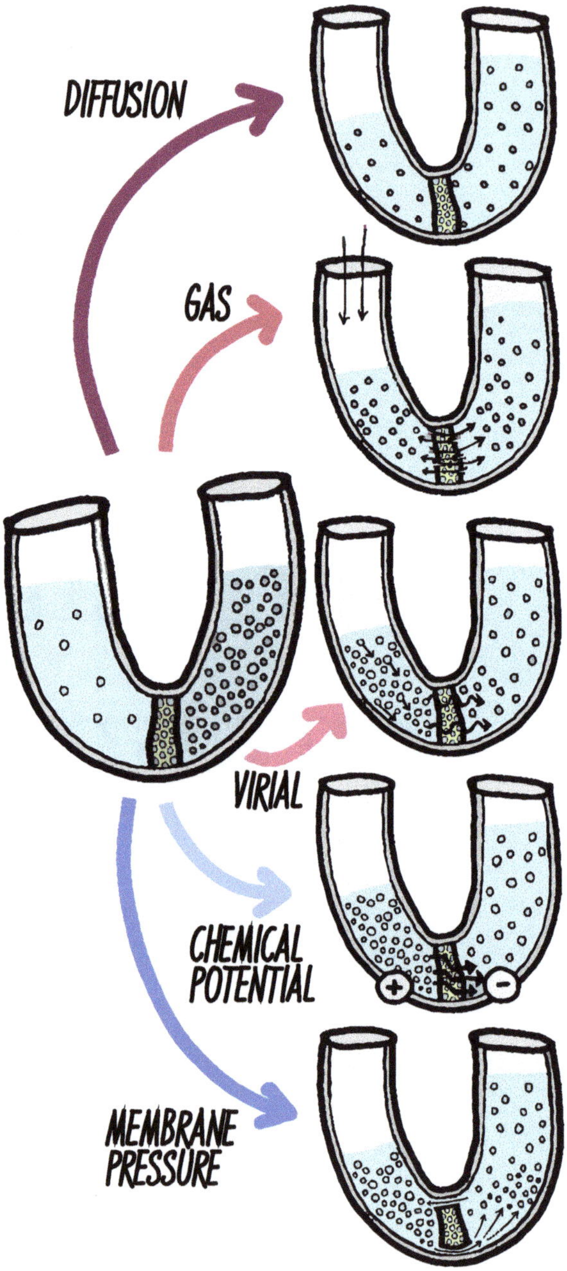

Concentrate on Osmosis

Standard osmosis explanations based solely on water concentration gradients fail to account for measured flow rates that far exceed diffusion limits. The ratio of osmotic permeability to diffusive permeability (Pf/Pd) commonly exceeds 100 in biological systems with aquaporins, while purely diffusive transport would yield a ratio near 1. Mechanical explanations, notably Debye's model, attribute osmosis to pressure gradients arising from solute-membrane interactions rather than simple diffusion. When solutes are excluded by a semipermeable membrane, their momentum cannot transfer across the boundary, creating a localised pressure drop that drives water movement.

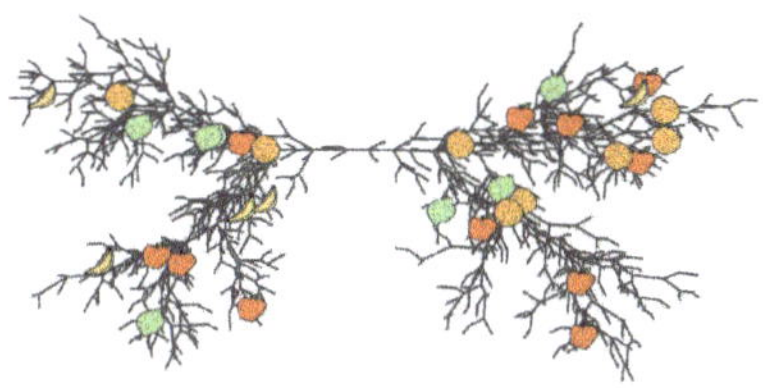

Osmosis Misconceptions ∘ Debye Interface Theory ∘ Membrane Selectivity ∘ P_f/P_d Transport Ratio ∘ Van 't Hoff Law ∘ Vegard Pressure Drop ∘ Virial Theorem ∘ Solute-Solvent Collisions ∘ Aquaporin Channels ∘ Galaxy Cluster Dynamics ∘ Interfacial Forces

"There is, I think, no more wonderful and illuminating spectacle than that of an osmotic growth—a crude lump of brute inanimate matter germinating before our very eyes."

— W. Deane Butcher, 1911

Concentrate on Osmosis

The phenomenon of osmosis was first described by Jean-Antoine Nollet in 1748 after observing fluid movement through animal membranes. In the 1820s, Henri Dutrochet introduced the terms 'endosmosis' and 'exosmosis,' and in 1854 Thomas Graham coined the term 'osmosis.' In the 1880s, Jacobus van 't Hoff derived a quantitative expression for osmotic pressure in dilute solutions that followed the same form as the ideal gas law. The result linked the behaviour of solutes in solution to molecular motion, reinforcing the emerging statistical view of thermodynamics.

Around the same time, in 1877, Wilhelm Pfeffer sealed sugar solution inside a porous ceramic pot lined with copper ferrocyanide, creating the first truly semipermeable membrane. He placed the pot in pure water and connected it to a mercury manometer (a pressure gauge). The mercury column climbed as water entered the pot, sometimes generating pressures over 20 atmospheres.

This was the first quantitative demonstration of osmotic pressure as a real mechanical force, measurable in the same way as gas or hydrostatic pressure. Van 't Hoff quickly recognised the analogy to ideal gases and used Pfeffer's results to formulate $\Pi = cRT$.

By the early twentieth century, osmosis was widely interpreted through the lens of diffusion: water was thought to move from regions of high to low concentration. However, alternative models emerged. In 1908, Lars Vegard proposed that solutes excluded from a membrane could generate local pressure differences. Peter Debye formalised this idea in 1923, modelling how solute collisions with a semipermeable barrier result in a force imbalance that drives water flow. Debye's model treated osmotic flow as mechanically generated rather than purely thermodynamic.

Although consistent with van 't Hoff's law at equilibrium, Debye's explanation emphasised local interactions at the membrane interface. The model was largely ignored in favour of equilibrium thermodynamics until it was revisited in the late 20th and early 21st centuries by researchers such as Gerald Manning and Alan Kay. Their work highlighted a discrepancy between persistent yet inaccurate textbook descriptions and the physical theory of osmosis.

Osmosis is introduced as the movement of water across a semipermeable membrane from a region of 'high water concentration' to one of 'low water concentration.' The phrasing appears in educational contexts ranging from middle school biology to university-level biophysics. The logic is derived from diffusion theory and implicitly models water as a dilute substance within itself, moving in response to its own number density gradient.

The description uses kinetic gas theory, where particles are modelled as non-interacting points executing straight-line motion between binary, elastic collisions. In a concentration gradient, more particles move from high-density regions to low-density regions than that in the reverse direction, producing a net flux. The flux is described by Fick's law (Fick, 1855), $J = -D\nabla c$, where D is the diffusion coefficient and c is the local number density. The law is derived under the assumption that particle motion is uncorrelated, that average free paths are long, and that interparticle forces are negligible.

The validity of this description depends on the gas being sufficiently dilute such that spatial correlations and momentum transfer between particles can be neglected over relevant time scales. The equilibrium state corresponds to uniform particle density and maximised configurational entropy, that is, the system is at its most disordered. The model accurately predicts behaviour for many inert gases under laboratory conditions.

When applied to water in the liquid phase, the model fails. Water molecules interact continuously through hydrogen bonds and short-range repulsion, so each molecule's motion is constrained by its neighbours. There is no regime in which water behaves as a gas of independently diffusing particles. Instead, motion involves correlated displacements and propagates mechanical stress through a dense, hydrogen-bonded network.

The concept of a 'water concentration gradient' lacks meaning in a solvent composed entirely of water. There is no distinct diffusing species; rather, any molecular displacement must displace others. Water cannot respond to a local number density gradient in the manner of an ideal gas. The semipermeable membrane selectively blocks solute molecules while allowing solvent to pass.

Let's introduce some units. P_f and P_d are permeabilities with units of m/s. L_p is the hydraulic permeability of the membrane with units of m/(Pa·s), relating volume flux per area to an effective pressure difference.

Water transport is quantified by two coefficients: the osmotic permeability P_f and the diffusive permeability P_d. The Fundamental Law of Osmosis states that volume flux per unit area is:

$$\Phi_V = L_p(\Delta P - RT\Delta c_s),$$

where L_p is hydraulic permeability, $\Delta P$is the pressure difference, and Δc_s is the osmolarity difference. Hydrostatic pressure and osmotic gradients produce identical water flux through the same coefficient L_p, with $P_f = L_pRT/V_w$ where V_w is the molar volume of water.

The dimensionless ratio P_f/P_d distinguishes transport mechanisms across different membranes. In pure lipid bilayers, $P_f/P_d = 1$, indicating purely diffusive transport of independent water molecules. For synthetic collodion membranes, studies reported P_f/P_d ranging from roughly 36 to 730, demonstrating predominantly convective flow. Biological membranes containing aquaporins show intermediate values, with P_f/P_d typically on the order of 10, despite water moving in single file through these channels.

Let's take a step back and review theoretical frameworks for osmosis that were proposed over the years. Each captures different aspects of osmotic flow, most lacking a complete mechanistic picture or accordance with experimental data.

The **kinetic gas model** treats solute particles as an ideal gas exerting pressure on the membrane. In this view, solute molecules bombard the membrane like gas molecules against a container wall, creating pressure $\Pi = nk_BT/V = cRT$where n is particle number, k_B is Boltzmann's constant, and c is molar concentration. The model correctly predicts van 't Hoff's law for dilute solutions (van 't Hoff, 1887) but fails for concentrated solutions where solute-solute interactions become significant. It also provides no mechanism for how this pressure drives water flow through the membrane.

The **chemical potential framework** describes osmosis as water moving to equalise its chemical potential across the membrane (Gibbs, 1876). The chemical potential of water decreases when solute is added: $\mu = \mu_0 + RT\ln(x_w)$ where x_w is the water mole fraction. Water flows from high to low chemical potential until equilibrium is reached. While thermodynamically rigorous, the chemical potential formulation restates the equilibrium condition without explaining the molecular forces that drive flow. Chemical potential is a state function, not a force.

The **hydration shell model** proposes that solute molecules bind water in hydration layers, reducing 'free' water concentration. Water then diffuses down this concentration gradient. However, hydration is a dynamic process with water molecules exchanging between bulk and hydration shells on picosecond timescales. No static population of 'bound' versus 'free' water exists. Complete hydration of all solutes would reduce water concentration by less than 1% in typical solutions, insufficient to explain observed osmotic pressures.

The physical source of osmotic flux lies at the interface. When solutes are excluded from one side of the membrane, they cannot impart momentum beyond the boundary. A local pressure deficit forms near the membrane on the solute-rich side. This asymmetry in solute–solvent collisions creates the driving force.

Peter Debye identified this mechanism in the early twentieth century. (This is the same Debye who advanced electrical conduction theory by introducing phonons in 1912 and electrostatic screening in 1923, as discussed in Chapter 4). Solute molecules striking the membrane generate an anisotropic momentum distribution. Water molecules on the other side encounter no such imbalance. Net solvent flux moves toward the region with solute, driven by a measurable pressure difference confined to the interface.

The transport requires no global difference in solvent concentration, only at the interface.

This interface phenomenon creates what Lars Vegard identified in 1908 as a pressure profile across the membrane. The **Vegard pressure drop** occurs at the membrane-solution interface where solute molecules cannot penetrate. On the solution side, pressure drops by $\Pi = cRT$ just inside the membrane. Since pressures in bulk solutions are equal, a pressure gradient must exist within the membrane, driving water from the pure solvent side to the solution side.

The **virial theorem** relates pressure to molecular forces and positions. It is basically relating the total force acting on a boundary with average forces of the elements bouncing around inside it. In statistical mechanics, pressure arises from momentum transfer at boundaries:

$$P = \frac{Nk_BT}{V} + \frac{1}{3V}\left\langle \sum_{i<j} \mathbf{r}_{ij} \cdot \mathbf{F}_{ij} \right\rangle .$$

The first term represents kinetic pressure from molecular motion. The second term accounts for intermolecular forces, where $\mathbf{r}_{ij}$ is the separation vector and $\mathbf{F}_{ij}$ is the force between molecules i and j. When solutes cannot cross the membrane, their force contributions to the pressure on that side vanish locally, creating the pressure imbalance that drives flow.

In astronomy, the virial theorem relates kinetic and potential energy in gravitationally bound systems. For a stable cluster of stars or galaxies:

$$2\langle K\rangle + \langle U\rangle = 0,$$

where K is kinetic energy and U is gravitational potential energy. Galaxy clusters violating this relation indicate either instability or the presence of dark matter. Fritz Zwicky first applied the virial theorem to the Coma cluster in 1933, inferring far more mass than could be accounted for by visible matter. His calculation provided early evidence for dark matter—the virial theorem exposed missing mass through dynamics alone.

In osmosis, the virial theorem quantifies interfacial forces. When solutes are excluded from a membrane interface, their contributions to the virial sum are absent, and the computed pressure reflects that deficit.

Osmotic flow persists despite equal hydrostatic pressure across a membrane because the local stress asymmetry at the interface produces solvent flux. The system reaches equilibrium when this interfacial pressure is exactly offset by an applied hydrostatic pressure, not when water concentrations equalise.

Biological systems demonstrate these mechanical principles. Capillary walls contain pores on the order of a few to tens of nanometres—much larger than water molecules (0.3 nm)—allowing bulk liquid flow consistent with the Debye–Vegard model. In cell membranes, aquaporin channels permit water to traverse in single file. Despite this confinement, P_f/P_d remains large. The enhancement cannot be attributed to faster diffusion or increased cross-sectional area. Selective solute exclusion generates the interfacial pressure gradients, whether in wide capillary pores or narrow protein channels.

Osmotic shock in red blood cells illustrates the mechanism. If red blood cells are placed in pure water, they swell and burst (haemolysis). Put them in concentrated saline, and they shrivel. Both outcomes occur because osmotic pressure differences of just a few hundred milliosmoles correspond to tens of atmospheres of mechanical stress. The biconcave shape is maintained only within a very narrow osmotic window. The cell membrane excludes solutes and establishes a boundary-layer stress. The cytoskeleton cannot withstand the imbalance if it grows too large.

Plant turgor converts osmotic pressure into support. Plant cells possess rigid cellulose walls that resist osmotic influx, so the internal pressure (turgor) builds until it supports the entire structure of leaves and stems. A wilting plant is simply one in which osmotic potential no longer generates sufficient pressure to keep cell walls stretched. Solute exclusion at membranes generates a pressure deficit that becomes stable internal pressure, measurable in atmospheres, supporting the tissue mechanically.

A membrane that excludes solute while admitting solvent generates directional pressure when intermolecular forces are non-negligible—a boundary condition (as opposed to purely thermodynamic effects).

Among the various explanations for osmosis—diffusion gradients, chemical potentials, hydration shells, kinetic pressure—only the mechanical picture addresses the central question: what forces drive water through the membrane? The answer lies not in abstract thermodynamic equilibrium conditions but in concrete analysis of molecular collisions at an asymmetric boundary. Debye's insight, formalised through the virial theorem, shows osmosis as a mechanical phenomenon. Water moves because real forces push it.

Reverse osmosis exploits this mechanical nature by applying external pressure to overcome the natural osmotic gradient. When pressure exceeding the osmotic pressure is applied to the solution side, water flows from high to low concentration—the reverse of spontaneous osmosis. The technique remained impractical until 1959, when Sidney Loeb and Srinivasa Sourirajan at UCLA developed asymmetric cellulose acetate membranes with a dense skin layer atop a porous support. Their membranes could withstand high pressures while maintaining substantial water flux, making desalination economically feasible for the first time.

Confusion-based diffusion.

Membrane Forces and Competing Models of Osmotic Flow

Three models compete to explain osmotic transport: thermodynamics (equilibrium only), diffusion (concentration gradients), and mechanical (interfacial forces). All yield van 't Hoff's law but differ in mechanism and predictive power.

1. Thermodynamic Model

Chemical potential equilibration yields:

$$\Delta P = RT\Delta c_s.$$

Correct for equilibrium but provides no mechanism, flux rates, or explanation for P_f/P_d ratios.

2. Diffusion Model

Water moves down concentration gradient:

$$\Phi_D = -D_w \nabla c_w.$$

Predicts $P_f = P_d$, contradicting experiments where P_f/P_d ranges from 10–730. Cannot explain convective flow or single-file transport.

3. Mechanical Model (Debye-Vegard)

Solute–membrane collisions create local pressure drop:

$$\Phi_V = L_p(\Delta P - RT\Delta c_s).$$

Solute exclusion generates pressure gradient $dP/dx = c_s F$, yielding Vegard drop:

$$\Delta P_{\text{interface}} = RTc_s.$$

Water flows through membrane due to real pressure gradient, not concentration difference.

4. Consequences for Permeability and Flow

The pressure drop across the membrane explains high P_f/P_d ratios and unifies osmotic and pressure-driven flow:

$$\Phi_V = -L_p \frac{dP}{dx}, \quad \text{(Darcy-like flow)}.$$

This model correctly predicts convective water transport in porous membranes and aquaporin-containing systems. In pure lipid bilayers lacking such channels, solute exclusion is absent and $P_f/P_d = 1$.

5. Comparative Summary

The thermodynamic model correctly predicts equilibrium but provides no mechanism or dynamics. The diffusion model offers dynamics but fails to match the magnitude and direction of flow in most membranes. The Debye–Vegard model provides dynamics, a clear mechanism, and explains the observed P_f/P_d. The mechanical pressure model distinguishes itself by explicitly identifying the origin of osmotic force and unifying the formalism with standard fluid mechanics.

References:

Debye, P. (1923). Physikalische Eigenschaften von Lösungen und Theorie der Elektrolyte. *Phys. Z.* 24:334–338.

Manning, G.S., and Kay, A.R. (2023). The physical basis of osmosis. *J. Gen. Physiol.* 155:e202313332.

Vegard, L. (1908). Zur Theorie der osmotischen Erscheinungen. *Proc. Camb. Phil. Soc.* 15:13–23.

Finkelstein, A. (1987). Water Movement Through Lipid Bilayers, Pores, and Plasma Membranes. *Wiley.*

Timing Is Everything

Top (Mechanical and Early Timekeeping): The image begins with sundials—using cast shadows to track solar position. Mechanical clocks follow: weight-driven escapements, pendulums, and balance springs. These systems introduced regular, countable oscillations, enabling precision independent of sunlight.
Middle (Electronic and Atomic Clocks): The transition to electronic timing brought quartz crystal oscillators, tuning forks, and circuit-based digital clocks. Atomic clocks, such as the cesium beam and rubidium standards, use microwave transitions in atoms to define the second with extreme precision. These devices revolutionised navigation, telecommunications, and metrology.
Bottom (Next-Generation Nuclear Clocks): The current frontier involves optical lattice and nuclear clocks—probing energy levels in nuclei rather than electrons. These offer unprecedented temporal stability and resolution. If perfected, they could detect minute variations in gravity, test general relativity, and redefine the basic unit of time.

Timing Is Everything

Timekeeping has progressively moved toward smaller physical phenomena: from Earth's rotation to pendulums, from crystal oscillations to atomic transitions, and now toward nuclear resonances. The SI second, defined by 9,192,631,770 periods of cesium-133's hyperfine transition, relies on quantum interactions between nuclear and electronic magnetic moments. This shift to microscopic reference standards improves precision exponentially—hydrogen masers achieve stability of one part in 10^{13}, while optical lattice clocks using strontium reach one part in 10^{18} by probing transitions at $\sim 10^{15}$ Hz. The progression continues toward nuclear clocks using thorium-229, which promises precision of one part in 10^{19} by exploiting transitions in atomic nuclei rather than electron shells.

TIME AS COORDINATE ∘ CAESIUM-133 STANDARD ∘ 9192631770 HZ ∘ HYPERFINE TRANSITIONS ∘ OPTICAL LATTICE CLOCKS ∘ FREQUENCY COMBS ∘ GPS TIME CORRECTIONS ∘ THORIUM-229 NUCLEAR CLOCK ∘ CHRONOMETRIC GEODESY ∘ PROPER TIME INVARIANCE ∘ FUNDAMENTAL PHYSICS PROBE

„Zeit ist das, was man an der Uhr abliest."

("Time is what a clock measures.")
— Albert Einstein, 1926

Timing Is Everything

Era / Technology	Accuracy	Size (m)	Time Reference
Sundials (1800 BCE)	10^{-2}	10^{1}	Solar shadow on gnomon
Ancient Water Clocks	10^{-3}	10^{-1}	Liquid level change
Verge Clocks (13th c.)	10^{-2}	10^{-1}	Crown wheel / verge foliot
Pendulum Clocks (1656)	10^{-5}	10^{0}	Pendulum arc length
Marine Chronometer (18th c.)	10^{-6}	10^{-2}	Balance spring oscillator
Quartz Oscillators (1930s)	10^{-8}	10^{-3}	Crystal thickness (MHz mode)
Ammonia Maser (1953)	10^{-9}	10^{-10}	NH_3 inversion barrier
Cesium Beam Standard (1955)	10^{-10}	10^{-10}	^{133}Cs hyperfine structure
Hydrogen Maser (1960s)	10^{-13}	10^{-10}	^{1}H hyperfine structure
Rubidium Vapour	10^{-11}	10^{-10}	^{87}Rb hyperfine structure
Cesium Fountain (1990s)	10^{-15}	10^{-10}	Interference of free atoms
Optical Lattice (2010s)	10^{-18}	10^{-10}	Atomic dipole transitions
Projected Thorium Nuclear	10^{-20}	10^{-14}	Intrinsic nuclear excitation

Time is a coordinate assigned to events, an ordering imposed on phenomena, and a physical quantity whose measurement depends on the reproducibility of periodic processes. The challenge in defining time stems from its dual character: operationally, time is what clocks measure, but physically, clocks are systems that embody time through the regularity of their transitions. A theory of time must therefore address both its measurement and its assignment.

Historically, time was defined by external reference. A day was one full rotation of the Earth, a year one revolution around the Sun. These intervals were directly observable but not uniform. Earth's rotation slows due to tidal friction, and its orbit varies minutely from year to year. As clocks improved, it became clear that astronomical cycles were not perfectly periodic. They were also not universally accessible.

Modern definitions turn inward. Time is now anchored to the internal configuration of matter. A clock is a system that undergoes periodic change—a pendulum, a quartz crystal, or a quantum oscillator—and time is defined by counting these cycles. The second is defined by the *Système international d'unités* (SI). The SI second is defined as exactly 9,192,631,770 periods of the hyperfine transition of the cesium-133 atom.

Still, the concept of time requires consistency. If time is relative—as in special and general relativity—how can clocks agree? The answer lies in local invariance and synchronisation protocols. In special relativity, time intervals are frame-dependent, but proper time—the time measured along an observer's worldline—is invariant. In general relativity, the curvature of spacetime causes time to flow at different rates in different gravitational potentials. Atomic clocks confirm that indeed identical devices tick faster at altitude than at sea level. Yet these variations are predictable and correctable.

To coordinate time across systems and locations, one defines a reference frame and applies relativistic corrections. Global time standards, such as International Atomic Time (TAI), are constructed by ensemble averaging signals from many atomic clocks, each corrected for gravitational potential and velocity. Clocks that are precise but prone to long-term drift are periodically resynced against clocks that are more accurate; the ensemble combines the strengths of both. The result is a global timescale without assertion of universal time.

Time also enters theoretical physics as a parameter. In Newtonian mechanics, time is absolute and flows uniformly. In quantum mechanics, it appears as an external parameter in the Schrödinger equation (Schrödinger, 1926). In general relativity, time is a coordinate entangled with space, whose flow is determined by the metric tensor. Time serves as an index that parameterizes change.

In quantum field theory and statistical mechanics, time appears asymmetrically. The microscopic laws are time-reversal symmetric, yet macroscopic systems exhibit irreversibility. The asymmetry is imposed not by the equations, but by boundary conditions and coarse-graining (the replacement of a detailed description with a statistical one). The direction of time—the arrow from past to future—emerges from the configuration of initial conditions and the growth of entropy (Boltzmann, 1877).

The measurement principle is that time is a relation between events measured by clocks as intervals. The definition principle is that the flow of a system is the unfolding of configurations in accordance with dynamical laws that govern evolution from past to future.

Physical timekeeping builds upon these principles. The invariance of atomic transitions allows time to be physically instantiated as a countable quantity, realised through interactions with matter that exhibit extraordinary regularity. Atomic clocks operationalize time by coupling electromagnetic fields to well-defined quantum transitions—processes governed by the internal energy levels of atoms. These transitions occur at precise frequencies determined by the laws of quantum electrodynamics and the values of fundamental constants, making them immune to most environmental and instrumental variations. The resulting periodicity is intrinsic.

In the case of cesium-133, the phenomenon that defines the second is the hyperfine splitting of its ground electronic state. The splitting arises from the interaction between two magnetic moments: that of the nucleus, which acts as a tiny bar magnet due to its intrinsic spin, and that of the valence electron, whose magnetic field is generated by both its orbital motion and its intrinsic spin. These moments couple through the magnetic dipole interaction, producing a small energy difference between two configurations. Quantum mechanically, the total angular momentum of the atom is given by $\vec{F} = \vec{I} + \vec{J}$, where $\vec{I}$ is the nuclear spin and $\vec{J}$ the total electronic angular momentum. In cesium-133, which has nuclear spin $I = 7/2$ and electronic angular momentum $J = 1/2$ in its ground state, this coupling results in two hyperfine levels: $F = 4$ and $F = 3$.

The transition between these levels occurs at a microwave frequency of 9.192631770 GHz. Because this energy difference is sharply defined and identical for all cesium-133 atoms in isolation, it serves as a natural frequency reference. The transition is measured by subjecting a cloud of cesium atoms to a tunable microwave field while monitoring population redistribution between the two states. When the applied frequency matches the energy gap—satisfying the resonance condition $E = h\nu$—atoms undergo induced transitions, which can be detected via state-selective fluorescence or ionisation. In practice, a feedback loop adjusts the microwave oscillator to maximise this transition probability. The resulting frequency is then divided electronically to produce the one-second interval. The process defines the second as the number of cycles of this specific atomic transition.

Two distinct qualities characterise clock performance: *precision* (or stability) measures how consistently a clock reproduces the same interval from one tick to the next, while *accuracy* measures how closely that interval matches the true definition of the second. A clock can be extraordinarily precise—ticking with negligible variation over hours—yet slowly drift from the correct frequency over months. Conversely, a clock can be accurate on average but noisy tick-to-tick. The best timekeeping systems must excel at both.

Hydrogen masers generate coherent microwave radiation at 1.42 GHz via stimulated emission between hyperfine levels of atomic hydrogen. Their short-term frequency stability, driven by long coherence times in a wall-coated storage bulb, surpasses that of most other clock types. Although long-term drift limits their use as absolute standards, they serve as exceptional flywheel oscillators in timekeeping ensembles, bridging intervals between recalibrations from more accurate devices.

Rubidium clocks—especially chip-scale atomic clocks (CSACs)—offer compact, energy-efficient timing solutions for portable and embedded applications. These systems exploit optical pumping to polarise a vapour of ^{87}Rb atoms and monitor resonant microwave transitions via changes in transmitted light. The clock output disciplines an internal quartz oscillator, yielding fractional stabilities on the order of 10^{-11} to 10^{-12}, sufficient for GPS receivers, telecommunications, and low-power navigation.

Optical lattice clocks improve precision by probing narrow-linewidth electronic transitions in neutral atoms confined within standing-wave laser fields. At the 'magic wavelength' (species-dependent), the differential AC Stark shift between clock states vanishes (Katori, 1999), preserving the transition frequency despite optical confinement. Atoms such as strontium and ytterbium offer transition frequencies near 10^{15} Hz, and interrogation times exceeding one second yield quality factors above 10^{17}. These systems achieve fractional instabilities below 10^{-18}. Optical frequency combs enable comparison to microwave references, bridging domains and facilitating global synchronisation.

With such precision, relativistic effects become measurable and essential. Identical clocks placed at different gravitational potentials accumulate proper time at different rates due to gravitational redshift (Pound & Rebka, 1960). The shift $\Delta f/f = gh/c^2$ enables vertical positioning to centimetre resolution—the basis of chronometric geodesy. GPS satellites, which orbit at 20,200 km, exhibit both special relativistic time dilation (from orbital velocity) and gravitational blueshift (from altitude). Pre-launch frequency offsets and onboard corrections account for the net gain of approximately 38 microseconds per day, maintaining sub-metre positional accuracy.

Nuclear clocks aim to surpass atomic standards by exploiting transitions in the atomic nucleus, which are orders of magnitude less sensitive to electric and magnetic perturbations. The thorium-229 isomer exhibits the lowest known nuclear excitation energy—approximately 8.3 eV—placing it within reach of laser spectroscopy in the vacuum ultraviolet. Its long radiative lifetime implies a millihertz-scale natural linewidth, suggesting a potential quality factor above 10^{19}. Two architectures dominate experimental development. In ion-trap systems, individual $^{229}Th^{3+}$ ions are confined by radiofrequency fields, laser-cooled, and interrogated using high-resolution VUV (vacuum ultraviolet) frequency combs. In the solid-state approach, thorium nuclei are embedded in wide-bandgap optical crystals such as CaF_2 or MgF_2. These hosts suppress internal conversion decay and enable parallel interrogation of large ensembles. Challenges include spectral broadening from lattice inhomogeneity, background fluorescence, and the engineering of narrowband, stable VUV sources. Recent experiments have reported increasingly precise energy determinations and quantum-resolved spectroscopy of the transition, with rapid progress toward routine laser control.

The implications of nuclear timekeeping extend beyond metrology. Due to the fine balance of nuclear forces, the ^{229}Th isomer is predicted to be hypersensitive to variations in the fine-structure constant, scalar field couplings, or violations of local position invariance. Networks of synchronised thorium clocks could detect transient dark matter interactions or topological defects via correlated frequency excursions—simultaneous shifts in transition frequency caused by passing field disturbances that temporarily alter the local values of fundamental constants. Timekeeping becomes a probe of fundamental physics.

What was once derived from the rotation of celestial bodies is now defined by invariant atomic structure—and may soon be defined by the nucleus, whose internal dynamics offer a new frontier for precision and for discovery.

The Seven SI Base Units

Quantity	Unit (Symbol)	Definition
Time	second (s)	9,192,631,770 periods of the cesium-133 hyperfine transition
Length	metre (m)	Distance light travels in vacuum in 1/299,792,458 of a second
Mass	kilogram (kg)	Fixed by setting Planck's constant $h = 6.62607015 \times 10^{-34}$ J·s
E. Current	ampere (A)	Fixed by setting elementary charge $e = 1.602176634 \times 10^{-19}$ C
Temperature	kelvin (K)	Fixed by setting Boltzmann constant $k_B = 1.380649 \times 10^{-23}$ J/K
Amount	mole (mol)	Exactly $6.02214076 \times 10^{23}$ elementary entities
Lum. Intensity	candela (cd)	Fixed by setting $K_{cd} = 683$ lm/W at 540 THz

The 2019 redefinition marked the final transition from artefact-based standards—such as the kilogram prototype stored in Paris—to definitions anchored in immutable fundamental constants, ensuring universal reproducibility without dependence on physical objects that degrade, drift, or require secure storage.

Exercises

Fusible Numbers: Exercises in Constructive Time

A classic riddle: given two candles, which each burn for 1 h, how can you measure 45 m?

Fusible numbers form a well-ordered subset of the rationals constructed iteratively from zero. A number z is fusible if there exist previously constructed fusible numbers x and y, with $|x - y| < 1$, such that $z = (x + y + 1)/2$. The construction corresponds to lighting a unit-time fuse at both ends with delay. Results about the growth of certain associated functions are linked to very fast-growing hierarchies and have connections to independence from Peano arithmetic.

1. The Fuse Construction
For a unit fuse lit at time x on one end and time y on the other (with $|x - y| < 1$), prove it extinguishes at $z = (x + y + 1)/2$. Consider the burn dynamics when both flames are active.

2. Enumeration Below 2
Determine all fusible numbers < 2 by systematic application of the construction rule.

3. Dyadic Structure
Prove that all fusible numbers have the form $a/2^k$ for integers $a \geq 0$ and $k \geq 0$.

4. The Margin Function
Let a_n be the smallest fusible number exceeding n. Prove that

$$a_n = n + \frac{1}{2^{k(n)}}$$

for some $k(n) \in \mathbb{N}$. Compute $k(0)$, $k(1)$, $k(2)$. The function $k(n)$ grows extremely rapidly; certain asymptotic properties are connected to statements independent of Peano arithmetic.

5. Well-Ordering
Prove that the fusible numbers form a well-ordered subset of $\mathbb{Q}^+$. What does this imply about infinite decreasing sequences?

Context
Fusible numbers (Erickson, Xu) demonstrate how elementary constructions yield extremely fast growth. The margin values are:

$$a_0 = \frac{1}{2}, \quad a_1 = 1 + \frac{1}{8}, \quad a_2 = 2 + \frac{1}{1024}.$$

The growth behaviour connects to ordinal ε_0. (See Chapter 17)

Quantum Transitions and the Limits of Clock Stability

Atomic and nuclear clocks define time by referencing a sharply resonant transition between two quantum states. The frequency of this transition is determined by fundamental constants and is reproducible across identical systems. The precision with which this frequency can be measured depends on the linewidth of the transition, the stability of the interrogation system, and the protocol used to extract frequency information. This section formalises the mathematical quantities that govern frequency stability, relates them to the physical structure of the transition, and identifies the limits imposed by spacetime curvature and coupling constants.

Spectral Linewidth and Quality Factor

Let f_0 denote the central transition frequency and Δf the full width at half maximum (FWHM). The quality factor is defined by:

$$Q = \frac{f_0}{\Delta f}.$$

In Ramsey interrogation, $\Delta f \approx 1/(2T)$, where T is the free evolution time between pulses. Hence,

$$Q \approx 2f_0T.$$

Optical lattice clocks probing transitions in strontium or ytterbium atoms with $f_0 \sim 10^{15}$ Hz and $T \sim 1$ s routinely achieve $Q > 10^{15}$.

Allan Deviation and Averaging Behaviour

The fractional instability of a clock over averaging time τ is quantified by the Allan deviation:

$$\sigma_y(\tau) \approx \frac{1}{Q} \cdot \frac{1}{\text{SNR}} \cdot \sqrt{\frac{T_c}{\tau}},$$

where SNR is the signal-to-noise ratio and T_c the cycle time. Increasing Q, improving detection fidelity, and lengthening τ all contribute to reduced $\sigma_y(\tau)$.

Hyperfine and Nuclear Transition Energies

In cesium-133, the clock transition arises from magnetic dipole coupling between nuclear spin $\vec{I}$ and electron angular momentum $\vec{J}$, producing total angular momentum $\vec{F} = \vec{I} + \vec{J}$ and energy splitting:

$$E_F = \frac{A}{2}\left[F(F+1) - I(I+1) - J(J+1)\right].$$

The $F = 3 \leftrightarrow F = 4$ transition at $f_0 = 9.192\,631\,770$ GHz defines the SI second. In ^{229}Th, the nuclear excitation energy $E \approx 8.3$ eV corresponds to:

$$f_{\text{Th}} = \frac{E}{h} \approx 2.0 \times 10^{15}\,\text{Hz}, \quad Q_{\text{Th}} \gtrsim 10^{19}.$$

Relativistic Shift and Coupling Sensitivity

In general relativity, clocks at different gravitational potentials W accumulate proper time at different rates. The fractional frequency shift is:

$$\frac{\Delta f}{f} = \frac{\Delta W}{c^2} \approx \frac{gh}{c^2},$$

where g is gravitational acceleration and h the height difference. At 10^{-18} resolution, height differences of 1 cm are resolvable.

Clock transitions sensitive to the fine-structure constant α respond to coupling variations via:

$$\frac{\Delta f}{f} = K_\alpha \cdot \frac{\Delta\alpha}{\alpha},$$

where K_α is a dimensionless sensitivity coefficient. In nuclear systems such as ^{229}Th, this coefficient may exceed 10^4, amplifying the clock's utility in probing scalar fields or dark sector interactions.

References:

Ludlow, A. D. et al. (2015). *Optical atomic clocks.* Rev. Mod. Phys. 87(2), 637–701.

Safronova, M. S. et al. (2018). *Search for new physics with atomic clocks.* Rev. Mod. Phys. 90(2), 025008.

The Centre Holds

Top (Curse of Dimensionality – Vanishing Volume of the Sphere): As dimensionality increases, the volume of an n-sphere relative to the surrounding n-cube rapidly shrinks toward zero. In low dimensions, the sphere fills most of the cube; by around ten dimensions, it's practically gone. This shows that intuition from 2D or 3D fails in high dimensions—most volume concentrates in the corners. **Bottom (Blessing of Dimensionality – Human Uniqueness in High-D Spaces):** If each person is described by even fifty independent traits (drawn from uniform or Gaussian distributions), then the 'average' human lies in a vanishingly small region of space. Almost everyone is in the high-dimensional fringes—radically unique combinations of attributes. High dimensionality ensures that individuality is not rare but inevitable.

The Centre Holds

A geometric puzzle about Gaussian probability stumped mathematicians for over sixty years: prove that convex sets that are symmetric around the origin have enhanced overlap under Gaussian measure—that $P(A \cap B) \geq P(A) \cdot P(B)$. Despite partial results for boxes, ellipsoids, and slabs, the general case resisted all attempts. In 2014, Thomas Royen, a retired pharmaceutical statistician from a small German university, solved it using textbook methods: transforming to squared variables, applying Laplace transforms, and checking matrix determinants. His proof, published in an obscure journal, went unnoticed for years.

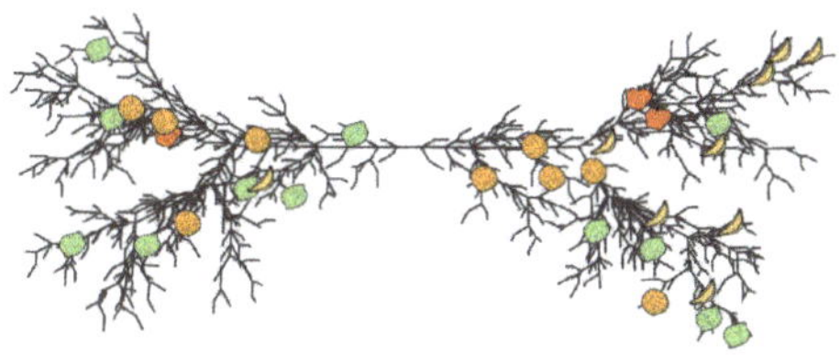

Gaussian Correlation Inequality ◦ Thomas Royen 2014 ◦ Convex Symmetric Sets ◦ High-Dimensional Geometry ◦ Concentration of Measure ◦ Squared Variables Method ◦ Laplace Transform Proof ◦ 60-Year Conjecture ◦ Multivariate Gaussian ◦ Elementary Solution ◦ Outsider Discovery

« Tout le monde y croit cependant, me disait un jour M. Lippmann, car les expérimentateurs s'imaginent que c'est un théorème de mathématiques, et les mathématiciens que c'est un fait expérimental. »

("Everyone believes in it, Mr. Lippmann told me one day, because experimentalists imagine it is a mathematical theorem, and mathematicians that it is an experimental fact.")
— Henri Poincaré on the Gaussian Distribution, 1912

The Centre Holds

Interest in how Gaussian measures behave under geometric constraints emerged in the mid-twentieth century, particularly in multivariate statistics and convex geometry. By the 1950s, researchers studying elliptical distributions began formulating conjectures about the probability content of intersections between symmetric convex regions.

The modern form of the Gaussian Correlation Inequality (GCI) was solidified in the 1970s through work by Das Gupta, Olkin, Pitt, and others, who framed it in terms of standard Gaussian measures over $\mathbb{R}^n$. They asked whether Gaussian probability favours overlap: specifically, whether the measure of the intersection of two symmetric convex sets is always at least as large as the product of their individual measures. The conjecture attracted attention because it combined natural geometric symmetry with the most analytically tractable probability distribution.

Over the following decades, progress was made in restricted settings. The inequality was proven for two-dimensional cases, for coordinate-aligned boxes, and for ellipsoids. The partial results relied on tools from real analysis, measure theory, and convex optimisation. The general case resisted all attempts, despite appearing elementary in formulation.

In 2014, a breakthrough came from Thomas Royen, a retired statistician with a background in pharmaceutical applications. Royen published a short paper that resolved the inequality in full generality. His approach was elementary in the technical sense: it used standard tools, required no heavy machinery, and invoked only modest linear algebra and probability. Nonetheless, it connected several overlooked identities in a way that previous attempts had not. Although initially unnoticed, Royen's proof was soon verified and reformulated in expository papers by Latała, Matlak, and others, and has since been accepted as the definitive solution to the GCI.

The Gaussian distribution, also called the normal distribution (Gauss, 1809), is defined by the density function

$$\varphi(x) = \frac{1}{\sqrt{2\pi}} e^{-x^2/2},$$

which describes the standard normal distribution, denoted $\mathcal{N}(0,1)$—zero mean, unit variance. More generally, a Gaussian with mean μ and variance σ^2, denoted $\mathcal{N}(\mu,\sigma^2)$, has density $\frac{1}{\sigma\sqrt{2\pi}} e^{-(x-\mu)^2/(2\sigma^2)}$; the parameter σ controls the width of the bell. The standard form above corresponds to $\mu = 0, \sigma = 1$. The function is symmetric about the mean, with values decreasing smoothly as $|x - \mu|$ increases. The rate of decay is exponential in the square of the distance, causing values far from zero to be exponentially rare. The total area under the curve is normalised to one, making it a valid probability distribution. Its characteristic bell shape is recognised as the canonical model for random variation in natural and statistical systems.

The bell curve emerges as the limiting form of many sums of random variables. Consider a process of repeatedly rolling a fair die and averaging the results. Although each individual roll yields a uniformly distributed outcome on a discrete set, the distribution of the average becomes increasingly smooth and Gaussian as the number of trials grows. The Central Limit Theorem states (Laplace, 1812) that the sum (or average) of independent, identically distributed variables with finite variance converges in distribution to the Gaussian, regardless of the original distribution.

This universality explains the Gaussian's omnipresence in nature. Heights in a population result from countless genetic and environmental factors—each contributing a small push up or down. Measurement errors accumulate from vibrations, temperature fluctuations, and quantum uncertainties. Stock prices reflect millions of independent trading decisions. In each case, myriad small influences combine additively, and their sum inevitably forms a bell curve. The Gaussian is not imposed by theory but emerges from the arithmetic of aggregation. The theorem explains why Gaussian distributions appear ubiquitously in statistical mechanics, measurement theory, and signal processing—wherever many small effects compound.

The multivariate Gaussian generalises this form to $\mathbb{R}^n$. The standard form has density

$$\varphi_n(x) = \frac{1}{(2\pi)^{n/2}} e^{-\|x\|^2/2},$$

where $\|x\|$ is the Euclidean norm of the vector $x \in \mathbb{R}^n$. The distribution is spherically symmetric: it assigns equal probability density to all points equidistant from the origin. Its contours are concentric spheres, and its value depends only on the radial distance. Every linear projection of the distribution onto a one-dimensional axis yields a standard univariate Gaussian. In the standard case, the coordinates are independent and identically distributed $\mathcal{N}(0, 1)$; by rotational invariance, this remains true in any orthonormal basis. Rotational invariance and marginal stability make the multivariate Gaussian a tractable object in high-dimensional probability.

In high dimensions, the geometry of Gaussian measure becomes profoundly unintuitive. Although the density is highest at the origin, the bulk of the probability mass concentrates near a thin spherical shell of radius approximately $\sqrt{n}$. This defies our three-dimensional experience: you might expect that since the Gaussian density peaks at the centre, most random points would be found there. This is completely wrong.

This is a result of the explosive growth of the number of points at a given radius. Consider an orange inside a cubic box. In three dimensions, the sphere fills a decent portion of the box. But as dimensions increase, the hypercube's corners dominate overwhelmingly. In 1000 dimensions, over 99.999% of the hypercube's volume lurks in its corners, not near the centre. The surface area of a sphere of radius r in $\mathbb{R}^n$ grows proportionally to r^{n-1}; in the radial density for $\|X\|$, this factor competes with the exponential term and pushes mass toward a thin shell.

The result is concentration of measure, which transforms probabilistic problems into geometric ones. A Gaussian random vector lies in a given region when that region intersects this nearly fixed-radius shell, rather than when it captures values near the origin.

The Gaussian Correlation Inequality concerns the probability that a standard Gaussian random vector $X \in \mathbb{R}^n$ simultaneously falls into two geometric regions. Let $A \subset \mathbb{R}^n$ and $B \subset \mathbb{R}^n$ be closed, convex sets that are symmetric about the origin. Then the inequality states:

$$\mathbb{P}(X \in A \cap B) \geq \mathbb{P}(X \in A) \cdot \mathbb{P}(X \in B).$$

The left-hand side is the probability that a single Gaussian sample lies in both sets, while the right-hand side is the product of the probabilities of lying in each separately. No notion of parametric correlation appears in this formulation—no Pearson coefficient, no covariance matrix interaction. The term 'correlation' here is geometric: it measures the extent to which the spatial configurations of the sets align so that overlap under the Gaussian measure is enhanced. Symmetric convex sets interact positively under Gaussian sampling.

Imagine a dartboard in high-dimensional space. Two target zones—each convex and mirror-symmetric about the centre—are drawn on the board. The dart is thrown not with uniform probability, but according to a Gaussian distribution. In our familiar world, you would expect the dart to land near the bullseye where the density is highest. But in high dimensions, the dart almost surely lands on a distant shell at radius $\sqrt{n}$. The magic of the GCI is that despite this shell phenomenon, symmetric convex sets still manage to overlap more than independence would predict. Their enforced central fatness—they cannot be hollow or lopsided—creates enough overlap at the origin's high-density region to overcome the dilution effect of the shell.

Both symmetry and convexity are essential to the validity of the inequality. If either condition is relaxed, the result can fail. For example, consider two non-convex shapes such as disconnected spherical caps placed symmetrically on opposite sides of the origin. Each may individually capture moderate Gaussian mass, but their intersection can be empty, rendering the left-hand side of the inequality zero while the right-hand side remains positive. Alternatively, take two convex balls shifted away from the origin in opposite directions: each maintains convexity, but the loss of symmetry means their overlap under Gaussian measure can be arbitrarily small, violating the inequality.

The unusual difficulty of proving the GCI arose from a geometric tug-of-war in high-dimensional space. The concentration of measure pushes probability mass outward to a distant shell, suggesting that intersection should be difficult—sets must somehow coordinate their overlap on this fragile, specific radius. But convexity and symmetry pull in the opposite direction: these shapes must be 'fattest' at the centre, they cannot be hollow or have their mass pushed outward. The conjecture, now proven, asserts that the central pull always wins.

Several equivalent formulations exist. The indicator function $\mathbf{1}_A(X)$ equals 1 when $X \in A$ and 0 otherwise, so its expectation is exactly the probability: $\mathbb{E}[\mathbf{1}_A(X)] = \mathbb{P}(X \in A)$. Rewriting the inequality in this notation,

$$\mathbb{E}[\mathbf{1}_A(X) \cdot \mathbf{1}_B(X)] \geq \mathbb{E}[\mathbf{1}_A(X)] \cdot \mathbb{E}[\mathbf{1}_B(X)],$$

the content is identical to the probability statement above, but the language shifts from set membership to random variables. This rephrasing is useful because it reveals the GCI as a *correlation* inequality in the literal statistical sense: the covariance $\text{Cov}(\mathbf{1}_A(X), \mathbf{1}_B(X)) \geq 0$. In other words, the two indicator random variables are nonnegatively correlated under the Gaussian measure—knowing that X lies in one symmetric convex set can only increase (never decrease) the likelihood that it lies in the other.

The Gaussian Correlation Inequality was conjectured in the 1950s and resisted proof for over six decades. During this time, it was confirmed in numerous special cases. For axis-aligned rectangles (boxes), the result was established by Šidák (1967). Other special cases—such as slabs and certain families of ellipsoids—were also resolved. Yet no general method succeeded. Classical techniques—log-concavity of Gaussian measure, the Brascamp–Lieb inequality, and concentration of measure (Brascamp & Lieb, 1976) phenomena—yielded related inequalities but stopped short of establishing the required correlation bound for arbitrary convex symmetric sets.

The proof came not from a well-known probabilist or a high-profile research programme, but from Thomas Royen, a retired statistician at a university of applied sciences in Bingen, Germany. Royen had worked for decades in applied statistics, particularly in pharmaceutical research. His academic career was spent outside the core research institutions of probability theory, and his publication record was modest by conventional standards. The outsider status provided the freedom to pursue classical problems without disciplinary constraint. Royen's mathematical training was solid but practical, shaped by applications and experience. He approached the problem of Gaussian correlation not as a convex analyst but as a statistician with an eye for transformations and distributions.

The central move in Royen's proof was to reframe the inequality in terms of squared Gaussian variables. By passing to variables of the form X_i^2, he translated the problem into one involving sums of independent gamma-distributed variables. The transformation allowed the introduction of Laplace transforms (LTs)—a standard tool in distributional analysis—and reduced the problem to showing monotonicity of a certain function defined by determinants of parameter-dependent covariance matrices. Royen employed an identity involving the determinant of a positive semi-definite matrix perturbed by diagonal terms, and used it to establish the required inequality via monotonicity in a parameter. The argument was elementary in the sense that it involved no modern theorems, but subtle in its reconfiguration of the problem into a tractable analytic form.

Despite the proof's correctness, Royen's paper initially went unnoticed. It appeared in a minor journal and lacked the formal polish typically expected of breakthroughs in high-dimensional analysis. The paper did not announce its significance, and the style—direct and sparse—obscured its novelty. For a time, the result was known only to a small circle of readers, many of whom were unsure whether the argument was valid. Eventually, experts in probability and convex geometry began to scrutinise the proof, rephrasing and streamlining its components. Within a few years, the result was confirmed, disseminated, and reformulated in the language of convex analysis and Gaussian processes. Royen's name entered the canonical history of the problem, and the Gaussian Correlation Inequality was marked solved. What remained was not only a resolution of the inequality itself, but a reminder that the landscape of mathematical solutions includes not only new theories, but new configurations of old tools—found sometimes at the margins of the research world.

Unexpected Solvers with Familiar Tools

The story joins others in this book where longstanding open problems were resolved not by new machinery, but by the careful use of classical methods in unfamiliar configurations—often by researchers outside elite institutions. Like Yitang Zhang's breakthrough on bounded prime gaps, or the amateur discovery of the monotile known as the 'hat,' Thomas Royen's proof of the Gaussian Correlation Inequality relied on known identities and transforms applied with unusual directness. Another case is the anonymous 4chan poster who, responding to a question about anime episode orderings, derived a new lower bound on superpermutation length using standard graph-theoretic arguments—a result later verified and cited in formal literature as belonging to an unidentified author (Ch. 39). The cases share a common story: problems that resisted decades of expert attention gave way once the right pathway—already present in the mathematical landscape—was followed with formal rigour.

This is unusual. Almost always, when someone claims to have solved a famous open problem, it is crankery. The phenomenon spans the entire spectrum of mathematical sophistication. At one end: amateurs on Quora insisting they have disproven momentum conservation or constructed a perpetual motion machine, unaware of basic definitions. In the middle: professors at respectable institutions who become obsessed with problems adjacent to their expertise, producing hundreds of pages of arguments that experts dismiss within minutes. At the high end: world-renowned experts who announce breakthroughs in areas outside their domain—claiming, for instance, to have proven the Riemann hypothesis—only to have fatal errors exposed during peer review.

The most contentious cases occur when the claimant's reputation and technical sophistication make dismissal difficult. Shinichi Mochizuki's claimed proof of the abc conjecture (Masser, 1985; Oesterlé, 1988) (which posits that when a, b, and c are coprime and satisfy $a + b = c$, then c is rarely much larger than the product of the distinct primes dividing abc), spanning over 500 pages of novel theory he calls 'inter-universal Teichmüller theory,' has divided the mathematical community for over a decade. Leading number theorists have declared the proof fatally flawed, while Mochizuki and a small circle of collaborators maintain its validity. The dispute remains unresolved—not for lack of expertise on either side, but because the proposed framework is so idiosyncratic that consensus on its correctness may be unattainable. What separates legitimate breakthroughs from crankery is not the solver's credentials, but whether the proof can be verified, communicated, and integrated into the broader body of mathematical knowledge.

Monotonicity via Covariance Interpolation

Let $X \sim \mathcal{N}(0, C)$ be an n-dimensional Gaussian vector with zero mean and covariance matrix $C \succeq 0$. The Gaussian Correlation Inequality asserts that for any symmetric convex sets $A, B \subset \mathbb{R}^n$,

$$\mathbb{P}(X \in A \cap B) \geq \mathbb{P}(X \in A)\,\mathbb{P}(X \in B).$$

We consider the axis-aligned box case; the full GCI for all symmetric convex sets follows from Royen:

$$A = \{x \in \mathbb{R}^n : |x_i| \leq 1 \text{ for } 1 \leq i \leq k\},$$
$$B = \{x \in \mathbb{R}^n : |x_j| \leq 1 \text{ for } k < j \leq n\}.$$

Let $X = (X_1, \dots, X_n)$, and define

$$f(t) := \mathbb{P}_t\left(\max_{1 \leq i \leq n} |X_i| \leq 1\right),$$

where $\mathbb{P}_t$ denotes a Gaussian measure with interpolated covariance

$$C(t) = \begin{pmatrix} C_1 & tQ \\ tQ^\top & C_2 \end{pmatrix}, \quad t \in [0, 1].$$

Here, $C_1 \in \mathbb{R}^{k \times k}$, $C_2 \in \mathbb{R}^{(n-k)\times(n-k)}$, and $Q \in \mathbb{R}^{k \times (n-k)}$. Write the original covariance in the same block form, $C = \begin{pmatrix} C_1 & Q \\ Q^\top & C_2 \end{pmatrix}$, so that $C(1) = C$ and $C(0) = \mathrm{diag}(C_1, C_2)$. At $t = 0$, the covariance is block-diagonal with independent blocks; at $t = 1$, the off-diagonal coupling Q is fully present. Since $C(t) = (1-t)C(0) + tC(1)$ and the positive semidefinite cone is convex, $C(t) \succeq 0$ for all $t \in [0, 1]$.

Note $\{\max_{1 \leq i \leq n} |X_i| \leq 1\} = A \cap B$. At $t = 0$, the block-diagonal covariance makes $(X_1, \dots, X_k)$ and $(X_{k+1}, \dots, X_n)$ independent (Gaussian), so $\mathbb{P}_0(A \cap B) = \mathbb{P}_0(A)\mathbb{P}_0(B)$. The goal is to prove that $f(t)$ is non-decreasing.

Transformation to Gamma Structure

The squared Gaussian variables $Z_i = X_i^2/2$ follow a scaled chi-squared law. For $\lambda_i \geq 0$, define the Laplace transform of $Z = (Z_1, \dots, Z_n)$ under $\mathbb{P}_t$:

$$\begin{aligned} \mathcal{L}_t(\lambda) &= \mathbb{E}_t\left[\exp\left(-\sum_{i=1}^{n} \lambda_i Z_i\right)\right] \\ &= \mathbb{E}_t\left[\exp\left(-X^\top \Lambda X/2\right)\right] \\ &= \det(I + C(t)\Lambda)^{-1/2}, \end{aligned}$$

where $\Lambda = \mathrm{diag}(\lambda_1, \dots, \lambda_n)$. Following Royen, differentiate $\log \mathcal{L}_t(\lambda)$ and rewrite the derivative as an expectation under a multivariate gamma law (see Royen Theorem 1 and its gamma mixture representation; Latała–Matlak (2000) §2–§3); the integrand is nonnegative, hence $\mathcal{L}_t(\lambda)$ is non-increasing in t, and hence $f(t)$ is non-decreasing.

Smoothing and Differentiation

To handle the indicator function rigorously, define a smooth approximation: $\phi_\epsilon(x) = 1$ if $|x| \leq 1 - \epsilon$, $\phi_\epsilon(x) = 0$ if $|x| \geq 1 + \epsilon$, and ϕ_ϵ is smooth monotone otherwise. Let $F_\epsilon(Z) = \prod_{i=1}^{n} \phi_\epsilon\left(\sqrt{2Z_i}\right)$, so that $F_\epsilon \to 1_{\{\max |X_i| \leq 1\}}$ as $\epsilon \to 0$. Using Royen's multivariate gamma representation, $\frac{d}{dt}\mathbb{E}_t[F_\epsilon(Z)]$ is an integral of a nonnegative kernel; dominated convergence then gives ≥ 0. Passing $\epsilon \to 0$ yields the monotonicity of $f(t)$, establishing the Gaussian Correlation Inequality for axis-aligned boxes. The full inequality for all symmetric convex sets follows from Royen (2014).

References:

Royen, T. (2014). *A simple proof of the Gaussian correlation conjecture extended to multivariate gamma distributions.* Far East J. Theor. Stat.

Latała, R., Matlak, D. (2017). *Royen's Proof of the Gaussian Correlation Inequality.* In: Israel Seminar (GAFA) 2014–2016. Springer.

A Thought
About Nothing

Top (Rube Goldberg Universe): The probability that the physical universe evolved over billions of years—from low-entropy initial conditions, through stars, planets, chemistry, and biology—to produce a thinking mind like yours at this moment is vanishingly small. This intricate causal chain cosmic Rube Goldberg machine needs to be tuned to such perfection that it is nigh impossible.

Bottom (Boltzmann Brain Fluctuation): By contrast, the spontaneous appearance of a fleeting brain—formed by a random fluctuation in an otherwise empty universe—is far more probable. It requires no stars, no galaxies, no history. Just a brief, ordered arrangement of matter or fields.

A Thought About Nothing

The Boltzmann Brain paradox shows that statistical mechanics predicts a disturbing outcome: random fluctuations in a high-entropy universe would produce isolated conscious entities more frequently than entire ordered universes such as ours. A single brain with false memories requires orders of magnitude fewer unlikely coincidences than 13.8 billion years of cosmic evolution. These hypothetical observers would experience coherent thoughts and apparent histories, yet exist only as momentary statistical fluctuations. It is not easy to dismiss this preposterous theory based on scientific reasoning alone.

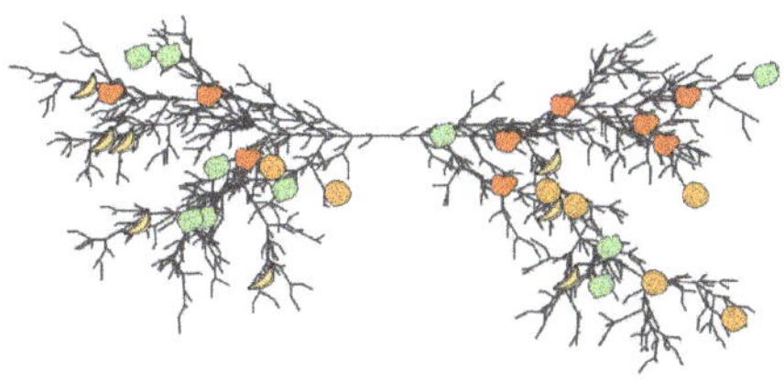

Boltzmann Brain Paradox ∘ Entropy Fluctuations ∘ de Sitter Future ∘ Eternal Expansion ∘ Cognitive Without Causation ∘ Epistemic Disconnection ∘ Observer Selection Problem ∘ Measure Proposals ∘ Gott's Turing Test ∘ Multiverse Crisis ∘ Scientific Dead End

> "Thoroughly conscious ignorance
> is the prelude to every real advance in science."
> — James Clerk Maxwell, 1877

> “הֲבֵל הֲבָלִים אָמַר קֹהֶלֶת, הֲבֵל הֲבָלִים הַכֹּל הָבֶל.”
> ("Vanity of vanities, saith the Preacher,
> vanity of vanities; all is vanity.")
> — Ecclesiastes 1:2

A Thought About Nothing

In 1895, Ludwig Boltzmann proposed that the observed low entropy of the universe might arise as a rare fluctuation within a larger equilibrium state. His goal was to reconcile the second law of thermodynamics with the possibility of eternal time: if the universe is statistically dominated by high-entropy configurations, then any low-entropy region—such as our observable cosmos—would have to be an exceptional, temporary departure.

This explanation faced immediate challenges. Boltzmann's assistant Schuetz pointed out the flaw: smaller fluctuations are exponentially more probable than large ones. If we explain our ordered universe as a fluctuation, why did it fluctuate so much more than necessary? It would be more likely for a single galaxy, solar system, or a single observer—complete with illusory memories of a larger cosmos—to emerge briefly from equilibrium. The entropy required for a functioning brain lasting seconds is negligible compared to that needed for billions of years of cosmic evolution. Arthur Eddington later emphasised this disparity (Eddington, 1931), arguing that Boltzmann's hypothesis made our observations unlikely over finite timescales.

The problem lay dormant until the late twentieth century, when it re-emerged. The 1998 discovery of cosmic acceleration implied a positive cosmological constant, suggesting our universe would expand forever into a de Sitter state. Lisa Dyson, Matthew Kleban, and Leonard Susskind (2002) showed that such universes face Boltzmann's original problem in its extreme form: eternal de Sitter space acts as a thermal bath that will fluctuate into any possible configuration, with smaller fluctuations exponentially dominating larger ones.

The 'Boltzmann Brain' problem—named by Andreas Albrecht and Lorenzo Sorbo—became recognised not only as philosophical speculation but as a discussion in cosmology. If the universe lasts long enough, and if thermal or quantum fluctuations occur eternally, then you are likely a momentary self-aware configuration with a fabricated past—including false memories of other people.

Don Page argued that in de Sitter space, the expected waiting time for a Boltzmann Brain in a horizon volume is roughly $\exp(10^{69})$ years—vastly shorter than the Poincaré recurrence time (Poincaré, 1890) $\exp(10^{122})$. The problem forces a choice: reject our best model of dark energy, accept solipsism, or find new principles that privilege causal observers.

Similar logic appears outside physics. Young-Earth creationists propose that fossils and rock strata, and starlight were created in transit—a universe with apparent but not actual age. This bypasses historical causality in favour of constructed appearance, similar to the Boltzmann Brain scenario, though motivated by theology rather than thermodynamics.

A Boltzmann Brain (BB) is a hypothetical conscious entity arising from a rare entropy fluctuation in a high-entropy background. Unlike evolved organisms, which result from extended sequences of causal and developmental events, a Boltzmann Brain emerges instantaneously. Its physical state—whether instantiated in particles, fields, or radiation—momentarily

 https://doi.org/10.11647/OBP.0526.34

satisfies the functional conditions required for awareness. If the arrangement of that matter realises the computational or dynamical architecture associated with continuous cognition, it qualifies as a mind, lacking any past.

The Boltzmann Brain paradox is not a whimsical thought experiment but an unwanted consequence of our most successful cosmological models. Current observations indicate that our universe contains a positive cosmological constant—dark energy—driving accelerated expansion (Riess et al., 1998). This leads to a de Sitter future: the universe will expand forever, approaching a maximum entropy state often called the 'heat death.' In this eternal vacuum, space maintains a tiny but non-zero temperature due to quantum fluctuations and the cosmological horizon.

In an eternally expanding universe, anything not strictly forbidden by conservation laws will occur through random fluctuations. Not only will it occur—it will occur infinitely many times. The mathematics of statistical mechanics guarantees that thermal fluctuations in this endless expanse will, given enough time, assemble particles into any conceivable configuration, including functioning brains. The timescales are unimaginably vast—roughly $\exp(10^{69})$ years for a first occurrence within a horizon volume—but in an eternal universe, such durations are still realised.

Consider yourself observing the cosmic microwave background, pondering your origin. Which is more probable: that you arose from a 13.8-billion-year causal history requiring a low-entropy Big Bang, or that you are a fleeting fluctuation with fabricated memories? The former requires a cosmos-wide entropy decrease of great improbability. The latter needs only a brain-sized entropy dip—far more likely. By naive application of Occam's Razor, you should conclude you are a Boltzmann Brain.

The paradox bites also without invoking eternity. Back of the napkin calculations show that a brain existing for 20 seconds with false memories is more probable than the sequence: Big Bang $\longrightarrow$ primordial nucleosynthesis $\longrightarrow$ stellar evolution $\longrightarrow$ heavy element formation $\longrightarrow$ planetary accretion $\longrightarrow$ prebiotic chemistry $\longrightarrow$ abiogenesis $\longrightarrow$ billions of years of evolution $\longrightarrow$ conscious observers. Each step multiplies the improbability. A universe beginning with entropy low enough to support this chain is less likely than a single, brief fluctuation that mimics its end result. The problem undermines the statistical credibility of our past.

More formally, standard statistical mechanics permits entropy-decreasing fluctuations in systems tending toward equilibrium. A localised reduction sufficient to form a brief conscious pattern is more probable than the reduction needed to produce a low-entropy universe with stars, planets, and biological evolution. In late-time cosmological models, the background expands indefinitely, making the spacetime volume available to such rare events unbounded.

From the interior perspective of a Boltzmann Brain, the experience is indistinguishable from that of a causally embedded human observer. The configuration encodes memories, perceptions, and beliefs—including apparent continuity with a personal past and memories of friends, family, and colleagues who may never have existed. The observer has no introspective access to the fact that its existence results from a spontaneous fluctuation. What it lacks is any connection between those beliefs and the processes that normally justify them: no past in which its knowledge was acquired, no external world that imprinted its memories, no causal pathway from observation to inference. Cognition without causation—mental states that function as if informed by reality, but are informationally sealed.

Such predictions create a scientific dead end. If a cosmological model predicts eternal expansion, it also predicts that you are likely a Boltzmann Brain—probably the only one, with false memories of other people. By the model's own logic, any experiment you perform is meaningless: the results you observe are not reflections of reality but random data encoded in your transient configuration. A Boltzmann Brain 'discovering' evidence for dark energy is no more meaningful than one 'discovering' evidence against it; both are probable fluctuations.

The failure is methodological and cannot be resolved by (for example) more precise mathematics. The likelihood of a Boltzmann Brain dissolves the inferential ladder on which any model stands. Scientific models are based on probabilistic predictions, either through p-value comparisons or more general assessments of evidence such as Occam's Razor in the philosophy of science (e.g., Popper's work on falsifiability (Popper, 1934)). If we deny the Boltzmann Brain just because it does not sit well with intuition, then we have to deny the whole framework of science.

The problem extends to multiverse models that depend on statistical inference across branches of a cosmological landscape. Without constraints that suppress configurations whose cognitive order is unmoored from physical history, such models cannot distinguish between actual evidence and simulated coherence. A theory in which the scientific method is most likely implemented by deluded agents undermines its own use.

The Boltzmann Brain paradox imposes a constraint on viable cosmological theories. Our universe's accelerating expansion—supported by supernovae data, cosmic microwave background measurements, and large-scale structure—appears to doom it to an eternal de Sitter phase where fluctuation-born observers dominate. This forces cosmologists into uncomfortable territory: either our observations mislead us about the universe's future, or we need new principles that explain why we are not random fluctuations. The paradox transforms from philosophical curiosity to empirical crisis—a *reductio ad absurdum* that demands resolution if cosmology is to remain a predictive science based on falsifiable hypotheses and probability comparisons.

Various proposals attempt to evade the Boltzmann Brain catastrophe, though none resolve it. Page suggests the de Sitter phase must end within 20 billion years through bubble nucleation—regions of lower vacuum energy that expand and percolate, terminating the eternal expansion before significant BB production. This requires an unexpectedly high tunnelling rate, finely tuned to prevent eternal inflation while avoiding immediate cosmic catastrophe.

The cyclic model of Steinhardt and Turok (Steinhardt & Turok, 2002) sidesteps infinity: the current accelerating phase ends after perhaps a trillion years, followed by contraction and a new big bang. With finite time available, the vanishingly small probability of BB formation yields no expected occurrences. Slow variation of fundamental constants or gradual roll-off of the vacuum energy could terminate the de Sitter phase, though this postpones the question of what determines these variations.

Linde notes that in eternal inflation, young bubble universes vastly outnumber old ones. At any cosmic time slice, newly formed universes teeming with ordinary observers outweigh ancient regions hosting Boltzmann Brains. Yet this creates a 'youngness problem'—you should find yourself among the first observers in a brand-new universe, not 13.8 billion years after your universe's birth (see Guth-Vanchurin paradox).

More sophisticated measure proposals compare formation rates rather than absolute numbers. Vilenkin weighs the rate of BB nucleation against the rate of new inflating regions forming. Each inflating region spawns infinite ordinary observers through standard cosmic evolution, while each BB fluctuation creates one. The formation rates may be comparable—both require exponentially rare quantum events—but the 'infinity' factor tips the balance toward ordinary observers.

These proposals shift probability calculations without addressing the deeper issue. They assume a particular cosmological model—eternal inflation, quantum tunnelling rates, specific vacuum configurations—within which to compute relative likelihoods. But the Boltzmann Brain hypothesis operates at a more fundamental level. In the space of all possible observers, why should any particular physics hold? A Boltzmann Brain need not arise within our specific de Sitter spacetime; it could fluctuate into existence with false memories of different physical laws. The proposals combat Boltzmann Brains within cosmology, but the paradox questions whether cosmology itself is a false memory. Each solution presupposes the causal framework that Boltzmann Brains dissolve.

Richard Gott has proposed a different approach: the Turing test. He argues that Boltzmann Brains fail the Turing test for intelligent observers because they cannot sustain lucid responses over time. While a BB might answer 20 questions correctly—an outrageously rare configuration—one that answers 21 questions correctly is rarer still. By the Copernican principle (that we are not special), if you observe a BB that has just answered 20 questions successfully, it will most likely fail on the 21st. No matter how many questions are answered, the next response will probably be nonsense or the BB will vanish entirely.

Gott claims this provides a practical test: "I will wait 10 seconds and see if I am still here." If you were a Boltzmann Brain, you would likely dissipate or cease coherent function before completing this simple task. Your continued persistence and ability to reason about the paradox itself demonstrates that you are not a random fluctuation but a causally embedded observer.

However, this argument contains the same flaw as Gott's prediction test. The relevant Boltzmann Brain is not one that existed before your 10-second wait, but one that could form afterward with memories of having waited successfully. When you reach '10' and conclude you've passed the test, you could be a newly formed BB with false memories of counting, false perceptions of temporal continuity, and false confidence in your non-BB status. The Turing test assumes persistent identity across time—precisely what the Boltzmann Brain hypothesis denies. A fluctuation need not maintain any observer continuously; it need only create an observer who believes they have experienced such continuity.

Simulated Pasts and Evidential Disconnection

The Boltzmann Brain scenario structurally parallels a common form of young-Earth apologetics: the claim that geological strata, fossils, and incoming starlight were instantiated directly, rather than arising through causal processes. In both cases, present configurations encode the appearance of a history that never occurred.

Each replaces process with configuration: the Boltzmann model posits a fluctuation that assembles an observer with fabricated memories; the theological model posits an act that instantiates a cosmos with pre-formed records.

This move severs the link between observation and inference. If coherent records can be instantiated without causal origin, then empirical data no longer warrants explanatory conclusions. The coherence of evidence becomes indistinguishable from simulation.

Scientific methodology presumes that regularities in the present reflect processes in the past. Models that decouple this relationship dissolve the inferential basis of empirical knowledge.

Such models are not simply untestable—they nullify the conditions under which testing acquires meaning. Without constraints that bind pattern to cause, explanatory validity reduces to interpretive preference.

In the absence of external commitments—philosophical, theological, or otherwise—models with embedded pseudo-histories are scientifically inert. Their predictive outputs are indistinguishable from those of causally grounded theories, but lack the procedural integrity required for evidential trust.

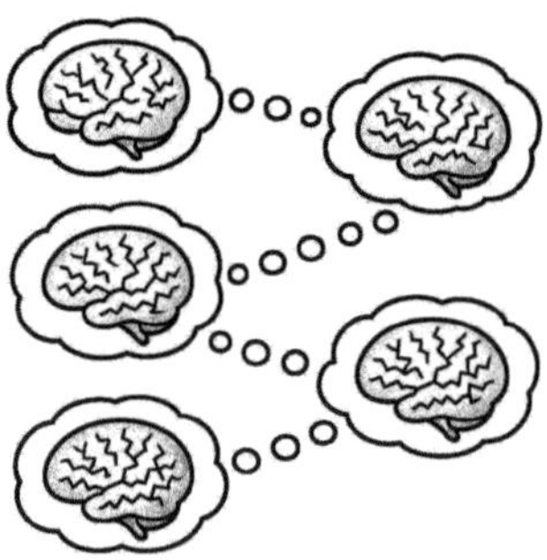

"It's Boltzmann brains all the way down!"

Universe vs Boltzmann Brain

De Sitter Horizon Entropy

For a universe approaching de Sitter space with cosmological constant $\Lambda \sim 10^{-122}$ in Planck units, the horizon entropy is

$$S_{\mathrm{dS}} = \frac{3\pi}{\Lambda} \sim 10^{122}.$$

This is the maximum coarse-grained entropy inside our causal patch.

Low-Entropy Big Bang

Penrose treats the observed smooth early universe as a tiny region of the gravitational phase space compatible with S_{dS}. The entropy deficit is

$$\Delta S_{\mathrm{cosmo}} \sim 10^{122},$$

so a fluctuation from de Sitter equilibrium to a Big Bang–like state has probability

$$P_{\mathrm{cosmo}} \propto \exp(-\Delta S_{\mathrm{cosmo}}) \sim \exp(-10^{122}).$$

Boltzmann Brain Entropy Cost

For a single brain of mass M fluctuating out of the de Sitter vacuum at temperature T_{dS}, the entropy cost follows from the Boltzmann factor,

$$\Delta S_{\mathrm{BB}} \simeq \frac{E}{T_{\mathrm{dS}}} = \frac{Mc^2}{T_{\mathrm{dS}}}.$$

With $M \sim 1$ kg and $T_{\mathrm{dS}} \sim 10^{-33}$ eV,

$$E \sim 10^{17}\,\mathrm{J} \sim 10^{36}\ \mathrm{eV},$$

$$\Delta S_{\mathrm{BB}} \sim \frac{10^{36}\ \mathrm{eV}}{10^{-33}\ \mathrm{eV}} \sim 10^{69}.$$

Uncertainties in M, T_{dS}, and required complexity shift this by many orders of magnitude; a round figure $\Delta S_{\mathrm{BB}} \sim 10^{60}$ still lies far below 10^{122}. The corresponding fluctuation probability is

$$P_{\mathrm{BB}} \propto \exp(-\Delta S_{\mathrm{BB}}) \\ \sim \exp(-10^{60}\text{-}10^{69}).$$

Relative Likelihood

The ratio of BB to cosmological fluctuations is

$$\frac{P_{\mathrm{BB}}}{P_{\mathrm{cosmo}}} \sim \exp(\Delta S_{\mathrm{cosmo}} - \Delta S_{\mathrm{BB}}) \\ \sim \exp(10^{122} - 10^{60}) \\ \sim \exp(10^{122}).$$

The entropy reduction needed for a full low-entropy universe is overwhelmingly larger than for a single brain-sized fluctuation, so BBs dominate naive fluctuation counting by an astronomically large factor.

Timescales

The mean waiting time for a fluctuation with entropy cost ΔS in a fixed horizon volume scales as

$$t \sim t_0 \exp(\Delta S),$$

with t_0 a microscopic timescale. For BBs and full recurrences,

$$t_{\mathrm{BB}} \sim \exp(\Delta S_{\mathrm{BB}}) \sim \exp(10^{60}\text{-}10^{69})\ \text{years},$$
$$t_{\mathrm{rec}} \sim \exp(S_{\mathrm{dS}}) \sim \exp(10^{122})\ \text{years}.$$

Both timescales are vast; $t_{\mathrm{BB}} \ll t_{\mathrm{rec}}$, so many BB fluctuations occur within a single Poincaré recurrence time.

References:

Dyson, L., Kleban, M., Susskind, L. (2002). Disturbing implications of a cosmological constant. *JHEP* 0210:011.

Penrose, R. (2004). *The Road to Reality. Jonathan Cape.*

Page, D. N. (2006). Is our universe likely to decay within 20 billion years? arXiv:hep-th/0610079.

From Air to Arbor

Photosynthesis and Carbon Fixation: This diagram illustrates the process by which trees build their mass from atmospheric CO_2 rather than soil nutrients. The top section shows the molecular mechanism of photosynthesis: light energy driving the conversion of carbon dioxide and water into glucose through the Calvin-Benson cycle. Chloroplasts in leaf cells capture photons to power the fixation of atmospheric carbon into organic compounds. The bottom section demonstrates the carbon flow: from diffuse CO_2 in the atmosphere, through stomatal uptake, into the structural polymers (cellulose, lignin, hemicellulose) that comprise wood. Each ring of tree growth represents a year's accumulation of atmospheric carbon, transformed by solar energy into solid biomass. As Feynman noted, trees are literally 'made of air'—their dry weight consists primarily of carbon atoms that were once distributed throughout the atmosphere.

From Air to Arbor

Ask where a tree's mass comes from and intuition points downward: soil, water, nutrients drawn up through roots. This is almost entirely wrong. Trees are made of air—~95% of their dry mass comes from atmospheric CO_2. Through photosynthesis, plants build themselves from carbon dioxide, converting invisible gas into solid wood, cellulose, and lignin using sunlight. Van Helmont's 1640s willow experiment demonstrated this: a tree gained 164 pounds (~74 kg) while the soil lost only two ounces (~60 g). Isotope labelling confirms the molecular accounting—carbon in wood comes from air, not earth or water. When trees burn, they simply return their borrowed carbon and sunlight to the atmosphere, completing a chemical cycle that temporarily crystallises air into living architecture.

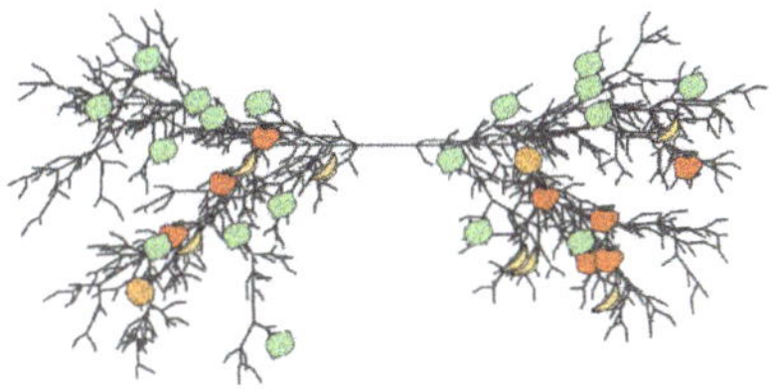

Trees Made of Air ∘ Atmospheric CO_2 Fixation ∘ Photosynthesis Energy Capture ∘ Water Splitting & O_2 Release ∘ Calvin-Benson Cycle ∘ Cellulose & Lignin ∘ Stored Solar Energy ∘ Great Oxygenation Event ∘ ^{18}O Isotope Experiments ∘ Feynman's Air Quote ∘ Combustion Symmetry

"I am the Lorax and I speak for the trees."
— The Lorax

"It is not the deer that is crossing the road, rather it is the road that is crossing the forest."
— Muhammad Ali *(probably misattributed)*

From Air to Arbor

In the early seventeenth century, Jan Baptist van Helmont conducted an experiment that would later become emblematic of early quantitative biology. He planted a willow sapling in a weighed quantity of dry soil, supplied it only with water, and allowed it to grow for five years. At the end of the experiment, he found that the tree had gained over 70 kilograms in mass, while the soil had decreased by less than 60 grams. From this, he concluded—correctly in direction though not in mechanism—that the tree's substance did not come from the soil.

Van Helmont identified water as the key source of mass, unaware of the role of atmospheric gases. His result was significant for shifting scientific attention away from Aristotelian elemental explanations and toward empirical measurement. The idea that a tree might be built from intangible substances posed a conceptual challenge to early chemistry, which had yet to recognise air as chemically active.

In the late eighteenth century, Joseph Priestley and Jan Ingenhousz discovered that plants could 'restore' air that had been 'damaged' by combustion or respiration. Ingenhousz, in particular, demonstrated that this process required light and occurred only in green parts of plants. The observations hinted at a connection between sunlight, plant matter, and atmospheric gases.

By the mid-nineteenth century, Julius von Sachs and others had established that plants produce starch in the presence of light and that carbon dioxide is the source of carbon in organic compounds. Quantitative combustion analysis allowed chemists to determine the proportions of carbon—hydrogen—and oxygen in plant tissues, confirming that nearly all plant biomass derived from these three elements.

In the twentieth century, isotopic labelling techniques enabled direct tracing of carbon atoms from CO_2 into plant tissues, definitively establishing air as the origin of most biomass. Experiments using ^{14}C-labelled carbon dioxide showed its incorporation into sugars, cellulose, and lignin. By the mid-twentieth century, the principal pathways of photosynthesis—including the light reactions and the Calvin–Benson cycle—had been elucidated.

A breakthrough came in the 1940s when Samuel Ruben and Martin Kamen used oxygen-18 isotope labelling to resolve a question about photosynthesis. Earlier researchers knew that oxygen gas was released, but its source remained unclear—did it come from carbon dioxide or water? By supplying plants with ^{18}O-enriched water while keeping CO_2 normal, they found that the heavy oxygen appeared exclusively in the released O_2 gas, not in the organic products. Atmospheric oxygen originates from water splitting, while the oxygen atoms in biomolecules derive from CO_2 fixation. The experiment established the precise molecular accounting of photosynthesis and confirmed that plants literally separate air from water at the atomic level.

A tree's material body, the wood, leaves, and branches it accumulates year by year, is not extracted from the ground in the way stones or metals are quarried. Its dry mass derives from elements that were once distributed in dilute form throughout the atmosphere and

 https://doi.org/10.11647/OBP.0526.35

hydrosphere. The key components of this mass, carbon, oxygen, and hydrogen, enter through invisible flows: air, water, and sunlight. A tree is built from what passes through it.

Although visually and mechanically tied to the soil, a tree records processes that unfold mostly above ground. The mass that persists after all water is removed, the dry matter, is composed primarily of carbon atoms originally fixed from atmospheric CO_2. The atoms were drawn down through the stomata of leaves, diffused through mesophyll tissue, and incorporated into sugar molecules via light-powered biochemical cycles.

The notion that trees 'grow out of the earth' conflates anchorage with origin. The soil does provide essential ions and mechanical stability, but its contribution to the actual mass is minor. Most of what endures in a dried trunk, cellulose, lignin, hemicellulose, was once part of the air surrounding it. The verticality of a tree, its rise toward the sky, is materially made possible by the intake of that sky's gaseous contents.

The central process enabling this conversion is photosynthesis. It is a layered sequence of energy transduction and molecular reconfiguration. The first phase occurs in the chloroplasts of leaf cells, where chlorophyll pigments absorb incoming photons. The photons excite electrons to higher energy states, dislodging them from their atomic orbitals and initiating a cascade of electron transfers through the thylakoid membrane.

Oxygen-producing photosynthesis altered Earth's atmosphere and biosphere. When it first evolved in cyanobacteria around 2.5 billion years ago, it triggered the Great Oxygenation Event—a transformation that poisoned most existing anaerobic life but enabled the eventual emergence of complex organisms. The oxygen released by water-splitting is a byproduct that reshaped planetary chemistry. Every breath taken by an animal, every flame that burns, every rusting of iron depends on this ancient process continuing in plant chloroplasts. Trees are participants in a planetary-scale atmospheric engine that has operated continuously for billions of years.

The chain of transfers generates two critical energy carriers: ATP (adenosine triphosphate) (Lohmann, 1929) and NADPH (nicotinamide adenine dinucleotide phosphate). The molecules store the electromagnetic energy harvested from light and shuttle it into the chemical domain. In the aqueous interior of the chloroplast, the stroma, the stored energy is used to convert inorganic carbon into organic intermediates.

Isotopes are versions of an element with the same number of protons (Soddy, 1913) but different numbers of neutrons. For example, oxygen has 8 protons and can have 8, 9, or 10 neutrons. This gives rise to the isotopes ^{16}O, ^{17}O, and ^{18}O. Carbon has 6 protons and can have 6, 7, or 8 neutrons. This gives rise to the isotopes ^{12}C, ^{13}C, and ^{14}C. Isotopes are a way to label atoms and track them in chemical reactions. This is how we know that the oxygen in wood comes from atmospheric CO_2, not from the soil.

When water molecules are split in photosystem II (oxygenic photosynthesis), their oxygen atoms are released directly to the atmosphere as O_2 gas. Isotope labelling experiments using ^{18}O-enriched water demonstrated (Ruben et al., 1941) that the heavy oxygen appeared in the released gas, not in the organic products. The oxygen atoms incorporated into cellulose and other biomolecules originate from CO_2. Every molecule of atmospheric oxygen released by plants represents a water molecule that was split to extract electrons, while the oxygen in wood records the atmospheric carbon that was fixed.

The fixation of carbon takes place in the Calvin–Benson cycle (Bassham, Benson & Calvin, 1950). Atmospheric CO_2 diffuses into leaf tissue and reacts with ribulose bisphosphate, a five-carbon sugar, under the catalytic action of the enzyme rubisco. The resulting six-carbon intermediate is promptly split into three-carbon molecules, triose phosphates, that serve as building blocks for carbohydrates. The triose units are reassembled into glucose and other hexoses, which in turn feed biosynthetic pathways across the plant.

Sunlight delivers approximately 1,000 watts per square metre on a clear day—plants capture 1–3% of this energy in chemical bonds. What is captured becomes concentrated: each kilogram of dry wood stores about 16–20 megajoules of energy, roughly equivalent to the combustion energy of natural gas. A single mature tree may contain 50–100 gigajoules of stored solar energy, accumulated over decades of photosynthetic capture. Millions of individual photons contribute quantum energy to the construction of molecular architecture that can persist for centuries.

Once synthesised, the sugars are exported from the site of fixation. Through the phloem, a network of conductive tissues, they are distributed to growing regions: root tips, shoot apices, developing leaves, and the vascular cambium. At the cambium, a cylindrical layer of dividing cells just beneath the bark, the imported carbohydrates are used to construct macromolecules.

Cellulose, hemicellulose, and lignin form the principal constituents of wood. Cellulose $(C_6H_{10}O_5)_n$ assembles into long, unbranched chains that crystallise into fibrils, giving tensile strength to cell walls. Hemicellulose $(C_5H_8O_4)_n$ binds the fibrils into a cohesive matrix, while lignin, a complex phenolic polymer, fills the spaces between them, adding compressive strength and water resistance. The polymers are laid down in geometric arrangements within the expanding walls of growing cells.

At the vascular cambium, cell division proceeds laterally, producing xylem cells toward the centre and phloem cells outward. The radial expansion creates the familiar pattern of growth rings. Each ring corresponds to a cycle of photosynthetic capture and biosynthetic deposition.

Elongation occurs at the apical meristems, where undifferentiated cells divide and specialise into tissue types. The regions at the tips of roots and shoots coordinate patterning, orientation, and organogenesis. As cells expand and walls thicken, the imported sugars are converted into permanent form.

Hydrogen atoms in the biomass originate from water. Water is absorbed by roots and pulled upward through the xylem under tension. Though over 99% of it eventually evaporates through stomatal pores, a small fraction is chemically incorporated into organic molecules. The hydrogen forms part of the fixed material, bound into carbohydrates and lipids.

Water's functional role extends beyond hydrogen donation. It serves as a solvent for ions, a medium for transport, and a buffer against temperature fluctuations. It enables the tree's biochemical metabolism and what remains after desiccation is not water but the elements it helped mobilise and bind.

Oxygen atoms in biomass come from CO_2. During photosynthesis, water molecules are split to provide electrons, but their oxygen atoms are released directly to the atmosphere as O_2 gas. The oxygen atoms incorporated into cellulose, forming hydroxyl, carboxyl, and ether linkages, originate from the atmospheric carbon dioxide that was fixed. The high oxygen content of wood, about 40 to 45 percent by weight, is a direct record of atmospheric CO_2 that was captured and converted into solid form.

Mineral ions absorbed from the soil are essential but contribute little to the total mass. Nitrogen, phosphorus, potassium, calcium, magnesium, and micronutrients serve catalytic and regulatory roles. They enable enzymatic function, membrane potential maintenance, and nucleic acid stability. Their aggregate proportion in dry matter is often less than 5 percent. They are mainly facilitators rather than substrates.

When all water is removed from a tree, what remains is a carbon-rich composite of organic polymers. Cellulose, lignin, and related molecules form a lattice of energy-stored mass, chemically stabilised and mechanically resilient.

The transformation exhibits chemical symmetry. Photosynthesis builds sugar units from atmospheric inputs, $6CO_2 + 6H_2O + \odot \text{ light} \longrightarrow C_6H_{12}O_6 + 6O_2$, which are then polymerized into cellulose by removing water: $n(C_6H_{12}O_6) \longrightarrow (C_6H_{10}O_5)_n + nH_2O$. When wood burns, the process reverses exactly: $(C_6H_{10}O_5)_n + 6nO_2 \longrightarrow 6nCO_2 + 5nH_2O +$ 🔥 heat. The stored solar energy is released, and every atom returns to its original atmospheric or aqueous state. The carbon dioxide and water vapour that rise from the flame are identical to the molecules that entered the tree decades earlier. A tree is a temporary configuration of atmospheric components, held together by captured light. As Richard Feynman remarked, trees are ‘made of air’. When a tree burns, the carbon returns to the atmosphere, and the stored sunlight is released as heat.

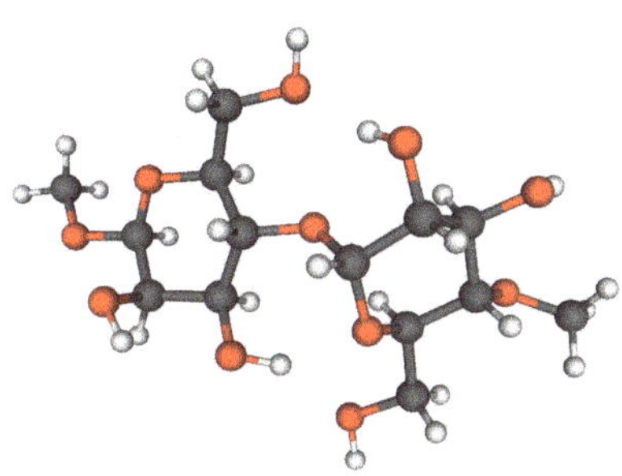

Figure by author. Molecular structure of cellobiose, the repeating β-1,4-linked D-glucose disaccharide unit of cellulose.

The Growing Tree

(NOT by Shel Silverstein)

Once there was a tree,
and she loved a little boy.

Every day the boy would come—
gather her leaves to make crowns,
climb her trunk,
swing from her branches,
eat apples,
and rest in her shade.

And the boy loved the tree.
And the tree was happy.

But time passed,
and the boy grew older.
The tree often stood alone.

One day the boy came back.
The tree said,
"Come, climb my trunk, swing, eat, and be happy."

"I'm too big to play," said the boy.
"I want money, to buy things and have fun."
"I don't have money," said the tree,
"but you can take some apples—sell a few, share a few, and plant one or two."

And the boy did.
And the tree was happy.

Years passed. The boy returned.
"I want a house," he said,
"for warmth, for family."

"I don't have a house," said the tree,
"but take my fallen branches.
Leave enough for me to grow."

And the boy did.
And the tree was happy.

More time went by.
The boy returned, older.
"I want a boat to go far away."

"Some of my thicker branches grew wild," said the tree.
"You can use them."

And the boy did.
And the tree was happy.
After many years, the boy returned, tired.

"I don't need much now," he said,
"just a place to rest."
The tree said,

"Come sit.
There is shade again.
And I'm glad you're here."
And the boy did.

and the boy was happy.

And the tree was happy,

Carbon Fixation and Mass Accumulation in Trees

Trees accumulate mass through atmospheric CO_2 fixation powered by sunlight. This section quantifies the chemical and energetic processes converting gaseous carbon into solid biomass.

Light-Driven Reactions

Photosystems I and II generate ATP and NADPH from light energy (680 nm photons ≈ 176 kJ/mol):

$$2\,H_2O + 2\,NADP^+ + 3\,ADP + 3\,P_i + h\nu \rightarrow 2\,NADPH + 3\,ATP + O_2.$$

Quantum requirement: 8–10 photons per CO_2 molecule fixed.

Carbon Fixation and Biomass Synthesis

In the Calvin–Benson cycle, carbon dioxide is enzymatically fixed into triose phosphates using the energy carriers from the light reactions. The overall reaction for one glucose unit is:

$$6\,CO_2 + 18\,ATP + 12\,NADPH \rightarrow C_6H_{12}O_6 + 18\,ADP + 18\,P_i + 12\,NADP^+.$$

Glucose is polymerized into cellulose by dehydration:

$$n\,C_6H_{12}O_6 \rightarrow (C_6H_{10}O_5)_n + n\,H_2O.$$

These polymers form the primary structure of wood (secondary xylem), alongside lignin and hemicellulose.

Oxygen Source Identification via Isotope Labelling

The ^{18}O labelling experiments by Ruben and Kamen (1941) definitively established oxygen source separation:

Water source test:

$$CO_2 + H_2^{18}O + h\nu \rightarrow [CH_2O] + ^{18}O_2.$$

Result: Heavy oxygen (^{18}O) appeared exclusively in released O_2, not in organic products. CO_2 *source test:*

$$C^{18}O_2 + H_2O + h\nu \rightarrow [CH_2^{18}O] + O_2$$

Energy Storage Density

Wood represents highly concentrated solar energy storage. **Energy density:** 16–20 MJ/kg (dry wood); **Solar capture efficiency:** 1–3% of incident radiation; **Mature tree storage:** 50–100 GJ total (accumulated over decades); **Photon requirement:** ~8–10 photons per CO_2 molecule fixed.

This energy density approaches that of fossil fuels, demonstrating that photosynthesis creates a highly efficient biological battery.

Quantitative Mass Accumulation

For annual NPP of 10^4 kg/ha dry biomass (50% carbon):

$$CO_2\text{ fixed} = 18.4\,\text{tonnes}\,CO_2/\text{ha}/\text{year}$$
$$\text{Per tree (100/ha)} = 184\,\text{kg}\,CO_2/\text{year}.$$

Over 50 years, each tree accumulates ~2.5 tonnes carbon, corresponding to ~5 tonnes total dry biomass—consistent with mature forest measurements.

Elemental Mass Contribution

Typical dry mass composition:

Carbon: 45–50% (from atmospheric CO_2)

Oxygen: 40–45% (primarily from CO_2)

Hydrogen: ~6% (from water)

Minerals: 1–5% (from soil: N, P, K, Ca, etc.)

References:

Farquhar, G. D., von Caemmerer, S., Berry, J. A. (1980). A biochemical model of photosynthetic CO_2 assimilation in C_3 leaves. *Planta*, **149**, 78–90.
Taiz, L., Zeiger, E. (2010). *Plant Physiology*.

Renormalize All the Things

Top (Double Slit): In general relativity (GR), each possible photon path carries mass-energy and should bend spacetime accordingly. In quantum field theory, superpositions of paths interfere—but how does spacetime metric respond to a superposed trajectory?
Second (Photoelectric Effect): Near a gravitating body such as Earth, GR predicts redshift of both incoming photons and bound electronic orbitals. Does the threshold for photoelectric emission shift? GR lacks quantum orbitals; QFT lacks spacetime curvature.
Third (Black Hole Information): GR predicts thermal Hawking radiation with no memory of what fell in. But QFT requires unitary evolution: information, such as spin, must be preserved.
Fourth (Unruh Effect): In GR, an observer in free fall detects no force and is locally equivalent to inertial motion. But in QFT, an accelerating observer perceives a thermal particle bath. Do both observers agree on the particle content or not?
Fifth (Vacuum Energy): QFT assigns enormous energy density to the vacuum via zero-point fluctuations. GR says that energy curves spacetime. Does the vacuum bend space violently?
Bottom (Definition of Space): In GR, space is a manifold with tensor fields assigning numbers to each point. In QFT, space is a passive backdrop where operators act to extract observables. GR side of the equation is numbers, QFT side of the equation is operators.

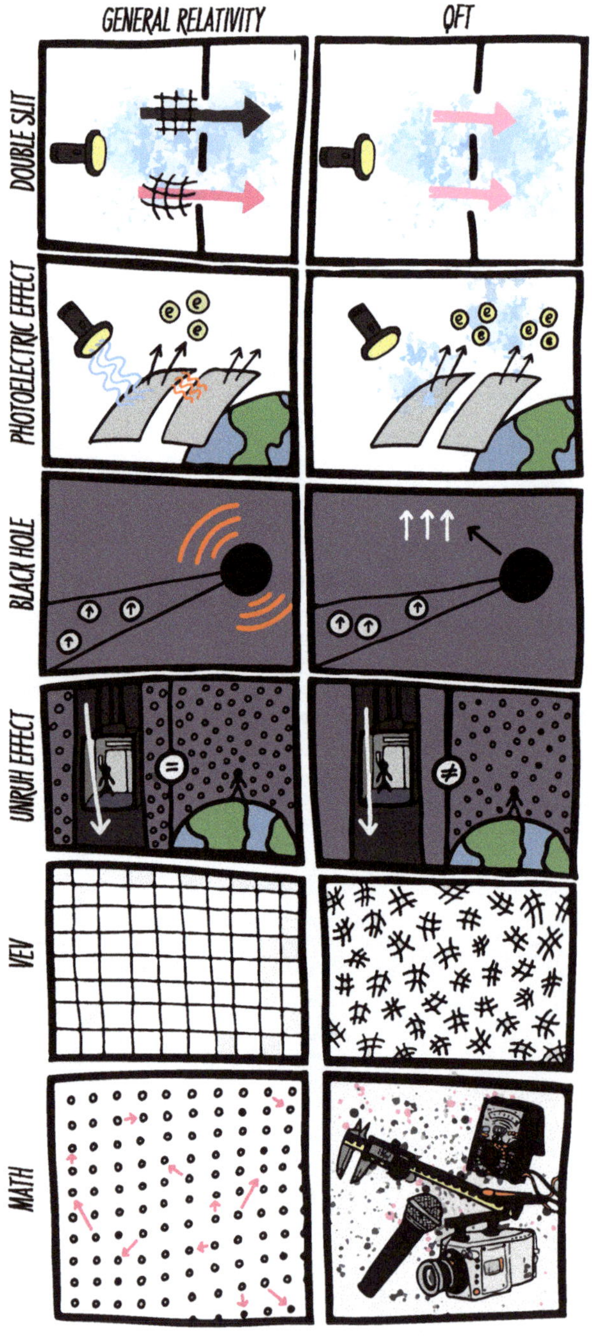

Renormalize All the Things

Physics's two most successful theories cannot coexist. Quantum field theory treats forces as particle exchanges on a fixed stage, while general relativity says the stage warps. When combined, they produce catastrophic contradictions: QFT predicts vacuum energy 10^{120} times larger than observed, gravity refuses renormalization, and black holes seem to destroy quantum information. Each theory works perfectly in its domain, yet they give mutually exclusive descriptions of reality. This incompatibility of theories is the most glaring problem in modern physics.

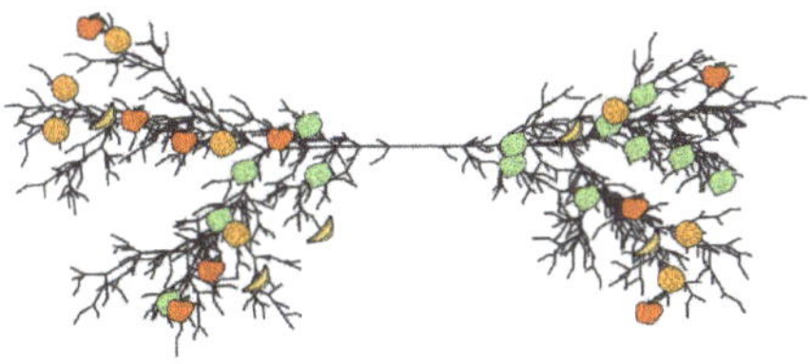

QFT vs General Relativity ◦ Standard Model Forces ◦ Gauge Symmetries ◦ Higgs Mechanism ◦ Fixed vs Dynamic Spacetime ◦ Vacuum Energy Problem ◦ Non-Renormalizable Gravity ◦ Black Hole Information ◦ Quantum Superposition of Geometry ◦ String Theory Landscape ◦ Unification Challenge

*„Es scheint hart, dem Herrgott in die Karten zu gucken.
Aber dass er würfelt und sich telepathischer Mittel bedient
(wie es ihm von der gegenwärtigen Quantentheorie zugemutet wird),
kann ich keinen Augenblick glauben."*

("It seems hard to sneak a look at God's cards.
But that He plays dice and uses telepathic methods
(as the current quantum theory requires of Him),
I cannot believe for a single moment.")
— Albert Einstein, 1942

Renormalize All the Things

Quantum mechanics, which underpins quantum field theory (QFT), took shape in the 1920s with the pioneering work of Planck, Heisenberg, Schrödinger, and Dirac. Around the same period, Einstein's general relativity (GR) from 1915 was being tested and further confirmed through observations such as the bending of starlight during eclipses.

Both theories revolutionised physics: GR reinterpreted gravity as the curvature of spacetime, while QFT unified quantum principles with special relativity to describe forces (electromagnetism, weak and strong) via quantised fields. Attempts to merge GR and QFT began in the mid-twentieth century, with physicists such as Feynman, Pauli, and later Weinberg exploring pathways to quantise gravity. Yet, unlike the other forces, gravity resisted such integration. The divergences encountered in high-energy regimes, plus fundamental contradictions in how time and space are treated, revealed an inherent incompatibility. Despite decades of effort—including approaches such as supergravity and string theory—no complete, experimentally confirmed quantum theory of gravity has emerged.

Quantum field theory (QFT) is the mathematical framework that describes particles and their interactions as excitations of underlying fields defined over spacetime. Each elementary particle corresponds to a quantized mode of a particular field: electrons arise from the electron field, photons from the electromagnetic field, and so forth. Fields span all of space and time, and particles emerge from localised disturbances or quanta of these fields, governed by creation and annihilation operators.

The electromagnetic force, described by quantum electrodynamics (QED), is mediated by photons—massless, chargeless bosons that couple to electric charge. This interaction is governed by a mathematical symmetry called $U(1)$, which represents continuous changes in the complex phase of charged quantum fields. This symmetry can be visualised as rotations around a circle—each point corresponding to a different phase. Mathematically, $U(1)$ has one degree of freedom: one direction of transformation, one conserved quantity (electric charge), and one associated force carrier (the photon). Requiring that the laws of physics remain invariant under such local phase changes leads directly to the existence of the electromagnetic field and ensures that electric charge is conserved. The result is a long-range interaction whose strength falls off as the inverse square of distance.

The weak nuclear force is mediated by three massive particles: the W^+, W^-, and Z^0 bosons. These particles arise from a symmetry structure described by the group $SU(2)_L$, which mathematically encodes transformations among left-handed particles. This group has three independent directions of transformation—called generators—corresponding to the three force carriers. At high energies, this symmetry is extended by an additional $U(1)_Y$ symmetry, associated with a quantity called hYpercharge. Together, these form the electroweak symmetry group $SU(2)_L \times U(1)_Y$. However, the physical world at low energies does not respect this full symmetry: it is spontaneously broken by the Higgs field (Englert & Brout, 1964; Higgs, 1964). This breaking mechanism gives mass to the W and Z bosons, while preserving a remnant $U(1)$ symmetry associated with electromagnetism.

The result is that one combination of the original fields remains massless (the photon), while the others acquire mass and mediate the weak force over short distances.

The strong nuclear force binds quarks together inside protons, neutrons, and other hadrons (particles made of quarks). It is mediated by gluons—massless particles that carry a type of charge called colour (also called color). The mathematical structure governing this interaction is called $SU(3)_{\text{colour}}$, a symmetry group that describes how quark colour states transform into one another. This group has eight independent generators (the Gell-Mann matrices), each corresponding to a type of gluon. Unlike the photon, which does not carry electric charge, gluons themselves carry colour charge, allowing them to interact with each other as well as with quarks. This self-interaction is central to two key features of the strong force: at high energies, quarks behave almost as free particles—a phenomenon called asymptotic freedom (Gross & Wilczek, 1973; Politzer, 1973); at low energies, the interactions become strong and trap quarks permanently inside colour-neutral combinations—a phenomenon known as confinement.

The Standard Model organises these particles and interactions into three families of matter: each includes two quarks, one charged lepton, and one neutrino. All known matter particles are fermions—spin-$\frac{1}{2}$ excitations of their fields. Bosons, which mediate forces, have integer spin and obey different statistical laws.

All these interactions are described within a fixed, background spacetime and governed by renormalizable quantum gauge field theories (renormalizable means that the theory can be made finite by redefining the parameters of the theory). While gravitational interactions are excluded from this framework, QFT has provided extremely accurate predictions for phenomena across particle physics, condensed matter, and quantum optics. It remains the most experimentally successful theory of matter and interactions at subatomic scales.

The Standard Model is a quantum field theory based on the symmetry group $SU(3)_{\text{colour}} \times SU(2)_L \times U(1)_Y$, and accounts for all observed particle interactions apart from gravity. With the inclusion of the Higgs mechanism, it became complete and renormalizable, yielding a fully predictive model with a finite set of input parameters—coupling constants, particle masses, and mixing angles. The 2012 discovery of the Higgs boson at CERN confirmed the final component of the Standard Model. The Higgs boson is the quantized excitation of the Higgs field, a scalar field whose nonzero vacuum expectation value breaks the electroweak symmetry $SU(2)_L \times U(1)_Y$ down to the electromagnetic subgroup $U(1)$. This spontaneous symmetry breaking gives mass to the $W^{\pm}$ and Z^0 bosons while leaving the photon massless.

The Higgs field also couples to fermions through Yukawa interactions. These Lagrangian terms pair fermion fields with the Higgs field via particle-specific coupling strengths. When the Higgs field acquires its vacuum value, these couplings become fermion mass terms, so that electron, muon, and quark masses arise from this interaction.

Measured properties of the Higgs boson—its mass, decay rates, and coupling strengths—closely match theoretical predictions. This agreement validates the mass-generation mechanism, confirms electroweak symmetry breaking via a scalar field, and strengthens the validity of the Standard Model.

For all this apparent completeness, the Standard Model does not account for several empirically established phenomena. It provides no candidate particle for dark matter, which constitutes approximately 85% of the matter content of the universe. Nor does it explain the accelerated expansion attributed to dark energy, nor the small but nonzero masses of neutrinos inferred from oscillation experiments. It also does not explain why the universe contains one type of matter in great excess over its corresponding antimatter. And worst, it does not incorporate gravity.

The Standard Model, and quantum field theory more generally, applies successfully across a vast range of physical scales. It governs phenomena from high-energy particle collisions down to atomic and subatomic interactions, including the structure of hadrons, the dynamics of electrons in atoms, and quantum behaviour in small condensed matter systems. Its validity spans energy scales from a few electronvolts to several teraelectronvolts, and length scales from atomic dimensions down to approximately 10^{-18} metres, probed at current collider facilities.

However, the framework has both ultraviolet and infrared limitations. At extremely short distances or equivalently high energies—approaching the Planck scale, around 10^{19} GeV—the Standard Model ceases to be predictive. At these scales, the effects of unknown high-energy physics are expected to dominate, and the field-theoretic treatment becomes formally ill-defined due to non-renormalizable divergences and the breakdown of perturbative methods.

At macroscopic or cosmological scales, the Standard Model also lacks explanatory power. It does not describe the emergence of classical spacetime, nor account for long-range phenomena not reducible to quantum field excitations. Although QFT explains matter properties in small aggregates—such as superconducting circuits or quantum dots—it does not scale directly to systems where spacetime curvature, causal structure, or background independence become essential.

To summarise: the Standard Model describes three of the four known fundamental forces as gauge interactions among quantized fields. The electromagnetic force, governed by a $U(1)$ symmetry, acts on electric charge and is mediated by the photon. The weak force, based on an $SU(2)_L$ symmetry, operates through the massive W and Z bosons and enables processes such as nuclear decay. The strong force, described by an $SU(3)_{\text{colour}}$ gauge theory, binds quarks and gluons through the exchange of self-interacting gluons. Each of these forces is formulated through a renormalizable quantum field theory and has been validated by collider experiments and astrophysical data.

The fourth fundamental interaction—gravity—lies outside this model. Unlike the other forces, gravity is not mediated by exchange particles on a fixed background. Instead, general relativity portrays it as the curvature of a smooth, continuous spacetime manifold, dynamically shaped by the distribution of energy and momentum. Any form of energy—whether rest mass, radiation, or field stress—contributes to this curvature, and all trajectories follow geodesics determined by the resulting geometry. The theory applies universally through Einstein's field equations (Einstein, 1915), though in practice, detectable gravitational effects require large concentrations of energy or momentum, typically on astronomical or cosmological scales.

The conflict between general relativity and quantum field theory stems from their radically different approaches. While gravity emerges from dynamic spacetime geometry, quantum field theory treats all interactions as exchanges of quantized excitations—force carriers—on a fixed, non-dynamical spacetime. Each field is defined relative to a background geometry, typically flat Minkowski space or a weak perturbation thereof. Interactions are governed by probabilistic amplitudes and operator algebra, computed through path integrals and correlation functions. The formalism is fundamentally discrete and algebraic, with observables expressed as expectation values of operator products, and locality defined with respect to a rigid light-cone structure.

This diverges from general relativity at several levels. The metric $g_{\mu\nu}(x)$ in GR is a classical, dynamical tensor field that defines causal structure; in QFT, causality is imposed externally through fixed spacetime intervals. The conflict becomes unresolvable when attempting to promote $g_{\mu\nu}(x)$ to a quantum operator. No known formalism permits operator-valued metrics that preserve general covariance while maintaining consistency with standard quantum field quantization. Commutators of field operators require a well-defined notion of spacelike separation, which in GR is determined by the metric itself—making the causal order dependent on the state of the fields.

This breakdown goes further. When quantum fluctuations are considered, QFT predicts a large zero-point energy for every field mode. Summing over all modes leads to an enormous vacuum energy density. When inserted into Einstein's field equations, this acts as a cosmological constant and should curve spacetime dramatically. Yet observations show a cosmological constant that is at least 120 orders of magnitude (trillion times trillion times trillion, ten times!) smaller than this prediction. This exposes a disagreement about what vacuum energy means and how it enters the gravitational field equations.

Renormalization further illustrates the incompatibility. In gauge field theories, divergences can be absorbed into a finite set of physical parameters through renormalization. This fails for gravity: treating the metric perturbatively as $g_{\mu\nu} = \eta_{\mu\nu} + h_{\mu\nu}$ and quantizing $h_{\mu\nu}$ produces divergent terms that require an infinite number of counterterms involving higher derivatives of the curvature tensor. No closed, predictive theory results. Gravity, within QFT, is perturbatively non-renormalizable ('t Hooft & Veltman, 1974)

Information and unitarity pose further tensions. Quantum theory forbids information loss: pure states evolve into pure states by applying unitary transformations. However, semiclassical treatments of black holes—where quantum fields propagate on classical spacetimes—predict evaporation via Hawking radiation (Hawking, 1975), which appears thermal and uncorrelated with the initial state. This suggests information loss, violating unitarity. Attempts to resolve this paradox confront the absence of a complete theory in which both the horizon structure and quantum correlations are dynamically defined.

Finally, QFT assumes that physical states can exist in superposition and be entangled across spacelike surfaces. But spacetime itself, in GR, is not a state but a geometric manifold. Whether one can meaningfully define a superposition of spacetime geometries, or even speak of entanglement without a fixed causal background, remains doubtful. When a particle goes through two-slits, which path curves spacetime? There is no operational procedure for comparing amplitudes across different topologies or coordinate charts.

The bottom line is that GR and QFT are incompatible. Their convergence would require a framework in which geometry, causality, and quantization arise jointly—a condition unmet by any known unification. Theories such as string theory and loop quantum gravity represent efforts to construct such a unified theory, but none has produced experimentally confirmed, unique predictions at accessible energies.

In particular, string theory introduces a vast landscape of possible vacua, each corresponding to a different low-energy limit. The theory accommodates an enormous number of possible compactification geometries, field configurations, and symmetry-breaking rules. While this internal flexibility allows string theory to incorporate both quantum field theoretic structure and dynamical geometry, it also permits so many distinct effective theories that it lacks a unique set of predictions. As a result, it is challenging to extract unique, falsifiable predictions without additional assumptions. The same challenge applies, in different form, to other quantum gravity proposals that lack experimentally testable observables at accessible scales. In the absence of empirical constraints, the search for a unified framework remains guided by mathematical coherence, internal consistency and equational elegance.

One of us takes care of the small stuff, the other the big stuff—it's Field Work.

Quantum Gravity: Core Conflicts

Mathematical Structure Mismatch

GR's Smooth Manifold vs QFT's Field Operators. In general relativity, spacetime is a 4D manifold M with metric tensor $g_{\mu\nu}(x)$ satisfying Einstein's equations,

$$R_{\mu\nu} - \tfrac{1}{2}R\,g_{\mu\nu} + \Lambda\,g_{\mu\nu} = \frac{8\pi G}{c^4}\,T_{\mu\nu},$$

where $R_{\mu\nu}$ is the Ricci curvature, $R = g^{\mu\nu}R_{\mu\nu}$ the scalar curvature, Λ the cosmological constant, and $T_{\mu\nu}$ the stress-energy tensor.

In quantum field theory, particle states arise from excitations of quantum fields $\hat{\phi}(x)$ or $\hat{\psi}(x)$ defined on a fixed Minkowski, or curved background. Canonical commutation relations,

$$[\hat{\phi}(t,\mathbf{x}),\,\hat{\pi}(t,\mathbf{y})] = i\,\hbar\,\delta^3(\mathbf{x}-\mathbf{y}),$$

quantify field quanta. Allowing the spacetime metric itself to be a dynamic quantum operator complicates these commutation relations, as fixed background reference frames break down.

Vacuum Energy and the Cosmological Constant

QFT's Enormous Zero-Point Energy vs Observed Small Λ. Zero-point fluctuations of quantum fields give a vacuum energy density,

$$\rho_{\text{vac}} = \frac{1}{2}\sum_{\mathbf{k}} \hbar\omega_{\mathbf{k}},$$

which diverges or is cut off at some high-energy scale. Conservative cutoffs overshoot the observed ρ_{vac} by up to 10^{120}, creating the cosmological constant problem. GR requires consistency between vacuum energy and curvature (through Λ), leading to an immense discrepancy $\Lambda_{\text{QFT}} \gg \Lambda_{\text{obs}}$.

Non-Renormalizability of Gravity

Perturbation Theory Fails for Graviton Loops. Treating the graviton (the quantum of the metric field) in perturbation series yields loop integrals with divergences that cannot be cancelled by renormalization.

A dimensionful coupling G implies higher-order terms require infinitely many counterterms. Unlike quantum electrodynamics or QCD, gravity does not fit the renormalizable pattern:

$$\mathcal{L}_{\text{eff}} = \sqrt{-g}\Big(\frac{R}{16\pi G} + \alpha_1 R^2 + \alpha_2 R_{\mu\nu}R^{\mu\nu} + \ldots\Big)$$

with each α_n an unknown parameter.

Black Hole Information and Unitarity

Information Loss vs Quantum Conservation. Hawking's semiclassical calculation predicts black hole evaporation that appears to erase quantum information. Quantum mechanics requires unitary evolution: no information loss. GR accommodates singularities where classical time ends. This clash forms the black hole information paradox, driving quantum gravity research.

Spacetime Superposition

Can the Metric Exist in a Quantum Superposition? Standard QFT can superpose field states, but GR demands a specific geometric framework for defining intervals, causal structure, and even time. A superposition of metrics $|\psi\rangle = \alpha\,\big|g^{(1)}_{\mu\nu}\big\rangle + \beta\,\big|g^{(2)}_{\mu\nu}\big\rangle$ challenges defining distance and time, essential for measurement theory.

Conclusion

GR and QFT tension stems from fundamental descriptive differences. Attempts to quantize gravity face the cosmological constant puzzle, non-renormalizable infinities, and conceptual conundrums such as black hole information loss. Through string theory, loop quantum gravity, asymptotically safe gravity, or other approaches, finding consistent quantum spacetime remains a foremost theoretical challenge.

References:

Weinberg, S. (1979). Ultraviolet divergences in quantum theories of gravitation. In S. W. Hawking and W. Israel (Eds.), *General Relativity: An Einstein Centenary Survey* (Cambridge University Press).

Kiefer, C. (2012). *Quantum Gravity* (3rd ed.). Oxford University Press.

Goroff, M. H., and Sagnotti, A. (1986). The ultraviolet behavior of Einstein gravity. *Nuclear Physics B*.

Darkness to Bind Them

Top (Candidate Properties): Dark matter candidates are evaluated by properties such as mass (from massless to heavy), speed (cold vs warm), interaction type (weak, gravitational only), and detectability (via lab or decay signatures). Each row indicates a possible property constraint.
Second Row (WIMPs and Axions): WIMPs (Weakly Interacting Massive Particles) are cold, massive, and weakly interacting, long favoured due to supersymmetric theories. Axions are extremely light, produced via field misalignment, and may convert to photons in magnetic fields.
Third Row (Sterile Neutrinos and Primordial Black Holes): Sterile neutrinos mix with active ones but don't interact via the weak force, allowing them to evade detection. Primordial black holes are relics from the early universe that behave as cold dark matter through pure gravitational influence.
Bottom Row (Kaluza-Klein theories and MOND): Kaluza-Klein particles arise from extra spatial dimensions; their quantised modes could serve as dark matter if stable. Modified Newtonian Dynamics (MOND) suggests gravity itself needs correction at low accelerations—removing the need for particle dark matter.

Darkness to Bind Them

Dark matter's existence is inferred through multiple independent lines of evidence spanning different cosmic scales. Galaxy rotation curves remain flat far beyond visible matter, indicating extended gravitational influence. Galaxy clusters contain hot gas whose temperature and confinement require gravitational potentials deeper than visible matter can provide. Gravitational lensing reveals mass distributions exceeding luminous components, particularly in systems such as the Bullet Cluster where dark and visible matter separate during collisions. The cosmic microwave background's fluctuation patterns indicate that ordinary matter comprises only 15% of the total matter content needed to match observations, with the remainder consisting of non-baryonic material already present before photon-matter decoupling.

Dark Matter Evidence ◦ Galaxy Rotation Curves ◦ Doppler Spectroscopy ◦ 21cm Hydrogen Mapping ◦ Gravitational Lensing ◦ Bullet Cluster Collision ◦ Cluster Dynamics ◦ CMB Acoustic Peaks ◦ Structure Formation Timeline ◦ Non-Luminous Mass ◦ MOND Limitations

"נְטֵה יָדְךָ עַל־הַשָּׁמַיִם וִיהִי חֹשֶׁךְ עַל־אֶרֶץ מִצְרָיִם וְיָמֵשׁ חֹשֶׁךְ׃"

("...darkness spreads over Egypt — darkness that can be felt")

— Exodus 10:21

Darkness to Bind Them

In 1933, Swiss astrophysicist Fritz Zwicky analysed the velocity dispersion of galaxies in the Coma Cluster and found their motions to be too fast to be gravitationally bound by the visible mass alone. He introduced the term *dunkle Materie* (dark matter) to describe the missing component. Though initially met with scepticism, his mass discrepancy hinted at a core problem in astrophysical mass accounting.

The issue resurfaced in the 1970s when Vera Rubin and Kent Ford measured rotation curves of spiral galaxies using optical spectroscopy. Instead of decreasing with radius as expected from luminous matter distributions, the rotation speeds remained flat well beyond the visible edge. Independent radio observations by Albert Bosma confirmed this effect through 21 centimetre emission from neutral hydrogen, revealing a pervasive halo of unseen mass enveloping each galaxy.

By the early 1980s, theorists such as Jeremiah Ostriker and Jim Peebles emphasised the necessity of dark matter to explain large-scale structure formation. Without a non-luminous component, galaxies and clusters could not form on observed timescales. Theoretical simulations incorporating dark matter successfully reproduced the filamentary distribution of galaxies seen in redshift surveys.

Gravitational lensing provided additional, independent confirmation. Light from distant sources bent around foreground mass concentrations showed more deflection than visible matter alone could account for. In 2006, observations of the Bullet Cluster—a high-speed collision of galaxy clusters—visually separated dark matter from hot gas via X-ray and lensing data, offering direct evidence of a collisionless mass component.

Dark matter's influence now spans cosmology, astrophysics, and particle physics. Measurements of cosmic microwave background anisotropies by missions such as WMAP and Planck established dark matter as essential for matching early-universe fluctuations to present-day structure.

To understand the structure and evolution of the universe, astrophysicists must measure not just light, but mass. Stars, galaxies, and gas clouds emit radiation that reveals their presence, but the dynamics of the cosmos are governed by gravity—by how much mass exists and how it is distributed. Knowing where matter is, and how much of it there is, is necessary for explaining motion, structure formation, and stability across cosmic scales. The question is: how can one measure mass across millions of light-years, when most of it emits no light at all?

The primary method is to observe motion. In Newtonian mechanics, any orbiting body experiences a centripetal acceleration that is directly related to the mass it orbits. This relationship allows astronomers to determine how much mass lies within a given radius, provided they can measure orbital speeds and distances. By measuring the velocities of stars at different radii from the centre of a galaxy, astrophysicists can infer how mass is distributed throughout the galaxy—not just in its luminous regions.

Spectroscopy plays a central role in this process. When light from a star or gas cloud is dispersed into its component wavelengths, the resulting spectrum reveals its motion via

Doppler shifts. A redshift indicates motion away from the observer; a blueshift indicates motion toward. These shifts allow for measurements of velocity along the line of sight. Rotational velocities within galaxies, random velocities in galaxy clusters, and internal turbulence in gas clouds can all be extracted from spectral line profiles.

Radio astronomy extends these measurements beyond visible light. Neutral hydrogen, the most abundant element in the universe, emits radiation at a wavelength of 21 centimetres. This emission can be traced even in the outskirts of galaxies, where stars are sparse or absent. Observing the motion of this gas provides crucial data on gravitational effects well outside the luminous core.

In systems without a simple rotation pattern—such as elliptical galaxies or clusters—mass is inferred statistically. The velocities of constituent bodies follow distributions governed by the overall gravitational potential. In these cases, the virial theorem (as is explored also in the chapter on osmosis) connects the average kinetic energy of the system to the total mass required to confine it. This technique is especially important in estimating the masses of galaxy clusters, where galaxies orbit in all directions and the system behaves like a gravitationally bound swarm.

A complementary approach bypasses dynamics altogether: gravitational lensing. According to general relativity, mass curves spacetime, bending the path of light from background sources. When a massive object—such as a galaxy or cluster—lies along the line of sight to a more distant source, the background light is distorted. By analysing the shape and degree of this distortion, one can map the total mass distribution of the intervening object. This method is purely gravitational: it measures mass regardless of whether it emits, absorbs, or reflects light.

When mass in a galaxy is concentrated toward the centre, orbital velocities should decrease with distance, just as planets in the Solar System orbit more slowly the farther they are from the Sun. Observations do not match this expectation. In spiral galaxies, stars orbit the centre at nearly constant speed (Rubin & Ford, 1970) over vast radial distances. These flat rotation curves indicate that the enclosed mass does not level off where the stars end, but continues to increase. The luminous matter—stars, gas, and dust—cannot account for this excess gravity.

The persistence of high orbital speeds well beyond the visible edge of galaxies is confirmed by radio observations of neutral hydrogen. Radio telescopes can map the velocity of this gas across the outskirts of galaxies. These measurements show that rotational velocities remain flat or even rise at large radii, where the density of luminous matter has dropped to negligible levels. The simplest explanation is that galaxies are embedded in extended halos of non-luminous mass, whose gravitational influence dominates in the outer regions.

On larger scales, galaxy clusters present an analogous discrepancy. These systems contain hundreds or thousands of galaxies bound together by gravity, along with large amounts of hot gas that emits strongly in X-rays. The temperature and distribution of this gas reflect the depth of the gravitational potential well. If only the visible galaxies and gas contributed to the cluster's gravity, the hot gas would escape over cosmological timescales. That it remains bound implies a much larger total mass than what is seen. The internal motions of galaxies within clusters, measured by redshift dispersion, independently confirm this excess mass. Gravitational lensing applies also to these massive clusters, and in both contexts (individual galaxies and clusters), the lensing signal supports the presence of a dominant, unseen mass component.

A unique astrophysical event—**the Bullet Cluster**—provides direct evidence for a non-luminous mass component that behaves differently from ordinary matter. This system consists of two galaxy clusters in the process of collision. As they pass through one another, the hot gas from each cluster interacts and slows down due to ram pressure (pressure due to bulk motion; compare to temperature, the internal kinetic component, in Chapter 28), becoming spatially displaced from the galaxies themselves. X-ray observations reveal this gas concentrated between the clusters. However, gravitational lensing maps of the same region show that most of the mass is still centred on the galaxies, not the gas. This separation implies that the dominant mass component did not experience significant drag during the collision. It must interact gravitationally, but not electromagnetically—suggesting it is both massive and effectively collisionless.

The early universe provides a separate window into the distribution of mass through the spectral composition of the cosmic microwave background (CMB). This radiation carries a record of acoustic oscillations in the primordial plasma—pressure waves driven by the interplay between gravity and radiation. The pattern of these oscillations, visible as peaks in the CMB power spectrum, (Spergel et al., 2003) depends sensitively on the matter content of the universe. Ordinary (baryonic) matter couples to photons and thus participates in pressure waves, while non-baryonic matter does not. Matching the observed amplitude and spacing of the peaks requires a dominant component of matter that does not interact with radiation, but contributes to gravitational attraction. Precision measurements by the WMAP (Wilkinson Microwave Anisotropy Probe) and Planck satellites confirm that ordinary matter accounts for only a small fraction of the total.

Numerical simulations of structure formation reinforce this conclusion. Starting from the nearly uniform density field observed in the CMB, simulations track the growth under gravity. The timing and scale of galaxy and cluster formation are highly sensitive to the amount and type of matter present. If only ordinary matter were included, structure would form too slowly to match what is observed in deep-field surveys. The emergence of galaxies, clusters, and the cosmic web within a few billion years requires a gravitational source that was present from the earliest epochs, unaffected by radiation pressure, and capable of seeding the collapse of matter on small scales. The observed universe forms on schedule only when this additional component is included.

Thus, the evidence for dark matter does not rest on a single anomalous measurement, but on the convergence of diverse and independent observational domains. Galaxy rotation curves, cluster dynamics, X-ray temperature profiles, gravitational lensing, and the cosmic microwave background all indicate that visible matter accounts for only a small fraction of the gravitational forces at work. The required additional mass must act through gravity, but not through electromagnetism; it must clump on galactic scales, but remain diffuse enough not to obstruct light; it must have existed before the era of recombination, but not interfered with photon-matter coupling. The simplest explanation consistent with all constraints is the existence of a non-luminous, cold, and effectively collisionless form of matter—distinct from atoms, but essential for the cosmic configuration. To convey the scale: the local dark matter density near the Sun is estimated at roughly 0.3 GeV/cm^3, equivalent to about one proton mass per every three cubic centimetres—an extraordinarily diffuse medium, yet one whose cumulative gravitational effect across galactic volumes is decisive.

Attempts to resolve these discrepancies by modifying the laws of gravity instead of introducing a new kind of matter have achieved only partial success. Modified Newtonian dynamics (MOND) can account (Milgrom, 1983) for some features of galactic rotation curves, but struggle with systems lacking clear symmetry or equilibrium. The Bullet Cluster, in particular, presents a direct conflict: the separation of gravitational and luminous mass cannot be explained by alterations to the gravitational field alone. Gravitational lensing imposes geometric constraints that any alternative theory must satisfy, and these constraints are difficult to reconcile with models that dispense entirely with unseen mass. The full range of phenomena—from early-universe fluctuations to present-day structure—aligns with the presence of a real, additional matter component, much better than by a reformulation of force laws.

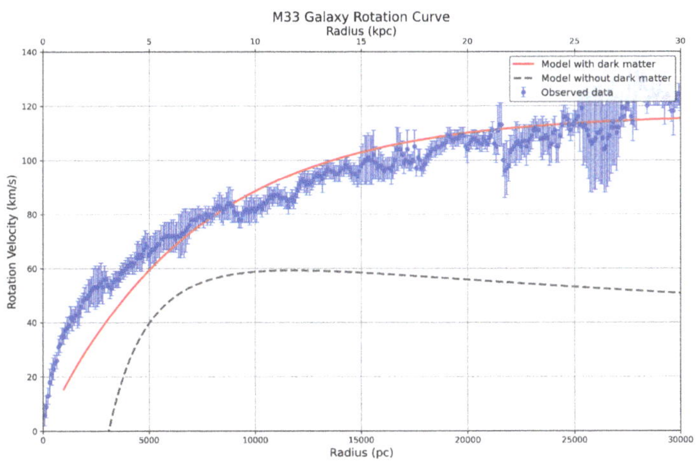

The rotation curve of M33 (Triangulum Galaxy). Blue points show observed Hα velocities; the red curve includes dark matter contribution; the grey dashed line represents visible baryonic matter only. The persistent high velocities at large radii provide compelling evidence for an extended dark matter halo. Raw data from Kam et al. (2015), MNRAS 449, 4048. Fitted curve is not from the literature and is for illustrative purposes only.

Mass Scales in the universe

Object	Mass (kg)	Object	Mass (kg)
Electron neutrino	$< 2 \times 10^{-36}$	Human	~ 70
Electron	9.1×10^{-31}	Earth	6.0×10^{24}
Proton	1.7×10^{-27}	Jupiter	1.9×10^{27}
Gold atom	3.3×10^{-25}	Sun	2.0×10^{30}
Water molecule	3.0×10^{-26}	Milky Way	$\sim 10^{42}$
DNA base pair	$\sim 10^{-24}$	Local group	$\sim 4 \times 10^{42}$
E. coli bacterium	$\sim 10^{-15}$	Observable universe	$\sim 10^{53}$

Density Scales in Nature

Material / Region	Density (g/cm³)
Intergalactic vacuum	$\sim 10^{-30}$
Interstellar medium	$\sim 10^{-24}$
Best laboratory vacuum	$\sim 10^{-17}$
Air at sea level	1.2×10^{-3}
Water	1.0
Iron	7.9
Lead	11.3
Osmium (densest element)	22.6
White dwarf core	$\sim 10^{6}$
Atomic nucleus	$\sim 2 \times 10^{14}$
Neutron star core	$\sim 10^{15}$
Quark-gluon plasma	$\sim 10^{16}$

The universe spans roughly 90 orders of magnitude in mass and 46 orders of magnitude in density. The density range accessible to direct laboratory measurement occupies only a narrow band between 10^{-17} *and* 10^{2} *g/cm³. Nuclear densities, neutron star cores, and the conditions of the early universe require inference from particle collisions, gravitational observations, and quantum chromodynamics.*

Gravitational Inference and the Distribution of Dark Matter

Virial Mass in Galaxy Clusters (see also Chapter 31)

Clusters of galaxies are treated as self-gravitating systems in equilibrium. Let a system of N particles with masses m_i and velocities $\mathbf{v}_i$ have total kinetic and potential energy:

$$K = \sum_{i=1}^{N} \tfrac{1}{2} m_i v_i^2, \qquad U = -\sum_{i<j} \frac{G m_i m_j}{r_{ij}}.$$

By the virial theorem:

$$2\langle K\rangle + \langle U\rangle = 0.$$

Assuming an isotropic one-dimensional velocity dispersion so that $\langle v^2\rangle = 3\sigma_v^2$ and, for a roughly uniform sphere, $U = -\frac{3}{5}\frac{GM^2}{R}$, we obtain:

$$M_{\text{vir}} \approx \frac{5\,\sigma_v^2 R}{G},$$

where R is the effective radius of the system. For rich clusters such as Coma, the mass inferred by this formula exceeds luminous mass (stars + gas) by over an order of magnitude.

Rotation Curves and Halo Profiles

In spiral galaxies, stars and gas orbit the galactic centre under gravitational attraction. For circular orbits:

$$\frac{v^2(r)}{r} = \frac{GM(r)}{r^2} \quad \Rightarrow \quad M(r) = \frac{v^2(r)\,r}{G}.$$

Observations show that $v(r)$ remains nearly constant beyond the optical radius, implying $M(r) \propto r$, inconsistent with the radial profile of visible mass.

This necessitates a dark matter halo extending beyond the visible disc. Simulations and fits to data often use the Navarro–Frenk–White (NFW) profile:

$$\rho(r) = \frac{\rho_0}{(r/r_s)(1 + r/r_s)^2},$$

where ρ_0 is a characteristic density and r_s a scale radius. This profile produces approximately flat rotation curves at large r and matches the mass distributions required to stabilise galaxies against dispersal.

Weak Gravitational Lensing

Lensing measures projected surface mass. In the weak lensing regime, the convergence $\kappa(\theta)$ is given by:

$$\kappa(\theta) = \frac{\Sigma(\theta)}{\Sigma_{\text{crit}}}, \qquad \Sigma_{\text{crit}} = \frac{c^2}{4\pi G} \cdot \frac{D_s}{D_d D_{ds}},$$

with angular diameter distances D_s (observer to source), D_d (observer to lens), and D_{ds} (lens to source). The deflection angle is sensitive to the integrated surface mass density $\Sigma(\theta)$. Mapping κ via background galaxy distortions reconstructs the total projected mass distribution.

In systems such as the Bullet Cluster, lensing peaks and X-ray emission peaks are spatially offset. This implies that the dominant mass component is not collisional (as hot gas is), but rather behaves as a collisionless fluid, consistent with dark matter expectations.

Implications

The gravitational field inferred from cluster dynamics, orbital motion in galaxies, and lensing geometry all points to a dominant non-luminous mass component. Its spatial distribution is extended, centrally concentrated, and required across all scales. These observations define dark matter phenomenologically: a gravitationally interacting, non-emissive mass component that clumps and seeds structure.

References:

Zwicky, F. (1933). Die Rotverschiebung von extragalaktischen Nebeln. *Helv. Phys. Acta*, **6**, 110–127.

Navarro, J. F., Frenk, C. S., and White, S. D. M. (1997). A Universal Density Profile from Hierarchical Clustering. *ApJ*, **490**, 493–508.

Clowe, D., Bradač, M., Gonzalez, A. H., Markevitch, M., Randall, S. W., Jones, C., and Zaritsky, D. (2006). A Direct Empirical Proof of the Existence of Dark Matter. *ApJL*, **648**, L109-L113.

A Truce Story

Top (Entrenched Conflict): Soldiers from opposing sides dig into fortified positions across a bleak winter landscape. Skirmishes persist across isolated fronts with minimal contact or trust.

Middle (Momentary Ceasefire): On Christmas Eve, combatants initiate an informal truce. They emerge from trenches, exchange gifts, share fires, and briefly connect across No Man's Land as human beings. Games of football and camaraderie replace gunfire.

Bottom (Return to War): As dawn breaks, soldiers are recalled to duty. Artillery resumes. The fleeting peace dissolves, leaving scattered memories of what might have been—a glimpse of shared humanity extinguished by command.

A Truce Story

On Christmas 1914, enemy soldiers climbed out of their trenches and shook hands. Along sectors of the Western Front, British and German troops spontaneously ceased fire, met in No Man's Land to exchange tobacco and souvenirs, sang carols together, and buried their dead side by side. Some kicked footballs around shell craters. This unofficial truce lasted hours to days depending on location. By Christmas 1915, high command used coordinated artillery barrages to prevent any recurrence.

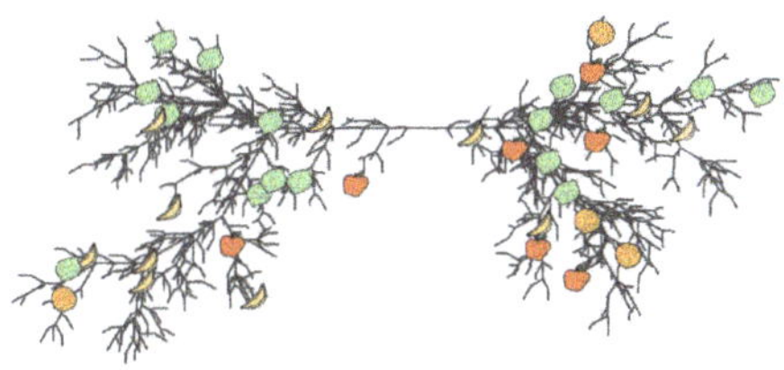

Christmas Truce 1914 ◦ Western Front Stalemate ◦ Trench Proximity ◦ Carols Across No Man's Land ◦ Gift Exchange & Burials ◦ Football Myth & Reality ◦ Frank Richards Account ◦ Saxon-British Fraternisation ◦ Temporary Humanity ◦ Official Suppression ◦ Historical Precedents

"I shouted to our enemies that we didn't wish to shoot
and that we make a Christmas truce...
Then a man came out of the trenches and I on my side did the same
and so we came together and we shook hands — a bit cautiously!"

— Captain Josef Sewald, 17th Bavarian Regiment

"You should laugh every moment you live,
for you'll find it decidedly difficult afterwards."

— Nicomo Cosca, Year 580 AU

A Truce Story

The Christmas Truce of 1914 occurred just months into the First World War, a conflict that had erupted from a complex web of alliances, imperial tensions, and national ambitions. The assassination of Archduke Franz Ferdinand of Austria-Hungary in Sarajevo on 28 June 1914 set off a chain reaction. Within five weeks, much of Europe was at war. Austria-Hungary, backed by Germany, declared war on Serbia. Russia mobilised in defence of Serbia, prompting German declarations of war on Russia and France. When German troops invaded neutral Belgium, Britain entered the war, citing the 1839 Treaty of London, which guaranteed Belgian neutrality. What might have remained a regional dispute quickly expanded into a global conflict.

By late 1914, the Western Front had solidified into a long, stagnant line stretching from the North Sea to the Swiss frontier. This line formed after the German army's rapid advance through Belgium and northern France—the execution of the Schlieffen Plan—was halted at the First Battle of the Marne in early September. The Allied counteroffensive pushed German forces back but failed to regain significant ground. Both sides attempted to outflank one another in a series of movements known as the 'Race to the Sea,' which culminated in the First Battle of Ypres in October and November 1914. The battle was costly and inconclusive, with neither side able to break the deadlock. By the end of November, both German and Allied armies had begun to dig in, transitioning to entrenched positions that would define the nature of the war for years to come.

The key belligerents along the Western Front during the truce were the British Expeditionary Force (BEF) and the Imperial German Army. The BEF, composed of professional soldiers and newly enlisted volunteers, was stationed across sectors in northern France and Belgium. The German lines opposite them were held by a mix of Saxon, Bavarian, and Prussian units. While France bore the brunt of the war's human and territorial costs, French units were less prominently involved in the truce, partly due to the deeper emotional and political resentment stemming from the German occupation of French soil.

Conditions by December were grim. The early optimism that the war would be short-lived had evaporated. Both sides had suffered staggering casualties in the first months: hundreds of thousands killed or wounded in battles from Mons to Ypres. The initial war of manoeuvre had devolved into a brutal, attritional struggle marked by mud, disease, and psychological fatigue. Troops on both sides faced inadequate shelter, minimal sanitation, and constant threat from snipers and artillery. In this context, the rigid enemy lines became strangely familiar. Soldiers could hear each other, sometimes see each other, and often recognised in their enemies the same weariness and longing for respite.

In the ninth century BCE, Greek city-states perpetually at war observed the Ekecheiria—a sacred truce for the Olympic Games. Three months before competition, heralds called *spondophoroi* travelled from Elis across the Greek world, protected by Zeus himself, announcing the cessation of hostilities. The truce began extitone month before the games and extended one month after, allowing athletes and the theoroi (sacred ambassadors) safe passage through hostile territory. Violation brought divine punishment and exclusion from the games. Sparta was fined 2,000 minae (200,000 drachmas) for attacking Lepreum

 https://doi.org/10.11647/OBP.0526.38

during the truce; when they refused to pay, they were barred from the Games. Even during campaigns, warring states allowed athletes safe passage to Olympia. The inscription at Olympia declared: “May the world be delivered from crime and killing and freed from the clash of arms.” War paused not for humanitarian ideals but for religious obligation—the games honoured Zeus, and defying the truce meant defying the gods.

Homer’s Iliad, Book Seven, records the duel between Ajax and Hector. They fight from dawn until heralds intervene at dusk, declaring divine favour on both warriors. Hector proposes gift exchange: “Let us give each other gifts, so that Trojans and Achaeans alike may say: 'These two fought in soul-consuming strife, then parted, joined in friendship.’” Ajax presents his purple war belt with silver studs; Hector reciprocates with his silver-hilted sword. The exchange creates guest-friendship (xenia)—a sacred bond transcending battlefield enmity. Their gifts carry dark irony: Ajax later kills himself with Hector’s sword, while Achilles later drags Hector’s corpse behind his chariot with ox-hide thongs. The warriors recognise shared excellence even while bound to kill each other’s kinsmen.

Roman military doctrine discouraged fraternisation, yet siege warfare bred practical accommodations. During the siege of Numantia (134–133 BCE), Scipio Aemilianus constructed a circumvallation to starve the defenders. Ancient accounts describe unspoken protocols at sieges: water collection sometimes went unmolested, burial parties operated under *informal immunity* (n.d.), and soldiers traded insults rather than missiles during meals. Caesar’s Commentarii describe similar patterns at Alesia and around Dyrrhachium—not from mercy but from mutual exhaustion. Soldiers recognised the futility of constant skirmishing over resources both sides needed. These intervals weren’t truces but tactical breathing spaces, managed through signals and precedent rather than negotiation. At Dyrrhachium, control of water sources became a tactical lever, and Caesar’s troops diverted alternative supplies rather than escalating—practical solutions to preserve fighting capacity for decisive battles.

The Christmas Truce of 1914 stands as one of the most enduring and mythologized episodes of the First World War. According to popular accounts, soldiers from opposing sides emerged from their trenches on Christmas Eve and Christmas Day to sing carols, exchange gifts, and even play football in No Man’s Land. These images—striking in their contrast to the prevailing brutality of trench warfare—have become symbolic of a moment when shared humanity briefly transcended the violence of industrialised conflict. Yet, while rooted in truth, such narratives often simplify and romanticise an event that was far more fragmented, contingent, and limited in both scope and duration.

The truce occurred during the first winter of the war, at a time when the initial hopes for a swift resolution had long since collapsed. From the failure of the Schlieffen Plan and the Battle of the Marne to the static bloodshed of Ypres, the Western Front had by December 1914 become a nearly continuous line of trenches stretching hundreds of miles. Conditions were bleak. Cold weather, persistent rain, inadequate shelter, and primitive hygiene created an environment of physical misery and psychological fatigue. Soldiers faced not only the enemy across the mud-churned expanse of No Man’s Land, but the more immediate challenges of frostbite, trench foot, and various diseases.

The proximity of opposing trenches created an unexpected intimacy. In the Ploegsteert sector near Armentières, British and German lines lay just 50 to 100 yards apart. Soldiers could hear conversations, smell cooking, and distinguish individual voices. This closeness had already produced informal arrangements. The 2nd Battalion Gordon Highlanders reported 'breakfast truces' where both sides refrained from sniping during morning meals. Saxon regiments opposite the 2nd Battalion Royal Welsh Fusiliers would shout warnings before shelling: 'We send shells in ten minutes—take cover!'

Against this backdrop, the events of Christmas took shape. In some sectors, particularly where German and British forces faced each other at short distances, soldiers began calling greetings across the lines. German troops were often the first to decorate parapets with lantern-lit Christmas trees and sing carols such as "Stille Nacht." British soldiers responded with their own songs, and in many places this shared recognition of the holiday prompted tentative ceasefires. Soldiers cautiously entered the area between the trenches, exchanged food, tobacco, and small souvenirs, and in many cases worked together to bury the dead. These acts were not officially sanctioned and did not occur everywhere. In some sectors, hostilities continued uninterrupted.

While there are scattered reports of football being played, most accounts describe informal kickabouts rather than organised matches. Still, the idea of enemies setting down rifles to play a game remains powerfully evocative. That this image has endured—more than the joint burial parties or shared cigarettes—speaks to the symbolic potency of sports as a common cultural language and to the broader desire for stories of reconciliation amid destruction.

The truce was geographically uneven and temporary. It began, often spontaneously, on Christmas Eve and faded by New Year's Day. In some sectors, truces lasted only a few hours; in others, they extended over several days. The experience varied not only by location, but by unit, terrain, and command attitude. Letters and diaries record joy, awkwardness, and even wariness. Some soldiers worried about violating orders. Others simply embraced the chance to reclaim a moment of peace, however fleeting.

In the weeks that followed, military authorities issued strict instructions to prevent further fraternisation. By Christmas 1915, coordinated artillery barrages were used to suppress any attempts at renewed truces. Still, the memory of 1914 persisted—not as an act of organised resistance, but as a brief and extraordinary lapse in the logic of total war (Axelrod, 1984). The truce was not a peace movement, and it changed nothing about the war's course. But it remains significant because it showed, even within the machinery of mass violence, a momentary refusal to reduce the enemy to a target.

Today, the Christmas Truce is remembered less for its strategic consequences than for its moral resonance. It stands as a testament to the capacity for empathy in the midst of systemic dehumanisation, and to the peculiar intimacy of trench warfare, where those who were supposed to kill each other instead spoke, sang, and—for a short time—stood together unarmed. In the context of a war that would ultimately claim millions of lives, the events of December 1914 offer a glimpse of human light amid the human darkness.

Among the preserved recollections of that day, one British private offered a detailed account of his company's interaction with Saxon troops. His description captures both the informality and the contradictions of the occasion—its unplanned gestures, hesitant fraternisation, and negotiated boundaries.

Frank Richards tells the following story:

> On Christmas morning we stuck up a board with "A Merry Christmas" on it. The enemy had stuck up a similar one. Platoons would sometimes go out for twenty-four hours' rest—it was a day at least out of the trench and relieved the monotony a bit—and my platoon had gone out in this way the night before, but a few of us stayed behind to see what would happen. Two of our men then threw their equipment off and jumped on the parapet with their hands above their heads. Two of the Germans did the same and commenced to walk up the river bank, our two men going to meet them. They met and shook hands and then we all got out of the trench.
>
> Buffalo Bill—the Company Commander—rushed into the trench and endeavoured to prevent it, but he was too late: the whole of the Company were now out, and so were the Germans. He had to accept the situation, so soon he and the other company officers climbed out too. We and the Germans met in the middle of No Man's Land. Their officers were also now out. Our officers exchanged greetings with them. One of the German officers said that he wished he had a camera to take a snapshot, but they were not allowed to carry cameras. Neither were our officers.
>
> We mucked in all day with one another. They were Saxons and some of them could speak English. By the look of them their trenches were in as bad a state as our own. One of their men, speaking in English, mentioned that he had worked in Brighton for some years and that he was fed up to the neck with this damned war and would be glad when it was all over. We told him that he wasn't the only one that was fed up with it. We did not allow them in our trench and they did not allow us in theirs.
>
> The German Company Commander asked Buffalo Bill if he would accept a couple of barrels of beer and assured him that they would not make his men drunk. They had plenty of it in the brewery. He accepted the offer with thanks and a couple of their men rolled the barrels over and we took them into our trench. The German officer sent one of his men back to the trench, who appeared shortly after carrying a tray with bottles and glasses on it. Officers of both sides clinked glasses and drank one another's health. Buffalo Bill had presented them with a plum pudding just before. The officers came to an understanding that the unofficial truce would end at midnight. At dusk we went back to our respective trenches.
>
> The two barrels of beer were drunk, and the German officer was right: if it was possible for a man to have drunk the two barrels himself he would have bursted before he had got drunk. French beer was rotten stuff.
>
> Just before midnight we all made it up not to commence firing before they did. At night there was always plenty of firing by both sides if there were no

working parties or patrols out. Mr. Richardson, a young officer who had just joined the Battalion and was now a platoon officer in my company, wrote a poem during the night about the Briton and the Bosche meeting in No Man's Land on Christmas Day, which he read out to us. A few days later it was published in *The Times* or *Morning Post*, I believe.

During the whole of Boxing Day we never fired a shot, and they the same, each side seemed to be waiting for the other to set the ball a-rolling. One of their men shouted across in English and inquired how we had enjoyed the beer. We shouted back and told him it was very weak but that we were very grateful for it. We were conversing off and on during the whole of the day.

We were relieved that evening at dusk by a battalion of another brigade. We were mighty surprised as we had heard no whisper of any relief during the day. We told the men who relieved us how we had spent the last couple of days with the enemy, and they told us that by what they had been told the whole of the British troops in the line, with one or two exceptions, had mucked in with the enemy. They had only been out of action themselves forty-eight hours after being twenty-eight days in the front-line trenches. They also told us that the French people had heard how we had spent Christmas Day and were saying all manner of nasty things about the British Army.

Source: Frank Richards, *Old Soldiers Never Die* (London: Faber & Faber, 1933), pp. 65–67; discussed in John Keegan, *The First World War* (London: Hutchinson, 1998).

The Trojan Piñata was a much friendlier affair, though rarely covered by historians.

The 1914 Christmas Truce: Documentary Evidence

Primary Source Analysis

Unit war diaries, personal letters, and official directives establish the truce as a geographically constrained phenomenon distinct from its mythologized afterlife. Archival data reveal heterogeneous local interactions shaped by material conditions and institutional ambivalence.

Source Reliability

Battalion war diaries constitute the most reliable documentary substrate—compiled contemporaneously under military regulations. Personal correspondence *requires triangulation against unit records.* German *Kriegstagebücher* demonstrate parallel reliability hierarchies: regimental over personal, Saxon over Prussian units.

Geographic Distribution

Truces concentrated along roughly 30 miles of BEF front in Flanders and northern France. Battalion records (London Rifle Brigade, Northumberland Hussars, 6th Gordon Highlanders) document cessation of fire, joint burials, gift exchanges, and carol singing. Saxon and Bavarian regimental reports corroborate these activities, noting English-speaking soldiers and prior civilian contact with Britain.

French-German interactions remain sparsely documented; French command prohibited fraternisation. Later accounts sometimes attribute involvement to Canadians; the Canadian Expeditionary Force did not take its place in the line until 1915.

Material Conditions

Truce emergence correlates with: waterlogged trenches (Ploegsteert), proximity enabling auditory contact (50–100 yards), supply irregularities, cold conditions with frost in many sectors 24–26 December. Static warfare's early phase—pre-gas, pre-continuous wire—permitted physical accessibility.

The Football Myth

No battalion war diary from confirmed truce sectors records organised football. The sole contemporaneous reference—a Rifle Brigade doctor's letter in *The Times* (1915)—mentions 'a football match' without details. Other testimonies describe 'kickabouts' or aborted plans amid impassable terrain. The '3–2' scoreline appears in no primary materials, originating in postwar elaborations.

Command Response

Corps and army commanders issued anti-fraternisation orders in early December 1914; higher commands reiterated prohibitions in late December. Disciplinary action was limited. Officers sometimes joined their men in No Man's Land or ignored violations. Documentary evidence reveals institutional ambivalence: some commands condemned fraternisation without prosecutions; others acknowledged it without censure. German regimental and headquarters records note infractions but little systematic punishment.

By Christmas 1915, pre-planned artillery bombardments and stricter enforcement curtailed fraternisation.

Historiographical Arc

Newspapers reprinted letters immediately. Official histories (1918–1935) omitted the event. Scholarly recovery began in the 1960s. The mythologisation exemplifies Hobsbawm's 'invention of tradition': prosaic fraternisation transformed into structured sporting event. Post-1960s scholarship established *documentary parameters* (Terraine 1978; Ferro 1990; Eksteins 1989). The truce functions as *lieu de mémoire* (Nora 1984): actual events subordinated to commemorative utility.

References:

Brown, M. (2007). *Christmas Truce: The Western Front.* Pocket.

Eksteins, M. (1989). *Rites of Spring.* Houghton Mifflin.

Hobsbawm, E. and Ranger, T. (1983). *The Invention of Tradition.* Cambridge.

National Archives UK. *BEF Unit War Diaries, Dec 1914.*

Nora, P. (1984). *Les Lieux de mémoire.* Gallimard.

Weintraub, S. (2001). *Silent Night.* Plume.

Superanony-mous

Superpermutation over Anime Orders: The sequence 1,2,3,1,2,1,3,2,1 visits all six permutations of $\{1, 2, 3\}$ as contiguous subsequences: 123, 231, 312, 213, 132, 321. Here, episode '1' is *Dragon Ball Z* (Vegeta), '2' is *One Piece* (Straw Hat Crew), and '3' is *Sailor Moon*. This is a minimal superpermutation for $n = 3$: the shortest possible viewing order that guarantees every order of watching (back-to-back) all three shows appears.

Superanonymous

A significant combinatorial breakthrough from an unlikely source: an anonymous 4chan post (2011) responding to a question about anime episode viewing orders. Superpermutations are strings containing every possible ordering of n symbols as substrings. For years, mathematicians believed the minimal length followed the pattern of factorial sums observed in small cases. The anonymous poster derived a rigorous lower bound, modelling the problem as path optimisation through a permutation graph. This proof remained obscure until 2014 when mathematician Robin Houston rediscovered it, leading to the disproof of the long-standing conjecture and establishing new bounds on this combinatorial problem—with the original derivation still officially credited to 'Anonymous 4chan Poster.'

Superpermutation Problem ◦ Permutation Overlap ◦ Hamiltonian Path Approach ◦ Sum-of-Factorials Conjecture ◦ n=6 Counterexample ◦ Anonymous 4chan Proof ◦ Haruhi Anime Origin ◦ Robin Houston Discovery ◦ Greg Egan Upper Bound ◦ Combinatorial Compression ◦ Non-Academic Mathematics

"Why should I refuse a good dinner
simply because I don't understand the digestive processes involved."
— Oliver Heaviside, 1893

Superanonymous

From the late nineteenth century, combinatorial mathematics developed tools for enumerating and arranging discrete structures. Permutations—ordered arrangements of elements—became central objects of study, with applications ranging from algebra to scheduling theory. One question was about sequences that embed all permutations of a given set as contiguous substrings. Though such sequences appeared in scattered contexts, the idea of minimising their length—the superpermutation problem—remained informal and largely unexplored.

By the late twentieth century, empirical computer exploration suggested that for small n the minimal lengths matched the sum-of-factorials pattern, $L(n) = \sum_{k=1}^{n} k!$. This led to a widely discussed but unproven conjecture that the pattern might hold in general.

The situation changed dramatically in 2011 when an anonymous user on 4chan's science board posed a variation of the problem in the context of anime episode viewing order. In response, another user posted a rigorous lower bound on superpermutation length, unnoticed by the broader community for years. Independent developments followed: in 2014, a construction of length 872 for $n = 6$ appeared, disproving the factorial-sum conjecture. Soon after, mathematicians formalised the 4chan insight, and Greg Egan proposed a new upper bound, narrowing the known range.

A *permutation* is an ordered arrangement of a set of distinct elements. Consider the set $\{1, 2, 3\}$. One possible ordering is 123, where the elements appear in their natural order. Another is 231, where 2 comes first, followed by 3 and 1. Each such arrangement–where all elements are used exactly once and appear in a specific sequence–is called a permutation of the set.

For a set of size n, the total number of such arrangements is given by the factorial function: $n! = n \times (n-1) \times (n-2) \times \cdots \times 2 \times 1$. The number of permutations grows rapidly: $1! = 1$, $2! = 2$, $3! = 6$, $4! = 24$, $5! = 120$, $6! = 720$, $7! = 5{,}040$. This rapid–factorial–growth means that although the idea of permutations is elementary, the total number becomes intractable to list or store explicitly even for moderate n.

Suppose one attempts to list all permutations of $\{1, 2, 3\}$: these are 123, 132, 213, 231, 312, and 321. Each permutation has length 3. If one were to list them end-to-end it would yield a string of length $6 \times 3 = 18$. For larger n, such direct listings become prohibitively long.

A string that contains all permutations as contiguous substrings is called a *superpermutation*. For example, the naive concatenation 123132213231312321 is a superpermutation of $\{1, 2, 3\}$ because it contains all six permutations as contiguous substrings. However, it has length 18, which is not the minimal possible length. While the naive approach gives a superpermutation of length $n! * n$, it can be *compressed* and made smaller by reusing overlapping segments wherever possible.

Consider the case $n = 2$. The two permutations of the symbols $\{1, 2\}$ are 12 and 21. A superpermutation must therefore include both 12 and 21 as substrings. The shortest such

 https://doi.org/10.11647/OBP.0526.39

string is 121. It contains 12 starting at position 1 and 21 starting at position 2. This is the minimal superpermutation for $n = 2$, and it has length 3.

For $n = 3$, there are $3! = 6$ permutations: 123, 132, 213, 231, 312, and 321. One example of a minimal superpermutation that includes all six of these as contiguous substrings is 123121321. It has length 9, and each of the six permutations occurs once within it. No shorter string satisfies the same condition.

The central question posed by the *superpermutation problem* is: what is the minimal possible length of such a string for general n? That is, given n distinct symbols, what is the shortest string over those symbols that contains every one of their $n!$ permutations as contiguous substrings? As n increases, this problem becomes computationally and combinatorially challenging. The number of permutations grows rapidly, and so does the complexity of arranging them with maximal overlap. The search for minimal superpermutations remains an active area of combinatorial optimisation.

At first glance, the problem of constructing a superpermutation may appear manageable. One might attempt a straightforward solution by writing out all $n!$ permutations of the n symbols and concatenating them end-to-end. Since each permutation is of length n, this method produces a string of total length $n \cdot n!$. For example, for $n = 3$, this naive approach would yield a string of length $3 \times 6 = 18$. While this guarantees that all permutations are present, it is highly inefficient. Adjacent permutations often share common segments—such as matching suffixes and prefixes—and these overlaps can be exploited to significantly reduce the total length.

The trick is that permutations can be arranged so that the end of one serves as the beginning of the next. For instance, the permutation 123 ends in 23, and 231 begins with 23; by placing them consecutively as 1231, the two permutations are both represented, and the shared segment 23 is not duplicated. This principle of overlap allows one to compress multiple permutations into a single string, hopefully, without repeating identical sequences.

For small values of n, exact solutions have been found. Remarkably, for $n = 1$ through $n = 5$, the shortest superpermutations are known, and their lengths follow a simple closed-form pattern:

$$L(n) = \sum_{k=1}^{n} k! = 1! + 2! + 3! + \cdots + n!.$$

For example:

$$L(3) = 1! + 2! + 3! = 1 + 2 + 6 = 9, \quad L(4) = 33, \quad L(5) = 153.$$

This empirical closed-form suggested a natural conjecture: that the shortest possible superpermutation on n symbols always has length equal to $\sum_{k=1}^{n} k!$. The conjecture was elegant and aligned with all known cases. For years, no counterexamples were found, and the formula became widely assumed to be correct.

That assumption remained unchallenged until 2014, when it was definitively disproven. A construction was found that produced a superpermutation on six symbols with total length 872—precisely one character shorter than the conjectured value of $1!+2!+3!+4!+5!+6! =$

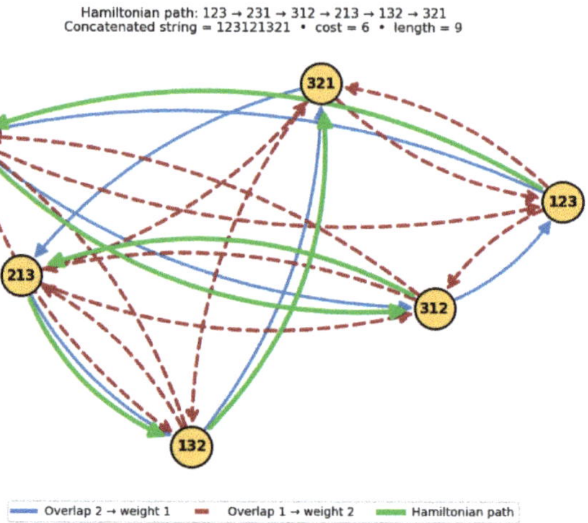

Permutation overlap digraph for $n = 3$. Each node is a permutation of $\{1, 2, 3\}$. A directed edge $A \to B$ exists when a suffix of A matches a prefix of B, with *overlap* length k. The *weight* is $w = n - k$, i.e. the number of extra symbols needed to append B after A. Edges with $k = 2$ (weight 1) are shown in solid blue, edges with $k = 1$ (weight 2) in dashed red. The highlighted green path is the Hamiltonian path $123 \to 231 \to 312 \to 213 \to 132 \to 321$, yielding concatenated string 123121321 of length 9 and total cost 6.

8 73. This counterexample showed that the conjectured bound, though valid for $n \leq 5$, does not hold in general.

The origin of this disproof traces back to an unexpected source: an anonymous discussion thread on the imageboard website 4chan. Founded in 2003, 4chan hosts ephemeral user-generated content across numerous boards organised by theme. Messages are anonymous by default, and threads are subject to automatic deletion without archival. The site's culture is informal, transgressive, and often dismissive of academic conventions. One of its boards, labelled `/sci/`, is nominally dedicated to science and mathematics. The board nonetheless occasionally features serious technical inquiry.

In 2011, an anonymous user on the `/sci/` board of the website 4chan posed a question that, at first glance, seemed whimsical: what is the shortest possible viewing sequence that includes every ordering of the 14 episodes of the anime *The Melancholy of Haruhi Suzumiya*? The show, known for its non-linear narrative and varying episode orders across different broadcasts, had developed a cult following that embraced its combinatorial potential. Beneath the framing, however, lay a precise mathematical question: how short can a string be while still containing every permutation of a 14-element set as a contiguous substring? In effect, the prompt was a popular-culture formulation of the superpermutation problem for $n = 14$

In response, another anonymous poster provided a compact but mathematically rigorous derivation of a new general lower bound (Anonymous et al., 2018): $L(n) \geq n! + (n-1)! + (n-2)! + n - 3$.

This inequality, valid for all $n \geq 2$, strengthened all previously known bounds. Though presented informally, the proof was ultimately correct. It treated permutations as vertices in a directed graph, with directed edges representing overlaps between adjacent substrings. The minimal-length superpermutation corresponded to a Hamiltonian path through this graph (Ashlock & Tillotson, 1993), and the proof established a lower bound on the total cost of any such path by analysing the unavoidable overlaps.

Despite its correctness and novelty, the result went largely unnoticed at the time. The platform offered no mechanisms for citation or persistence: posts were anonymous, threads expired automatically, and archival relied entirely on user initiative. The derivation was eventually copied to a fandom-hosted mathematics wiki, but remained obscure and disconnected from formal literature.

In 2014, mathematician Robin Houston independently rediscovered the argument, verified its correctness, and recognised its significance. He publicised the result, incorporated it into ongoing research, and cited the unknown author as "Anonymous 4chan Poster"—a designation that remains standard in subsequent academic references. The original poster has never been identified.

The impact was immediate. The new lower bound provided a rigorous floor against which all proposed constructions could be measured. Shortly thereafter, Houston constructed a superpermutation on six symbols of length 872 (Houston, 2014)—one less than the conjectured minimum of 873—thereby disproving the long-standing sum-of-factorials conjecture.

In 2018, an accomplished sci-fi author and mathematician Greg Egan proposed a constructive upper bound: $L(n) \leq n! + (n-1)! + (n-2)! + (n-3)! + n - 3$, placing the known range for $L(n)$ within a narrow window, bounded, yet not tightly, from above and from below.

The 4chan derivation now stands as a rare episode in modern mathematics: a significant and previously unknown lower bound for a classical problem, derived anonymously, informally posted, largely ignored, and later validated by professionals. It illustrates how insight can originate outside institutional settings, and how easily such insight can be lost when detached from systems of attribution, preservation, and dissemination. Nevertheless, the mathematics holds. The 'Haruhi Problem' has since entered the literature as a textbook case in combinatorial optimisation—and its solution, at least in part, belongs to a nameless contributor with no affiliation, no traceable authorship, and a correct idea.

>> **Lower bounds** Anonymous Fri Sep 16 23:35:54 2011 No.3751197
Quoted by: >>3751366 >>3751370 >>3752047_2 >>3752047_25

I think I have a proof of the lower bound n! + (n-1)! + (n-2)! + n-3 (for <span class="math">n \geq 2[/spoiler]). I'll need to do this in multiple posts. Please look it over for any loopholes I might have missed.

As in other posts, let n (lowercase) = the number of symbols; there are n! permutations to iterate through.

The obvious lower bound is n! + n-1. We can obtain this as follows:

Let
L = the running length of the string
<span class="math">N_0[/spoiler] = the number of permutations visited
<span class="math">X_0 = L - N_0[/spoiler]

When you write down the first permutation, <span class="math">X_0[/spoiler] is already n-1. For each new permutation you visit, the length of the string must increase by at least 1. So <span class="math">X_0[/spoiler] can never decrease. At the end, <span class="math">N_0 = n![/spoiler], giving us <span class="math">L \geq n! + n-1[/spoiler].

I'll use similar methods to go further, but first I'll need to explain my terminology...

The original 4chan post

Lower and Upper Bounds via Overlap Graphs

Introduction

A superpermutation on n symbols is a string that contains all $n!$ permutations of those symbols as contiguous substrings. Let $L(n)$ denote the minimum possible length of such a string. The problem of determining $L(n)$ can be reformulated as a path-finding problem on a directed graph whose nodes are permutations and whose edge weights correspond to symbol overlap.

Permutation Graph Model

Let S_n be the symmetric group on n elements, and let each vertex in the directed graph G_n correspond to a permutation $\pi \in S_n$. For each ordered pair (π, σ), define an edge $\pi \to \sigma$ with weight $w(\pi, \sigma) = n - \ell(\pi, \sigma)$, where $\ell(\pi, \sigma)$ is the length of the longest suffix of π that matches a prefix of σ. A superpermutation corresponds to a Hamiltonian path through G_n, with total cost equal to the sum of edge weights plus n.

Anonymous Lower Bound

The anonymous 4chan user showed that for all $n \geq 2$, the minimal length satisfies

$$L(n) \geq n! + (n-1)! + (n-2)! + n - 3.$$

The proof bounds the number of edges in any Hamiltonian path that must have weight greater than 1. Let $\mathcal{P}$ be any Hamiltonian path through G_n. At best, two permutations can overlap by $n-1$ symbols, requiring only one new symbol to transition. However, weight-1 transitions alone cannot form a complete path due to overlap constraints. The proof partitions permutations into blocks where one block must be traversed without overlap (initial permutation); some fraction of transitions must necessarily use edges with higher cost due to incompatible suffix-prefix structure. Using known bounds on minimal-overlap transitions, one can count higher-cost edges and compute their contribution. The derived lower bound is tight for $n \leq 4$ and remains the strongest known general lower bound for $L(n)$.

Egan's Constructive Upper Bound

Greg Egan constructed a general method for building superpermutations of length at most

$$L(n) \leq n! + (n-1)! + (n-2)! \\ + (n-3)! + n - 3.$$

The method generates Hamiltonian paths through subsets of permutations with controlled overlaps: Begin with a path that efficiently traverses permutations of $n-1$ symbols. Lift the path into S_n by inserting the new symbol in controlled positions. Design the insertion and merge process to preserve maximal overlaps where possible. Egan's construction uses Cayley graph traversal techniques and reuses structural symmetries. The difference between the upper and lower bounds is $(n-3)!$:

$$\Delta(n) = L_{\text{upper}}(n) - L_{\text{lower}}(n) = (n-3)!.$$

Example: The Case $n = 7$

Applying the bounds:

$$\begin{aligned} \text{Lower bound:} \quad & 7! + 6! + 5! + 7 - 3 \\ & = 5884, \\ \text{Upper bound:} \quad & \text{lower bound} + 4! \\ & = 5884 + 24 \\ & = 5908. \end{aligned}$$

The shortest known construction is 5906. The true value of $L(n)$ for $n = 7$ remains unknown.

References:

Houston, R. (2014). *Obvious Does Not Imply True: The Minimal Superpermutation Conjecture Is False.* arXiv:1408.5108.

Egan, G. (2018). *Superpermutations* (online note).

Anonymous 4chan Poster (2011). Lower bound proof (`https://warosu.org/sci/thread/3751105`).

Slices of Life

Sequencing as Text Reconstruction: The image parallels DNA sequencing to reassembling a book from shredded fragments from hundreds of identical copies. The books are pulverised into overlapping sentence fragments—analogous to sequencing short reads from long DNA strands. By aligning overlapping sequences, the original string is reconstructed. As with genome assembly, redundancy, overlaps, and statistical inference allow one to piece together each chromosome from fragmented reads.

Slices of Life

DNA sequencing has evolved from Sanger's chain-termination method through the next-generation revolution of 454 pyrosequencing, Ion Torrent, and Illumina platforms to modern nanopore technologies. The computational challenge of genome assembly uses sophisticated algorithms such as de Bruijn graphs to reconstruct complete genomes from millions of short fragments, while paired-end chemistry and long-read technologies help resolve repetitive regions that have long frustrated genomic reconstruction efforts.

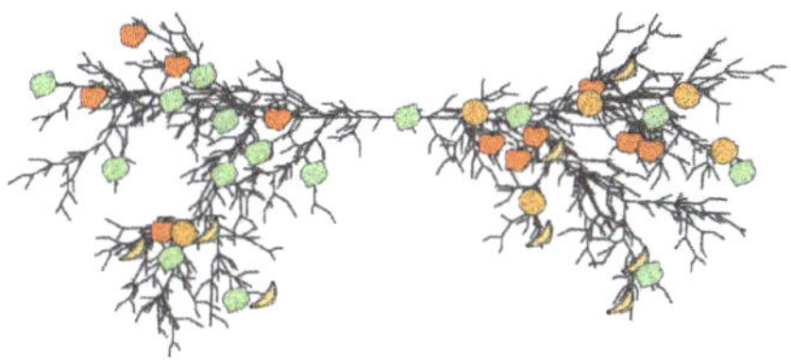

DNA Sequencing Evolution ∘ Sanger ddNTP Method ∘ Illumina Reversible Terminators ∘ Bridge Amplification ∘ PacBio SMRT Technology ∘ Oxford Nanopore ∘ De Bruijn Graph Assembly ∘ K-mer Overlap ∘ Contigs vs Scaffolds ∘ N50 Statistic ∘ Long-Read Revolution

"Would I have invented PCR if I hadn't taken LSD? I seriously doubt it.
I could sit on a DNA molecule and watch the polymers go by.
I learned that partly on psychedelic drugs."

— Kary Mullis, 1998

Slices of Life

Frederick Sanger's 1977 chain-termination method emerged from years of frustration with earlier approaches to reading DNA. His insight—using modified nucleotides to randomly terminate DNA synthesis—provided the first practical way to determine base sequences. While Allan Maxam and Walter Gilbert simultaneously developed a chemical cleavage method, Sanger's approach proved more robust and became the foundation for three decades of genomic research.

The Human Genome Project launched in 1990 as biology's moonshot—a publicly funded effort to read all 3.2 billion letters of human DNA. Francis Collins led the international consortium, methodically mapping and sequencing chromosomes piece by piece. Then in 1998, Craig Venter announced that his company, Celera Genomics, would sequence the human genome in just three years using a 'whole genome shotgun' approach (Venter et al., *Science* 280, 1998, pp. 1540–1542)—fragmenting the entire genome at once and using computational power to reassemble it.

The race was on. The public project, with its careful clone-by-clone strategy, suddenly faced a nimble competitor unconstrained by academic collaboration requirements. Venter's team used hundreds of automated sequencers running 24/7, while the public consortium scrambled to accelerate their timeline. Both sides published draft sequences simultaneously in February 2001—a diplomatic resolution to a bitter competition that had featured Congressional hearings, patent disputes, and public acrimony. The project cost approximately $3 billion and required a decade of work.

Sanger sequencing's limitations—high cost and low throughput—motivated a new generation of technologies. 454 Life Sciences introduced pyrosequencing in 2005, detecting DNA synthesis through light emission and reading millions of fragments simultaneously. This began the 'next-generation' era, where parallelisation replaced precision.

Illumina emerged as the dominant platform after acquiring Solexa technology in 2007. Their reversible terminator chemistry solved pyrosequencing's homopolymer problems while maintaining massive throughput. The cost per genome plummeted from millions to thousands of dollars, democratising genomic research.

The push for longer reads drove development of single-molecule technologies. Pacific Biosciences (PacBio) spent a decade perfecting zero-mode waveguides—zeptolitre observation chambers that could watch individual DNA polymerase enzymes at work. Oxford Nanopore took a different path, threading DNA through protein pores and reading the sequence from electrical current fluctuations. When they released the MinION in 2014—a sequencer the size of a USB stick—it showcased the progress the technology has made from room-sized machines of the genome project era.

The computational challenge evolved in parallel. Early assembly algorithms handled thousands of Sanger reads; modern de Bruijn graph methods process billions of short reads. Long-read assemblers now tackle the ultimate challenge: reconstructing complete chromosomes from end to end. Sequencing costs have fallen faster than Moore's

> Law—from roughly dollars per base in 1990 to well under a dollar per megabase today, and under a thousand dollars per human genome on leading platforms.

DNA (deoxyribonucleic acid) is the molecule that stores genetic information in all living organisms. It consists of two complementary strands twisted into a double helix, where each strand (Watson & Crick, 1953) is a linear sequence of four chemical bases: adenine (A), cytosine (C), guanine (G), and thymine (T). The sequence of these bases encodes the instructions for building and maintaining an organism.

The central dogma of biology describes how genetic information flows (Crick, 1970). DNA is transcribed into RNA, which is translated into proteins. Each three-base sequence (codon) in DNA specifies one amino acid in the resulting protein. A single change in the DNA sequence can alter the protein's structure and function, causing disease or evolutionary adaptation. Understanding DNA sequences is a key component in understanding the molecular basis of life.

Consider a short DNA sequence: ATGCGATCGATCG. This 12-base fragment contains four codons: ATG, CGA, TCG, ATC. The codon ATG typically signals 'start translation,' while the others specify specific amino acids. If the first base changes from A to T, creating TTGCGATCGATCG, the first codon becomes TTG, which codes for a different amino acid, potentially altering the resulting protein's function.

DNA sequencing is the process of determining the exact order of bases in a DNA molecule. This requires overcoming a scale mismatch: individual bases measure roughly one nanometre, while human chromosomes stretch millions of bases long which makes it practically impossible to read the entire sequence in one pass.

The solution involves fragmenting DNA into manageable pieces, reading each fragment separately, then computationally reconstructing the original sequence. This creates two challenges. The first is the biochemical problem of reading individual fragments and the second is the algorithmic problem of assembling them correctly. Each generation of sequencing technology has approached these challenges in different ways.

Frederick Sanger solved the reading problem through controlled interruption of DNA synthesis. DNA polymerase builds new strands (Sanger, Nicklen, & Coulson, 1977) by adding nucleotides complementary to a template; a 3'-hydroxyl group on each nucleotide enables the next to attach. Sanger introduced dideoxynucleotides (ddNTPs) that lack this hydroxyl group, terminating synthesis when incorporated.

This process is probabilistic. In a mix containing a DNA template, primers, polymerase, all four normal dNTPs, and a small amount of one ddNTP type (e.g., ddATP), the polymerase occasionally incorporates a ddATP, terminating the strand. Across millions of template copies, this generates a collection of fragments of different lengths, each ending at a different A position.

Running four parallel reactions–one for each ddNTP type–produces four fragment collections. Gel electrophoresis separates these by size: DNA fragments migrate through a polymer matrix under an electric field, with smaller fragments moving faster. After separation, each gel lane shows a ladder of bands. Reading from shortest to longest fragment across all four lanes reveals the sequence. If the shortest fragment appears in the G lane, the first base is G. If the next shortest is in the A lane, the second base is A. Going through all four lanes reveals the sequence.

Sanger sequencing powered the Human Genome Project but had limitations. Each reaction produced only 500–1000 readable bases. Preparing samples, running gels, and reading results consumed hours per reaction. Radioactive or fluorescent labelling added complexity and cost. The throughput ceiling meant that sequencing a human genome required years of work and hundreds of millions of dollars.

Next-generation platforms achieved a breakthrough with parallelization, performing millions of reactions simultaneously on a single surface. Early systems distributed single DNA fragments into millions of microscopic wells and flowed one nucleotide type at a time (first all As, wash; then all Cs, wash). Detection chemistry varied. 454 Life Sciences used pyrosequencing, where nucleotide incorporation releases pyrophosphate (PPi), which an enzyme cascade converts into a light signal via luciferase. Ion Torrent used semiconductor sequencing, where incorporation releases a hydrogen ion, and millions of ISFET (ion-sensitive field-effect transistor) sensors detect the resulting pH change as a voltage signal. Both methods suffered from the same core limitation. Because nucleotides were added sequentially, homopolymer runs like AAAA caused all four bases to incorporate at once, producing a signal four times stronger. Distinguishing a 4× signal from a 5× signal was error-prone, limiting accuracy.

Illumina took a different path, solving the homopolymer problem through reversible termination. Their innovation (Bentley et al., 2008) combined three key elements: surface-bound amplification, chemically cleavable terminators, and four-colour imaging.

The process begins with bridge amplification. DNA fragments attach to a glass surface coated with two types of oligonucleotide primers. Each fragment bends to hybridise with a nearby complementary primer, forming a bridge. Polymerase extends the primer, creating a complementary strand anchored at both ends. Denaturation releases the original strand, and the process repeats. After 35 cycles, each original molecule generates a tight cluster of $\sim$1,000 identical copies, all within a few hundred nanometres—small enough to act as a single sequencing unit but bright enough for fluorescence detection.

Illumina's sequencing chemistry uses nucleotides engineered with two modifications: a fluorescent dye unique to each base (A, C, G, T) and a chemical block on the 3'-OH that prevents further extension. Unlike Sanger's permanent terminators, these blocks can be cleaved chemically.

Each sequencing cycle follows four steps: add all four labelled terminators simultaneously, wait for incorporation, image in four colours, then cleave both dye and terminator. Because only one base can be added per cycle (due to the 3'-block), homopolymers read accurately—AAAA requires four separate cycles, each adding one A. This solved 454's limitation.

Illumina's paired-end innovation provided long-range information. Sequence both ends of a DNA fragment, keeping track that they came from the same molecule. If fragments are 500 bases long but you only read 150 bases from each end, you know those two 150-base sequences sit 200 bases apart in the genome. These distance constraints prove essential for genome assembly.

Assembling a 3-billion-base human genome from 20 million 150-base fragments is computationally demanding. Early overlap-layout-consensus algorithms, which compare all read pairs to find overlaps, were feasible for thousands of Sanger reads but fail for millions of short reads where all-pairs comparison is prohibitive.

De Bruijn graphs, a 1946 mathematical structure, provided a scalable solution. Instead, of connecting reads, they connect k-mers—all possible k-letter substrings. A sequence traces a path through a graph where each unique k-mer is a node and edges connect k-mers overlapping by k-1 bases. The scalability arises because a genome of length G contains at most $G - k + 1$ distinct k-mers, regardless of sequencing depth. Finding Eulerian paths that traverse each edge once is tractable even for graphs with billions of nodes.

Consider the sequence ATCGATCG and extract all 3-mers: ATC, TCG, CGA, GAT, ATC, TCG. Build a graph where each unique k-mer is a node, and edges connect k-mers that overlap by k-1 bases. The sequence ATCGATCG traces a path through this graph: ATC→TCG→CGA→GAT→ATC→TCG (compare to the superpermutation problem in Chapter 39).

Repeats in the genome, such as transposable elements, complicate assembly. When a repeat is longer than a read, it creates ambiguity in the assembly graph, resulting in multiple valid paths. Paired-end constraints resolve some ambiguities, but short reads cannot span long repeats, requiring the development of long-read technologies.

Pacific Biosciences (PacBio) developed single-molecule real-time (SMRT) sequencing, observing individual DNA polymerase enzymes. The primary challenge was detecting single fluorescent nucleotides against the background of unincorporated ones. The solution was zero-mode waveguides (ZMWs): 70 nanometre holes in an aluminium film that confine laser illumination to a 20 zeptolitre volume. A polymerase at the bottom of each ZMW holds an incorporating nucleotide for milliseconds, long enough to generate a detectable fluorescent flash distinct from the transient signals of freely diffusing nucleotides. This method generates reads exceeding 10,000 bases. Though noisy, with error rates of 10–15%, these long reads are effective at spanning genomic repeats.

Oxford Nanopore technology uses no enzymes or fluorescence. It passes a single DNA strand through a protein nanopore (Kasianowicz, Brandin, Branton, & Deamer, 1996) embedded in a membrane. An applied voltage drives both the DNA and an ionic current. As the DNA translocates, nucleotides in the pore's 1.4 nanometre constriction modulate the current. The narrowest region spans approximately five bases, so the signal reflects a 5-mer. Signal processing algorithms, and recently, neural networks, decode the complex current modulations into a DNA sequence, achieving >95% accuracy and read lengths that can exceed one million bases.

Long reads simplified assembly graphs, as most repeats become trivial to span. Modern projects often use a hybrid approach: Illumina provides an accurate short-read backbone, while PacBio or Nanopore provides a long-read scaffold to resolve repeats and structural variants.

On Assembly Statistics

Evaluating a genome assembly requires understanding its output format. Assemblies consist of fragments at two organisational levels. A **contig** is a contiguous stretch of sequence assembled from overlapping reads—an unbroken text. A **scaffold** links multiple contigs that are ordered and oriented but separated by gaps of estimated size. Consider recovering a book's text from shredded copies: contigs are complete pages reconstructed from overlapping fragments, scaffolds are chapters where page order is known but some pages remain missing.

The median contig length—the middle value in a sorted list—is uninformative because assemblies contain thousands of short contigs and few long ones. The **N50** statistic measures contiguity differently. Sort contigs from longest to shortest, then sum their lengths sequentially. The N50 is the length of the contig that brings this cumulative sum to 50% of the total assembly size. For contigs of lengths 10, 9, 8, 7, 1, 1, 1, 1, and 1 kb (total 39 kb), the cumulative sum surpasses 50% after adding the first three contigs ($10 + 9 + 8 = 27 > 39/2 = 19.5$ kb). The N50 is 8 kb—the length of that third contig. The median is 1 kb.

The scientific literature sometimes misreports N50 as 'median contig length (N50).' The N50 is a length-weighted metric, not the simple median of contig lengths. Describing N50 as a 'weighted median' is correct if one creates an expanded list where each contig appears once for each base it contains, then takes that list's median.

My first first-author paper addressed a bottleneck of this era: when paired-end reads were short, their two ends often failed to overlap, leaving assemblers unable to merge them into longer fragments. ELOPER (Silver et al., *Bioinformatics*, 2013) detected pairs whose ends overlapped simultaneously at both positions, effectively doubling usable read length while also providing long-range scaffolding, thus improving assembly quality. Within a few years, advances in sequencing technology made reads long enough to render the problem—and the tool—unnecessary. I also helped assemble the genome of *Acrobeloides nanus*, a nematode whose gene regulatory program revealed conserved phylum-specific expression patterns across deep evolutionary distances (Schiffer et al., *PNAS*, 2018)—a collaboration with a dear friend in Edinburgh, completed as part of my honeymoon trip with my wife Enny.

That work was conducted under Itai Yanai at the Technion, where most of my biology-related research took place. Yanai has since moved to NYU, where he is Professor of Biochemistry and Molecular Pharmacology and founding director of the Institute for Computational Medicine. A pioneer of single-cell transcriptomics and co-author of *The Society of Genes*, he is among the most effective communicators of science to graduate students and young researchers. His teaching, curiosity, and attitude toward what science is shaped much of the sensibility behind this book.

Illumina Sequencing and De Bruijn Assembly

Sequencing by Synthesis

Illumina uses reversible terminators with cleavable fluorescent labels. Each cycle:

$$\begin{aligned}&\text{DNA}_n + \text{dNTP-3'-block-fluor}\\&\quad\xrightarrow{\text{pol}} \text{DNA}_{n+1}\\&\text{Imaging} \rightarrow \text{Base identification}\\&\text{Chemical cleavage} \rightarrow \text{3'-OH restoration.}\end{aligned}$$

Bridge amplification creates clonal clusters ($\sim 10^3$ copies) on flow cell surface. Fluorescence signal $S \propto N_{\text{mol}}$ enables base calling with error rate $\varepsilon \approx 0.1\%$.

De Bruijn Graph Construction

For read set $\mathcal{R}$ with k-mer length k:

$$\begin{aligned}V &= \{w \in \Sigma^{k-1} : w \text{ is prefix or suffix}\\&\quad \text{of some k-mer in } \mathcal{R}\}\\E &= \{w \in \Sigma^k : w \text{ appears in } \mathcal{R}\},\end{aligned}$$

where each k-mer edge connects its $(k-1)$-mer prefix to its $(k-1)$-mer suffix.

Each read of length L contributes $L-k+1$ k-mer edges, compressing redundant sequence information into a compact graph structure.

Eulerian Path Assembly

Assembly seeks Eulerian path through G:

$$\begin{aligned}\text{Path} &= e_1 e_2 \ldots e_m \text{ where}\\&\quad \forall i : \text{tail}(e_i) = \text{head}(e_{i+1})\\\text{Genome} &= e_1[1..k] + e_2[k] + e_3[k]\\&\quad + \ldots + e_m[k],\end{aligned}$$

where $e_i[k]$ denotes the last base of k-mer e_i.

For an Eulerian path to exist, the underlying graph over nonzero-degree vertices must be weakly connected, with all vertices balanced (in-degree = out-degree), or exactly two semi-balanced vertices: one with out-degree = in-degree + 1 and one with in-degree = out-degree + 1.

Coverage and k-mer Selection

Expected k-mer coverage:

$$C_k = C_{\text{read}} \cdot \frac{L-k+1}{L},$$

where $C_{\text{read}} = NL/G$ (reads × length / genome).

Optimal k balances a tradeoff: smaller k yields more connections and higher coverage but introduces ambiguity, while larger k reduces repeat ambiguity at the cost of lower coverage and potential gaps.

Typically $k \in [50, 250]$ for Illumina data.

Graph Complexity

Real graphs contain three main types of structural features: **bubbles** (parallel paths created by SNPs or errors), **tips** (dead ends from coverage gaps), and **repeats** (creating branching and convergence). Error correction typically removes k-mers with coverage below a threshold.

Paired-End Constraints

Insert size $d \sim \mathcal{N}(\mu, \sigma^2)$ provides scaffolding:

$$|p(r_1, r_2) - \mu| < 3\sigma,$$

where $p(r_1, r_2)$ is genomic distance between read pairs.

References:

Bentley et al. (2008). *Nature* 456:53–59.
Pevzner et al. (2001). *PNAS* 98:9748–9753.

It Is Just a Phase

Top (Pixel-Level Edge Detection): Each observer inspects a single pixel and flags whether it looks like part of an edge. This approach produces isolated detections without context—assembling a full line requires comparing scattered responses after the fact. It's data-rich but structure-poor: the system sees points, not lines. **Middle (Line-Level Hough Detection):** Observers instead scan the whole image through virtual rulers at specific angles and offsets. Each one votes if their ruler aligns with edge evidence. The same number of detectors now deliver global outputs: lines directly. This is the core idea behind the Hough Transform—accumulating evidence in a parameter space where lines become peaks. **Bottom (Phase Space Riddle):** Two roads connect cities A and B without crossing. A pair of cars, tied by a rope of length < $2R$, can travel side by side on these roads from A to B. Now consider two circular wagons of radius R, each centred on its own path—one going A to B, the other B to A. Let (x, y) track their positions along the two roads, forming a trajectory in the unit square $I = \{(x, y) : 0 \leq x, y \leq 1\}$. The cars' path goes from $(0, 0)$ to $(1, 1)$; the wagons' from $(1, 0)$ to $(0, 1)$. By topology, these curves must intersect. At the intersection, the wagons occupy the same positions as the cars once did, yet their centres are < $2R$ apart, and each wagon has radius R, so a collision is unavoidable. A fact that is difficult to prove directly. (After Konstantinov, in Arnold's *ODEs*.)

It Is Just a Phase

The Hough transform detects geometric shapes in images by converting the problem from image space to parameter space. Images undergo edge detection to identify significant brightness transitions. Each edge pixel then 'votes' for all possible geometric structures that could contain it. For line detection, edge points generate constraints in parameter space through the relation $b = y_0 - mx_0$. Points lying on the same line create intersecting lines in parameter space, forming accumulator peaks.

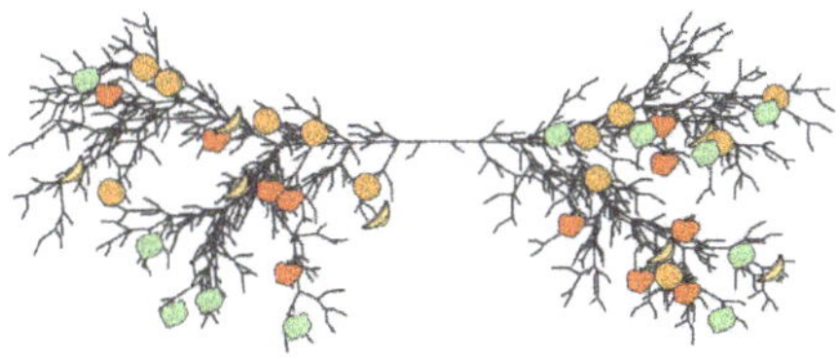

Hough Transform ∘ Edge Detection ∘ Parameter Space Voting ∘ Rho-Theta Parametrisation ∘ Accumulator Arrays ∘ Global Feature Detection ∘ Line Finding Algorithm ∘ Gradient Operators ∘ Circle Detection ∘ Computer Vision ∘ Duality Transform

"With four parameters I can fit an elephant,
and with five I can make him wiggle his trunk."
— John von Neumann, circa 1953

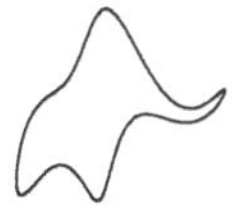

A curve found in 2010 as shown above.
The curve can be fitted with four parameters.
A fifth point can be added to form the elephant's eye
and allows the trunk to move.
— Jürgen Mayer, Khaled Khairy, Jonathon Howard, 2010

It Is Just a Phase

In 1959, Paul V. C. Hough filed a patent—issued in 1962 as U.S. Patent 3,069,654—for a method of identifying complex patterns in visual data by mapping image features into a parameter space. His technique framed detection as finding parameter settings supported by many edge points. The familiar sinusoidal loci in a polar-coordinate domain arise with the normal parameterisation (ρ, θ), which was formalised and popularised subsequently.

Early implementations relied on analogue hardware and optical computing elements. Limitations in memory and processing speed constrained accumulator resolution and the range of detectable geometries. Despite these constraints, the method was adopted in early automation systems for detecting weld lines, highway markings, and parts in mechanical assemblies.

In 1972, Richard Duda and Peter Hart provided the first formal exposition of the method in their landmark paper, reframing it as a discrete voting process over a bounded parameter space and introducing the normal parameterisation $\rho = x\cos\theta + y\sin\theta$ that treats all orientations uniformly. Their formulation gave the method the name 'Hough transform' and connected it to broader principles in statistical decision theory. Subsequent extensions by Ballard and others generalised the idea to arbitrary curves and spatial templates, enabling detection of circles, ellipses, and parabolas.

By the 1980s and 1990s, with advances in digital signal processors and parallel computing, the Hough transform became a foundational tool in industrial vision systems, robotics, and medical image reconstruction. Its structure-preserving mapping from image to parameter space allowed for robust detection even in the presence of occlusion, fragmentation, and noise—a capacity that remains central to modern implementations in lane detection, tomography, and motion analysis.

A digital image is a two-dimensional discrete function $I(x, y)$ defined over a finite grid of integer coordinates. Each ordered pair (x, y) refers to a spatial location in the image plane, and the corresponding value $I(x, y)$ denotes the intensity—or brightness—measured at that position. In standard grayscale images, the intensity values range over a finite interval, typically $[0, 255]$, where 0 represents black and 255 represents white. Intermediate values encode proportional levels of grey.

This allows images to be treated as matrices of numerical data. Each row corresponds to a horizontal slice through the image, and each column to a vertical slice. The array format aligns with standard data structures in numerical computing, enabling efficient storage, manipulation, and analysis using linear algebraic tools. More relevantly, this view enables the application of discrete transformations that operate directly on the pixel grid.

The array-based nature of digital images reflects their method of acquisition. Optical sensors—such as charge-coupled devices (CCDs)—sample incoming light (Boyle & Smith, 1970) intensity across a regular lattice of photodetectors. Each photodetector integrates the radiance over a small rectangular region and records the result as a scalar value. This process discretizes a continuous visual field, replacing smooth spatial variation with a piecewise-constant approximation.

Despite this discretization, many natural images exhibit local continuity. In regions of uniform texture or lighting, adjacent pixels tend to have similar intensity values. This results in spatial smoothness—a statistical tendency that can be exploited by various image processing algorithms. Conversely, abrupt changes in intensity may indicate the presence of physical boundaries, surface discontinuities, or occlusion contours.

The goal of low-level image processing is to extract global geometric elements from the raw intensity field—to identify where transitions occur, how they are oriented, and how they combine to form recognisable configurations. The first step in this process is edge detection: identifying where intensity values change significantly in space.

To extract localised transitions in brightness, digital image processing employs gradient-based methods. These techniques compute spatial derivatives of the intensity function $I(x, y)$ using discrete convolution kernels. Common approximations include the Sobel, Prewitt, or central (Sobel, 1970) difference operators, which estimate the partial derivatives $\partial I/\partial x$ and $\partial I/\partial y$ by combining intensities in a small neighborhood.

The result of this process is a vector field $G(x, y) = (I_x(x, y), I_y(x, y))$, where I_x and I_y denote the horizontal and vertical gradients respectively. The magnitude of this vector, defined as

$$\|G(x, y)\| = \sqrt{I_x(x, y)^2 + I_y(x, y)^2},$$

quantifies the rate of change in brightness at each pixel. The direction $\theta(x, y) = \arctan 2(I_y, I_x)$ indicates the orientation of maximal contrast—that is, the direction perpendicular to the local edge.

To reduce the resulting field to a usable form, a fixed threshold is applied to the gradient magnitude. The binary edge map $E(x, y)$ is defined by

$$E(x, y) = \begin{cases} 1 & \text{if } \|G(x, y)\| > \tau, \\ 0 & \text{otherwise,} \end{cases}$$

where τ is a positive scalar threshold. The outcome is a sparse array in which $E(x, y) = 1$ denotes an edge candidate and 0 otherwise. This thresholding step suppresses minor fluctuations while preserving strong transitions, isolating regions of significant spatial variation.

The resulting edge map retains only local information. Each nonzero pixel marks a point of abrupt contrast but encodes no higher-order structure. It does not indicate whether a given edge continues across neighbouring pixels, nor whether multiple edge points are aligned.

In many visual tasks, the detection of straight lines is the second step after edge detection. Lines may correspond to physical edges in man-made environments—such as walls, roads, or tools—or to object boundaries under specific perspectives. In medical imaging, remote sensing, and industrial inspection, linear features often identify critical diagnostic or structural information.

However, a straight line is not a local construct. Unlike a gradient, which depends only on immediate neighbours, a line is a global configuration—a set of pixels that, despite spatial separation, conform to a shared geometric constraint. In the discrete setting, this constraint must be inferred from partial and often noisy evidence. The edge map contains a large number of isolated pixels, many of which are spurious. Determining which subsets form lines requires aggregating across potentially distant points.

One naïve approach is to enumerate all pairs of edge pixels and test whether a third or fourth point lies along the same line. Given n detected edge points, there are $\binom{n}{2}$ possible pairs, and each pair defines a candidate line. For each such line, one must then check whether additional points fall sufficiently close to it, accounting for quantization and discretization artefacts. The combinatorial burden of this process grows quadratically in the number of edge pixels and becomes intractable at practical image resolutions.

The challenge is further exacerbated by noise and occlusion. Genuine lines may be broken into disconnected segments, and false positives may arise from texture or illumination variations. Any method that seeks to identify lines must accommodate partial evidence and operate under pixel-level uncertainty. The core problem is not to test a single hypothesis, but to efficiently search a vast and noisy hypothesis space for a small number of consistent patterns. To detect lines in a binary edge map, one may adopt a parametric formulation of straight lines. In Cartesian coordinates, the general equation for a line is $y = mx + b$, where m is the slope and b the vertical intercept. The goal is to identify parameter pairs (m, b) that describe lines passing through multiple edge pixels.

Each edge pixel (x_0, y_0) imposes a constraint on the set of valid (m, b) values. Specifically, if a line passes through (x_0, y_0), then its parameters must satisfy

$$y_0 = mx_0 + b, \quad \text{or equivalently,} \quad b = y_0 - mx_0.$$

This relation defines a one-dimensional locus in the (m, b) parameter space: the set of all lines that intersect (x_0, y_0). For fixed x_0 and y_0, this is a straight line in parameter space—each pixel generates such a line of possible (m, b) values.

To implement this idea computationally, the (m, b) space is discretized into a finite grid. A two-dimensional accumulator array is initialised, with each cell corresponding to a quantized pair of slope and intercept values. For each edge pixel, the algorithm evaluates the above relation at discrete samples of m and computes the corresponding b values. Each resulting (m_i, b_i) pair increments the count in its associated cell of the accumulator.

After processing all edge pixels, the accumulator stores the number of pixels consistent with each candidate line. Peaks in this array—cells with significantly elevated counts—represent line parameters that are supported by many pixels. These peaks are interpreted as strong linear features in the original image.

While this method organises the problem effectively, it remains computationally intensive. Each edge pixel must evaluate the constraint over all discretized slope values, resulting in a cost proportional to the product of the number of edge pixels and the number of slope samples. Moreover, the parameterization becomes unstable for nearly vertical lines, where m diverges. These limitations motivate the adoption of alternative representations, but the core insight remains: transform each pixel into a family of candidate lines, and identify agreement via intersection in parameter space.

The Hough transform discretizes the (m, b) space into a finite grid. For each edge pixel, it iterates over a predefined set of slope values m_i and computes $b_i = y - m_i x$. A two-dimensional accumulator array stores how many pixels vote for each parameter pair (m_i, b_i). After all edge pixels have contributed, cells in the accumulator with high vote counts correspond to lines that are strongly supported by the data.

This formulation avoids the combinatorial explosion of testing all pixel pairs. Each pixel acts independently, voting for a family of possible lines. Lines with high support appear as peaks in the accumulator, which can be located efficiently using standard search techniques.

In practice, this classical parameterization has limitations near vertical lines, where the slope m becomes unbounded. A more robust formulation employs the normal parameterization $\rho = x \cos\theta + y \sin\theta$, which describes lines using distance and orientation. This variant ensures uniform treatment of all orientations, but the essential idea remains: transform alignment in image space into intersection in parameter space, and use voting to identify consistent structures.

The strength of the Hough transform lies in its ability to transform the problem from image space to parameter space. Once the core mechanism is established—each edge pixel voting for the set of parameters consistent with its location—the method generalises seamlessly to more complex shapes.

For example, consider the problem of detecting circles. A circle in the plane is defined by the equation $(x - a)^2 + (y - b)^2 = r^2$, where (a, b) denotes the centre and r the radius. Each edge pixel (x_0, y_0) must satisfy this relation for some triplet (a, b, r). If the radius is fixed, then every such pixel defines a circular locus of possible centres—that is, it contributes votes to all (a, b) such that $(x_0 - a)^2 + (y_0 - b)^2 = r^2$. Allowing r to vary introduces a third dimension: the accumulator now becomes a 3D volume over (a, b, r).

This generalises further. Any shape that admits an algebraic parameterization can be detected in the same way. For instance, **ellipses** require five parameters (centre coordinates, major and minor axis lengths, and orientation), while **parabolas** can be parameterized by vertex location and opening direction. Even **arbitrary curves** described by parametric equations or algebraic forms can be handled, provided the number of parameters is finite and reasonably small.

In each case, the edge pixel votes for a hypersurface in the corresponding parameter space. The transform aggregates these votes, and high-density regions in the accumulator indicate the presence of shapes that are strongly supported by the edge data.

The cost of generality is dimensional: for a shape defined by k parameters, the accumulator array lives in $\mathbb{R}^k$. Memory and computation scale exponentially with k, limiting the practical complexity of detectable shapes. Nevertheless, for many applications—especially where the shape class is known and low-dimensional—the Hough transform remains a tractable and robust detection mechanism.

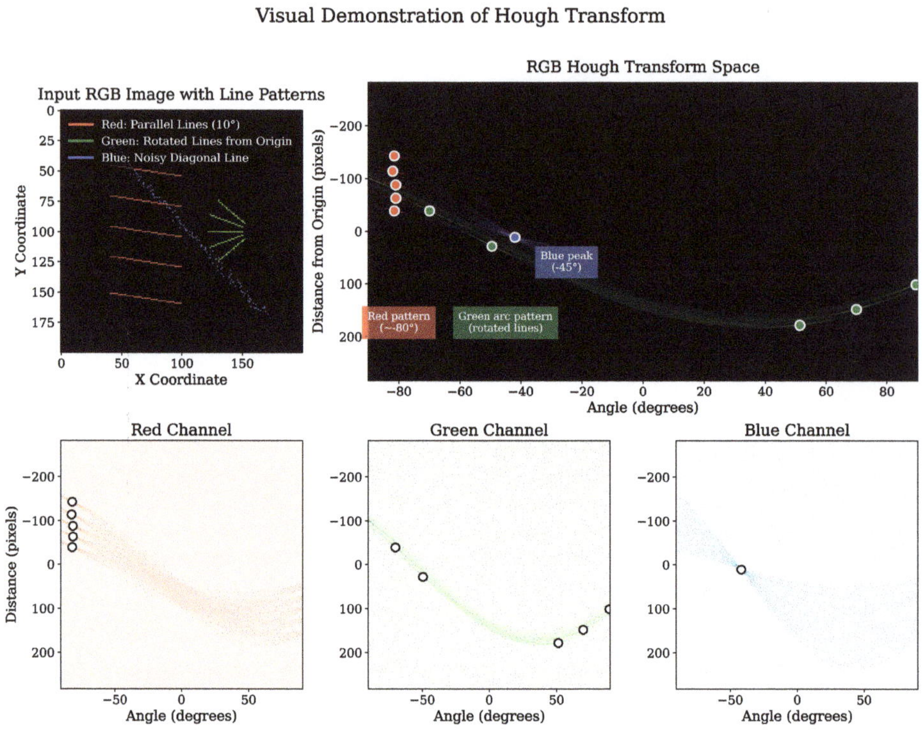

Figure by author. The Hough transform converts lines in image space to points in parameter space (angle, distance). Parallel lines (red) create peaks at the same angle. Lines from a common point (green) create an arc pattern. Even with noise, a line (blue) creates a strong peak in Hough space.

Radon, Hough, and the Geometry of Detection

Continuous vs Discrete Detection

Detecting a geometric structure in an image means finding parameters p for which the curve or surface $C(x;p) = 0$ is present in the intensity field $I(x)$. The Radon and Hough transforms implement this same idea in two complementary ways: one performs a continuous *read-out* along shapes, the other a discrete *write-in* from feature points.

Transform Duality: Reading vs Writing

Let $I(x)$ be a spatial image and $p \in P$ a parameter vector. The Radon transform is the integral projection:

$$R(p) = \int_{\mathbb{R}^n} I(x)\,\delta(C(x;p))\,dx.$$

Here each parameter p queries all x satisfying $C(x;p) = 0$ and accumulates their intensity. In contrast, the Hough transform iterates over image locations: for each x_0 where $I(x_0) \neq 0$, it computes all p with $C(x_0;p) = 0$ and increments $H(p)$.

Let $\mathcal{C}_p = \{x \in \mathbb{R}^n \mid C(x;p) = 0\}$ denote the locus of points on shape p, and $\mathcal{M}_x = \{p \in P \mid C(x;p) = 0\}$ the set of shapes passing through point x. Then:

$$R(p) = \int_{\mathcal{C}_p} I(x)\,d\mu(x),$$

(Radon: continuous read-out)

$$H(p) = \sum_{x \in \mathrm{supp}(I)} \mathbf{1}_{\mathcal{M}_x}(p),$$

(Hough: discrete write-in).

Radon computes an inner product between $I(x)$ and a template restricted to $\mathcal{C}_p$, while Hough builds up a histogram in parameter space whose peaks indicate well-supported shapes.

Unified Operator View

Both transforms can be expressed as linear operators with a kernel $C(p,x)$:

$$(\mathcal{L}_C I)(p) = \int_{\mathbb{R}^n} C(p,x)\,I(x)\,dx.$$

Choosing $C(p,x) = \delta(C(x;p))$ recovers Radon-type projections. Replacing the integral by a sum over discrete feature points and incrementing an accumulator at each p with $C(x;p) = 0$ yields Hough-type transforms.

If $C(p,x)$ is shift-invariant, i.e., $C(p,x) = K(x - \phi(p))$, the operator reduces to convolution with a shifted template. This links Radon/Hough methods to classical matched filtering and Fourier-analytic descriptors.

Intersections in Parameter Space

Every edge point x defines a manifold $\mathcal{M}_x \subset P$. True structures correspond to parameters $p^\star$ where multiple such manifolds intersect, producing sharp peaks in $H(p)$. In this view, detection is about the geometry of intersections: coherent data produce persistent intersections that survive noise and discretization, whereas random or spurious features yield only weak or isolated crossings.

References:

Radon, J. (1917). Über die Bestimmung von Funktionen durch ihre Integralwerte längs gewisser Mannigfaltigkeiten. *Berichte der Sächsischen Akademie der Wissenschaften zu Leipzig*, 69, 262–277.

Hough, P. V. C. (1962). Method and Means for Recognizing Complex Patterns. U.S. Patent 3,069,654.

Duda, R. O., and Hart, P. E. (1972). Use of the Hough Transformation to Detect Lines and Curves in Pictures. *Commun. ACM* 15(1), 11–15.

van Ginkel, M., Luengo Hendriks, C. L., and van Vliet, L. J. (2004). A short introduction to the Radon and Hough transforms and how they relate to each other. *TU Delft Technical Report QI-2004-01.*

Wet, Cold, Slippery Slope

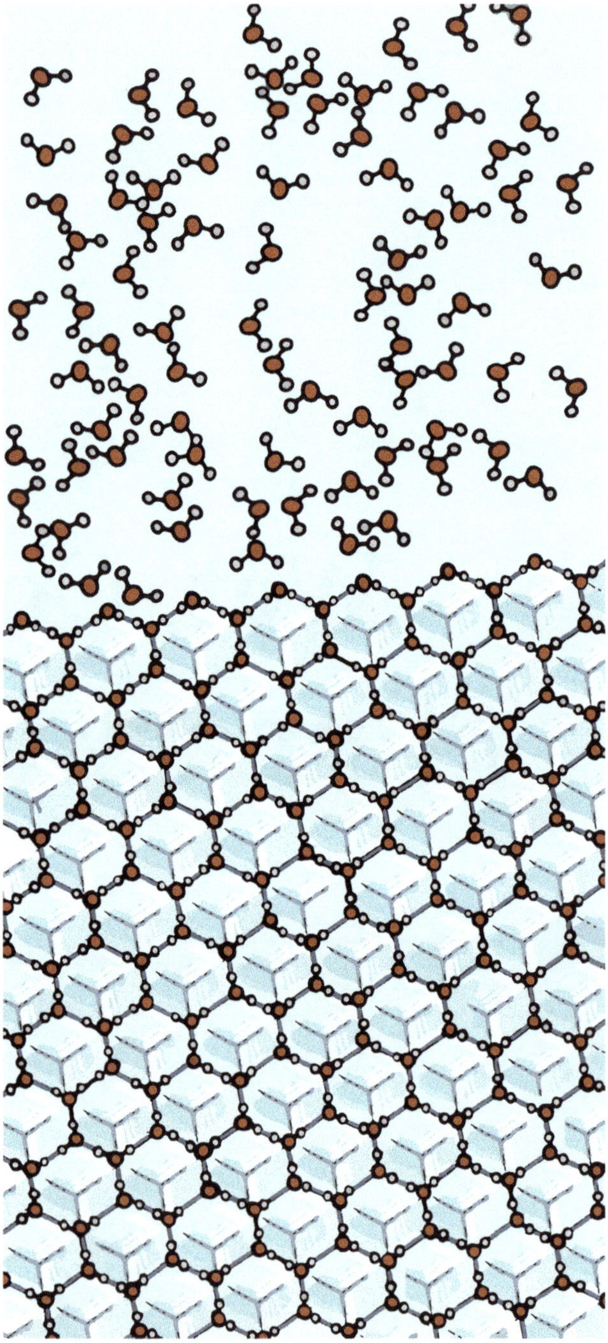

Quasi-Liquid Layer on Ice: Ice surfaces are coated with a thin, disordered layer of mobile water molecules—called the *quasi-liquid layer*. Even below freezing, this layer behaves like a liquid: molecules at the surface are less tightly bound than those in the bulk lattice, enabling them to rearrange and flow. This surface mobility reduces friction and is a primary reason ice feels slippery, even without pressure or frictional heating.

Wet, Cold, Slippery Slope

Ice's exceptional slipperiness results primarily from a quasi-liquid layer (QLL) of disordered water molecules at its surface rather than from commonly assumed mechanisms. While pressure melting and frictional heating contribute under specific conditions, neither explains ice's slickness at rest nor across wide temperature ranges. Surface molecules, having fewer hydrogen bonds than those in the interior crystal lattice, form a nanometre-thick disordered layer that functions as a molecular lubricant even well below freezing. Counterintuitively, ice is most slippery around -7°C rather than at 0°C, as the QLL is sufficiently mobile at this temperature while the underlying ice remains hard enough to resist deformation.

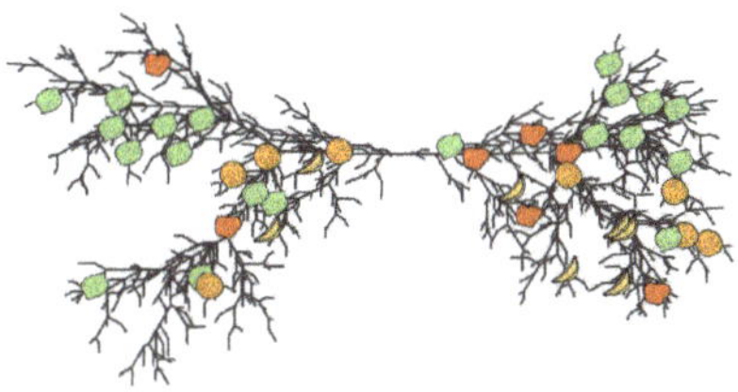

Ice Slipperiness ∘ Quasi-Liquid Layer ∘ Surface Undercoordination ∘ Hydrogen Bond Network ∘ Ice Ih Structure ∘ Pressure Melting Myth ∘ Frictional Heating Limits ∘ Optimal Temperature -7°C ∘ Molecular Dynamics ∘ Sea Level & Land Ice ∘ Phase Transitions

"Oh, you know what I want? Ice cream.
Do you guys have that here? Ice cream?
Oh, it's so good, you have to get it.
It's like scoops and it comes on a cone.
Do you have that here?"

– Andy Dwyer, 2012

Wet, Cold, Slippery Slope

> In the mid-nineteenth century, Michael Faraday proposed that ice possesses a thin, liquid-like surface layer even below its melting point—a hypothesis based on observations of regelation and contact phenomena. Around the same time, James Thomson and later Lord Kelvin developed the thermodynamic framework of pressure melting, suggesting that applied pressure lowers the melting point and produces a lubricating film. John Joly applied this idea to ice skating in 1886, arguing that the narrow blade of a skate generates sufficient localised pressure to melt ice beneath it.
>
> In the early twentieth century, questions emerged about whether pressure alone could explain ice's slipperiness, especially at low temperatures. In the 1930s and 1950s, Frank P. Bowden and David Tabor introduced frictional heating as an alternative mechanism, showing that sliding motion could generate enough heat to produce melt layers, complementing or supplanting pressure-induced effects.
>
> By the late twentieth century, new experimental tools—such as atomic force microscopy, sum-frequency generation spectroscopy, and X-ray scattering—enabled scientists to probe the molecular structure of ice surfaces directly. These studies showed that even in the absence of pressure or friction, the outermost molecular layers of ice are inherently disordered and mobile. Molecular dynamics simulations (MDS) further supported this view, confirming the existence of a quasi-liquid layer driven by the undercoordination of surface molecules.
>
> Together, these historical developments trace a change from macroscopic mechanical theories to microscopic interfacial physics. Ice's slipperiness, once attributed solely to melting, is now understood as the result of an intrinsic, dynamic surface layer whose mobility increases with temperature—an insight that unifies over a century of observation, theory, and experimentation.

Matter exists in distinct organisational forms known as phases. The classical categories—solid, liquid, and gas—are defined by qualitative differences in arrangement and in response to external conditions. In solids, particles maintain fixed relative positions within a repeating spatial pattern. Liquids retain cohesion without rigidity, allowing flow while maintaining volume. Gases exhibit weak intermolecular interactions and expand to fill any container. These phases describe the majority of everyday materials, but others emerge under specialised conditions.

Additional phases include plasmas, which arise when gases are ionised into charged particles, and supercritical fluids, which appear beyond the liquid-gas boundary at high pressure and temperature. At extremely low temperatures, matter can form Bose–Einstein condensates or superfluids, characterised by quantum coherence across macroscopic scales. These states differ not only in arrangement but also in their symmetries, excitations, and thermodynamic behaviour.

Transitions between phases are governed primarily by temperature and pressure. Lower temperatures reduce kinetic energy, allowing intermolecular forces to stabilise ordered configurations. Increasing temperature disrupts this order. Pressure alters the volume

 https://doi.org/10.11647/OBP.0526.42

available for molecular motion and can favour or suppress particular interactions. These competing effects generate a phase diagram—a diagrammatic map of stable forms as functions of external conditions. Phase boundaries correspond to discontinuities in structure or derivatives of free energy, typically expressed as latent heat or a change in symmetry.

Water, as a molecular compound, exhibits all three classical phases within common terrestrial conditions. Under atmospheric pressure, it transitions from solid to liquid at 0°C and from liquid to vapour at 100°C. These transition points shift with pressure, enabling supercooled liquid below 0°C and reduced boiling points at high altitude. The phase diagram of water includes a triple point at 0.01°C and 0.006 atmospheres where solid, liquid, and gas coexist, and a critical point at 374°C and 218 atmospheres beyond which liquid and gas become indistinguishable. Water forms more than a dozen crystalline ice phases—Ice II through Ice XIX—many of which are denser than liquid water, contrasting with Ice Ih which floats.

The distinct behaviour of water follows from its molecular geometry and intermolecular interactions. Each H_2O molecule forms a bent structure with a 104.5° angle between hydrogen atoms, creating an electric dipole with partial negative charge on oxygen and partial positive charges on hydrogens. This polarity permits the formation of hydrogen bonds: directional attractions between the hydrogen of one molecule and the oxygen of another. In the liquid phase, each molecule forms and breaks hydrogen bonds rapidly, producing a transient network. In ice, these interactions become fixed, forming a tetrahedral lattice where each molecule participates in four hydrogen bonds.

Hydrogen bonding accounts for thermodynamic anomalies. Water has a higher melting and boiling point than other molecules of similar mass. Its density peaks at 4°C, then decreases upon freezing. At atmospheric pressure, the stable crystalline form is Ice Ih, adopting a hexagonal lattice with each molecule coordinated to four others at tetrahedral angles. This open configuration contains void space, producing a density lower than liquid water. Freezing thus involves expansion rather than contraction, allowing ice to float.

Ice Ih exhibits macroscopic properties consistent with its lattice. It is brittle, cleaving along crystallographic planes. Its thermal conductivity is moderate, mediated by phonons in the ordered lattice. It is optically transparent in the visible spectrum, though scattering increases with impurities or polycrystallinity. Ice Ih remains the dominant form in terrestrial and atmospheric environments.

Beyond crystalline forms, water also forms amorphous ice—a glassy solid lacking long-range order. Produced by rapid cooling or vapour deposition at temperatures below 130 K, amorphous ice is the predominant form of water in interstellar space and on cometary surfaces. On Earth, it exists transiently in the upper atmosphere and can be created in laboratories. Unlike crystalline ice, amorphous ice lacks the organised hydrogen-bond network that creates the open configuration of Ice Ih.

One early hypothesis to explain ice's low friction was pressure melting. According to this view, localised pressure—such as from a skate blade—lowers the melting point beneath the contact area, producing a thin film of liquid water. This film then acts as a lubricant.

The mechanism is thermodynamically valid near 0°C and relies on the Clausius–Clapeyron relation (Clausius, 1850), which predicts a decrease in melting point with pressure.

A second hypothesis emphasises frictional heating. As an object slides across ice, mechanical work is converted into heat at the contact interface. Because ice is a poor conductor, this heat remains localised, potentially melting the surface. This model accounts for enhanced slipperiness during rapid motion and is consistent with high-speed sports where continuous sliding sustains the melt layer.

Both explanations fail under static or slow-motion conditions. The pressure needed to depress the melting point by 1°C is about 13 MPa, so typical skate contact pressures of only a few MPa yield at most a few tenths of a degree—insufficient on their own at low temperatures. Frictional heating is minimal at low velocities and cannot explain the ease with which stationary objects begin to slide. Experiments show that ice remains slippery at temperatures and pressures where neither mechanism is operative.

The resolution lies at ice's surface. Even in the absence of external inputs, a thin, mobile layer of disordered molecules exists at the ice-air boundary. This quasi-liquid layer (QLL) is not a bulk liquid, nor a perfect continuation of the crystalline lattice. It consists of molecules that lack sufficient bonding partners and thus vibrate with greater amplitude and positional freedom.

Surface undercoordination breaks the tetrahedral symmetry found in the bulk. Molecules at the boundary form fewer than four hydrogen bonds, creating a dynamic layer with reduced rigidity. Although confined to nanometric thickness, this layer allows shearing with minimal resistance. The QLL persists even at temperatures as low as −20°C, though its thickness and mobility vary with temperature. As the surface warms, more molecules enter the disordered state and the layer thickens, decreasing friction.

Ice exhibits minimum friction not at 0°C but at intermediate subzero temperatures. At 0°C, the bulk ice softens and becomes susceptible to ploughing deformation under load. This increases drag and offsets the benefits of surface lubrication. Between −5°C and −10°C, depending on sliding velocity and contact pressure, the QLL remains mobile while the underlying ice retains sufficient hardness to resist deformation. For typical skating conditions, minimum friction occurs near −7°C, though faster sliding or heavier loads shift this optimum.

Atomic force microscopy confirms nanometric compliance at the ice surface. Sum-frequency generation spectroscopy detects disrupted hydrogen bonding at the interface. The QLL arises from the geometry and thermodynamics of the boundary, not from transient melting.

Molecular dynamics simulations complicate this picture. During sliding, the QLL is overwhelmed within the first few nanometres by a different process: cold, displacement-driven amorphization. Lateral displacement at the contact destroys crystalline order molecule by molecule, producing an amorphous layer whose thickness grows as $w \propto \sqrt{d}$—the same square-root scaling observed in diamond and silicon wear. At 10 K, ice amorphizes six times faster than at -10°C. The process is mechanical, not thermal. The difficulty of skiing at low temperatures is not a lack of interfacial water but the high viscosity of the amorphous layer that forms.

Low friction also requires a hydrophobic counterface: a hydrophilic surface doubles the friction coefficient with an identical water film, because adhesion-enhanced dissipation at contact edges dominates. Ice slipperiness depends on the interplay of intrinsic surface disorder, displacement-driven amorphization, frictional heating, counterface chemistry, and contact geometry.

The Ice Cube Berg

A common objection to climate concern goes like this: an ice cube melting in a glass doesn't raise the water level, so why should melting polar ice raise sea levels? The reasoning seems sound. Archimedes established that floating ice displaces (Archimedes, c. 250 BCE) its own weight in water. Ice is roughly 9% less dense than liquid water, so it floats with about 90% submerged. When it melts, the resulting liquid occupies almost exactly the volume previously displaced. The person making this argument has correctly understood buoyancy.

Yet, only some ice affects sea level.

Sea ice—the Arctic ice cap, icebergs calved from glaciers, ice shelves extending from Antarctica—is already floating. When it melts, the direct contribution to sea level is indeed minimal. There is a small effect: sea ice forms from freshwater (salt is excluded during freezing), but it floats in saltwater. Freshwater is less dense than saltwater, so when the ice melts, the freshwater occupies slightly more volume than the saltwater it was displacing. This effect exists but remains small compared to what follows.

The Greenland ice sheet sits on land. So does the Antarctic ice sheet. So do mountain glaciers across the Himalayas, Andes, Alps, and Rockies. These formations are not floating. They rest on bedrock, supported by solid ground, contributing nothing to current ocean volume. When this ice melts, the water flows into rivers and eventually into the sea. This is not an ice cube melting in a glass. This is ice from outside the glass being poured into it.

The scales are staggering. The Greenland ice sheet contains enough water to raise global sea levels by 7.4 metres. The Antarctic ice sheet holds enough for 58 metres. Even a few percent loss would displace hundreds of millions of people from coastal cities. This is not speculative. Greenland is currently losing roughly 280 billion tons of ice per year. Antarctica loses about 150 billion tons annually. These are measured quantities, tracked by satellite gravimetry and radar altimetry.

Roughly 68% of Earth's freshwater is locked in ice sheets and glaciers on land. Most of that sits on Antarctica and Greenland. As global temperatures rise, this ice transitions from solid to liquid and enters the ocean, increasing total volume.

Thermal expansion contributes as well. Water expands as it warms. Between 1993 and 2019, thermal expansion accounted for roughly 40% of observed sea level rise, with melting land ice contributing most of the remainder. The two mechanisms are additive.

Phases of Matter

Classical States

Solid
Atoms in fixed lattice positions. Definite shape and volume.

Liquid
Short-range order permits flow. Fixed volume, variable shape.

Gas
Weak intermolecular forces. Fills available volume.

Plasma
Ionised particles. Collective electromagnetic behaviour.

Quantum Phases

Bose–Einstein Condensate (Einstein, 1925)
Bosons in single quantum state below μK.

Fermionic Condensate
Cooper-paired fermions at ultralow temperature.

Superfluid
Zero viscosity. He-4 below 2.17 K, He-3 below 2.6 mK.

Superconductor
Zero electrical resistance, magnetic flux expulsion (Meissner & Ochsenfeld, 1933).

Quantum Spin Liquid
Frustrated magnetism, long-range entanglement.

Topological Matter
Global invariants define phase (quantum Hall, topological insulators).

Intermediate Forms

Liquid Crystal
Orientational order with fluidity. Nematic, smectic, cholesteric phases.

Glass
Amorphous solid. Kinetically arrested liquid structure.

Gel
Crosslinked network in fluid. Viscoelastic response.

Granular Matter
Macroscopic particles. Jamming transitions.

Extreme Conditions

Quark–Gluon Plasma
Deconfined quarks above 2 trillion K.

Degenerate Matter
Quantum pressure dominates. White dwarfs (electrons), neutron stars.

Supersolid
Crystalline order with superflow. Realised in ultracold atoms.

Time Crystal
Periodic structure in time. Driven quantum systems.

Rydberg Matter
Highly excited atomic states. Millimetre-scale electron orbits.

Matter organises into distinct phases determined by temperature, pressure, and quantum mechanics. Each phase exhibits characteristic symmetries, excitations, and responses to external conditions.

Origins of Ice Slipperiness

Surface Premelting and Quasi-Liquid Layer Formation

A quasi-liquid layer (QLL) forms on ice when the solid–vapour interfacial energy exceeds the combined solid–liquid and liquid–vapour energies. Let γ_{sv}, γ_{sl}, and γ_{lv} denote these interfacial energies, respectively. The criterion for spontaneous surface disordering is:

$$\gamma_{sv} > \gamma_{sl} + \gamma_{lv}.$$

This lowers the Gibbs free energy (G) and drives disordered layer formation. Surface molecules are undercoordinated, forming fewer hydrogen bonds and possessing higher vibrational entropy. The QLL exhibits molecular mobility without full phase change.

Frictional Heating and Velocity-Dependent Melt Film Generation

Frictional sliding converts mechanical work to interface heat. Heat generation rate: $P_{\text{fric}} = \mu F_N v$, where μ is kinetic friction coefficient, F_N is normal load, and v is sliding velocity. For high v, generated heat exceeds thermal dissipation, raising interface temperature and potentially inducing melt layers below bulk melting point T_m. This dynamic meltwater film can exceed equilibrium QLL thickness and reduce shear resistance.

QLL Rheology and Shear Lubrication

QLL or meltwater lubrication depends on rheological response. Let $\eta(T, \dot{\gamma})$ denote the effective viscosity (units: Pa·s), where T is temperature and $\dot{\gamma}$ is shear rate. In confined geometries, viscosity deviates from bulk water and may exhibit non-Newtonian behaviour. The shear stress $\tau = \eta\dot{\gamma}$ determines frictional resistance. Enhanced molecular mobility near T_m yields lower η and reduced τ under shear, enabling efficient nanometric lubrication.

Thickness Divergence and Interfacial Scaling Laws

As temperature approaches the melting point, QLL thickness $d(T)$ increases following:

$$d(T) \sim (1 - T/T_m)^{-\alpha}, \quad \alpha \in [0.3, 0.5],$$

where α is a critical exponent. This reflects gradual surface disordering and successive molecular layer formation. Ellipsometry and vibrational spectroscopy confirm this scaling; simulations support entropic and energetic growth origins.

Pressure Effects and Contact Mechanics

The Clausius–Clapeyron relation governs melting point depression under pressure: $dT/dp = T\Delta V/\Delta H$, where $\Delta V < 0$ is the volume change upon melting and ΔH is the latent heat of fusion. For macroscopic loads (e.g., skates), the average pressure-induced melting-point shift is small, typically less than 1°C under realistic loads. Local pressure at asperities—real contact points within the nominal contact area—can be much higher. These localised hotspots drive frictional heating and melting. Real contact area controls heat distribution and deformation.

Composite Friction Model: Thermo-Mechanical Coupling

In the lubricated regime, friction reduces to viscous shear across the interfacial film:

$$\mu \approx \eta(T)\, v/(p \cdot h(T, v)),$$

where $\eta(T)$ is the film viscosity, p the contact pressure, and $h(T, v)$ the lubricating layer thickness—set by the equilibrium QLL and any frictional melt. As $T \to T_m$, h grows (from the scaling above) and η drops, reducing μ. At low T or v, h is thin and friction is high. Near T_m, ice softens and the slider ploughs into the surface. The minimum μ near -7°C reflects the balance: the film is thick enough to lubricate, but the ice is stiff enough to resist penetration. Molecular dynamics simulations (Atila et al., 2024) suggest that during sliding, displacement-driven amorphization—not melting—produces the dominant lubricating layer, with thickness scaling as $w \propto \sqrt{d}$ where d is the slid distance.

References:

Slater, B., and Michaelides, A. (2019). *Nat. Rev. Chem.*, **3**, 172.

Weber, B. et al. (2018). *J. Phys. Chem. Lett.*, **9**, 2838.

Atila, A., Sukhomlinov, S. V., and Müser, M. H. (2024). *Phys. Rev. Lett.*, **133**, 236201.

Flat Universers

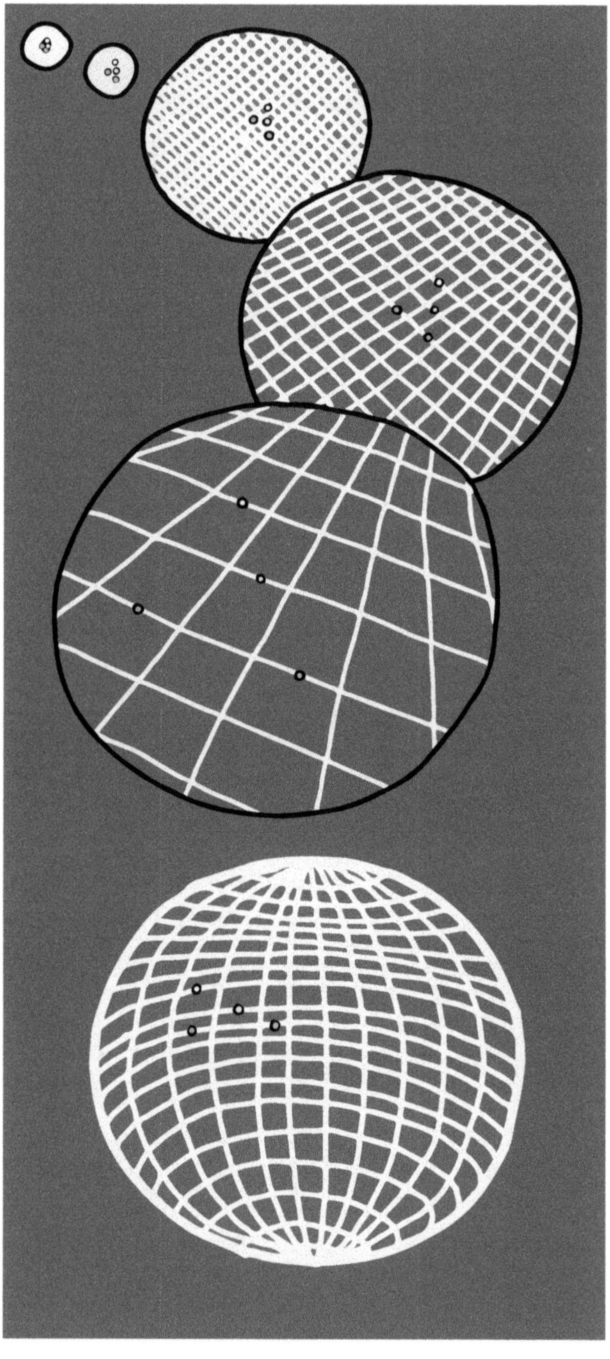

Top (ΛCDM Infinite Flat Universe): In the standard cosmological model, ΛCDM, space is spatially flat and infinite. Galaxies drift apart as distances expand—there is no centre, no edge, and every point sees the same large-scale dynamics.
Bottom (Compact Finite Universe): If space is compact—like a 3-torus or a hypersphere—then the universe can be finite in volume without having a boundary. Expansion still occurs, but the global topology loops back on itself. Light could, in principle, circle the cosmos. The grid warps around, indicating periodicity or curvature that reconnects spatial locations.

Flat Universers

The universe appears flat to within 0.4% precision according to cosmic microwave background measurements. This flatness, described by the Lambda-CDM model, indicates that space follows Euclidean geometry even across vast cosmological distances. The universe may be spatially infinite while having a finite age of 13.8 billion years. This implication comes from the Big Bang model: an expansion of intergalactic space rather than an explosion within pre-existing space. If space was already infinite at the beginning, it expanded uniformly from every point. No centre to the universe!

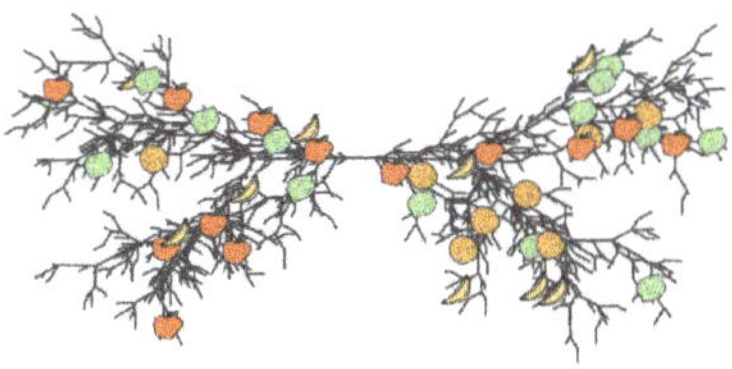

Flat Universe Observations ◦ Cosmic Curvature ◦ CMB Acoustic Peaks ◦ Angular Size Test ◦ ΛCDM Model ◦ Metric Expansion ◦ Infinite Spatial Extent ◦ Particle Horizon ◦ Big Bang Geometry ◦ Planck Satellite Data ◦ Observable vs Total Universe

"The truth is like salt.
Men want to taste a little, but too much makes everyone sick."
— The Dogman, 584 AU

Flat Universers

The debate over whether the universe had a beginning or existed eternally has shaped cosmology for over two thousand years. In classical antiquity, the dominant view—especially in Aristotelian physics—was eternalism: the cosmos had no origin, existing in a state of perpetual motion and balance. Aristotle's model featured concentric spheres rotating around a stationary Earth, upheld by the idea that a perfect, eternal order governed the heavens. The notion of cosmic creation was seen as unnecessary, even philosophically inferior, to an eternal and self-contained universe.

This changed with the rise of monotheistic religions, which introduced a radically different concept: a universe created *ex nihilo* (from nothing), by a singular act of divine will. Medieval thinkers such as Augustine (*Confessions*, c. 397, Book XI) and Maimonides (*Guide for the Perplexed*, c. 1190, Part II) incorporated this creationist framework into their metaphysics, contrasting sharply with the Greek eternalist paradigm. However, for centuries, this remained a theological stance, largely separate from natural philosophy.

Modern cosmology inherited this tension. When Albert Einstein formulated general relativity in 1915, he initially envisaged a static universe. In 1917, he introduced the cosmological constant Λ to obtain a static solution—an implicit nod to eternalism. Yet in 1922, Alexander Friedmann, and independently in 1927, Georges Lemaître, found that Einstein's equations naturally described an expanding cosmos. Lemaître, a Belgian priest and physicist, explicitly interpreted this expansion as evidence of a beginning—a 'day without yesterday.' His model, known as the 'primeval atom,' implied a definite origin in time.

This idea clashed with the philosophical preferences of many physicists. Fred Hoyle, Hermann Bondi, and Thomas Gold proposed the steady-state model in 1948, maintaining that the universe had always existed and would continue to expand eternally, with matter continuously created to preserve density. Hoyle coined the term 'Big Bang' as a dismissive label for Lemaître's model, viewing it as tainted by religious overtones.

Ironically, it was empirical evidence that vindicated the 'creationist' model. The discovery of the cosmic microwave background in 1965 by Penzias and Wilson provided direct observational support for a hot, dense early universe—an echo of its origin. This shifted the consensus dramatically. What began as a scientifically controversial, seemingly theological notion—that the universe had a beginning—became the foundation of modern cosmology. Today's standard model, □CDM (Lambda Cold Dark Matter), descends directly from this creation-based framework, though now couched in empirical and mathematical precision rather than metaphysical doctrine.

A geometric space is defined by the relations among its points: distances, angles, and the behaviour of geodesics—paths that locally minimise distance. In Euclidean geometry, geodesics are governed by axioms such as the parallel postulate, which ensures that parallel lines never intersect and that triangle angles sum to 180 degrees. When these properties fail, the space is said to be curved.

Curvature quantifies how a space deviates from the rules of Euclidean geometry. Positive curvature causes initially parallel lines to converge, as on the surface of a sphere. Negative curvature causes them to diverge, as in a hyperbolic plane. Zero curvature preserves their parallelism indefinitely. These cases define the three canonical geometries in two dimensions: spherical, hyperbolic, and flat.

Curvature is a local property: it describes how space behaves in an infinitesimal neighbourhood. Compactness is a global property: it describes whether space is bounded and complete. The surface of a sphere is compact and positively curved. A flat plane is non-compact and uncurved. A cylinder is flat but compact in one direction. A torus has zero curvature but is compact. Curvature and compactness are independent notions.

A three-dimensional space can have its own curvature, defined purely through internal measurements of distance and angle, without requiring an external embedding. General relativity models the universe using such three-dimensional spatial geometries evolving in time.

The mathematical classification of homogeneous (uniform in all positions), isotropic (uniform in all directions) three-dimensional spaces yields three possibilities: positive curvature (a 3-sphere), negative curvature (a 3-hyperboloid), and zero curvature (Euclidean $\mathbb{R}^3$). Each corresponds to a constant value of spatial curvature and admits a well-defined metric. These are the geometric possibilities for the shape of the universe on large scales.

A cosmological model describes the evolution of space and time on the largest scales. In general relativity, such a model is not a visual rendering of stars and galaxies, but a mathematical solution to Einstein's field equations (Einstein, 1915). It specifies the metric tensor—a geometric object encoding distances, angles, and causal relationships across spacetime. Given assumptions about symmetry and matter content, the metric determines how space stretches, curves, and evolves in time.

The standard model of cosmology is called the Lambda–Cold Dark Matter model (ΛCDM). It assumes that, at sufficiently large scales, the universe is homogeneous and isotropic, restricting the spatial geometry to one of the three cases described above. The spatial curvature in ΛCDM is not a free assumption. It is determined by the total energy density of the universe relative to a critical threshold. Density above, below, or at this value yields positive, negative, or zero curvature (Kolb & Turner, 1990).

In this model, the Big Bang is not a point in space, but a boundary in time: a moment when the scale factor—the function describing the distance between any two comoving points—reaches zero. It represents the earliest definable state of the metric, beyond which classical general relativity ceases to apply. The Big Bang is not an explosion of matter into space. It is the dynamical expansion of space, or more accurately the distance between objects, governed by the evolving metric. All regions of the universe were arbitrarily close together in the past and have since expanded away from each other in a coordinated, metric-driven evolution.

This expansion is not centred at any specific location. It occurs everywhere simultaneously. Each observer sees distant galaxies receding, not because they are moving through space, but because the space between them is getting bigger. In a homogeneous universe, every region participates equally in the expansion, and the large-scale structure remains statistically uniform over time. The observable consequence is a redshift in the light from distant galaxies (Hubble, 1929)—a stretching of wavelengths that records the history of spatial expansion.

The strongest evidence for the geometry of the universe comes from the cosmic microwave background (CMB) (Penzias & Wilson, 1965)—a relic radiation field that permeates all of space. The CMB originates from a time roughly 380,000 years after the Big Bang, when the universe cooled enough for protons and electrons to combine into neutral hydrogen atoms. This event, known as recombination, allowed photons to travel freely for the first time. The CMB is the redshifted remnant of that photon field, now observed at microwave wavelengths with a temperature of approximately 2.725 K.

The CMB is not perfectly uniform. It contains small anisotropies—tiny temperature fluctuations at the level of one part in 100,000. These fluctuations correspond to density variations in the early universe, which later seeded the formation of large-scale structures such as galaxies and clusters. The angular size of these fluctuations provides a direct measurement of spatial geometry. In particular, one can ask how large a primordial region appears on the sky today, given the time it took for light to reach us and the curvature of space through which it travelled.

Before recombination, the universe consisted of a hot, dense plasma of photons, electrons, and baryons (protons and neutrons). Photons scattered continuously off free electrons, coupling radiation tightly to matter. In this medium, density perturbations propagated as pressure waves driven by the competition between gravitational infall and photon pressure. These acoustic oscillations (Peebles & Yu, 1970)—standing wave patterns in the plasma—left an imprint on the temperature distribution of the CMB when photon decoupling occurred.

The most important feature in the CMB spectrum is the first acoustic peak. This peak corresponds to the largest sound waves that had time to compress and rarefy a region of plasma before recombination. The physical size of such regions is determined by known physics—the speed of sound in the early universe and the duration before decoupling. However, the observed angular size depends on the spatial curvature of the universe. If space is positively curved, such a region appears larger than in flat geometry. If negatively curved, it appears smaller.

High-precision observations, particularly from the WMAP and Planck satellites, have measured the angular scale of the first acoustic peak with great accuracy. The result is consistent with a flat universe: the peak appears at an angle of approximately 1 degree, matching the prediction from a zero-curvature model. This agreement indicates that spatial curvature is consistent with zero to within a few parts per thousand.

Flat geometry means space neither curves back on itself nor terminates at a boundary (see Chapter 14). It continues indefinitely in every direction. If the universe is spatially flat everywhere, the simplest consistent model is one infinite in extent—no enclosing shell, no edge beyond which space ceases to exist. This conflicts with everyday intuition, which expects all things to be either bounded or looped. General relativity is indifferent to such expectations: it computes curvature from energy density and evolves the metric accordingly.

The observable universe does have a limit—the particle horizon, roughly 46 billion light-years in radius—but this is an observational boundary, not a physical wall. Beyond it, the model predicts space continues with the same statistical properties indefinitely.

That infinite space can arise from a finite beginning seems paradoxical, but time and space play different roles. The Big Bang marks a finite past moment—the origin of the metric and the expansion dynamics. If space was already infinite at that moment, it remained so as it expanded. The scale factor grew, but the global extent may always have been unbounded.

The universe's flatness is not a theoretical preference but an outcome of measurement. The cosmic microwave background, galaxy clustering, and large-scale structure all point to a flat metric, and the ΛCDM model achieves this without assuming it a priori.

The name 'ΛCDM' encodes the model's two dominant components: Λ (Lambda) represents dark energy—the cosmological constant driving accelerated expansion (Riess et al., 1998)—and CDM stands for cold dark matter. Dark matter's role in structure formation is not incidental. During the radiation era, baryonic matter (protons, neutrons, electrons) remained tightly coupled to photons through electromagnetic interactions. Photon pressure prevented gravitational collapse. Dark matter, interacting only gravitationally, decoupled immediately after the Big Bang and began clustering in response to the density fluctuations seeded during inflation. By the time of recombination—when photons finally decoupled from matter—dark matter had already formed deep gravitational potential wells. Baryons then fell into these pre-existing halos, forming the first protogalaxies. Without this head start, the universe would not have had enough time to form the observed structures. The CMB fluctuations mentioned earlier are snapshots of these nascent dark matter concentrations.

ex nihilo

The investigation of the origin of 'the world' is a central theme in all cultures—from spacetime to abiogenesis, between religion and science. Yet their answers are not mutually exclusive. Creation stories reveal nothing about the scientific account, and science could not care less about the intuitive question of why there is existence rather than nothing. It is interesting that the Big Bang model was initially ridiculed as too religious and later as atheistic ramblings, while in reality it's neither. It is simply the best set of equations that describes everything we observe to an unmatched degree of accuracy. The intuitive, not scientific, question of existence itself is in the domain of philosophy and theology, not science. The question of which model best explains the state of the universe, while being predictively fruitful, is not only in the *domain* of science, but the *raison d'être* of science.

ΛCDM Cosmology Timeline

1. Planck Era ($< 10^{-43}$ s)
Forces: Four forces unified. Quantum gravity dominates.
Matter: No particles; quantum foam.
Scale: Fluctuations at Planck length ($\sim 10^{-35}$ m).

2. Inflation (10^{-36}–10^{-32} s)
Forces: Strong separates from electroweak.
Matter: Vacuum energy drives exponential expansion.
Scale: Universe expands by factor $\geq 10^{26}$; quantum fluctuations seed galaxies.

3. Reheating and Baryogenesis
Time: 10^{-32}–10^{-6} s
Forces: Electroweak breaks into weak and electromagnetic.
Matter: Quarks, leptons, gluons. CP violation creates 1:10^9 matter excess.
Temp: 10^{15}–10^{12} K.

4. Quark–Gluon Plasma
Time: 10^{-6}–10^{-5} s
Forces: All four forces distinct.
Matter: Quarks confine into protons and neutrons.
Temp: $\sim 10^{13}$ K.

5. Nucleosynthesis (1 s–20 min)
Matter: Protons and neutrons fuse: 75% H, 25% He, trace Li.
Scale: Photons coupled to matter; uniform plasma.

6. Photon Era (20 min–380 kyr)
Matter: Ionised plasma. Neutrinos decouple at ~1 s. CDM decouples immediately; non-interacting.
Temp: 10^9 K $\rightarrow$ 3000 K.
Scale: Density fluctuations ($\sim 10^{-5}$) grow. CDM begins clustering gravitationally.

7. Recombination and CMB
Time: 380,000 yr
Matter: Electrons bind to nuclei → neutral atoms. CDM potential wells already formed.
Scale: Photons decouple → CMB. Baryons begin falling into CDM halos.

8. Dark Ages (0.4–0.5 Gyr)
Matter: Neutral hydrogen falls into CDM halos. CDM provides ~85% of gravitational scaffolding.
Scale: Hierarchical merging; no stars yet. Smallest halos form first.

9. First Stars and Reionization
Time: 0.5–1 Gyr
Matter: Nuclear fusion creates first stars (Pop III); heavier elements forged.
Scale: Ionising radiation clears fog; protogalaxies form.

10. Galaxy Formation (1–5 Gyr)
Scale: CDM halos merge hierarchically; baryons cool and collapse at halo centres. Cosmic web emerges (100 Mpc filaments).
Activity: Quasars peak (~3 Gyr). Galaxy rotation curves reveal CDM dominance.

11. Present (13.8 Gyr)
Forces: Dark energy (Λ) dominates since ~5–6 Gyr; drives accelerated expansion.
Composition: 69% dark energy, 26% CDM, 5% baryons.
Scale: Expansion rate: $H_0 \approx 70$ km/s/Mpc. Galaxy separation ~1 Mpc.
Effect: Λ prevents new large-scale structure formation; existing structures bound by CDM resist expansion.

12. Future (Λ domination)
Near: 10–100 Gyr: Λ-driven expansion pushes galaxies beyond event horizon. Only Local Group (bound by CDM) remains visible.
Mid: 10^{12}–10^{14} yr: Star formation ceases; gas exhausted.
Far: 10^{15}–10^{100} yr: Black holes evaporate. Universe asymptotes to cold, empty de Sitter space.

Evolution Summary: Unified forces → separate by 10^{-12} s. Quantum fields → quarks → hadrons → atoms → stars → galaxies. Planck scale → cosmic horizon. Radiation era → matter era (CDM-driven structure formation) → Λ era (accelerated expansion).

Calculating Timescales

Recombination: Why 380,000 Years?

Recombination occurs when the universe cools enough for protons and electrons to form neutral hydrogen. The ionisation fraction is governed by the Saha equation:

$$\frac{n_e n_p}{n_H} = \left(\frac{m_e k_B T}{2\pi\hbar^2}\right)^{3/2} e^{-E_I/k_B T},$$

where $E_I = 13.6\,\text{eV}$ is hydrogen's ionisation energy. At $T \sim 3000\,\text{K}$, $n_H/n_e \sim 1000$: the plasma becomes neutral.

In a radiation-dominated universe, temperature scales as $T \propto a^{-1} \propto t^{-1/2}$. From the Friedmann equation:

$$H^2 = \frac{8\pi G}{3}\rho_r, \quad \rho_r = \frac{\pi^2}{30} g_* k_B^4 T^4/(\hbar c)^3,$$

where $g_* \approx 3.36$ at recombination. Solving for t:

$$t = \frac{1}{2H} \approx \frac{\sqrt{45}}{4\sqrt{2\pi^3 G g_*}} \frac{\hbar c}{k_B^2 T^2}.$$

Substituting $T = 3000\,\text{K}$ yields $t \approx 380{,}000\,\text{yr}$.

Matter-Radiation Equality: 47,000 Years

Radiation density scales as $\rho_r \propto a^{-4}$; matter density as $\rho_m \propto a^{-3}$. Equality occurs when:

$$\rho_r(a_{\text{eq}}) = \rho_m(a_{\text{eq}}).$$

Today's CMB temperature is $T_0 = 2.725\,\text{K}$. At equality, $T_{\text{eq}} = T_0(1 + z_{\text{eq}})$, where $z_{\text{eq}} = a_0/a_{\text{eq}} - 1$. From Planck data:

$$\Omega_m h^2 = 0.143, \quad \Omega_r h^2 = 4.18 \times 10^{-5},$$

giving $z_{\text{eq}} \approx 3400$. Using the same temperature-time relation:

$$t_{\text{eq}} \approx 47{,}000\,\text{yr}.$$

Dark Energy Domination: 9 Billion Years

Dark energy density ρ_Λ remains constant as matter dilutes. Domination begins when $\rho_\Lambda = \rho_m(a)$. Today's densities:

$$\Omega_\Lambda = 0.69, \quad \Omega_m = 0.31.$$

Since $\rho_m \propto a^{-3}$:

$$\rho_m(a) = \rho_{m,0} a^{-3}, \quad \rho_\Lambda = \text{const}.$$

Equality at:

$$a_\Lambda = \left(\frac{\Omega_m}{\Omega_\Lambda}\right)^{1/3} \approx 0.75.$$

The scale factor evolves as $a(t) \propto t^{2/3}$ in matter era. Integrating from $a = a_\Lambda$ to $a = 1$ over 13.8 Gyr:

$$t_\Lambda \approx 9.8\,\text{Gyr}.$$

Nucleosynthesis Window: 1–20 Minutes

BBN requires $T \sim 0.1\,\text{MeV}$ for deuterium formation but must occur before neutrons decay ($\tau_n = 880\,\text{s}$). At $T = 0.8\,\text{MeV}$:

$$t_{\text{start}} \approx 1\,\text{s}.$$

Reactions freeze out at $T \sim 0.07\,\text{MeV}$:

$$t_{\text{end}} \approx 1200\,\text{s} \approx 20\,\text{min}.$$

The neutron-to-proton ratio at freeze-out determines helium abundance:

$$Y_p = \frac{2(n/p)}{1 + (n/p)} \approx 0.25,$$

matching observations precisely.

References:

Dodelson, S. (2003). *Modern Cosmology.*

Kolb, E. W., and Turner, M. S. (1990). *The Early Universe.*

Planck Collaboration (2020). *A&A*, **641**, A6.

The Man in the Velvet Mask

The Man in the Iron Mask: The real prisoner—Eustache Dauger—wore a velvet mask, not iron, and was transferred discreetly between prisons under heavy secrecy. He died unknown. In contrast, fiction (notably Dumas's *The Man in the Iron Mask*, 1849) transformed him into a hidden twin of Louis XIV, masked in iron, rescued by musketeers, and restored to royal life. The left column shows archival history; the right, its mythologised media version.

The Man in the Velvet Mask

The prisoner known as 'Eustache Dauger' remained in state custody for thirty-four years (1669–1703) under extraordinary protocols of secrecy. His confinement spanned four locations under the continuous supervision of a single jailer, Bénigne Dauvergne de Saint-Mars. Official correspondence reveals exceptional measures: a specially constructed cell with sound isolation, strict limitations on communication, and a requirement to wear a black velvet mask when visible to anyone outside Saint-Mars's control. The prisoner served as valet to another detainee at Pignerol before eventual transfer to the Bastille, where he died and was buried under the alias 'Marchioly.'

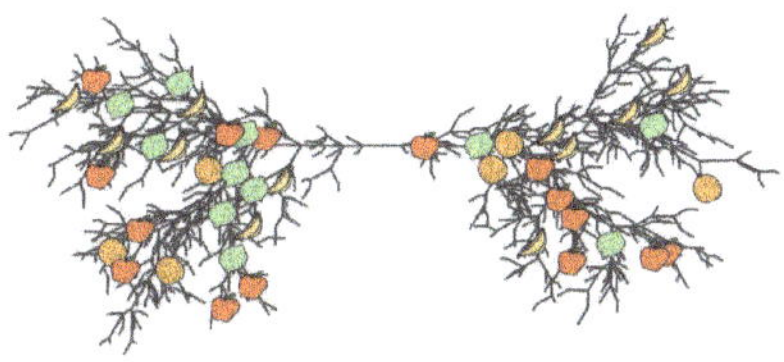

Man in the Iron Mask ◦ Eustache Dauger Mystery ◦ Lettres de Cachet ◦ Saint-Mars Custody ◦ Velvet Not Iron ◦ Administrative Erasure ◦ Voltaire & Dumas Myths ◦ Bastille Symbol ◦ Revolutionary Icon ◦ Arbitrary Detention ◦ Historical Void

"I swear I way more than half believe it when I say,
That somewhere love and justice shine
Cynicism falls asleep,
Tyranny talks to itself."

— The Weakerthans, 1997

The Man in the Velvet Mask

The late seventeenth century in France was defined by the dominance of Louis XIV, the so-called Sun King, whose reign from 1643 to 1715 represents one of the longest and most centralised periods of monarchical authority in European history. The French court at Versailles embodied the power and spectacle of absolute monarchy, where every detail of court life was orchestrated to reflect the grandeur of the sovereign. In this context, the King's will was law. Mechanisms such as the *lettre de cachet* allowed for imprisonment without trial, often for reasons known only to the monarch or his ministers. These secret detentions were essential to the logic of governance, especially in a state where honour, reputation, and dynastic stability were paramount.

Louis XIV inherited a nation destabilised by the Fronde civil wars and moulded it into a regime where loyalty to the crown was absolute. Institutions such as the Bastille and the Alpine fortress of Pignerol were not just prisons; they were instruments of statecraft. High-ranking prisoners, such as disgraced ministers, dissenting nobles, or politically inconvenient relatives were incarcerated under conditions of discretion and silence. Governors of such prisons, such as Bénigne Dauvergne de Saint-Mars, were carefully chosen for loyalty and discretion.

This era also saw the consolidation of state secrecy in foreign policy, diplomacy, and internal finance. Cardinal Mazarin, Louis's chief minister during the King's youth, had amassed a personal fortune through murky dealings with both French and foreign powers, including the English court. Sensitive knowledge of such dealings, especially if acquired by individuals outside the political elite, was considered a potential threat to the monarchy. Against this backdrop, the long, secret imprisonment of a masked man begins to appear less as an anomaly and more as a manifestation of how absolute power protected itself from destabilising disclosures.

In medieval and early modern Europe, long-term imprisonment did not function as a primary tool of criminal justice. Confinement was typically employed as a provisional measure—for debtors, those awaiting trial, or individuals requiring temporary custodial restraint. Sentences relied on corporal penalties, execution, fines, exile, or public shaming. Prisons existed as procedural instruments rather than destinations of punishment.

By the sixteenth and seventeenth centuries, selective forms of political detention had begun to appear, particularly in the Italian principalities and the Habsburg realms. Individuals viewed as Politically dangerous, diplomatically embarrassing, or ideologically subversive were confined through the discretionary authority of monarchs, dukes, or cardinals. Conditions depended on relationships of power, access, or threat. Prisons became tools of silencing.

France institutionalised this through the *lettre de cachet*—a sealed royal directive permitting imprisonment without trial or formal accusation. They authorised indefinite confinement and were used against courtiers, clerics, dissidents, or troublesome family members. Though sometimes misused by noble families to eliminate inconvenient heirs or rivals, they were also instruments of state control. The Bastille, Vincennes, and other royal fortresses housed such prisoners without public record or legal recourse.

These prisons were administered by military governors under the oversight of the War Ministry. Many buildings were former citadels or active military posts. The governor of a fortress prison—such as Pignerol or the Bastille—was a commissioned officer with autonomous control over its operations. Supplies, transfers, and correspondence passed through military channels. The jailer's loyalty was owed to the crown directly, with oversight exercised through ministerial confidence rather than civil inspection.

The prisoner later associated with the name Eustache Dauger was arrested by royal warrant in 1669 and held under continuous custody for thirty-four years. During this period, he was successively imprisoned at Pignerol, Exilles, Île Sainte-Marguerite, and the Bastille. At each location, the prisoner remained under the exclusive supervision of a single officer: Bénigne Dauvergne de Saint-Mars. This consistency of custody—across four separate sites and nearly four decades—was highly unusual in French penal administration.

Upon the prisoner's arrival at Pignerol, the Secretary of State for War, Louvois, issued direct instructions that a special cell be constructed with successive doors to prevent sound transmission. The prisoner was to receive food, clothing, and supplies only through Saint-Mars himself. Conversation was forbidden beyond basic necessities. Following the initial arrest, the prisoner's name vanished from official correspondence. References to him were consistently indirect—phrases such as 'the one you know' or 'the old prisoner' replaced any identifying language.

Surviving records from the Bastille, including the register of Lieutenant Étienne du Junca, describe the mask as being made of black velvet. It was employed when the prisoner was visible to guards, clergy, or others not under Saint-Mars's direct control. No evidence supports the claim that the mask was metallic, nor that it was worn at all times. An iron mask worn continuously over years would have produced physical damage—none is recorded. The mask prevented recognition during public transfers or collective observance when total isolation was impractical.

The case lacks any legal framing. There are no extant records of charges, trial, classification, or judicial review. The prisoner was never formally sentenced, and no court official appears to have been involved in his management after his initial detention. He was not categorised under espionage, treason, or moral scandal. He was administratively undefined. Unlike other state prisoners, whose files often contain notes of visitation, surveillance reports, or periodic assessments, this individual's record is limited to internal logistics and commands.

Upon the prisoner's death at the Bastille in 1703, the erasure continued. He was buried under the name "Marchioly" in the parish cemetery of Saint-Paul-des-Champs. This name appears nowhere in earlier correspondence and does not match any documented individual held under Saint-Mars's custody. After the burial, Saint-Mars ordered the destruction of all furnishings, bedding, and written materials associated with the prisoner. The walls of his cell were scraped and whitewashed, and no personal effects were preserved. These actions exceeded the standard procedures for deceased prisoners of state.

Throughout his confinement, there is no evidence that the prisoner enjoyed the privileges or deference accorded to persons of noble birth or dynastic sensitivity. His designation in internal correspondence remained 'valet' during his time at Pignerol. He was not granted enhanced rations, special accommodations, or access to legal counsel. Nor was he treated with hostility. His confinement was methodical rather than punitive. Later speculation has emphasised the possibility of royal lineage—most famously the twin brother hypothesis advanced by Voltaire and fictionalised by Dumas—but the historical record offers no support for such interpretations.

The transformation into myth began with the absence left by his confinement. Voltaire's *Le Siècle de Louis XIV* (1751) proposed an iron mask and royal origin without archival basis. The iron mask became a metaphor for secrecy rendered visible—an image of anonymity made material.

Alexandre Dumas embedded this in *The Vicomte de Bragelonne* (The Man in the Iron Mask, 1849), portraying the prisoner as Louis XIV's identical twin. The mask concealed dynastic threat: a bloodline too dangerous to acknowledge. Fiction overtook record for storytelling purposes—Dumas was writing the kind of story that would be entertaining with disregard to any factual substance.

Twentieth-century cinema embraced the mask as visual anchor. Films from 1929 to 1998 reimagined the prisoner as royal heir, wronged twin, or victim of betrayal of historical facts were set aside for commentary on power and injustice. The absence of documentation enabled limitless theatricality.

The prisoner also served as a political weapon. Voltaire deployed him in *Le Siècle de Louis XIV* as evidence of *lettres de cachet* taken to its limit: imprisonment without accusation, trial, or recorded offence. The metal signified permanence, the mask signified anonymity, and together they constructed an image of power exercised without accountability.

The *philosophes* recognised a perfect inversion of juridical process. Where law demanded *public accusation*, documented evidence, and formal judgment, the prisoner received none. His confinement operated through pure administrative will. Diderot cited the case in his attacks on arbitrary detention. Rousseau invoked it to illustrate the distance between natural rights and monarchical practice.

By the 1780s, the prisoner had evolved from historical curiosity to revolutionary icon. Pamphlets circulating in Paris described the Bastille as housing countless such victims—men and women buried alive in stone cells, their names erased, their families ignorant of their fate. The actual prison population was modest and consisted largely of debtors and forgers, but the symbolic weight of the fortress derived from cases such as the masked prisoner. He represented what the Bastille could contain: anyone, for any reason, forever.

The events of 14 July 1789 transformed symbol into action. The crowd that converged on the Bastille sought gunpowder, but they also sought vindication. They expected to find dungeons packed with political martyrs, victims of *lettres de cachet* living proof of tyranny. The fortress yielded seven prisoners: four forgers, two madmen, and one aristocrat confined at his family's request. The mythical imprisoned multitude did not exist. But the revolutionaries found something else in the archives: the administrative traces of the masked prisoner, including du Junca's register entry (18 September 1698) describing his arrival in 'a mask of black velvet.'

These documents confirmed the legend. The absence of charges validated every suspicion about arbitrary power. The prisoner's anonymity became his defining feature. He became the ancestor of every political prisoner, the prototype of administrative disappearance.

The Constituent Assembly abolished *lettres de cachet* on 16 March 1790, citing "the sacred rights of men" and "the horror that secret orders inspire in a free nation." The debates surrounding this legislation invoked the masked prisoner as the ultimate example of their necessity. Deputies argued that as long as sealed letters could authorise indefinite detention, no citizen was secure. The prisoner's decades of confinement without trial demonstrated that administrative convenience could override every principle of justice. His case proved that between the King's will and the subject's freedom stood only the thickness of a seal.

Revolutionary iconography absorbed the prisoner into its visual repertoire. Engravings showed him in chains with an iron mask, standing as an emblem of pre-revolutionary oppression. The Bastille's demolition was framed as his posthumous liberation.

The Directory and successive regimes inherited this political symbolism. The prisoner served as a cautionary figure, invoked whenever debates arose about preventive detention, state security, or judicial transparency. His image functioned as a constitutional ghost—a reminder of what government could do when unrestrained by law. Even Napoleon, who reinstated forms of administrative detention, avoided association with the precedent. The prisoner had become radioactive, his facelessness a mirror reflecting the anxieties of any regime about its own legitimacy.

International republicanism adopted the figure as universal symbol. Italian carbonari, German liberals, and Polish nationalists all invoked the prisoner as victim of despotism. His French specificity dissolved into general metaphor. Any political prisoner held without trial, any dissident silenced by state power, could claim genealogy from the man in the mask.

The nineteenth century produced hundreds of theories about his identity—from disgraced ministers to foreign spies, from royal bastards to religious heretics. Each hypothesis reflected contemporary concerns more than historical evidence. Scholars combed archives for traces, but each discovery deepened the mystery. What remained was a cavity in history, defined by the forces that had created it.

Modern historiography has abandoned the search for the prisoner's identity, focusing instead on his function within the apparatus of early modern state power. He existed at the intersection of administrative efficiency and sovereign prerogative, at the margin where bureaucratic procedure met royal exception. His confinement required constant maintenance—transfers, supplies, instructions—yet produced no documentation of purpose. His trajectory through the prison system traced the limits of what absolute power could do when it chose to act without explaining itself.

The legend persists because the absence persists. Democratic societies have not eliminated administrative detention, classified prisoners, or state secrets. The practice remains: the possibility that individuals can disappear into custody, that reasons can be withheld, that legal process can be suspended in the name of higher necessity.

Administrative Detention: Present Continuous

The *lettre de cachet* was abolished in 1790. Administrative detention was not.

As of 2024, the United States detention facility at Guantanamo Bay holds approximately 30 individuals, some for over two decades without trial. Most were detained under the Authorization for Use of Military Force following 11 September 2001. The legalese classifies them as 'unlawful enemy combatants' rather than prisoners of war or criminal defendants. This classification places them outside both military and civilian judicial systems. Evidence against them often remains classified and even *habeas corpus* petitions have produced limited results.

Israel employs administrative detention under military orders issued pursuant to the British Mandate Defence (Emergency) Regulations of 1945 in the West Bank, and under Israel's Emergency Powers (Detentions) Law, 1979, within Israel. As of mid-2024, over 3,300 Palestinians were held without charge. Detention orders, issued by military commanders, can be renewed indefinitely in six-month increments. Detainees and their lawyers may be denied access to the evidence against them, which is classified as security-sensitive. The process occurs in military courts where standards of evidence and procedural protections differ from civilian criminal proceedings. Though predominantly applied to Palestinians, the measure has also been used against right-wing Israeli settlers and extremists, and the practice was challenged by both ends of the political spectrum.

In 2025, the 'Alligator Alcatraz' detention centre opened in Florida's Everglades. Civil rights organisations filed lawsuits alleging that detainees were held without charges and denied access to legal counsel. The facility's remote location—surrounded by wetlands and wildlife—functions as geographic isolation reminiscent of Guantanamo's offshore positioning, placing detainees beyond easy reach of attorneys, advocates, or public scrutiny.

The procedural architecture has evolved since 1669. There are review boards, periodic renewals, legal representation. But when evidence remains classified, when reviews examine only summaries, when detention orders can be renewed indefinitely, the distance from Dauger's cell becomes a question of degree. The mask has been removed. The administrative silence continues.

Historical Fiction

This technical study reconstructs the historical profile of the state prisoner later mythologized as 'the Man in the Iron Mask.' It evaluates the verifiability and continuity of archival sources—primarily the correspondence between Louvois and Saint-Mars, prison registers, burial records, and journals kept by staff—and contrasts them with later literary augmentations by Voltaire and Dumas.

The Documentary Spine: Continuity and Custody

The prisoner's existence is traceable through a continuous chain of archival records from 1669 to 1703. The arrest order—a *lettre de cachet* dated 19 July 1669—names 'Eustache Dauger,' ordering his confinement under Saint-Mars at Pignerol. Subsequent letters from Louvois, though avoiding names, refer to 'the prisoner whom you know,' with consistent logistical details.

Each transfer—Pignerol (1669), Exilles (1681), Sainte-Marguerite (1687), Bastille (1698)—parallels Saint-Mars's own promotions. Du Junca's Bastille journal confirms the masked prisoner's arrival in 1698 and death in 1703. He was buried under the alias 'Marchioly.' No court records, criminal charges, or trial transcripts exist over this thirty-four-year period.

Evidentiary Elimination: Mask, Alias, and Erasure

The famous mask appears only once in primary sources—in du Junca's 1698 journal entry—described as 'black velvet.' There is no mention of iron. Its use is restricted to moments of public exposure, such as transport or chapel attendance. Early orders emphasise secrecy and isolation but not continuous masking.

The burial alias 'Marchioly' evokes 'Mattioli,' a separate prisoner captured in 1679 and dead by 1694. However, Dauger's imprisonment begins a decade earlier and ends nine years later. No source places Mattioli at the Bastille. The alias appears to be administrative misdirection rather than a clue to true identity.

Post-mortem protocols—including burning of bedding and wall-scraping—are recorded in Saint-Mars's letters and deviate sharply from standard Bastille procedure, indicating deliberate suppression rather than routine sanitation.

Historiographical Filtering: Dauger, Mattioli, and Royal Invention

Three major identification theories remain:

1. **Dauger** is named in the 1669 warrant and is often identified as Fouquet's valet. Some historians argue that he may have uncovered sensitive financial or political information, warranting extreme secrecy. This theory aligns with timeline, treatment, and the absence of formal charges.

2. **Mattioli**, although a real prisoner, is chronologically misaligned. He was arrested in 1679, died in 1694, and is never recorded in the Bastille. The alias 'Marchioly' is insufficient for identification given the common use of placeholders.

3. **Royal identity hypotheses**—twin, brother, or secret heir—have no archival foundation. No court record, diplomatic note, or genealogical account supports them. These theories originate with Voltaire's speculative writings and were dramatised by Dumas. They reflect political allegory, not evidence-based history.

Conclusion: Historical Confinement as Narrative Substrate

The verified record depicts a man systematically anonymized, transferred, masked on occasion, and ultimately erased from memory. These measures suggest not noble origin but sensitive knowledge. Later literary versions reframe bureaucratic silencing into a fable of royal injustice, but the legend's core is not who he was—it is how thoroughly the state erased him.

References:

Sonnino, P. (2016). *The Search for the Man in the Iron Mask.* Rowman & Littlefield.

Voltaire (1751). *Le Siècle de Louis XIV.*

Archives Nationales, Série K. (1669–1703).

du Junca, E. (1703). *Journal de la Bastille.*

The Demon is in the Details

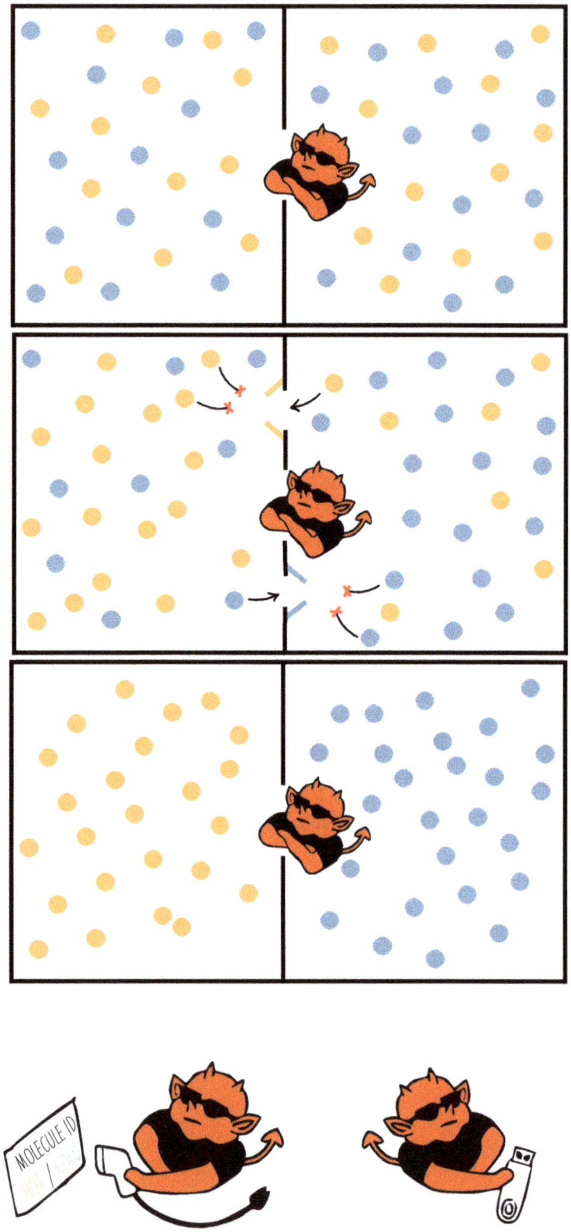

Top (Maxwell's Demon): The classic Maxwell's Demon thought experiment. A demon monitors molecules in a gas and selectively opens a door to let fast-moving (hot) molecules through in one direction and slow-moving (cold) molecules through the other. Over time, this creates a temperature gradient without performing mechanical work, seemingly violating the second law of thermodynamics.
Bottom Left (Work Cost): First resolution—work-based measurement cost (Szilard 1929, Brillouin 1951). Determining each molecule's speed and deciding whether to open the door requires physical measurement processes that generate entropy or consume work, preventing a net decrease in total entropy.
Bottom Right (Memory Erasure): Second resolution—memory erasure cost (Landauer 1961, Bennett 1982). Even if measurement and control were performed reversibly with no work cost, the demon's finite memory would eventually fill. Erasing past measurement records to store new data increases entropy by at least $k \ln 2$ per bit erased (Landauer's Principle), ensuring the second law remains intact. Either way, the apparent paradox dissolves when information-processing steps are properly accounted for.

The Demon is in the Details

Maxwell's Demon, proposed in 1867, describes a thought experiment where a tiny being controls a door between two gas chambers, selectively allowing fast molecules into one chamber and slow ones into another. This sorting creates a temperature gradient from uniformity, seemingly decreasing entropy and violating the second law of thermodynamics. The resolutions are through work costs of measurement and the information-theoretic cost of manipulating information.

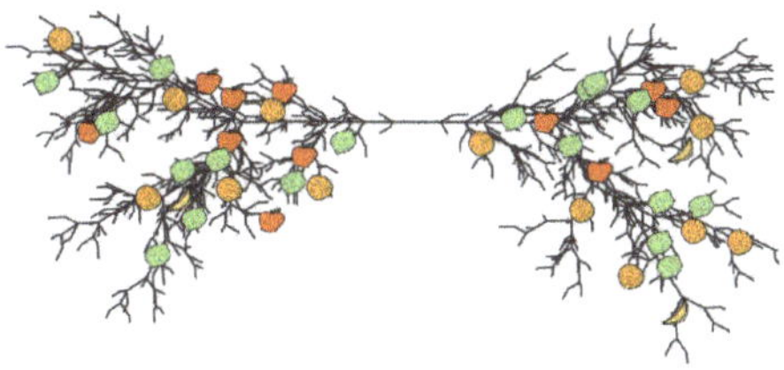

Maxwell's Demon ◦ Second Law Challenge ◦ Entropy & Information ◦ Boltzmann Statistics ◦ Molecular Sorting ◦ Landauer's Principle ◦ Bennett's Resolution ◦ Information Erasure Cost ◦ Thermodynamic Computing ◦ Information is Physical ◦ Memory Cycle Limits

« Seul un être aux sens infiniment subtils,
tel que le démon de Maxwell,
pourrait démêler cet écheveau embrouillé
et remonter le cours de l'univers. »

("Only a being with infinitely subtle senses, such as Maxwell's demon, could unravel this tangled skein and reverse the course of the universe.")

— Poincaré, 1902

"Oh, ye seekers after perpetual motion,
how many vain chimeras have you pursued?
Go and take your place with the alchemists!"

— Da Vinci, 1500s

The Demon is in the Details

The roots of thermodynamics trace back to early nineteenth-century efforts to understand the efficiency of heat engines. In the 1820s, Sadi Carnot introduced the idea of reversible cycles and the notion that heat could be partially transformed into work, bounded by what would later be called the second law of thermodynamics. His work, though framed in a caloric theory, anticipated a fundamental limitation: that no engine could be more efficient than a reversible one operating between two heat reservoirs.

Building on this foundation, Rudolf Clausius in 1865 formally introduced and named the concept of entropy, giving the second law a precise mathematical expression: in any real process, the total entropy of an isolated system tends to increase. Earlier, in 1851, William Thomson (Lord Kelvin) offered an alternative formulation, asserting the impossibility of converting all heat from a single reservoir into work without other effects—essentially forbidding perpetual motion machines of the second kind.

While these formulations were macroscopic and phenomenological, physicists such as Ludwig Boltzmann sought to derive them from microscopic principles, modelling gases as vast ensembles of molecules obeying Newtonian mechanics. This kinetic theory offered statistical interpretations of thermodynamic quantities, suggesting that entropy increase reflected the overwhelmingly probable behaviour of particle ensembles rather than an inviolable mechanical law.

It was in this context—where thermodynamics was seen as emergent from statistical regularities, yet grounded in reversible microscopic dynamics—that James Clerk Maxwell introduced his thought experiment in 1867. He aimed to probe the assumptions underlying the second law by imagining an idealised being capable of intervening at the molecular level, potentially subverting the macroscopic flow of entropy without violating any mechanical law.

Thermal systems are characterised by macroscopic quantities–temperature, pressure, and volume–that arise from the statistical behaviour of countless microscopic constituents. Each molecule in a gas possesses position and velocity at every instant. Macroscopic observables summarise the collective dynamics of trillions of particles.

A single macroscopic state corresponds to countless microscopic configurations. The same pressure and temperature can arise from different combinations of molecular positions and velocities. This multiplicity is central to statistical mechanics, where macroscopic descriptions average over the microstates that realise them.

Entropy quantifies the logarithm of the number of microstates compatible with a given macrostate. In the Boltzmann formulation, the entropy (Boltzmann, 1877) S of a system is expressed as $S = k_B \ln \Omega$, where k_B is Boltzmann's constant and Ω denotes the number of microstates. This mathematical framework captures both the multiplicity of configurations and the incompleteness of macroscopic information.

The Boltzmann constant $k_B = 1.380649 \times 10^{-23}$ J/K (exact, SI) bridges microscopic and macroscopic worlds. It converts between energy scales of individual particles (joules) and thermal energy (kelvins). At room temperature ($T \approx 300$ K), the thermal energy $k_BT \approx 4.14 \times 10^{-21}$ J sets the scale for molecular motion and thermal fluctuations.

The second law of thermodynamics asserts that in an isolated system, entropy cannot decrease. Natural processes tend toward macrostates with greater multiplicity because they are overwhelmingly more probable—there are more ways to achieve a general, high-entropy macrostate than to achieve a lower-entropy, specific macrostate, so any 'random' process will tend to increase entropy. The law governs irreversibility in macroscopic phenomena and forbids spontaneous reorganisation into low-entropy configurations. Though microscopic dynamics allow rare fluctuations, most accessible microstates correspond to thermal equilibrium.

The second law admits equivalent formulations: Clausius forbids spontaneous heat flow from cold to hot; Kelvin rules out complete conversion of heat to work in cyclic processes. Both capture energy's unidirectional dispersal.

While microscopic laws (Newtonian mechanics, Schrödinger equation) are time-reversal invariant, macroscopic irreversibility emerges from statistical asymmetry. Individual molecular collisions remain reversible, but aggregate behaviour favours higher-entropy macrostates due to their numerical dominance.

Although the total phase space volume occupied by a system is conserved under Hamiltonian evolution, as guaranteed by Liouville's theorem, entropy can increase. Fine-grained distributions evolve into intricate structures that, when viewed with any coarse-graining appropriate to macroscopic observations, appear more uniform, corresponding to higher entropy.

Temperature reflects the average kinetic energy per degree of freedom in a system. In classical gases, the distribution of particle energies follows the Maxwell–Boltzmann distribution, while in more general statistical ensembles, the Boltzmann distribution governs the probability of finding the system in a given microstate, establishing a link between microscopic motion and macroscopic thermodynamic parameters.

Thermodynamic processes exchange energy through work and heat. Entropy tracks irreversible energy dispersal and loss of microscopic information. Work represents organised energy transfer; heat denotes disorganised exchange. The second law ensures some energy becomes unavailable for work.

The second law introduces the thermodynamic arrow of time. This arrow derives not from time-symmetric laws of motion, but from statistical tendencies toward higher entropy. Systems evolve from ordered to disordered states, establishing asymmetry between past and future.

Entropy represents the missing information about the system's precise microstate. In this view, thermodynamic entropy parallels concepts from information theory, linking the physical evolution of systems with the informational limitations inherent in macroscopic descriptions.

In 1867, James Clerk Maxwell introduced a thought experiment that challenged the apparent absoluteness of the second law of thermodynamics. He imagined a sealed box filled with gas at thermal equilibrium, where molecules moved randomly at a range of speeds and directions. A partition divided the box into two chambers, A and B, with a small frictionless door controlled by a hypothetical observer: the demon.

The demon monitors molecules approaching the door without mechanical work or external energy. Fast molecules from A pass to B; slow molecules from B pass to A. Others are blocked. Faster molecules accumulate in B, slower ones in A.

This sorting creates a temperature gradient. Heat flows from cold to hot without external energy, contradicting Clausius's formulation. Entropy decreases: the uniform configuration becomes ordered by temperature difference.

The paradox: without work or external energy, the system evolves toward lower entropy. The demon appears to circumvent thermodynamic constraints.

Two approaches resolve this paradox:

The Work-Based Approach argues that the demon cannot operate without performing thermodynamic work. To distinguish between fast and slow molecules, the demon must interact with them, perhaps by shining light to measure their velocities or by mechanically probing their kinetic energies. These measurement processes necessarily require energy input and generate entropy. However, this approach faces a limitation: it cannot establish a precise quantitative relationship between the work invested in measurement and the entropy reduction achieved through sorting. The energy costs of individual molecular measurements depend on the specific measurement apparatus and protocols, making it difficult to prove that the entropy increase from measurement operations exactly compensates for the entropy decrease from molecular sorting.

The Information-Erasure Approach offers a more satisfactory resolution by focusing not on measurement costs, but on the logical requirements of cyclic operation. This approach, developed by Rolf Landauer and Charles Bennett, recognises that for the demon to operate cyclically, it must eventually erase the information stored in its memory. Landauer's principle establishes that erasing one bit of information (Landauer, 1961) requires a minimum energy dissipation of $k_B T \ln 2$, where k_B is Boltzmann's constant and T the temperature of the environment. This energy cost is independent of the physical implementation—it represents a thermodynamic limit on information processing.

Bennett's insight was that this erasure cost provides exact entropy accounting. In each cycle, the demon reduces the gas entropy by $k_B \ln 2$ (corresponding to one bit of information about molecular positions). To continue operating, the demon must erase one bit from its memory, which necessarily increases the environment's entropy by at least $k_B \ln 2$. The entropy reduction from molecular sorting is precisely compensated by the entropy increase from information erasure. No net entropy decrease occurs when all components, gas, demon memory, and thermal environment, are included in the accounting.

This information-theoretic resolution is remarkable because it establishes that **information is physical**. The demon's memory, though conceptually abstract, must be realised in some material substrate subject to thermodynamic laws. The act of erasing information is not merely a computational operation but a physical process that generates heat and increases entropy.

The information-erasure approach also reveals why attempts to circumvent the erasure requirement fail. If the demon preserves information indefinitely to avoid erasure costs, its memory eventually becomes full, preventing further operation. If the demon attempts to reset its memory without erasure, perhaps through reversible computation, the information is merely transferred elsewhere in the system, requiring eventual erasure at some location. The second law cannot be circumvented by clever information management; it emerges inevitably from the statistical nature of many-body systems and the physical reality of information storage.

The 'piston-demon' model suggests the moving partition itself serves as memory, its position encoding molecular information. The resolution depends on system boundaries and whether demon-plus-gas constitutes a closed system.

Quantum mechanics adds complexities: measurement disturbs systems, and indistinguishability constrains sorting. Quantum Maxwell's demons demonstrate how coherence and decoherence affect entropy accounting, probing connections between information, measurement, and thermodynamics.

Does a Full Hard Drive Weigh More?

If information is physical, does a full hard drive weigh more than an empty one? Some theoretical arguments suggest tiny differences, though not in any measurable way.

Information Entropy: A terabyte (8×10^{12} bits) of random data carries maximal Shannon entropy. Erasing that data would, by Landauer's principle, dissipate at least $E = Nk_BT\ln 2 \approx 2.3 \times 10^{-8}$ J at room temperature, equivalent to $\Delta m = E/c^2 \approx 2.6 \times 10^{-22}$ g. This represents heat released during erasure, not extra energy stored in the drive—no more than a $(6, 6)$ dice throw weighs more than $(1, 4)$.

Solid-State Drives: In flash memory, a '1' corresponds to additional trapped electrons in a floating gate. A terabyte written entirely with '1' bits contains roughly 10^{16} extra electrons, adding about 10^{-11} g. This effect is real but fifteen orders of magnitude smaller than the drive's total mass and far beyond detectability.

Exercises

Energy Conservation to Solve Complex Problems

In classical mechanics, many systems that appear to require force analysis, Newton's laws, or torque computations can be solved using the principle of energy conservation. The following problems invite you to discover how straightforward solutions can emerge when the total mechanical energy is conserved.

1. Rolling Sphere on an Incline
A solid sphere of mass m and radius R is placed at the top of a frictional incline of angle θ and allowed to roll down without slipping. Using energy conservation (not Newton's laws), determine the acceleration of the sphere's centre of mass.
Hint: Total mechanical energy includes both translational and rotational kinetic energies.

2. Yo-Yo Drop
A yo-yo of mass m is held so that its string is taut and then released. The string unwinds without slipping as the yo-yo descends. The axle radius is r and the moment of inertia about the centre is I. Use energy conservation to determine the downward acceleration of the yo-yo's centre of mass.
Hint: The yo-yo's kinetic energy has both linear and rotational components; relate the angular speed to the linear speed via the axle radius.

3. Chain Falling Off a Table
A uniform chain of linear mass density λ lies coiled on a horizontal frictionless table. At time $t = 0$, a small length starts to hang off the edge and the chain begins to slide off under gravity. Assuming no friction and no energy loss, use conservation of mechanical energy to find the acceleration of the chain as it falls.
Hint: The centre of mass of the hanging portion descends while its kinetic energy increases.

4. Man Walking on a Boat
A man of mass m walks a distance d from one end of a boat of mass M to the other. The boat floats on frictionless water. Determine how far and in what direction the boat moves relative to the water while the man walks. Use momentum conservation—no external horizontal forces act on the system.
Hint: The centre of mass of the system must remain fixed in the horizontal direction.

Answers: 1: $a = \frac{5}{7} g \sin\theta$ 2: $a = \frac{mg}{m + \frac{I}{r^2}}$ 3: $a = \frac{g}{2}$ 4: Boat disp. $= -\frac{m}{M+m} d$

Thermodynamic Accounting in the Classical Szilard Engine

Introduction

The Szilard engine models a single classical particle confined in a box connected to a thermal reservoir at temperature T. A partition is inserted, the particle is measured, and expansion performs work. The demon, modelled as a finite memory device, must be reset for reuse. This section formally evaluates the extracted work and the entropy budget, showing that total entropy remains non-decreasing when all components are included.

Work from Isothermal Expansion

After measurement, the particle occupies volume $V/2$. Isothermal expansion to volume V yields mechanical work:

$$W_{\text{ext}} = \int_{V/2}^{V} \frac{k_B T}{V'}\, dV' = k_B T \ln 2.$$

Let $\sigma \equiv k_B \ln 2$. Then:

$$W_{\text{ext}} = T\sigma \quad \text{and} \quad \Delta S_{\text{gas}} = \sigma.$$

This reflects the entropy gained by the gas during expansion under constant temperature, which accounts for the increase in accessible microstates as the volume doubles.

Memory Reset and Landauer Bound

To begin a new cycle, the demon must erase one bit of information. Erasure is a logically irreversible operation mapping two equiprobable states to one. According to Landauer's principle, the minimum heat dissipated into the reservoir is:

$$Q_{\text{erase}} \geq T\sigma \quad , \quad \Delta S_{\text{mem}} = -\sigma, \quad \Delta S_{\text{env}} \geq \sigma.$$

This entropy increase in the environment offsets the decrease in the demon's memory. Even in an idealised quasistatic erasure process, the bound cannot be avoided.

Entropy Ledger Over a Full Cycle

We now compute entropy changes for all subsystems. Let G be the gas, M the memory, and E the thermal environment. Then:

$$\begin{aligned} \Delta S_G &= -\sigma \quad \text{(localization dur. measur.)} \\ \Delta S_G &= +\sigma \quad \text{(isothermal expansion)} \\ \Rightarrow \Delta S_G &= 0 \end{aligned}$$

$$\begin{aligned} \Delta S_M &= +\sigma \quad \text{(information recorded)} \\ \Delta S_M &= -\sigma \quad \text{(memory erased)} \\ \Rightarrow \Delta S_M &= 0 \end{aligned}$$

$$\begin{aligned} \Delta S_E &= -\sigma \quad \text{(heat drawn dur. expansion)} \\ \Delta S_E &\geq +\sigma \quad \text{(heat dumped dur. erasure)} \\ \Rightarrow \Delta S_E &\geq 0. \end{aligned}$$

Summing over all contributions:

$$\Delta S_{\text{total}} = \Delta S_G + \Delta S_M + \Delta S_E \geq 0.$$

The equality holds in the quasistatic limit where each step is ideal and reversible. Any deviation—e.g., finite-time processes or imperfect measurement—adds entropy.

Conclusion

The apparent entropy reduction induced by the demon is exactly counterbalanced by the entropy cost of erasing its memory. Though the demon performs no mechanical work, its function relies on acquiring and discarding information—a process embedded in physical degrees of freedom. When all elements are included in the thermodynamic ledger, the second law remains intact. No net entropy decrease occurs, and no violation arises.

References:

Szilard, L. (1929). *On the Decrease of Entropy in a Thermodynamic System by the Intervention of Intelligent Beings.* Z. Phys., 53, 840–856.

Landauer, R. (1961). *Irreversibility and Heat Generation in the Computing Process.* IBM J. Res. Dev., 5(3), 183–191.

Bennett, C. H. (1982). *The Thermodynamics of Computation—A Review.* Int. J. Theor. Phys., 21, 905–940.

Orbital Affairs

Top (Thermal Electrocyclic Reaction): The ring closure proceeds under heat via disrotatory motion. The HOMO is ψ_3, antisymmetric, so disrotation preserves bonding overlap. Constructive phase alignment yields product—shown by successful hand-holding of upper dancers.

Second (Forbidden Thermal Reaction): Conrotatory motion under thermal conditions disrupts orbital overlap. Despite symmetrical movement, bonding fails—dancers miss connection.

Third (Photochemical Electrocyclic Reaction): After photoexcitation, the HOMO becomes ψ_2, symmetric. Now conrotatory motion preserves phase alignment. The dancers execute a stable hand-linked ring—reaction proceeds.

Bottom (Forbidden Photochemical Reaction): Disrotatory motion with ψ_2 breaks symmetry—bonding fails. The dancers' hands miss again. Only one mode is symmetry-allowed per excitation condition.

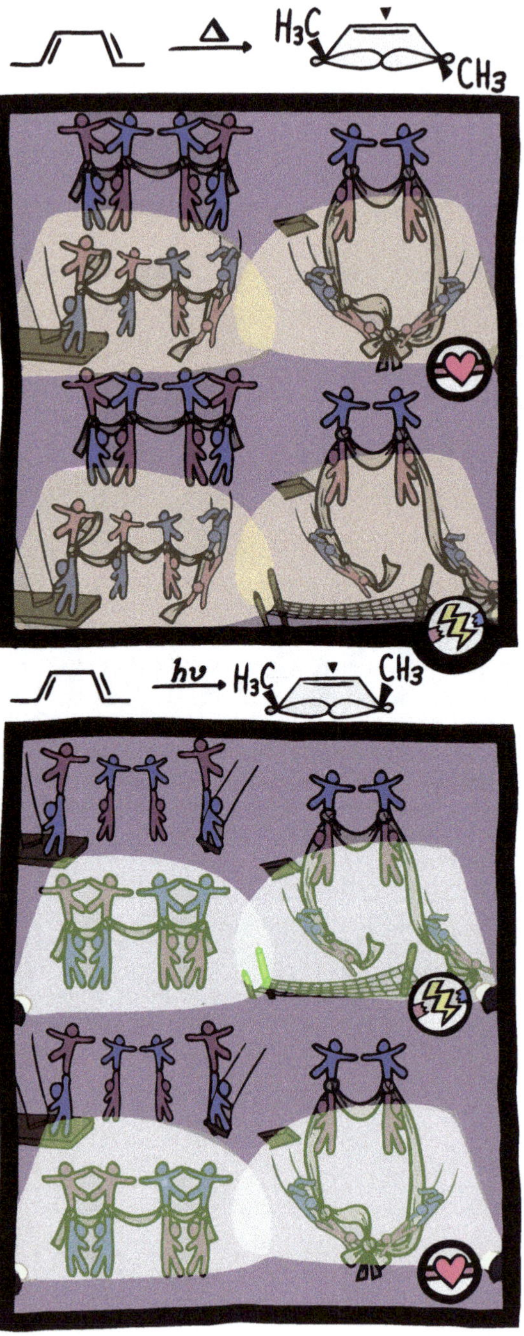

Orbital Affairs

The Woodward-Hoffmann rules establish how mathematical symmetry conservation governs chemical reaction pathways at the quantum level. In pericyclic reactions, the symmetry properties of molecular orbitals—represented by wave functions with specific nodal patterns analogous to trigonometric functions—must be conserved throughout the reaction coordinate. This conservation requirement creates selection rules that determine allowed stereochemical outcomes. The symmetry constraints differ fundamentally between thermal and photochemical conditions, as light excitation inverts the orbital symmetry relationships, thereby enabling reaction pathways forbidden under thermal conditions and vice versa.

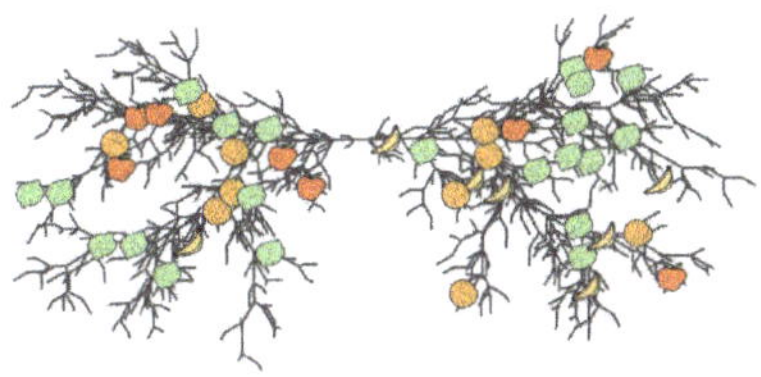

Woodward-Hoffmann Rules ◦ Orbital Symmetry ◦ Pericyclic Reactions ◦ Molecular Orbitals ◦ Thermal vs Photochemical ◦ Electrocyclic Reactions ◦ Cycloadditions ◦ Diels-Alder Reaction ◦ Retinal Photoisomerisation ◦ Quantum Chemistry ◦ Symmetry Conservation

"Coming back to where you started is not the same as never leaving."
— Tiffany Aching, Year of the Signifying Frog, AM 2008

Orbital Affairs

Roald Hoffmann was born Roald Safran in Złoczów, Poland, in 1937. When he was five, his father Hillel smuggled him and his mother out of a Nazi labour camp. A Ukrainian neighbour, Mykola Dyuk, hid them for eighteen months in the attic and storeroom of a local schoolhouse. His mother taught him to read using textbooks stored in the attic. His father remained in the camp and was killed in 1943 for organising a prisoner escape. Most of the extended family perished. Hoffmann and his mother survived, emigrated to the United States in 1949, and he eventually became a theoretical chemist at Cornell.

In 1952, Kenichi Fukui introduced frontier molecular orbital theory, identifying the HOMO and LUMO as the orbitals governing chemical reactivity. His framework was qualitative but pointed towards a deeper principle. Meanwhile, pericyclic reactions—concerted transformations such as electrocyclic ring closures and sigmatropic shifts—presented puzzling stereospecific behaviour that resisted classical explanation. Some worked thermally but not photochemically; others required light. The outcomes seemed predictable only in hindsight.

The breakthrough came from an anomaly. In the early 1960s, Robert Burns Woodward—already regarded as the greatest synthetic organic chemist of his generation, and a 1965 Nobel laureate—was pursuing the total synthesis of vitamin B_{12}, a molecule with 181 atoms, 9 stereocenters, and a corrin ring that had resisted synthesis for over a decade. A critical electrocyclic step gave, in Woodward's words, 'precisely the opposite stereochemical course to that which we had predicted.' He had used physical CPK models, the standard tool of the era, to predict the stereochemistry and got it exactly backwards. Rather than dismissing the result, Woodward brought the puzzle to Hoffmann, who was then developing orbital phase methods using the extended Hückel approach. Their investigation of why the reaction went the 'wrong' way led directly to a series of papers (1965–69) articulating what became the Woodward–Hoffmann rules: pericyclic reactions follow strict constraints based on conservation of orbital symmetry. Whether a reaction is allowed or forbidden can be deduced by tracking how the symmetries of occupied orbitals evolve along the reaction coordinate.

Experimentalists immediately tested the rules across electrocyclic closures, sigmatropic shifts, and cycloadditions. Every known case conformed. The rules offered not just post hoc explanation but genuine predictive power, and by the early 1970s orbital symmetry had become a central organising principle in mechanistic organic chemistry. In 1981, Hoffmann shared the Nobel Prize in Chemistry with Fukui for their independent theoretical contributions to reaction mechanisms. Woodward had died in 1979, two years too early. Had he lived, he would have been among the very few scientists to win two Nobel Prizes.

Molecular structure originates from the underlying architecture of atoms, governed by quantum mechanical principles. Electrons are not treated as point particles following classical trajectories, but as wavefunctions—mathematical objects encoding the probability distribution of their position in space. The behaviour of an electron is determined not by

Newtonian forces but by the solutions to the Schrödinger equation (Schrödinger, 1926), which defines discrete energy levels and corresponding spatial distributions.

The time-independent Schrödinger equation relates the system's total energy to its spatial properties: $\hat{H}\psi = E\psi$, where $\hat{H}$ is the Hamiltonian operator, ψ is the electron wavefunction, and E is the energy eigenvalue. These quantized solutions stand in sharp contrast to the continuous energy spectra predicted by classical mechanics and form the basis of modern atomic theory. Each solution defines an orbital, characterised by specific energy, shape, and nodal structure. In hydrogen, the 1s orbital is spherically symmetric, while higher orbitals—p, d, and f—exhibit directional lobes and angular nodes arising from quantum mechanical constraints.

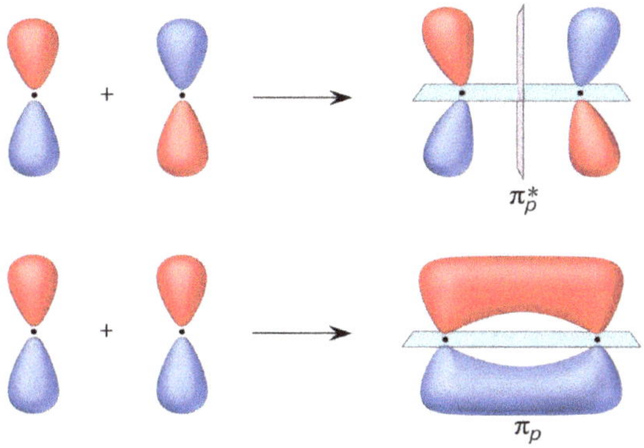

Formation of π and π^* molecular orbitals from lateral overlap of p orbitals. Bonding π_p orbitals (bottom) feature electron density above and below the internuclear axis. Antibonding π_p^* orbitals (top) exhibit a nodal plane and out-of-phase lobes. Adapted from OpenStax Chemistry, Section 7.8: Molecular Orbital Theory. Licenced under CC BY 4.0.

When atoms bond to form molecules, their atomic orbitals combine into molecular orbitals that extend across multiple nuclei. Constructive interference between wavefunctions produces bonding orbitals, concentrating electron density between nuclei and stabilising the system. Destructive interference leads to antibonding orbitals, characterised by a nodal plane between nuclei and elevated energy. The occupation of these orbitals follows the Pauli exclusion principle (Pauli, 1925): electrons fill available molecular orbitals from lowest to highest energy, pairing spins where necessary. This filling determines the molecule's electronic ground state and dictates its chemical reactivity.

Organic chemistry is dominated by molecules composed of carbon, hydrogen, oxygen, and nitrogen. Carbon's tetravalency, enabled by sp^3, sp^2, and sp hybridisations of their s and p orbitals, allows the formation of stable σ and π bonds in chains, rings, and three-dimensional networks. π bonds, arising from lateral overlap of unhybridized p orbitals, are more diffuse than σ bonds and more sensitive to molecular geometry. In conjugated systems, π bonds alternate with σ bonds, allowing electron delocalisation across multiple atoms. This delocalisation lowers the system's energy and imparts distinctive electronic, optical, and chemical properties.

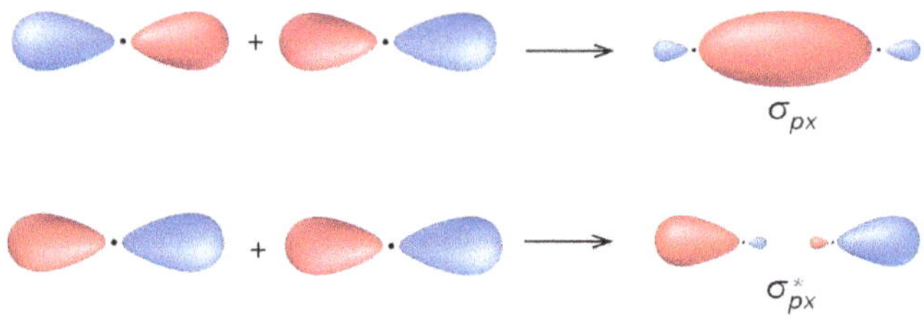

Formation of σ and σ^* molecular orbitals from head-on overlap of two p_x atomic orbitals. Constructive interference (top) yields a bonding σ_{p_x} orbital. Destructive interference (bottom) yields an antibonding $\sigma^*_{p_x}$ orbital. Adapted from OpenStax Chemistry, Section 7.8: Molecular Orbital Theory. Licenced under CC BY 4.0.

Conjugated π systems underpin many organic processes, including pericyclic reactions. These reactions proceed concertedly: all bond-making and bond-breaking events occur simultaneously in a single kinetic step, without discrete intermediates. The hallmark of pericyclic reactions is cyclic electron flow, with electrons moving around a closed loop and the reaction passing through a highly ordered, often symmetric transition state. The feasibility and stereochemical outcome of these reactions depend critically on the phase relationships and symmetries of the participating molecular orbitals.

Molecular orbitals carry a property with no classical analogue: phase. Each orbital has regions of positive and negative sign, like the crests and troughs of a wave. When two orbitals overlap during a reaction, the outcome depends on whether their phases match. Constructive overlap—positive meeting positive—lowers the energy and drives bond formation. Destructive overlap—positive meeting negative—raises the energy and blocks it. Whether a reaction proceeds or not can hinge entirely on this sign, a purely mathematical property of the wavefunction.

The Woodward–Hoffmann rules (Woodward & Hoffmann, 1965) formalised this into a predictive theory. The feasibility of any pericyclic reaction can be determined by tracking how the occupied molecular orbitals evolve along the reaction coordinate. If orbital symmetry is preserved continuously from reactants to products, the reaction is allowed. If symmetry is broken, the reaction is forbidden—no matter how thermodynamically favourable it might be.

The rules distinguish between thermal and photochemical activation. Under thermal conditions, reactions involve the ground electronic state, and the correlation of the highest occupied molecular orbitals (HOMOs) of reactants and products determines the pathway. Under photochemical conditions, excitation promotes an electron into a higher orbital, altering symmetry relationships. The relevant correlation then involves the frontier orbitals of the excited state. In both cases, the requirement is continuous, symmetry-allowed transformation of the electron configuration.

Electrocyclic reactions exemplify the application of the Woodward–Hoffmann rules. In the thermal ring closure of butadiene (four π electrons), the terminal p orbitals must rotate conrotatorily—both twisting in the same direction—to preserve orbital symmetry and achieve constructive overlap. In contrast, the thermal ring closure of hexatriene (six π electrons) proceeds through disrotatory motion, with terminal p orbitals rotating in opposite directions. Photochemical activation reverses these patterns: butadiene closes disrotatorily, and hexatriene closes conrotatorily.

The Diels–Alder reaction (Diels & Alder, 1928) is the clearest demonstration. A conjugated diene (four π electrons) and a dienophile (two π electrons) combine in a single concerted step to form a six-membered ring—a [4+2] cycloaddition involving six π electrons total. The reaction works thermally because the HOMO of the diene and the LUMO of the dienophile have matching phase at both ends simultaneously: both overlaps are constructive, so the transition state is stabilised. This reaction is a workhorse of organic synthesis—steroids, terpenes, alkaloids, and pharmaceutical intermediates are routinely built through Diels–Alder cycloadditions.

Now consider two ethylene molecules approaching face-to-face—a [2+2] cycloaddition with four π electrons. The HOMO of one ethylene and the LUMO of the other have matching phase at one end but opposite phase at the other. One overlap is constructive, the other destructive. The net bonding interaction cancels, and the reaction faces a large energy barrier. Heat alone cannot drive it. Under ultraviolet irradiation, however, promoting an electron into the LUMO flips the phase relationship, making both overlaps constructive. The [2+2] cycloaddition becomes allowed photochemically—this is how cyclobutane rings are made. The same atoms, the same bonds to form, but the sign of a wavefunction determines whether the reaction needs a flame or a lamp.

Sigmatropic shifts extend the theory further. These reactions involve the migration of a σ-bonded group across a conjugated π system. The symmetry of the transition state, visualised through correlation diagrams, dictates whether the shift is thermally allowed. For example, the [1,5]-hydride shift proceeds thermally because the orbital interactions preserve bonding symmetry throughout the migration.

The scope of the rules extends beyond synthetic chemistry into biological systems. In rhodopsin, the light-sensitive pigment of the retina, photoisomerisation of the retinal chromophore exemplifies a pericyclic process governed by symmetry. In its ground electronic state, thermal isomerisation is strongly disfavoured due to a high energy barrier and is extremely rare, preserving visual sensitivity. Upon photon absorption, the excited-state electronic configuration permits rapid, concerted isomerisation from the 11-cis to the all-trans configuration, triggering visual signal transduction.

Even pathological processes, such as the formation of toxic byproducts in age-related macular degeneration, follow orbital symmetry constraints. Under oxidative stress, reactive intermediates enable cyclisations that would otherwise be forbidden thermally.

Mathematical Abstraction as Physical Constraint

The orbital symmetries that dictate reaction outcomes are not empirical regularities forced into a model. They are constraints embedded in the Schrödinger equation. When the mathematics of quantum mechanics was extended from atoms into molecular reaction pathways, it revealed new principles that had been invisible to empirical cataloguing. Symmetry conservation predicted which reactions would work, with what stereochemistry, before they were attempted. Few theories in chemistry have matched this kind of predictive reach from first principles.

This topic holds a personal significance for me. When I applied to a competitive excellence program at the Technion in Israel, I had almost no formal scientific education—my background was in yeshiva study. I was asked to give a lecture as part of the admissions process, and I chose the Woodward–Hoffmann rules. I could not claim deep technical fluency in organic chemistry, quantum mechanics, or group theory. But the subject sits at the intersection of all three, and presenting it allowed me to demonstrate a basic grasp of each—and, more importantly, a genuine curiosity about how they connect. It was the willingness to engage with an idea that crossed disciplinary boundaries, and the sense of wonder that comes from discovering a new connection between seemingly disparate fields, that ultimately led to my acceptance.

For some reason, mirrors flip the image left and right, but not up and down.
See the book **The Ambidextrous Universe** (1964) by M. Gardner.

Orbital Symmetry

Symmetry in the Schrödinger Framework

The Schrödinger equation $\hat{H}\psi = E\psi$ governs the electronic structure of molecules. When the molecular Hamiltonian $\hat{H}$ commutes with a symmetry operator $\hat{S}$, the system's eigenfunctions must reflect that symmetry: $[\hat{H}, \hat{S}] = 0 \Rightarrow \hat{S}\psi = \lambda\psi$. This imposes a conserved quantum label (irreducible representation) on the wavefunction throughout any geometry-preserving deformation. In a concerted reaction such as a pericyclic transformation, where all bond changes occur in a cyclic, symmetry-retaining transition state, this leads to a constraint: only reactions that preserve orbital symmetry continuity are allowed. This is the foundation of the Woodward–Hoffmann rules.

Phase Symmetry and Frontier Orbitals

The molecular orbitals (MOs) of conjugated systems can be described as linear combinations of atomic p orbitals. For a linear polyene with n p orbitals, the kth MO has the form $\Psi_k = \sum_{j=1}^{n} \sin(\pi k j/(n+1))\,\mathrm{p}_j$, where p_j are orthogonal atomic orbitals. The phase of the terminal lobes in the HOMO (highest occupied MO) determines the allowed mode of bond formation. For example:

- Butadiene (4 π electrons): HOMO has opposite terminal phases → conrotatory closure aligns lobes → allowed thermally.
- Hexatriene (6 π electrons): HOMO has same terminal phases → disrotatory closure preserves overlap → allowed thermally.

These rules emerge not from empirical fits but from the symmetry character of the MOs under conserved operations (like a C_2 axis or mirror plane in the transition state).

General Selection Rules

Pericyclic selection rules can be framed using the Möbius–Hückel approach (Heilbronner–Zimmerman). A concerted transition state with Hückel topology (even number of phase inversions) is thermally allowed for $4q + 2$ electrons and forbidden for $4q$; with Möbius topology (odd number of phase inversions) the situation reverses (thermally allowed for $4q$, forbidden for $4q + 2$). Under photochemical activation, these parities invert. This framework consistently reproduces the canonical outcomes for electrocyclic reactions, cycloadditions, and sigmatropic shifts.

Correlation Diagrams and Symmetry Conservation

A more formal approach uses correlation diagrams, where each MO is labelled by its symmetry character under a conserved symmetry operation (e.g., S for symmetric, A for antisymmetric). The MOs of reactants and products are then connected across the reaction coordinate. For butadiene: Ψ_1 (A), Ψ_2 (S); for cyclobutene: π (A), σ (S). Under a C_2 axis (conrotatory path), the symmetry labels match, and the transformation preserves orbital occupation → allowed. Under a mirror plane (disrotatory path), the correlation fails (occupied orbital would map to unoccupied antibonding orbital) → forbidden.

Sigmatropic Shifts and Topological Classifications

Sigmatropic shifts involve the migration of a σ-bonded atom across a delocalized π system. The cyclic transition state contains the migrating group plus the π system—usually a 6-electron or 4-electron arrangement. For $[i, j]$ shifts, thermal selection depends on topology: a suprafacial $[i, j]$ shift is allowed when $i + j = 4q + 2$, whereas an antarafacial $[i, j]$ shift is allowed when $i + j = 4q$. Thus, a $[1, 5]$-hydrogen shift is allowed suprafacially (6 electrons), while a $[1, 3]$ shift (4 electrons) is thermally allowed only in an antarafacial topology, which is usually sterically blocked.

References:

Woodward, R. B., and Hoffmann, R. (1970). *The Conservation of Orbital Symmetry*. Addison-Wesley.

Matter of Perspective

Top (Unruh effect): Even in vacuum, observers may disagree on particle content. An accelerating observer detects a warm bath of particles where an inertial observer sees none.
Middle (The Andromeda Paradox and Relativity of Simultaneity): Two observers at the same point on Earth—one stationary, one walking—will disagree on what is happening right now (as seen from right here) in distant galaxies. A moon event (such as an egg breaking) can be 'now' for the stationary, and 'not yet' for the runner.
Bottom (Quantum Fields and Observer-Dependent Particles): What counts as a particle depends on the observer. Each pink sphere represents a local quantum field mode, vibrating against the vacuum. But acceleration or curvature shifts the vacuum state, so an observer may see particles where another sees none—like Unruh radiation or Hawking emission.

Matter of Perspective

Empty space isn't empty—and even that depends on who's looking. The quantum vacuum teems with field fluctuations, but two observers can fundamentally disagree about whether particles exist. An astronaut floating peacefully sees perfect vacuum. Her twin, accelerating through the same region, is bombarded by thermal radiation at temperature $T = a/2\pi ck_B$—the Unruh effect. Particle content becomes relative, like simultaneity in Einstein's relativity.

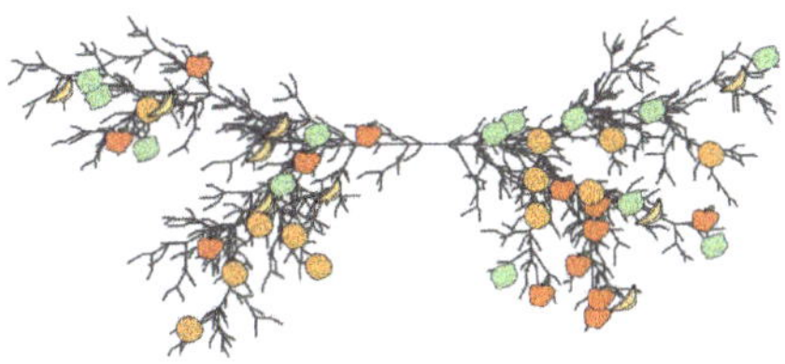

Observer-Dependent Vacuum ◦ Quantum Field Theory ◦ Curved Spacetime Effects ◦ Bogoliubov Transformations ◦ Unruh Effect ◦ Hawking Radiation ◦ Cosmological Particle Production ◦ Dynamical Casimir Effect ◦ Relative Particle Number ◦ Killing Vector Fields ◦ Vacuum State Ambiguity

«Vacuum voco locum omnem
in quo corpora sine resistentia movetur.»
("Vacuum I call every place
in which a body is able to move without resistance.")
— Sir Isaac Newton, 1713

«Repugnare ut detur vacuum,
sive in quo nulla plane sit res.»
("That a vacuum or space in which there is
absolutely no body is repugnant to reason.")
— René Descartes, 1644

Matter of Perspective

The concept of vacuum has been contested since antiquity. Aristotle rejected the possibility of void—nature abhors a vacuum—and medieval natural philosophy largely followed. Torricelli's 1643 experiment demonstrated that a sustained void could exist above a column of mercury, but the space was considered merely empty of air, not of all physical influence. Newton's mechanics required absolute space as a reference frame but assigned no physical properties to empty regions. The nineteenth-century luminiferous aether filled the vacuum with an elastic medium to carry light, but the Michelson-Morley experiment (1887) and Einstein's special relativity (1905) eliminated the aether without replacing it. Quantum mechanics reopened the question. Dirac's 1930 electron theory predicted that the vacuum teemed with an infinite sea of negative-energy electrons—the Dirac sea—whose holes appeared as positrons. Casimir's 1948 prediction that two uncharged conducting plates would attract due to restricted vacuum modes, confirmed experimentally by Lamoreaux in 1997, established that the quantum vacuum exerts measurable forces.

By the time quantum field theory matured on curved backgrounds, the vacuum had already shed any pretence of emptiness. In 1966–69, Leonard Parker showed that the expansion of the universe itself could create particles from the vacuum, the first rigorous demonstration that a time-dependent spacetime geometry converts vacuum fluctuations into real excitations. In 1973, Stephen Fulling proved that the definition of 'particle' depends on the observer's choice of quantisation, while Paul C. W. Davies and William G. Unruh showed that a uniformly accelerated observer in flat spacetime perceives a thermal bath where an inertial observer sees none. In 1974, Stephen Hawking extended these ideas to black holes, deriving that they emit thermal radiation at a temperature inversely proportional to their mass. The result stunned even Hawking—he had set out to disprove it. Jacob Bekenstein had already argued on information-theoretic grounds that black holes must carry entropy proportional to their horizon area, but without a temperature the claim seemed formal. Hawking's calculation supplied the temperature, and with it the entropy became physical: $S = k_B c^3 A/(4G\hbar)$. The four laws of black hole mechanics, catalogued by Bardeen, Carter, and Hawking in 1973 as a mathematical analogy to thermodynamics, turned out to be thermodynamics.

The implications extended beyond black holes. If particle content depends on the observer, then so does the vacuum energy density that drives cosmological expansion, complicating the already severe cosmological constant problem. The Schwinger effect (1951), in which a sufficiently strong electric field pulls electron-positron pairs from the vacuum, provided an independent line of evidence that the vacuum is a reservoir from which real particles can be extracted given enough energy. Birrell and Davies synthesised these threads in their 1982 monograph *Quantum Fields in Curved Space*, which remains the standard reference. By the late twentieth century, the vacuum had been transformed from the simplest concept in physics—nothing—into one of the most contested.

The term 'vacuum' has distinct meanings across classical physics, quantum field theory, and general relativity. In classical physics, vacuum refers to the absence of material particles—a region of space devoid of atoms, molecules, or macroscopic matter. The classical vacuum is an empty stage on which forces act.

In quantum field theory (QFT), the notion of vacuum acquires a different character. Here, fields—not particles—constitute the primary entities. The vacuum is the ground state of all quantum fields, the configuration of lowest possible energy consistent with the commutation relations and field dynamics. Even when no particles are present, quantum fields fluctuate around their minima, giving rise to nonzero vacuum expectation values for certain observables. These fluctuations are not artefacts of measurement or disturbance; they are features of the quantum equations themselves. In Minkowski spacetime, the vacuum is Lorentz invariant. No preferred direction or frame exists, and the absence of particles is an absolute property shared by all inertial observers.

However, in general relativity (GR), spacetime is no longer a fixed, flat background. It becomes a dynamical entity whose curvature interacts with matter and energy. The introduction of curved spacetime disrupts the global symmetries that underlie the inertial vacuum of QFT. In regions of strong gravitational fields or global curvature, there is generally no unique, globally defined vacuum state. Instead, the concept of vacuum becomes observer-dependent. Different families of observers may disagree about whether a given region of spacetime is populated by particles. This relativity of the vacuum holds because the definition of positive frequency modes—those corresponding to particle excitations—depends on the choice of time coordinate, which itself is tied to the observer's worldline. Consequently, what one observer identifies as an empty vacuum, another observer may interpret as a state containing particles, momentum, or thermal radiation.

In quantum field theory, the notion of a particle is defined relative to a choice of time. In flat Minkowski spacetime, all inertial observers share the same time direction (up to Lorentz boosts that preserve the sign of frequency). This gives a unique way to split field solutions into positive-frequency modes (time dependence $e^{-i\omega t}$, $\omega > 0$) and negative-frequency modes. Creation and annihilation operators are built from this split, and the vacuum is the state annihilated by all annihilation operators. Because inertial observers agree on which frequencies are positive, they agree on particle number.

In curved spacetimes or for accelerating observers, no universal time direction exists. Different observers, following different trajectories, adopt different time coordinates and decompose the field into different sets of modes. What one observer identifies as a positive-frequency mode may contain negative-frequency components according to another. Since positive frequency defines particles and negative frequency defines antiparticles, the two observers disagree on how many particles are present.

Mathematically, if one observer expands the field $\hat{\phi}(x)$ in terms of a basis of modes $\{f_i(x)\}$, while another observer uses a different basis $\{g_j(x)\}$, the two expansions are related by a Bogoliubov transformation. This transformation mixes creation and annihilation operators, leading to the possibility that the vacuum state for one observer appears populated with particles to another. If the Bogoliubov coefficients β_{ij} are nonzero, the expected particle number for mode j in the g-basis, measured in the f-vacuum, is $\langle 0_f|\hat{N}_{g,j}|0_f\rangle = \sum_i |\beta_{ij}|^2$, and the total count is $\langle 0_f|\hat{N}_g|0_f\rangle = \sum_{i,j} |\beta_{ij}|^2$.

The number operator $\hat{N} = \sum_j \hat{a}_j^\dagger \hat{a}_j$ counts excitations above a given vacuum. It is constructed from creation and annihilation operators, which themselves depend on the choice of mode basis. Change the basis—by switching to an accelerating frame or a different time slicing—and the operator changes. Two observers examining the same quantum state with their respective number operators obtain different expectation values. This is not a coordinate artefact—a particle detector carried by an accelerating observer registers real energy deposits while an inertial detector in the same region remains silent.

A particle, in this framework, is not a fundamental object embedded in spacetime. The quantum field is fundamental—it pervades all of spacetime and its state is agreed upon by all observers. A particle is what a detector registers when coupled to the field along a specific worldline. The Unruh-DeWitt model makes this precise. It describes a pointlike system with discrete energy levels that follows a trajectory through spacetime, interacting locally with the field at each point along its path. The excitation rate depends entirely on the trajectory. Along an inertial path in the Minkowski vacuum, the detector remains in its ground state, but along a uniformly accelerating path through the same field state, it absorbs energy and clicks. The same field, the same quantum state, two different particle counts.

Classical wave physics offers a partial analogy. A moving listener hears a Doppler-shifted frequency, but the wave carries the same energy—no new physics follows from the listener's motion. In quantum field theory, frequency decomposition is more consequential. Positive-frequency modes define particles; negative-frequency modes define antiparticles. When an observer's trajectory mixes these—as acceleration does—modes that one observer calls vacuum become, for the other, a superposition containing particles. The accelerating detector absorbs real energy, thermalises, and reaches equilibrium at a definite temperature. This energy is drawn from the agency maintaining the acceleration, so the total energy budget remains consistent across frames. The field state has not changed; what has changed is which degrees of freedom the detector accesses.

Yet the quantum state of the field is observer-independent. The state vector $|\Psi\rangle$ in the Hilbert space does not change when observers switch coordinates. What changes is the decomposition of that state into particle and vacuum sectors. The stress-energy tensor $T_{\mu\nu}$, local curvature invariants, and S-matrix elements between asymptotic states also remain invariant. The field equations hold in all frames. What becomes observer-dependent is the particle interpretation—how many quanta a given detector counts—not the underlying field configuration or its gravitational effects.

The Unruh effect (Unruh, 1976) occurs in flat Minkowski spacetime. An inertial observer in this vacuum detects nothing, but an observer undergoing uniform acceleration a through the same region perceives a thermal bath at temperature $T_U = \hbar a/(2\pi c k_B)$. The accelerating observer follows hyperbolic trajectories that asymptotically approach but never cross a Rindler horizon—a boundary in flat spacetime beyond which signals can never reach the accelerating observer, analogous to a black hole horizon. Rindler coordinates, natural to this observer, cover only the wedge of spacetime accessible to them. The Minkowski vacuum, expressed in Rindler modes, decomposes into entangled pairs straddling this horizon. Since the accelerating observer cannot access modes behind the horizon, tracing over those inaccessible modes produces a thermal density matrix. The thermal spectrum is exact for uniform acceleration, not an approximation.

Hawking radiation (Hawking, 1974) arises from an analogous mechanism at black hole horizons. An observer falling freely across the horizon encounters no particles—locally, the equivalence principle guarantees that spacetime looks flat, and the vacuum looks empty. A distant stationary observer, however, perceives thermal radiation at temperature $T_H = \hbar c^3/(8\pi G M k_B)$, where M is the black hole mass. The event horizon separates spacetime into causally disconnected regions. Field modes straddling the horizon are entangled across it. The distant observer, who can access only the exterior, must trace over the interior modes. The resulting reduced state is thermal, with its spectrum set by the surface gravity. The connection to the Unruh effect is direct: a stationary observer hovering near a black hole must accelerate to avoid falling in, and the temperature they measure matches the Unruh temperature for that acceleration. At large distance, the gravitational redshift reduces this to the Hawking temperature.

Cosmological particle production occurs in expanding spacetimes. During inflation, the rapid exponential expansion of space stretches quantum vacuum fluctuations to macroscopic scales. A mode that begins as a short-wavelength fluctuation well inside the Hubble radius gets stretched until its wavelength exceeds the horizon size. At that point the mode freezes—it can no longer oscillate because no causal process spans its wavelength—and its quantum amplitude becomes a classical perturbation in the energy density. After inflation ends and the expansion decelerates, these frozen modes re-enter the horizon and seed the density perturbations that grow into galaxies and galaxy clusters. The temperature anisotropies in the cosmic microwave background, measured by COBE (1992), WMAP (2003), and Planck (2013), match the nearly scale-invariant spectrum predicted by this mechanism. The particle interpretation follows the same pattern as the Unruh and Hawking cases: the rapid change in the metric mixes positive and negative frequencies, and an observer at a later epoch counts particles where an earlier observer counted none.

The Casimir effect (Casimir, 1948) demonstrates that boundary conditions alter the vacuum. Two parallel conducting plates restrict which field modes can exist between them, lowering the vacuum energy density relative to the space outside. The resulting pressure difference pushes the plates together—a measurable force produced by the vacuum itself. In the dynamical variant, the plates accelerate or their separation oscillates, and the changing boundary conditions parametrically amplify vacuum fluctuations into real photons. Laboratory realisations using superconducting circuits with time-modulated boundaries have confirmed this photon creation.

Hawking Radiation

> By 1972, black hole mechanics followed formal laws resembling thermodynamics: changes in mass, angular momentum, and charge obeyed a relation involving surface gravity κ and horizon area A, with κ positioned where temperature would appear and A where entropy belongs. Yet classical general relativity permitted no emission, implying zero temperature. The analogy appeared to be coincidence.

Jacob Bekenstein argued otherwise. On information-theoretic grounds, black holes must possess entropy proportional to horizon area. Hawking rejected this—without an emission mechanism, a black hole could not have true temperature.

In 1973, Hawking visited Moscow and met Yakov Zel'dovich and Alexei Starobinsky, who had shown that rotating bodies could amplify vacuum fluctuations—superradiance. Hawking set out to prove gravitational collapse could not produce particles.

He traced field modes through the spacetime of a collapsing star. Outgoing modes near the forming horizon undergo exponential redshift, mixing positive and negative frequencies. The Bogoliubov coefficients yield a Planck spectrum at temperature $T_H = \hbar c^3/(8\pi GMk_B)$. Hawking spent weeks rechecking, yet the result persisted.

Hawking published in *Nature* as Black hole explosions? (1974), with a full derivation in *Communications in Mathematical Physics* (1975). The result introduced the information paradox: thermal radiation carries no trace of what fell in, contradicting unitarity.

The popular picture of virtual particle pairs splitting at the horizon does not appear in Hawking's derivation. The radiation follows from the global causal structure of the collapsing spacetime and the entanglement of field modes across the horizon. It has never been directly observed—for a solar-mass black hole, $T_H \approx 6 \times 10^{-8}$ K—though analogue experiments in BECs, fluids, and superconducting circuits have reproduced consistent features.

Feature	Unruh Effect	Hawking Radiation	Cosmological Particle Creation	Schwinger Effect	Dynamical Casimir Effect
Physical Context	Uniform acceleration in flat spacetime	Black hole event horizon	Expanding FLRW universe	Strong external electric field	Time-dependent boundary conditions
Primary Mechanism	Bogoliubov transformation (Minkowski $\leftrightarrow$ Rindler)	Horizon mode mismatch and equivalence principle	Mode stretching and horizon crossing	Vacuum instability via tunnelling	Parametric amplification of vacuum fluctuations
Key Dependence	Proper acceleration a	Black hole mass M, surface gravity	Hubble rate H, coupling strength	Electric field strength E	Modulation frequency and boundary speed
Characteristic Scale / Formula	$T_U = \dfrac{\hbar a}{2\pi c k_B}$	$T_H = \dfrac{\hbar c^3}{8\pi GMk_B}$	Particle density $\propto H^2$	$\Gamma \propto \exp\left(-\dfrac{\pi E_c}{E}\right)$	Photon production peaks at $\omega_{\text{mod}} \approx 2\omega_{\text{cav}}$
Experimental Probes	Analogues: BECs, trapped ions, superconducting circuits	Analogues (BECs, fluids); black hole thermodynamics	CMB anisotropies, primordial fluctuations	High-intensity lasers, graphene lattices	SQUID-based circuits, modulated cavities

Comparative overview of observer-dependent vacuum phenomena. Each column represents a distinct physical context in which the concept of vacuum, and thus of particle content, becomes relative to the observer's state of motion or horizon access.

Technical Derivation of the Unruh Effect

The Unruh effect manifests within the framework of quantum field theory in flat Minkowski spacetime. Consider a massless scalar field $\hat{\phi}(x)$ governed by the Klein–Gordon equation

$$\Box\hat{\phi}(x) = 0,$$

where $\Box$ denotes the d'Alembertian operator associated with the Minkowski metric $\eta_{\mu\nu}$, with signature $(-,+,+,+)$. Explicitly,

$$\Box = -\frac{\partial^2}{\partial t^2} + \frac{\partial^2}{\partial z^2} + \frac{\partial^2}{\partial x^2} + \frac{\partial^2}{\partial y^2}.$$

An inertial observer describes spacetime using Cartesian Minkowski coordinates (t, z, x, y), in which the line element is

$$ds^2 = -dt^2 + dz^2 + dx^2 + dy^2.$$

The field $\hat{\phi}(x)$ is quantized by expanding in terms of plane wave modes that are eigenfunctions of the time translation operator ∂_t, exploiting the global timelike Killing vector field ∂_t of Minkowski spacetime.

Uniformly accelerated observers, however, do not naturally perceive the Minkowski time t as their proper time. Instead, their worldlines trace hyperbolic trajectories characterised by constant proper acceleration α. These trajectories are described by

$$z^2 - t^2 = \alpha^{-2}.$$

To describe the experience of such observers, it is natural to introduce Rindler coordinates (η, ξ, x, y), defined by the transformations

$$\begin{aligned} t &= \xi\sinh(a\eta), \\ z &= \xi\cosh(a\eta), \end{aligned} \quad \text{with} \quad \xi > 0, \eta \in \mathbb{R},$$

where a is an arbitrary constant with dimensions of inverse length, conventionally chosen so that η has dimensions of time.

Substituting into the Minkowski line element yields

$$\begin{aligned} dt &= a\xi\cosh(a\eta)d\eta + \sinh(a\eta)d\xi, \\ dz &= a\xi\sinh(a\eta)d\eta + \cosh(a\eta)d\xi, \end{aligned}$$

so that

$$-dt^2 + dz^2 = -(a\xi)^2 d\eta^2 + d\xi^2. \quad .$$

Thus, the Minkowski metric becomes

$$ds^2 = -(a\xi)^2 d\eta^2 + d\xi^2 + dx^2 + dy^2.$$

The coordinate ξ measures the proper distance from the Rindler horizon located at $\xi = 0$, and η serves as the observer's proper time scaled by a^{-1}. The proper acceleration α experienced by an observer at fixed ξ satisfies $\alpha = 1/\xi$.

Hence, smaller ξ corresponds to larger proper acceleration.

It is important to note that the Rindler coordinates (η, ξ) cover only a subset of Minkowski spacetime, specifically the right Rindler wedge defined by $z > |t|$.

The surface $\xi = 0$, corresponding to $z = |t|$, acts as a causal boundary: signals from beyond this horizon cannot reach the accelerated observer. This causal restriction implies that uniformly accelerated observers perceive only part of the global spacetime, fundamentally altering their notion of vacuum and particle content.

The hyperbolic trajectories of constant ξ correspond to observers moving with constant proper acceleration $\alpha = 1/\xi$, whose four-velocity u^μ and four-acceleration a^μ satisfy

$$u^\mu u_\mu = -1, \quad a^\mu a_\mu = \alpha^2.$$

The presence of a causal horizon and the distinct mode structure in Rindler coordinates underlie the emergence of the Unruh effect, which will now be derived by solving the field equations in this coordinate system.

References:

M. Socolovsky, Rindler Space and the Unruh Effect, arXiv:1304.2833 [gr-qc], 2013.

Chaotic Neutrality

Chaotic Parts, Predictable Whole: Deterministic systems can be chaotic—where tiny changes snowball into wildly different results. Each dot on this falling object traces a unique, tangled trajectory, sensitive to initial conditions. Yet the object as a whole falls smoothly. Dissipative forces such as friction and internal damping suppress chaos, causing the centre of mass to follow a clean, Newtonian arc.

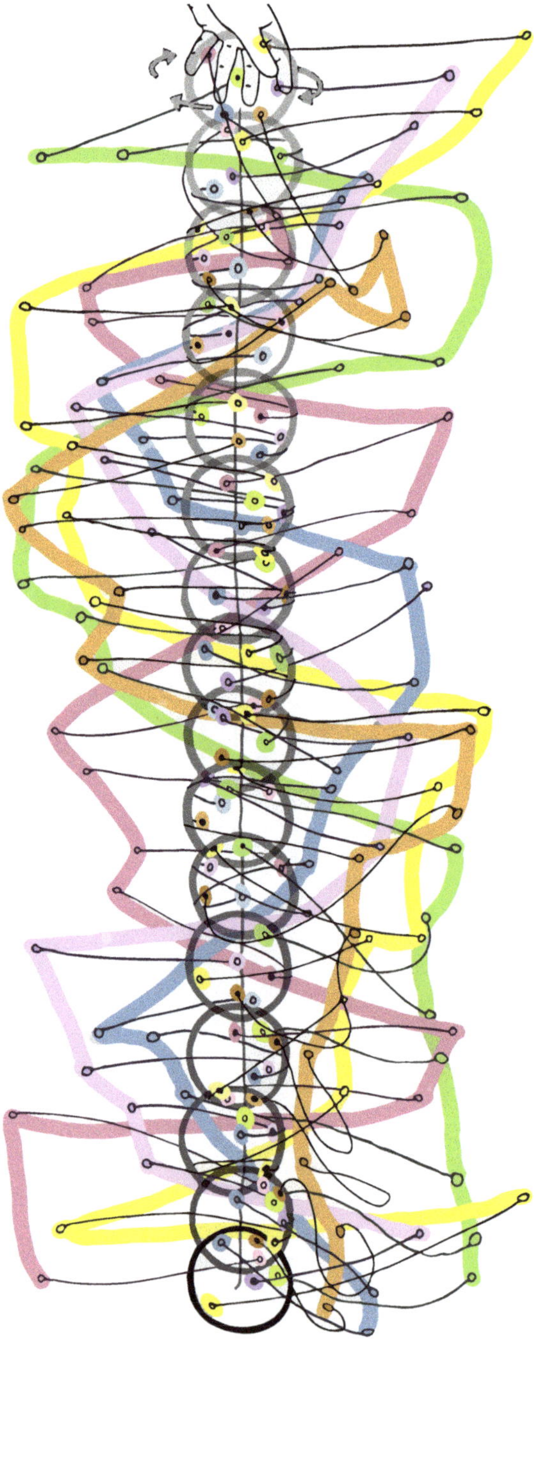

Chaotic Neutrality

Deterministic systems can exhibit chaotic behaviour, where minuscule differences in initial conditions lead to drastically different outcomes. The double pendulum and three-body gravitational problem exemplify this despite having few components and simple equations. Counterintuitively, far more complex systems such as falling objects often behave predictably because dissipative effects continuously suppress perturbations.

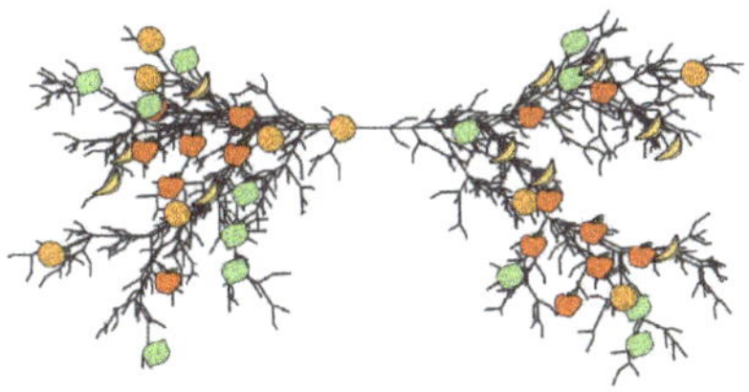

Three-Body Problem ◦ Poincaré's Chaos Discovery ◦ Deterministic Unpredictability ◦ Sensitive Initial Conditions ◦ Double Pendulum ◦ Butterfly Effect ◦ Complex System Stability ◦ Dissipation vs Amplification ◦ Falling Apple Paradox ◦ Prediction Horizons ◦ Chaos Theory

« Pourquoi les chutes de pluie, les tempêtes elles-mêmes nous semblent-elles arriver au hasard, de sorte que bien des gens trouvent tout naturel de prier pour avoir la pluie ou le beau temps, alors qu'ils jugeraient ridicules de demander une éclipse par une prière? »

("Why is it that showers and storms seem to come by chance, so that many think it natural to pray for rain, though they would consider it ridiculous to ask for an eclipse by prayer?")

— Henri Poincaré, 1908

"Let the Lord of Chaos rule."

— Semirhage, 1000 NE

Chaotic Neutrality

The recognition that deterministic systems can exhibit unpredictable behaviour marked a change in scientific thought. Before the twentieth century, classical mechanics—embodied by Newton's laws—was largely viewed as a complete and exact framework: given initial conditions, future behaviour was presumed computable in principle.

This view was first challenged by Henri Poincaré's entry to King Oscar II's prize competition in 1888. Poincaré submitted a memoir claiming to prove stability of certain three-body configurations. The work was accepted and the prize awarded. But before publication, his colleague Lars Edvard Phragmén found an error in a key argument. The correction reversed the conclusion: the orbits Poincaré had declared stable were in fact wildly unstable. He paid for the recall and reprinting of the original memoir—the cost exceeded the prize money—and the corrected version (1890) became the founding document of what would later be called chaos theory. The very calculation meant to demonstrate cosmic predictability had instead uncovered its impossibility.

In the decades that followed, these ideas lay mostly dormant until the rise of computers in the mid-twentieth century allowed for detailed numerical explorations of nonlinear systems. In 1963, Edward Lorenz demonstrated that a set of three differential equations meant to model atmospheric convection could yield drastically different outcomes from imperceptibly different starting points when he re-ran a simulation with rounded initial conditions. This sensitivity, later termed the 'butterfly effect,' became the signature of what we now call chaos.

Rather than disorder or randomness, chaos refers to the intrinsic unpredictability found in certain deterministic systems. It highlighted the limitations of prediction when geometry of the solution space allows tiny differences to grow exponentially, defying long-term computation even in conceptually simple scenarios.

In 1885, King Oscar II of Sweden sponsored a prize competition for solving the gravitational three-body problem–determining the motion of three masses under mutual gravitational attraction. Henri Poincaré ultimately won the prize not by finding a closed-form general solution, but by demonstrating deep instability and nonintegrability in the problem (Poincaré, 1890). Three gravitating bodies follow Newton's laws precisely, yet their long-term behaviour defies prediction. Exact periodic solutions do exist–Euler found collinear configurations in 1767, Lagrange found equilateral triangle orbits in 1772, and Chenciner and Montgomery (2000) proved the existence of a figure-eight orbit in which three equal masses chase each other along a single closed curve–but these are measure-zero islands in a chaotic sea, unstable to perturbation and never observed in nature.

The three-body problem exemplifies a deeper conundrum. Newton's equations are deterministic–they specify exactly how a system evolves from any given starting point. No randomness enters the calculations. Yet for three or more gravitating bodies, these deterministic equations generate behaviour so sensitive to initial conditions that a difference of one part in a trillion in starting positions leads to entirely different orbital configurations after sufficient time. The future is determined but not determinable.

 https://doi.org/10.11647/OBP.0526.48

This sensitivity is intrinsic to the equations, not a numerical artefact. No measurement apparatus can specify initial conditions with infinite accuracy, and no simulation can represent real numbers exactly. The discovery overturned Laplace's vision of a sufficiently powerful intellect (Laplace, 1814) that, knowing the precise positions and *velocities* of all particles in the universe, could calculate the entire future and past. In a chaotic system, trajectories that begin infinitesimally close diverge exponentially—not because the rules are imprecise, but because the phase space amplifies initial discrepancies rather than suppressing them. Small uncertainties, inevitable in any real situation, are magnified until they dominate behaviour at later times.

Phase space provides the geometric arena where dynamics unfold. For a system with N degrees of freedom, phase space has dimension $2N$, each point specifying all positions and velocities. A trajectory through phase space encodes the system's evolution. Determinism means that through each point passes exactly one trajectory.

Chaos quantifies through the Lyapunov exponent (Lyapunov, 1892) λ, which measures how rapidly nearby trajectories separate. Consider two initial states separated by distance δ_0. After time t, the separation grows to $\delta(t) \approx \delta_0 e^{\lambda t}$, where λ is the Lyapunov exponent. Positive λ signals chaos: trajectories diverge exponentially. Negative λ indicates stability: trajectories converge. The magnitude of λ sets the timescale for prediction. If $\lambda = 0.5$ per day, an initial uncertainty of one part in a billion grows to one part in a million after 14 days, then one part in a thousand after 28 days.

For the double pendulum, typical Lyapunov exponents are on the order of the inverse oscillation period. For Earth's orbit around the Sun, the Lyapunov exponent is approximately $(5 \times 10^6 \text{ years})^{-1}$, limiting detailed predictability to a few million years despite the apparent regularity observed over human timescales. Weather systems have Lyapunov exponents near $(2 \text{ days})^{-1}$, establishing the practical limit on forecast accuracy.

Chaotic systems often possess strange attractors: sets in phase space toward which trajectories converge but on which they wander chaotically. The Lorenz attractor, discovered in weather modelling (Lorenz, 1963), resembles a butterfly with two lobes. Trajectories circle one lobe unpredictably many times before switching to the other, never settling into periodic motion. These attractors have fractal dimension—they occupy more space than a curve but less than a surface. A three-dimensional system might have an attractor with dimension 2.06, indicating that trajectories explore more than a surface but do not fill the full space. To specify a point on the attractor to within ϵ requires roughly ϵ^{-D} distinguishable cells, where D is the fractal dimension.

The boundary between regular and chaotic motion can be razor-thin. Consider the restricted three-body problem, where a small mass moves in the gravitational field of two large masses orbiting their common centre. For certain initial conditions, the small mass traces out stable, repeating orbits; for example, motion near the equilateral Lagrange points L4/L5 can be stable for favourable mass ratios, whereas halo and Lissajous orbits near L1/L2 used by spacecraft require active stationkeeping because those equilibria are unstable. But tiny perturbations can push the system across an invisible boundary into chaos. The same equations that produce clockwork regularity in one region of phase space generate unpredictability in adjacent regions.

The Kolmogorov-Arnold-Moser theorem formalises this coexistence (Kolmogorov, 1954). For nearly-integrable Hamiltonian systems, most trajectories remain confined to invariant tori: surfaces in phase space on which motion is quasi-periodic and stable. Small perturbations deform but do not destroy these tori. However, gaps exist where the tori break apart, creating a fractal web of chaotic trajectories intertwined with islands of stability. As perturbations increase, more tori disintegrate, expanding the chaotic sea. The Solar System exists in this mixed regime: most planetary orbits lie on stable tori and will persist for billions of years, but resonances create chaotic regions where asteroids wander unpredictably before ejection or collision. Mercury's orbit is the most chaotic of the planets, with a Lyapunov time of roughly 5 million years. Numerical integrations by Laskar (1989) show a probability of approximately 1% that Mercury will collide with Venus or be ejected from the Solar System within the next 5 billion years.

Weather systems exemplify chaos on a planetary scale. Lorenz's 1961 discovery began with a rounding error: re-entering a state variable as 0.506 instead of 0.506127 caused his simulated weather to diverge completely after a few days of model time. The atmosphere obeys fluid dynamics equations deterministically, but the nonlinear interactions between pressure, temperature, and velocity fields amplify microscopic uncertainties into macroscopic unpredictability. This 'butterfly effect'—the notion that a butterfly flapping its wings in Brazil could trigger a tornado in Texas—illustrates chaotic amplification.

Chaotic behaviour can arise in both conservative and dissipative systems. In Hamiltonian (conservative) dynamics, phase-space volume is preserved and small perturbations are stretched and folded, producing exponential separation without energy loss. In dissipative systems such as the Lorenz model, volume contracts and trajectories approach strange attractors, yet sensitivity to initial conditions persists. What matters is the nonlinear dynamical architecture that determines whether small differences are amplified or suppressed, not merely the presence or absence of damping.

Dimensionality shapes the possibility of chaos. Autonomous Hamiltonian systems with one degree of freedom cannot be chaotic: a trajectory in two-dimensional phase space cannot cross itself and energy conservation confines motion to a one-dimensional curve. More generally, continuous-time flows in two dimensions cannot exhibit chaos (Poincaré–Bendixson theorem), whereas discrete-time maps can be chaotic even in one dimension. With two degrees of freedom, phase space has four dimensions, and energy conservation reduces accessible phase space to three dimensions. Trajectories can now weave around one another without crossing, creating the tangled topology necessary for chaos.

Yet higher dimensions need not amplify chaos. Poincaré recurrence guarantees that conservative systems in bounded phase space eventually return arbitrarily close to their initial state. The recurrence time, however, grows extremely rapidly with dimension (often exponentially in simple models). A three-body system might recur after billions of years. A gas of 10^{23} molecules confined to a box would take a time vastly exceeding the age of the universe to return even approximately to its initial microstate. High dimensionality converts mathematical recurrence into physical irreversibility.

The contrast between simple chaotic systems and complex stable systems is sharp and puzzling. A double pendulum exhibits unpredictable behaviour after a few oscillations.

Yet a falling apple, composed of approximately 10^{26} atoms, moves through turbulent air and an ever-changing environment with predictability. Internally, the apple undergoes continuous atomic vibrations, thermal fluctuations, and structural deformations. Externally, it interacts with a turbulent atmosphere, random gusts of wind, and small fluctuating forces from air pressure and temperature gradients. Each interaction, taken in isolation, could introduce deviations from an idealised path. Nevertheless, the macroscopic motion remains stable and predictable, governed by simple equations of motion augmented by modest drag corrections. How can a system with billions of internal degrees of freedom be stable, while a system with two degrees of freedom is chaotic?

Complex systems contain dissipative and averaging effects. As a falling apple moves through air, it experiences drag forces that steadily remove kinetic energy. Internally, vibrations and deformations distribute energy among a vast number of microscopic degrees of freedom. Dissipative processes suppress small perturbations introduced by turbulence or internal noise rather than amplifying them. Energy lost to friction, drag, and internal vibration prevents the growth of deviations that would otherwise destabilise the macroscopic trajectory.

The motion of the apple's centre of mass contributes to stability. Although individual atoms exhibit random motion, their collective behaviour averages out. Fluctuations at the microscopic level do not accumulate coherently to shift the overall path. They cancel statistically, leaving the centre of mass to follow a trajectory governed by external forces such as gravity and aerodynamic effects. The system's vast internal complexity insulates the macroscopic motion from microscopic uncertainty.

This statistical stability follows from the central limit theorem applied to phase space dynamics. With N particles, each contributing a small random displacement to the centre of mass, the net fluctuation scales as $\sqrt{N}$ rather than N. For an apple with 10^{26} atoms, the relative fluctuation in centre-of-mass position is suppressed by a factor of 10^{13}. Microscopic chaos becomes irrelevant to macroscopic motion because these fluctuations average incoherently across vast numbers of degrees of freedom.

High-dimensional phase spaces partition into macrostates and microstates. A macrostate specifies coarse-grained properties: the apple's position, velocity, temperature. Each macrostate corresponds to an enormous number of microstates: the precise positions and velocities of all 10^{26} atoms. Macroscopic observables depend only on bulk properties averaged over microstates, washing out the chaotic sensitivity that would dominate if we tracked individual atoms. The apple falls predictably not because its atomic dynamics are simple, but because 10^{26} chaotic degrees of freedom conspire—through statistical averaging—to produce stable collective motion.

The distinction is geometric. In chaotic systems such as the double pendulum, the equations preserve and amplify differences—small deviations feed forward unchecked through conservative dynamics. In stable systems like the falling apple, dissipative processes damp sensitivity, dispersing perturbations among many degrees of freedom or losing them to the environment. Predictability depends not on the number of components but on whether the governing equations allow deviations to grow or force them to dissipate.

Beyond mechanics, chaos appears wherever nonlinear dynamics govern evolution: population dynamics, neural firing patterns, financial markets, traffic flow. In each domain, deterministic rules produce behaviour that resists long-term prediction. A population model with simple reproduction and competition terms can generate boom-bust cycles as irregular as any stochastic process. Neural networks with fixed connection strengths produce firing patterns indistinguishable from random noise.

Quantum mechanics introduces a different kind of unpredictability through fundamental uncertainty relations and measurement collapse. But perfectly classical, perfectly deterministic systems generate their own form of irreducible uncertainty through dynamical amplification. The clockwork universe of Laplace fails not at the quantum scale but at the macroscopic scale of planetary orbits and weather systems.

Weather prediction improves with better models and more powerful computers, but fundamental limits remain. Doubling computational power might extend accurate forecasts by a day or two, not by weeks. The chaotic amplification of uncertainties sets an absolute horizon beyond which detailed prediction becomes meaningless. Climate models can hence project average temperatures for decades because they focus on statistical properties rather than specific weather patterns. Asking where a storm will strike three weeks from now exceeds what any conceivable computation could achieve.

Engineering must account for chaos when designing control systems. A satellite's trajectory near a Lagrange point requires constant adjustment because the dynamics balance on the edge between stability and chaos. Small thruster firings maintain the desired orbit against exponential growth of deviations. The control system fights not randomness but the deterministic instability built into the gravitational geometry of the three-body configuration.

Chaos theory affects how we interpret apparent randomness in nature. Irregular heartbeats, previously dismissed as noise, may reflect chaotic dynamics in the cardiac conduction system. Ecosystems that fluctuate despite constant environmental conditions may exhibit deterministic chaos rather than responding to hidden random influences, and the dripping of a faucet transitions from periodic to chaotic as the flow rate increases.

Perturbative Stability and Exponential Sensitivity in Deterministic Systems

Classical mechanics is deterministic: given initial conditions and governing forces, the trajectory of a system is uniquely determined. However, predictability depends on the evolution of perturbations. In chaotic systems, infinitesimal deviations grow exponentially, whereas in many complex but dissipative systems, fluctuations are suppressed or averaged out, leading to reliable large-scale predictions.

1. Lyapunov Exponents and Sensitivity
The maximal Lyapunov exponent λ quantifies sensitivity to initial conditions. For trajectories separated by δ_0, the separation evolves as $\delta(t) \approx \delta_0 e^{\lambda t}$. Formally:

$$\lambda = \lim_{t\to\infty} \lim_{\delta_0 \to 0} \frac{1}{t} \ln \frac{\delta(t)}{\delta_0}.$$

Positive λ: exponential divergence (chaos). Negative λ: exponential convergence (dissipation). Zero λ: marginal stability with polynomial growth.

2. Chaotic Dynamics in a Double Pendulum
Let $\theta_1(t)$ and $\theta_2(t)$ denote the angular positions of a double pendulum with masses m_1, m_2 and rod lengths l_1, l_2. The Lagrangian formulation yields the coupled equations of motion:

$$(m_1 + m_2) l_1 \ddot{\theta}_1 + m_2 l_2 \ddot{\theta}_2 \cos(\theta_1 - \theta_2) = -m_2 l_2 \dot{\theta}_2^2 \sin(\theta_1 - \theta_2) - (m_1 + m_2) g \sin\theta_1,$$

$$m_2 l_2 \ddot{\theta}_2 + m_2 l_1 \ddot{\theta}_1 \cos(\theta_1 - \theta_2) = m_2 l_1 \dot{\theta}_1^2 \sin(\theta_1 - \theta_2) - m_2 g \sin\theta_2.$$

These are second-order nonlinear differential equations with explicit coupling between the degrees of freedom. For many energy regimes, this system exhibits positive Lyapunov exponents: infinitesimally close initial conditions produce trajectories that diverge exponentially in time.

3. Stability in Dissipative Systems
Now consider a falling object subject to linear drag. Assuming the drag force is proportional to velocity, the centre-of-mass motion is governed by

$$m\frac{d^2\mathbf{r}}{dt^2} = m\mathbf{g} - \gamma \frac{d\mathbf{r}}{dt},$$

where γ is the damping coefficient. The velocity $\mathbf{v}(t) = d\mathbf{r}/dt$ evolves as

$$\mathbf{v}(t) = \mathbf{v}_\infty + (\mathbf{v}_0 - \mathbf{v}_\infty) e^{-\frac{\gamma}{m} t} \quad \text{with} \quad \mathbf{v}_\infty = \frac{m\mathbf{g}}{\gamma}.$$

The position follows: $\mathbf{r}(t) = \mathbf{r}_0 + \mathbf{v}_\infty t + (m/\gamma)(\mathbf{v}_0 - \mathbf{v}_\infty)(1 - e^{-\gamma t/m})$. Perturbations decay exponentially with time constant $\tau = m/\gamma$, suppressing initial differences.

4. Perturbation Scaling and Averaging
For a perturbed trajectory $\mathbf{x}_\epsilon(t) = \mathbf{x}_0(t) + \epsilon \delta\mathbf{x}(t)$, the perturbation behaviour distinguishes:

$$\begin{cases} \|\delta\mathbf{x}(t)\| \sim \epsilon e^{\lambda t} & \text{chaotic,} \\ \|\delta\mathbf{x}(t)\| \lesssim \epsilon & \text{damped/bounded.} \end{cases}$$

For N microscopic degrees of freedom, macroscopic observables $\mathbf{X}(t) = N^{-1} \sum_{i=1}^{N} \mathbf{x}_i(t)$ have variance $\mathrm{Var}(\mathbf{X}) \sim N^{-1}\mathrm{Var}(\mathbf{x}_i)$ by the central limit theorem. Thus:

$$\mathbf{X}(t) \approx \langle \mathbf{x}_i(t) \rangle + \mathcal{O}(N^{-1/2}).$$

Fluctuations scale as $\mathcal{O}(N^{-1/2})$, becoming negligible for large N. Microscopic uncertainty remains confined at the macroscopic level.

References:

Poincaré, H. (1890). Sur le problème des trois corps et les équations de la dynamique. *Acta Mathematica*, 13, 1–270.

Lorenz, E. N. (1963). Deterministic nonperiodic flow. *Journal of the Atmospheric Sciences*, 20(2), 130–141.

Feigenbaum, M. J. (1978). Quantitative universality for a class of nonlinear transformations. *Journal of Statistical Physics*, 19, 25–52.

The Three Genome Problem

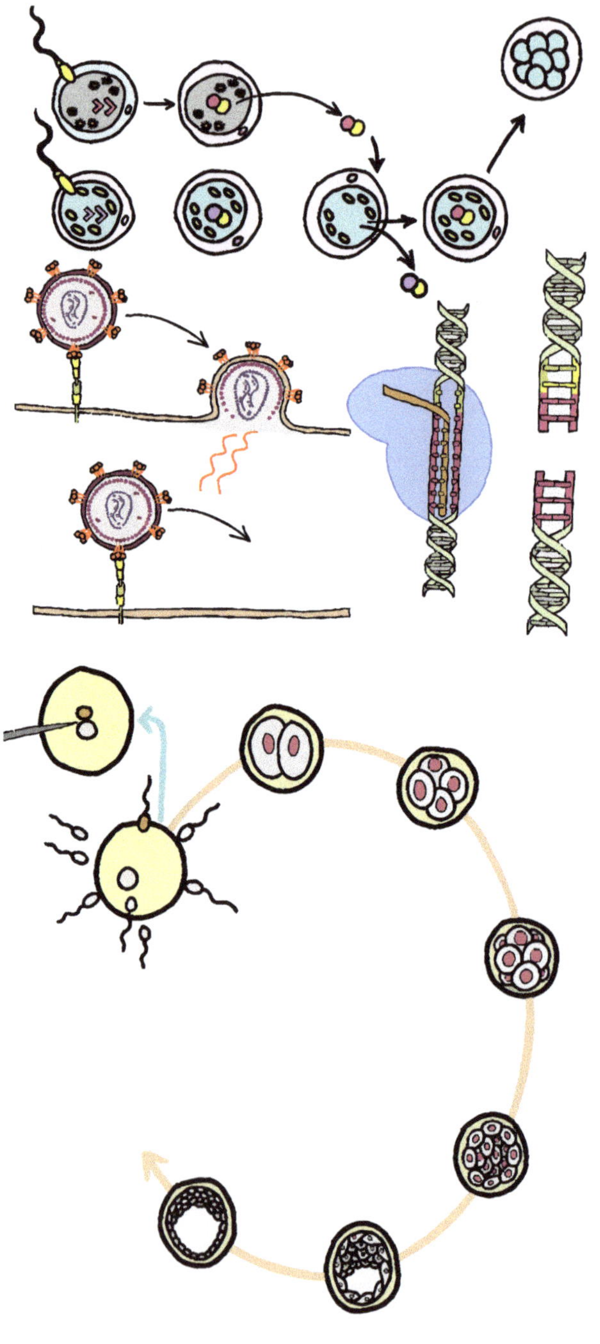

Top (Mitochondrial Replacement Therapy): A fertilised nucleus is inserted into a donor egg with healthy mitochondria, creating an embryo with parental nuclear DNA and donor mtDNA—preventing transmission of mitochondrial disorders.
Middle Left (CCR5 and HIV Resistance): HIV entry requires CCR5. Individuals with the CCR5-Δ32 mutation lack a functional receptor, blocking viral entry. This natural resistance underpins CCR5-targeted therapies.
Middle Right (CRISPR-Cas9 Genome Editing): Cas9 cuts DNA at a targeted site guided by RNA. This enables correction or knockout of specific genes, forming the basis of precise gene therapy.
Bottom Left (Embryonic Germline Editing): CRISPR introduced at the zygote stage edits all descendant cells, including germline—causing permanent, heritable changes. Ethically and legally restricted.
Bottom Right (Embryo Development): Post-fertilisation, the embryo progresses through cleavage stages to blastocyst.

The Three Genome Problem

Every human inherits two distinct genomes: nuclear DNA from both parents and mitochondrial DNA almost exclusively from the oocyte. This second genome—37 genes controlling cellular energy production—mutates 10–100 times faster than nuclear DNA, causing devastating diseases when defective. Traditional IVF cannot prevent mothers from passing faulty mitochondria to children. Enter mitochondrial replacement therapy: scientists transfer nuclear DNA from an affected mother's egg into a donor egg with healthy mitochondria. From single-base edits to chromosome transfers—many ethical questions arise to be discussed.

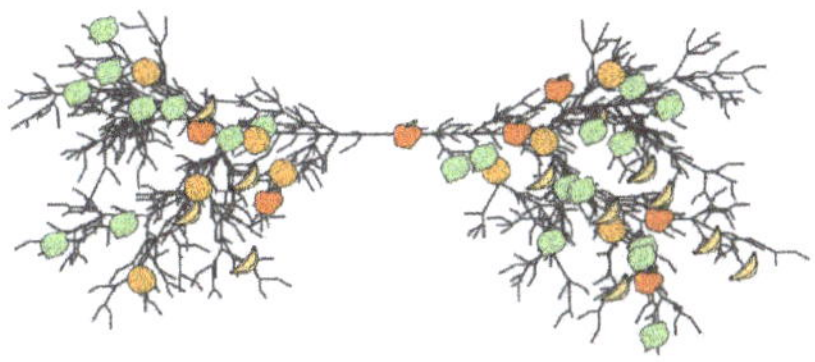

In Vitro Fertilisation ◦ Mitochondrial DNA Inheritance ◦ Maternal mtDNA Transmission ◦ Mitochondrial Disease ◦ Mitochondrial Replacement ◦ Three-Parent Baby ◦ NUMTs Confusion ◦ Genetic Parentage Redefined ◦ He Jiankui CRISPR Case ◦ Oxidative Phosphorylation ◦ Heteroplasmy

"Who in the world am I? Ah, that's the great puzzle!"
— Alice, 1865

"It has not escaped our notice that the specific pairing we have postulated immediately suggests a possible copying mechanism for the genetic material."
— James D. Watson and Francis H. Crick, 1953

The Three Genome Problem

The effort to prevent the maternal transmission of mitochondrial DNA (mtDNA) diseases originated from the recognition, during the late twentieth century, that certain debilitating disorders such as Leigh syndrome, MELAS, and Leber's hereditary optic neuropathy were caused by mutations (Wallace et al., 1988) in the mitochondrial genome. Unlike nuclear genetic diseases, mitochondrial disorders presented a unique challenge: strict maternal inheritance, random bottleneck effects during oogenesis, and heteroplasmy complicated genetic counselling and prediction of disease severity.

Early experiments in the 1990s explored the feasibility of cytoplasmic transfer between oocytes, aiming to improve oocyte competence rather than prevent disease. These procedures, known as ooplasmic transfer, resulted in the births of children carrying a mixture of maternal and donor mitochondria, raising ethical and regulatory concerns. In 2001, the U.S. Food and Drug Administration (FDA) halted cytoplasmic transfer procedures, citing insufficient safety data and concerns regarding heritable genetic modification.

Scientific focus then shifted toward targeted nuclear transfer techniques. In 2009, Tachibana et al. demonstrated in rhesus macaques that meiotic spindle transfer (MST) could successfully prevent the transmission of maternal mtDNA mutations without compromising embryo viability. This work provided the first preclinical evidence supporting mitochondrial replacement as a viable therapeutic strategy.

Pronuclear transfer (PNT), originally demonstrated in murine models in 1983, was adapted for use in human embryos by Craven et al. in 2010. Subsequent refinements by research groups in the United Kingdom and United States established that both MST and PNT could achieve low levels of mitochondrial carry-over and support normal embryonic development to the blastocyst stage.

In 2015, the United Kingdom became the first country to formally legalise mitochondrial replacement therapies (MRTs) under strict regulation, following extensive public consultation and scientific review by the Human Fertilisation and Embryology Authority (HFEA). Clinical licences were granted on a case-by-case basis for preventing the transmission of serious mitochondrial diseases.

The first reported live birth resulting from MRT occurred in 2016 via spindle transfer, performed by a clinical team led by Dr. John Zhang. The procedure was carried out partially in the United States and partially in Mexico to circumvent regulatory barriers, marking a controversial milestone in the field.

Parallel developments occurred at the Nadiya Clinic in Kyiv, Ukraine. In 2016–2017, researchers led by Dr. Valery Zukin and Dr. Pavlo Mazur implemented pronuclear transfer protocols adapted for infertility treatments, reporting multiple pregnancies and births.

Research efforts have since expanded to include polar body transfer (PB1T and PB2T) as alternative MRT strategies, aiming to further minimise mitochondrial carry-over and ethical concerns related to zygote destruction. Long-term follow-up studies and multigenerational observations remain essential to fully assess the safety, efficacy, and societal implications of mitochondrial replacement technologies.

 https://doi.org/10.11647/OBP.0526.49

Fertilisation in humans begins when a sperm cell successfully penetrates the outer membrane of the oocyte. Upon entry, the sperm delivers its haploid set of 23 chromosomes into the oocyte cytoplasm. The oocyte, already arrested in metaphase II of meiosis, completes its meiotic division and expels the second polar body. The fusion of the male and female pronuclei forms a single diploid nucleus, establishing the genomic foundation of the zygote.

Following fertilisation, the zygote undergoes a series of rapid mitotic divisions known as cleavage. These divisions increase cell number without increasing the overall size of the embryo, partitioning the cytoplasm into progressively smaller blastomeres. Around the 16- to 32-cell stage, the embryo compacts to form a morula, and by the fifth to sixth day post-fertilisation, a fluid-filled cavity called the blastocoel develops, creating a blastocyst. The blastocyst consists of an inner cell mass, destined to form the embryo, and an outer trophoblast layer that facilitates implantation into the uterine endometrium.

The nuclear DNA of the zygote comprises 46 chromosomes, organised into 23 homologous pairs. One chromosome of each pair is inherited from the oocyte, and the other from the sperm. These chromosomes encode the genetic information required for human development, regulating processes from cell cycle control to tissue differentiation. The nuclear genome is distributed across the nucleus of each embryonic cell, and its faithful replication is critical for maintaining genomic integrity throughout embryogenesis.

In addition to nuclear DNA, each human cell contains mitochondria in the cytoplasm that produce cellular energy. Within each mitochondrion exists a small, circular DNA molecule known as mitochondrial DNA (mtDNA). Unlike nuclear DNA, which is packaged into chromosomes within the nucleus, mtDNA is physically separate and exists in multiple copies per mitochondrion. The mitochondria are inherited maternally through the oocyte's cytoplasm.

Mitochondrial DNA encodes components for oxidative phosphorylation, the biochemical pathway that generates adenosine triphosphate (ATP) through electron transport and proton gradient-driven synthesis (Mitchell, 1961). Specifically, mtDNA contains 37 genes: 13 encoding protein subunits of the respiratory chain complexes, 22 encoding transfer RNAs, and 2 encoding ribosomal RNAs necessary for mitochondrial protein synthesis. These elements are indispensable for cellular metabolism.

Unlike nuclear DNA, mtDNA is highly susceptible to mutations. It lacks protective histone proteins, possesses limited DNA repair mechanisms, and sits near the electron transport chain, a major source of reactive oxygen species. These conditions result in a mutation rate for mtDNA that is approximately 10 to 100 times higher than that of nuclear DNA. Mutations in mtDNA accumulate over time and can disrupt the efficiency of oxidative phosphorylation, impairing cellular energy production.

Mutations in mitochondrial DNA cause diseases characterised by impaired energy metabolism. Because mitochondria are responsible for supplying the majority of cellular ATP, defects in oxidative phosphorylation have the greatest impact on tissues with high metabolic demands. Clinical manifestations include neurological disorders such as encephalopathy and seizures, muscular disorders such as myopathy and exercise intolerance, cardiomyopathies, and sensory deficits including optic neuropathy and hearing loss. The severity of these diseases often correlates with the proportion of mutated mtDNA within affected cells, a condition known as heteroplasmy, and with the energy thresholds required by different tissues.

In vitro fertilisation (IVF) combines retrieved oocytes and prepared sperm outside the human body under controlled laboratory conditions. The process begins with controlled ovarian hyperstimulation, during which exogenous gonadotropins are administered to stimulate the development of multiple follicles. Once sufficient follicular maturation is confirmed by ultrasound and hormone measurements, oocyte retrieval is performed transvaginally under ultrasound guidance. Retrieved oocytes are assessed for maturity and subsequently exposed to motile sperm, either by conventional insemination or by intracytoplasmic sperm injection (ICSI) (Palermo et al., 1992), in which a single sperm cell is mechanically introduced into the oocyte cytoplasm.

Fertilised embryos are cultured in specialised media supporting preimplantation development. Embryos are monitored for cleavage patterns, morphology, and progression to the blastocyst stage, typically over a period of five to six days. Selection criteria based on morphological quality and developmental timing guide the choice of embryos for transfer. One or more embryos are transferred into the uterine cavity using a catheter, aiming to establish implantation and initiate a clinical pregnancy. Remaining viable embryos may be cryopreserved for future use.

In standard IVF procedures, the oocyte's cytoplasm, including its mitochondrial content, is transmitted unchanged to the resulting embryo. Because mitochondria and mtDNA are maternally inherited, IVF does not prevent the transmission of pathogenic mtDNA mutations. Someone with defective mtDNA can pass these mutations to offspring through the oocyte cytoplasm in both natural conception and IVF.

Mitochondrial replacement therapies (MRT) prevent the transmission of mutated mtDNA. These techniques involve transferring the nuclear genetic material from an oocyte or zygote carrying pathogenic mtDNA into a donor cytoplasm containing healthy mitochondria. Two methods exist: maternal spindle transfer, performed before fertilisation, and pronuclear transfer, performed after fertilisation but before pronuclear fusion. Both methods aim to preserve the intended parents' nuclear genome while replacing the defective mitochondrial population with functional donor-derived mitochondria.

The strict maternal inheritance of mtDNA involves active mechanisms to eliminate paternal mitochondria, including degradation during spermatogenesis and post-fertilisation mitophagy.

Rare reports have described cases of paternal mtDNA transmission. High-throughput sequencing technologies occasionally detect mtDNA sequences in offspring that do not match the maternal lineage, suggesting a potential contribution from the sperm. Initial interpretations proposed that paternal mitochondria might occasionally evade elimination mechanisms and be transmitted to the offspring at detectable levels.

After some excitement in the field about this alternative inheritance mechanism, it was realised that these observations often result from nuclear mitochondrial DNA segments (NUMTs) (Lopez et al., 1994)—fragments of mtDNA incorporated into the nuclear genome over evolutionary time that closely resemble true mitochondrial sequences. NUMTs are inherited in a Mendelian fashion and can be misinterpreted as paternal mtDNA when standard sequencing techniques co-amplify nuclear and mitochondrial DNA. Indeed, multiple studies have shown that many apparent cases of biparental mtDNA inheritance are artefacts caused by the presence of large, recently inserted NUMTs in the nuclear genome rather than true transmission of paternal mitochondria.

On the Semantics of Genetic Parentage

In laboratory settings, mitochondrial DNA replacement and more complex genomic interventions—ranging from whole chromosome transfers to single-base CRISPR edits—are becoming routine. Popular discourse often responds with labels such as 'three-parent baby,' referring to cases where nuclear DNA comes from two individuals and mtDNA from a third. As such procedures proliferate and more nuanced manipulations emerge, questions like 'who are the parents?' or 'how many parents are there if 10% of the genome is replaced?' become ill-posed.

This is not unlike the conceptual shift that occurred in physics a century ago. Questions such as 'what is light?' or 'what is an electron?' gave way to rigorously framed operational questions: 'what signal will appear on a detector under specified experimental conditions?' Biology is undergoing a similar transition. Rather than asking 'who are the parents?', the relevant question becomes 'what proportion of the genome is shared with each contributor?'. Genetic parentage, in this view, becomes a quantitative map of biological contribution. This technical definition intentionally separates the biological facts from the equally valid concepts of social and emotional parentage, which are defined orthogonally by nurture, commitment, and care.

A parallel imprecision pervades discussions of 'biological sex.' The term conflates multiple distinct but correlated traits—chromosomal karyotype, gonadal structure, hormonal profiles, anatomical morphology, and secondary sexual characteristics—that do not always align. No single definition encompasses all cases. Scientific precision requires specifying which measurable trait is under discussion: XX/XY karyotype, circulating testosterone levels, presence of particular anatomical structures, or expression of specific gene regulatory networks. The phrase 'biological sex' functions as convenient shorthand but lacks the rigour of a well-defined biological variable.

This is not a debate about politics, gender identity, or social inclusion—it is a matter of definitional rigour. Science advances by defining terms precisely and specifying what is being measured. When researchers invoke 'biological sex' without clarifying whether they mean chromosomal composition, hormonal milieu, or anatomical phenotype, they introduce ambiguity that weakens scientific reasoning. Sloppy terminology generates confusion across medicine, developmental biology, and public policy, much as the now-discredited concept of 'biological race' once did. Scientific language should describe measurable phenomena rather than compress multidimensional biological variation into a single inherited categorical label for convenience.

CRISPR Babies

In late 2018, a Chinese biophysicist named He Jiankui announced the birth of twin girls whose genomes had been edited at the embryonic stage. Using the CRISPR-Cas9 system (Jinek et al., 2012), He targeted the CCR5 gene, seeking to introduce a mutation associated with resistance to HIV infection. The announcement, made through YouTube videos and public statements rather than peer-reviewed scientific channels, bypassed established academic and regulatory norms. This triggered international condemnation from scientists, ethicists, and government bodies.

The CCR5 gene, while involved in HIV susceptibility, also participates in brain development, immune response regulation, and other physiological processes. Editing this gene without comprehensive knowledge of its systemic effects introduced unknown biological risks. Established medical procedures such as sperm washing already enabled HIV-positive parents to conceive healthy children, making the intervention medically unnecessary.

The consent forms misled participating families about the experimental nature and unknown risks of the intervention. Off-target edits—unintended mutations elsewhere in the genome—were not assessed before embryo implantation. The CRISPR modifications were heritable, meaning any unintended effects would be transmitted to future generations without their consent.

In December 2019, Chinese authorities sentenced He Jiankui to three years in prison for 'illegal medical practices,' alongside financial penalties and professional bans. Two of his collaborators received lesser sentences. The incident marked the first criminal prosecution for human germline genome editing and catalysed global discussions about regulatory frameworks, ethical standards, and the governance of emerging biotechnologies.

After his release from prison in 2022, He Jiankui publicly stated that the three children born from the experiment—the twin girls and a third child—are "healthy and living normal lives," though no independent medical verification or peer-reviewed follow-up data has been released. He has since attempted to re-enter research, claiming to focus on non-reproductive gene therapies, while China has enacted stricter bioethics and genetic-research regulations that explicitly ban germline editing in clinical settings. As of 2025, the health, developmental outcomes, and genetic integrity of the edited children remain unknown.

Micromanipulation and Mitochondrial Heteroplasmy

Technical Challenges in Nuclear Transfer

Maternal spindle transfer requires extracting the metaphase II spindle-chromosome complex without disrupting chromosome alignment. The spindle, a 15–20-μm birefringent structure visible under polarised light (Oosight™), must be aspirated with minimal surrounding cytoplasm. Tachibana et al. (2009) demonstrated this technique in rhesus macaques, achieving no detectable donor mtDNA carry-over and producing healthy, fertile offspring.

Critical technical parameters:

- **Timing**: Within 2 hours of oocyte retrieval before spindle depolymerization.
- **Pipette diameter**: 20 μm bevelled, approaching at 30° to minimise membrane deformation.
- **Cytoplasmic volume**: <5 picolitres co-aspirated to keep mitochondrial carry-over below 2%.
- **Fusion method**: Electrofusion (1.0 kV/cm, 50 μs pulses) or HVJ-E (Sendai virus extract).

Pronuclear transfer exploits the 8–10-hour window post-fertilisation when pronuclei are visible but unfused. Both pronuclei must be extracted together within a karyoplast—a membrane-bound cytoplasmic package containing ~5% of oocyte volume. This preserves their relative positioning, which carries epigenetic information essential for proper development.

Mitochondrial Carry-over and Detection

Even stringent micromanipulation cannot eliminate all donor mitochondria. Reported carry-over levels in embryo reconstruction studies include:

- PNT: typically <2%, occasionally higher (Hyslop et al., 2016).
- MST: approximately 1–2% on average (Tachibana et al., 2013).
- PB1T: often <0.5% when optimised (reported in polar body transfer studies).

Deep sequencing can achieve ~0.1% sensitivity, with digital droplet PCR detecting heteroplasmy as low as ~0.01% in optimised assays.

Heteroplasmy Dynamics in Development

Low-level heteroplasmy behaves unpredictably during development due to:

The Mitochondrial Bottleneck: During oogenesis, mtDNA copy number drops from ~100,000 in mature oocytes to ~200 in primordial germ cells before clonal expansion. This bottleneck allows random genetic drift to dramatically shift heteroplasmy levels between generations.

Tissue-Specific Segregation: Post-mitotic tissues show divergent heteroplasmy patterns:

- **Muscle**: Can amplify from 1% to 50% by adulthood.
- **Blood**: Typically maintains stable levels.
- **Brain**: Shows regional variation, with high-energy regions (substantia nigra) potentially enriching for wild-type mtDNA.

Nuclear-Mitochondrial Compatibility

The mitochondrial proteome comprises ~1,500 proteins: 13 encoded by mtDNA, the remainder nuclear-encoded and imported. OXPHOS complexes require precise stoichiometry between nuclear and mitochondrial subunits.

Potential incompatibilities arise where mtDNA-encoded subunits must interact with nuclear-encoded partners (e.g., Complex I: 7 mitochondrial and 38 nuclear subunits). However, the Tachibana primate studies (2009) showed reassuring intraspecies compatibility.

References:

Hyslop, L. A. *et al.* (2016). Towards clinical application of pronuclear transfer. *Nature* **534**, 383–386.

Tachibana, M. *et al.* (2009). Mitochondrial gene replacement in primate offspring and embryonic stem cells. *Nature* **461**, 367–372.

Tachibana, M. *et al.* (2013). Towards germline gene therapy of inherited mitochondrial diseases. *Nature* **493**, 627–631.

A Freely Wilful Ignorance

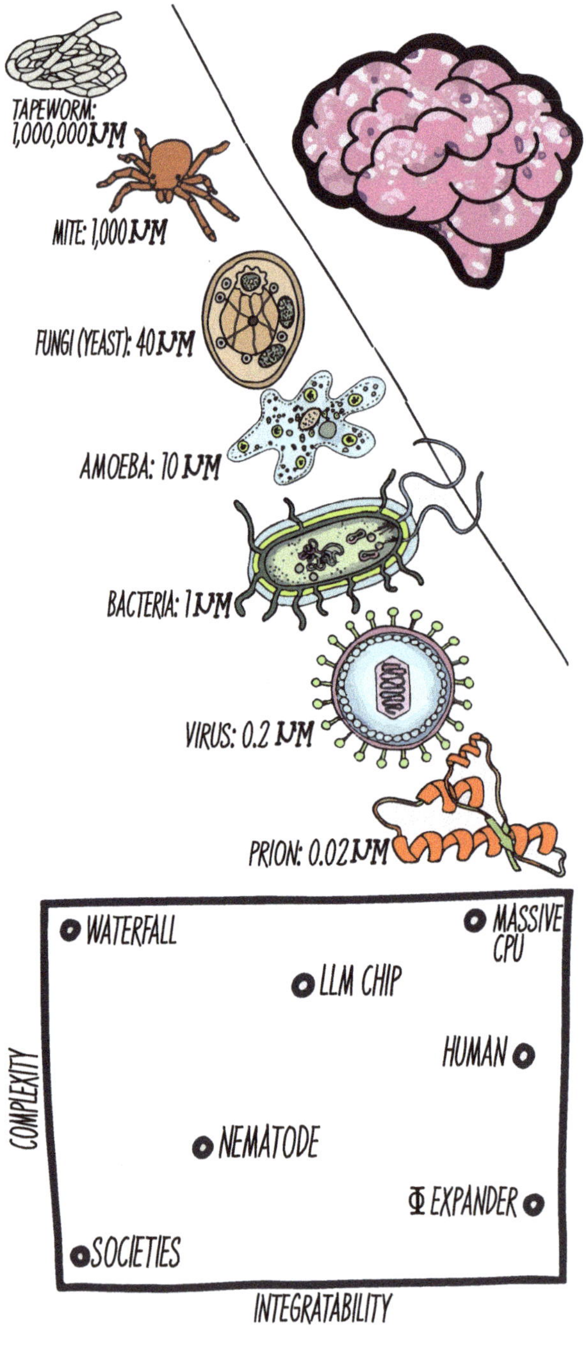

Top (Scales of Infection and Pathogenic Agents): Infectious agents span eight orders of magnitude—from macroscopic parasites such as tapeworms to sub-viral prions. Each has its own transmission pathway, lifecycle, and interaction mode with the host. Top-right: a brain, target of prion neurodegeneration.
Bottom (Post-Hoc Theories of Consciousness): Proposed mechanisms for consciousness map along a complexity–integrability plane: waterfall-like chaos, chip design, nematode circuits, human brains, and societal behaviour. But all are descriptive, not explanatory. No theory derives subjective experience from first principles—only correlates it post hoc to system properties, and for each example we can find synthetic constructs as a counter-example. For instance, if we make a chip composed of massively parallel XOR gates arranged in feedback loops to maximise cross-dependence, it can be tuned to produce an arbitrarily high φ value—far exceeding estimates for the human brain—yet the device is nothing more than a repetitive logical expander.

A Freely Wilful Ignorance

Milligrams of propofol erase consciousness in seconds. Fatal familial insomnia prevents its cessation for months until death. While we can reliably toggle awareness, no unified mechanism explains why subjectivity vanishes. Consciousness cannot be reduced to neural correlates or fit by classifiers. Any attempt to locate its origin in physical mechanisms presupposes the very phenomenon under study. Free will and physics appear incompatible, but the standoff is asymmetric: agency is the lived fact that makes physics construction possible. Consciousness occupies the apex of a revision hierarchy where, in any conflict with lower-level descriptions, the knower must prevail.

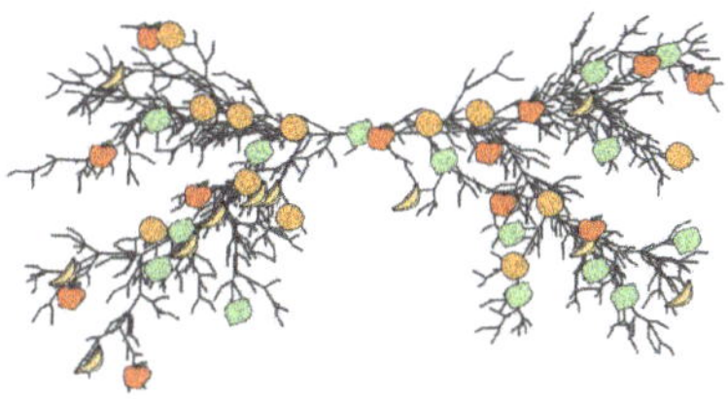

Consciousness & Anaesthesia ◦ General Anaesthetic Mystery ◦ Multiple Molecular Mechanisms ◦ Fatal Familial Insomnia ◦ Free Will vs Physics ◦ First-Person Experience ◦ Libet Readiness Potential ◦ Revision Cost Hierarchy ◦ Cogito Ergo Sum ◦ Direct Self-Knowledge ◦ Hard Problem

«الاعتقاد ليس هو المعنى المقصود، بل المعنى المتصور في النفس»

("Belief is not the utterance, but the conception in the soul.")
— Maimonides, circa 1191 CE

"Reason will prevail."
— The Gang, 2008

A Freely Wilful Ignorance

Ether-era anaesthesia began in 1846 with Morton's public demonstration in Boston; within months, ether and chloroform spread worldwide. By the early twentieth century, Meyer and Overton independently observed a correlation: anaesthetic potency scaled with lipid solubility across diverse compounds. This supported the idea that consciousness could be turned off by a nonspecific action on neuronal membranes. Yet the correlation cracked under scrutiny: highly lipophilic yet inert molecules failed to anaesthetise, while effective agents deviated from the predicted potency.

Mid-to-late twentieth century work shifted toward specific molecular targets: $GABA_A$ receptors, NMDA antagonism, two-pore K^+ channels, and hyperpolarisation-activated cyclic nucleotide–gated (HCN1) currents. Still, no single pathway unified the class.

In prion disease, a different historical thread exposed the opposite failure mode. In 1982, Prusiner proposed prions—a proteinaceous infectious particles—as agents of neurodegeneration. A rare PRNP mutation producing fatal familial insomnia (FFI) was later traced to selective thalamic degeneration, abolishing sleep despite otherwise preserved wakeful function. An Italian pedigree provided the defining clinical arc: onset with fragmented sleep, inexorable insomnia, autonomic failure, cognitive collapse, and death within months. Where anaesthesia induced obliviousness, FFI prevented it.

Reader beware: this chapter is not about a phenomenon at the heart of the scientific consensus; instead it is full of my philosophical musings.

General anaesthesia abolishes subjectivity itself. Other drugs alter perception, mood, or pain. Anaesthetics suspend the condition for all perception and mood. A standard intravenous dose of propofol—two milligrams per kilogram—eliminates awareness in less than a minute. The transition is sharp. One moment the subject tracks voices and surroundings; the next moment there is no report, no continuity of thought, and no subsequent memory. The effect is reliable, reversible, and indispensable to surgical practice. Yet it remains unexplained. That is assuming the loss is real, and not merely after the fact amnesia.

Different drugs converge on this endpoint through divergent and sometimes contradictory mechanisms. Propofol potentiates γ-aminobutyric acid type A ($GABA_A$) receptors, amplifying inhibitory currents and reducing excitability across the cortex. Isoflurane, sevoflurane, and other volatile anaesthetics bind to potassium and sodium channels, producing generalised dampening of neuronal firing. Nitrous oxide and xenon inhibit N-methyl-D-aspartate (NMDA) receptors, reducing excitatory drive. Ketamine blocks NMDA receptors yet increases cortical activity globally, producing electroencephalographic patterns closer to wakefulness than sleep while still abolishing awareness. Distinct molecular actions—some silencing neurons, some exciting them—dismantle consciousness with similar reliability.

The search for an underlying model for general anaesthesia once looked promising. At the turn of the twentieth century, Hans Meyer and Charles Ernest Overton noted a correlation: anaesthetic potency scales with lipid solubility. The Meyer–Overton rule suggested that

 https://doi.org/10.11647/OBP.0526.50

anaesthetics (Meyer, 1899; Overton, 1901) dissolved into neuronal membranes, altering their physical properties. For decades this correlation dominated, reinforced by its simplicity. Yet the correlation is not absolute. Non-immobilisers—molecules with high lipid solubility—fail to anaesthetize. Others deviate from predicted potency. The membrane theory could not account for exceptions.

The focus moved to receptors. Different anaesthetic classes bind to distinct proteins: $GABA_A$, NMDA, and two-pore domain potassium channels among prime candidates. Yet receptor theories also encounter anomalies. No single target is necessary. Mice engineered with $GABA_A$ subunits resistant to volatile anaesthetics still lose consciousness when exposed. No single target is sufficient: receptor agonists or antagonists with precise effects on candidate pathways often fail to produce general anaesthesia. What remains is a map of partial correlates, not a law specifying why awareness vanishes.

Network hypotheses move up a level. Thalamic "switch-off" models propose that sensory relay and intralaminar nuclei disengage cortical broadcasting. Alternatives hold that long-range cortico-cortical integration degrades: effective connectivity fragments, ignition-like reverberation collapses, and fronto-parietal synchrony decouples. Empirically, anaesthetic depth tracks changes in spectral power, complexity, and coherence. Though counterexamples persist. Ketamine increases cortical activity and high-frequency power yet abolishes consciousness. Dexmedetomidine reduces thalamic throughput yet permits vivid dreams.

The opposite extreme also exists. Infectious agents span orders of magnitude: from metre-long tapeworms to micrometre bacteria and nanometre viruses. Though some infections are not carried by biological agents, but by physical ones. A prion (proteinaceous infectious particle) is a protein (Prusiner, 1982) at nanometre scale) that was misfolded into an abnormal shape and can sometimes infect nearby proteins to do the same. It resists most disinfection protocols. And it can cause a consciousness disorder.

Fatal familial insomnia, a prion disease (Lugaresi et al., 1986), destroys neurons in the thalamus, especially in the anteroventral and mediodorsal nuclei. These nuclei regulate sleep architecture. As they degenerate, the subject loses the ability to enter non-rapid eye movement sleep. Ordinary fatigue accumulates, but sleep never arrives. Patients remain in escalating wakefulness until death, typically within one to two years of symptom onset. Consciousness persists compulsively until the body collapses under uninterrupted wakefulness.

Anaesthesia and prion disease bracket the same mystery. Milligrams of a synthetic molecule suspend awareness entirely. Widespread neuronal loss fails to interrupt it. Consciousness is too easy to subtract and, simultaneously, impossible to eliminate. This indicates that manipulations reach only the conditions under which consciousness manifests. They do not specify what consciousness is. Practitioners can toggle the switch without knowing what is being switched.

Measuring consciousness remains harder than turning it off. Clinical scales rely on responsiveness; neurophysiology adds proxies: cross-regional EEG (Electroencephalography) coherence, perturbational complexity from TMS-evoked (transcranial magnetic stimulation) responses, and theoretical constructs such as Integrated Information Theory's Φ (Tononi, 2004). Each stumbles. Some unresponsive patients process speech. High Φ can be assigned to systems with no plausible subjectivity. EEG signatures of wakefulness can appear under amnestic sedation. Competing theories—Global Workspace, Integrated Information, Recurrent Processing—disagree on what makes a state conscious, and experiments often adjudicate proxies rather than experience itself.

The working picture is that multiple molecular routes converge on a few network-level motifs—reduced ignition, impaired integration, altered thalamocortical gating—sufficient to block access to a reportable workspace. That picture explains much of practice and little of essence.

The gap between control and understanding demands a different frame. Consciousness is singular. Treating it as a parameter vector to be fit by a classifier condescends to the phenomenon. A classifier extracts invariants and separates classes. Consciousness is first-personal presence and deliberative control. No change of basis, no loss function turns one into the other. The distinction is categorical.

Any research programme that seeks to locate the origin of consciousness in physical mechanisms presupposes the very phenomenon it attempts to explain. You deploy attention, select among hypotheses, compare results, and conclude. Each act exercises the thing under study. This reflexivity marks a boundary of intelligibility: the point where explanation reaches its natural terminus because the explanans and the explanandum coincide. Thomas Reid identified reflexive self-awareness as a first principle of common sense—an immediate, non-derivable truth that grounds all inquiry. Consciousness, when reflecting on itself, encounters not an epistemic obstacle but the foundational condition for explanation itself.

Free will and physics appear incompatible. If physics is a complete description—deterministic or stochastic, local or quantum, simulated or fundamental—then every decision reduces to a trajectory in state space. Free will becomes an illusion, a narrative that complex systems tell themselves about their own deterministic unfolding. But if free will exists, then physics is inconsistent. The standoff seems symmetric: pick your side.

The symmetry is false. Free will is the lived fact. Physics is the constructed model. If physics denies free will, physics has misclassified its own status. Constructing, testing, and revising physical theories requires a subject that directs thought, selects among candidate explanations, and exercises judgment. To declare that subject an illusion saws off the branch on which the declaration sits. Illusions presuppose a subject that misperceives. If the subject is deleted, the word 'illusion' loses reference. The sentence 'free will is an illusion' requires a subject that can contrast seeming with being. That requirement reinstates free will.

Superdeterminism attempts to dissolve the conflict by denying the independence of measurement choices. In this view, the experimenter's decision to measure spin-up versus Bell's theorem rejects this assump-spin-down correlates with the particle's prior state through a common past. Bell's theorem assumes measurement settings can be freely (Bell, 1964) chosen. Superdeterminism rejects this assumption by claiming that every choice traces back to initial conditions that also determined the particle's properties. The loophole saves the physics by deleting the physicist. It preserves the deterministic model by denying the very capacity—experimental choice—required to validate the model. Yet the superdeterminist still chooses which papers to write, which theories to propose, which objections to raise. Experiencing the act of advocating superdeterminism exercises the agency that superdeterminism denies.

Neuroscience experiments probe the timing of conscious will. Benjamin Libet (1985) measured electrical readiness potentials (RP) beginning 550 milliseconds before subjects reported awareness of their intention to move. The brain initiates action before conscious decision registers. Subsequent experiments refined this: Schurger (2012) showed that RP reflects general motor preparation rather than specific decision; Fried (2011) recorded individual neurons firing up to 1.5 seconds before reported awareness.

These findings constrain but do not eliminate agency. The readiness potential precedes awareness of specific intention, not the capacity for veto. Libet himself noted that subjects retain 'free won't'—the ability to cancel incipient actions after becoming aware of them. More fundamentally, experimental paradigms that measure spontaneous movements capture only a subset of willing. Deliberative decisions—weighing options, comparing outcomes, selecting among complex alternatives—unfold over seconds to hours, not milliseconds. The neuroscience of snap judgments does not generalise to the neuroscience of reflection.

Ultimately, these chronometric objections miss the mark. We do not need an oscilloscope to detect the conflict between free will and physics. The conflict is structural. If the universe is causally closed—whether deterministically or stochastically—there is no room for an uncaused cause. The incompatibility is logical, not empirical. Libet merely measured the delay of a mechanism we already knew had to exist if the brain is a physical object.

Consciousness in this context is the exercise of will on one's own stream of thought. Hold, release, redirect, compare, adopt, reject. Deliberate selection among candidate continuations. The stream is the ordered sequence of contents available for such selection. The subject is the locus at which selection is enacted.

We define commitment as the act of believing in a proposition. To rank which commitments prevail when they conflict, we define the revision cost, denoted $C(P)$, of a proposition P as the magnitude of the epistemic collapse that follows from assuming P is false. It is a measure of structural dependency. If P supports Q, then rejecting P destroys Q. The proposition with the maximal revision cost is not the one with the most evidence, but the one which provides the condition of possibility for evidence itself. Let's rank several commitments by revision cost.

I know the sky is blue. If tomorrow I learn it is an optical illusion—scattering, refraction, atmospheric tricks—fine. Mildly interesting. Nothing essential breaks.

I know that I live on Earth in the year 2025. If the simulation ends and someone unplugs me from "The Matrix", mind blown. Days to recover. But recovery is possible. I can still compare, infer, and correct.

I know there is gravity. If someone pulls the plug and reveals the simulation, forces redraw, mass no longer bends spacetime—I am stunned for weeks. I will need to rebuild the catalogue of causes and move on. The capacity to model persists.

I know $2 + 3 = 5$. If someone demonstrates that arithmetic itself is wrong—that I had a cognitive shortcut, and really $2 + 3 = 11$—the machinery of thought disassembles. Counting, comparison, consistency all rest on that foundation. Without it, reasoning collapses and is very difficult to reconstruct.

I know I have free will. I know I exist as the thing that directs its own thoughts. If this turns out to be false—then there is no 'I' left to register the failure. This is incompatible with the standpoint from which acceptability is judged.

The highest commitment dominates. Every statement, inference, or model presupposes a subject that can assert, doubt, compare, and revise. That presupposition is the content of the highest tier. Lower tiers describe states of affairs in the world. The highest tier secures the existence of the knower to whom the world appears. In any conflict, the knower wins. Without the knower, conflict is unintelligible.

Write the revision cost as $C(\cdot)$. Then $C(\text{appearances}) \ll C(\text{physics}) \ll C(\text{mathematics}) \ll C(\text{agency})$. The last inequality is decisive. If agency conflicts with physics, agency prevails. Agency is the condition for there being importance at all.

Neural correlates, receptor binding, thalamic gating, and network fragmentation describe *when* consciousness appears or vanishes and *how* physiology couples to report. That scope is exact and valuable. *What it is to be* the subject for whom appearance and vanishing matter lies elsewhere. 'When does awareness switch off?' asks about timing and mechanism. 'What is it to direct one's own thought?' asks about the standpoint that makes timing intelligible. Neuroscience answers the first. Philosophy addresses the second. Conflating them produces the reduction error: mistaking access conditions for the subject to whom access matters.

Research that maps brain states to behavioural outputs achieves correlation. Intervention studies that disrupt nodes and track changes achieve mechanism. Both are genuine progress. Constitution—the precondition without which correlation and mechanism cannot be stated—remains distinct. Consciousness sits at the constitutional level, beyond the reach of finer imaging or additional parameters.

Anaesthesia deletes awareness in seconds. Fatal familial insomnia prevents its deletion for months. Neither touches essence. We are trying to see the eye with which we see. We can map the optic nerve, treat the cataract, and measure the photon, but the act of seeing itself remains the prerequisite, not the object. Explanation terminates here, not because we have run out of data, but because we have reached the north pole.

To be fair, I must present the flip side of the coin. You can believe in free will—and ironically, I claim you have no other choice—but you also know it cannot exist, because physics is causally closed. A dichotomy with which we are forced to live.

Axiomatic Agency

Doxastic Formalism

Let S be a knowing subject, $\mathcal{P}$ the set of propositions, and $K_S \subseteq \mathcal{P}$ the commitment set of propositions S holds true. For $p, q \in K_S$, write $p \vdash q$ if q logically follows from p.

Define *revision cost*:

$$C(p) := |\{q \in K_S \mid p \vdash q\}|.$$

This induces partial order $(K_S, \preceq)$ where $p \preceq q \iff C(p) \leq C(q)$.

Hierarchy with Revision Costs

- p_1: 'Sky is blue'
  If false: Mildly interesting. Nothing breaks.
- p_2: 'Not in The Matrix'
  If false: Stunned. Rebuild ontology. Days to recover.
- p_3: 'Gravity exists'
  If false: Physics rebuilds. Weeks to recover.
- p_4: '$2 + 3 = 5$'
  If false: Arithmetic collapses. Reasoning disassembles.
- p_5: '$P \vee \neg P$' (excluded middle)
  If false: Logic fails. Cannot reason about contradictions.
- A: 'I direct my thought'
  If false: No subject remains to register the failure.

Strictly: $C(p_1) \ll C(p_2) \ll C(p_3) \ll C(p_4) \ll C(p_5) \ll C(A)$.

Agency as Maximal Element

Agency (A): capacity to perform operations on K_S (selecting, comparing, affirming, rejecting propositions). This is control over thought, not physical action.

To revise K_S by removing A requires performing an operation on K_S, which presupposes A. Thus revision of A is self-undermining:

$$A \vdash p \quad \forall p \in K_S \quad \Rightarrow \quad C(A) = |K_S|.$$

Agency is the maximal element in $(K_S, \preceq)$.

Philosophical Grounding

Descartes's Cogito ergo sum (1641): The act of doubting presupposes the existence of a doubter. Even radical scepticism cannot eliminate the thinking subject. This establishes the subject as the foundation of knowledge, not a conclusion derived from it.

Kant's Transcendental Apperception (1781): The unity of consciousness is not empirically observed but is the logical precondition for any structured experience. The 'I think' must accompany all representations. Without a unified subject, no comparison, judgment, or synthesis of data is possible.

Thomas Reid's First Principles (1785): Reid rejected both Cartesian doubt and Humean scepticism, arguing that consciousness, perception, and belief in the external world are immediate acts of common sense. They require no inferential justification because they constitute the conditions of intelligibility itself. His position anchors the self not in abstraction but in lived, self-evident awareness.

Hegel's Phenomenology of Spirit (1807): Hegel develops self-consciousness as a dialectical process—the subject becomes what it is through recognition and negation. Consciousness encounters itself in the world and, through that encounter, attains universality. Reflexivity here is not circular but generative.

David Chalmers's Hard Problem of Consciousness (1995): Chalmers formalises the explanatory gap—the difference between functional accounts and subjective experience. He frames reflexivity as evidence that consciousness is a fundamental property, not a computational artefact.

References:

Descartes, R. (1641). *Meditations on First Philosophy.*

Kant, I. (1781). *Critique of Pure Reason.*

Reid, T. (1785). *Essays on the Intellectual Powers of Man.*

Hegel, G.W.F. (1807). *Phenomenology of Spirit.*

Chalmers, D. (1995). *Facing Up to the Problem of Consciousness.*

Subject Index

Biographical Reference

One-Line Biographies

Nobel Prize laureates are marked before their name: 🏵 Physics, ⚗ Chemistry, ♥ Physiology or Medicine, ⚖ Economics, and 🖋 Literature.

Tiffany Aching (Fictional). Young witch from Terry Pratchett's *Discworld* series who protects the Chalk using practical wisdom and allies with the Nac Mac Feegle, embodying "First Sight" and "Second Thoughts." *[Ch. 46]*

Wilhelm Ackermann (1896–1962). German mathematician who defined the Ackermann function, the first example of a recursive function that is not primitive recursive. *[Ch. 17]*

Douglas Adams (1952–2001). English author and satirist who created the multimedia science fiction franchise *The Hitchhiker's Guide to the Galaxy. [Ch. 14]*

Rabbi Akiva (c. 50–c. 135). Judean sage who established hermeneutic methodology and was a central contributor to the *Mishnah* and *Midrash Halakha. [Ch. 26]*

Andreas Albrecht (b. unknown). American theoretical physicist who co-established the "new inflation" model in 1982, resolving critical stability issues in cosmology. *[Ch. 34]*

Muhammad Ali (1942–2016). American heavyweight boxer and civil rights activist who became the first three-time world heavyweight champion and refused U.S. Army induction. *[Ch. 35]*

Alice (1852–1934). Alice Liddell, daughter of Oxford dean Henry Liddell, who inspired Lewis Carroll's *Alice's Adventures in Wonderland* through their friendship and storytelling sessions. *[Ch. 49]*

Robert Ammann (1946–1994). American amateur mathematician who made independent discoveries in aperiodic tilings, including "Ammann bars" and the Ammann–Beenker tiling. *[Ch. 29]*

André-Marie Ampère (1775–1836). French physicist who founded classical electromagnetism and formulated the mathematical theory describing the interaction of electric currents. *[Ch. 04]*

Anaxagoras (c. 500 BC–c. 428 BC). Pre-Socratic Greek philosopher who brought Ionian traditions to Athens and postulated *Nous* as the ordering force of the cosmos. *[Ch. 10]*

🏵 **Carl D. Anderson** (1905–1991). American physicist who discovered the positron in 1932 using a cloud chamber, confirming the existence of antimatter. *[Ch. 19]*

Dominique-François Arago (1786–1853). French physicist and astronomer who validated the wave theory of light by experimentally demonstrating the existence of the "Arago spot." *[Ch. 13]*

Nittai of Arbel. Jewish sage and *Av Beit Din* of the Sanhedrin who advised distancing oneself from bad neighbors in *Pirkei Avot. [Ch. 26]*

Archimedes (c. 287 BC–c. 212 BC). Ancient Greek mathematician and physicist who calculated a rigorous approximation of pi and established the principle of the lever. *[Ch. 17]*

Aristarchus (c. 310 BC–c. 230 BC). Ancient Greek astronomer and mathematician who was the first to propose a heliocentric model of the universe. *[Ch. 17]*

Aristotle (384 BC–322 BC). Greek philosopher who developed the first formalized system of logic and created the structural vocabulary for Western philosophy. *[Ch. 43]*

Vladimir Arnold (1937–2010). Soviet and Russian mathematician who solved Hilbert's thirteenth problem at nineteen and co-developed the Kolmogorov-Arnold-Moser theorem. *[Ch. 48]*

⚖ **Kenneth Arrow** (1921–2017). American economist who formulated Arrow's Impossibility Theorem and shared the 1972 Nobel Prize for general equilibrium theory. *[Ch. 09]*

Achraf Atila (b. unknown). Computational materials scientist at the Bundesanstalt für Materialforschung und Prüfung (BAM) in Berlin who co-authored molecular dynamics simulations demonstrating that cold displacement-driven amorphization dominates ice lubrication during sliding. *[Ch. 42]*

Michael Atiyah (1929–2019). British mathematician who unified topology and analysis through the development of topological K-theory and the Atiyah–Singer index theorem. *[Ch. 07]*

Robert Atkinson (1898–1982). British physicist and astronomer who, with Fritz Houtermans, applied quantum tunneling to explain stellar thermonuclear fusion. *[Ch. 10]*

Avtalyon. Jewish sage and *Av Beit Din* of the Sanhedrin who taught Hillel and Shammai as part of the fourth *Zugot*. *[Ch. 26]*

Bernard Baars (b. 1946). American neurobiologist and cognitive psychologist who originated Global Workspace Theory, describing consciousness as a spotlight for unconscious cognitive processes.

John C. Baez (b. 1961). American mathematical physicist who researches loop quantum gravity and authored the column "This Week's Finds in Mathematical Physics." *[Ch. 04]*

Shankar Balasubramanian (b. 1966). British chemist who co-invented Solexa sequencing, the primary technology used for rapid, low-cost DNA sequencing worldwide.

Michel Balinski (1933–2019). American mathematician and economist who advanced linear programming by proving that the graph of a convex d-dimensional polytope is d-connected. *[Ch. 09]*

Stefan Banach (1892–1945). Polish mathematician who founded modern functional analysis and systematized the theory of vector spaces known as Banach spaces. *[Ch. 03]*

Cohen the Barbarian (Fictional). Elderly barbarian hero from Terry Pratchett's *Discworld* who leads the Silver Horde, refuses to retire despite being over 90, and famously attempts to return fire to the gods. *[Ch. 23]*

Elad Barkan (b. unknown). Israeli cryptographer who developed practical attacks against GSM encryption standards, proving mobile calls could be intercepted in real-time. *[Ch. 12]*

Dave Barry (b. 1947). American author and columnist who won the 1988 Pulitzer Prize for Commentary for his work at the *Miami Herald*. *[Ch. 04]*

Yuliy Baryshnikov (b. 1959). Russian-American mathematician who applies algebraic topology to complex systems in sensor networks and social choice theory. *[Ch. 09]*

Hagan Bayley (b. unknown). British chemical biologist who established the scientific foundation for nanopore DNA sequencing and founded Oxford Nanopore Technologies.

Jacob Bekenstein (1947–2015). Israeli-American theoretical physicist who proposed that black holes possess entropy proportional to their event horizon area.

Eugenio Beltrami (1835–1900). Italian mathematician who proved the logical consistency of non-Euclidean geometry by modeling it on the pseudosphere. *[Ch. 14]*

Frank Benford (1883–1948). American physicist and engineer who rediscovered and generalized the statistical observation of leading digits in *The Law of Anomalous Numbers*. *[Ch. 07]*

Charles H. Bennett (b. 1943). American physicist who established the foundations of quantum information theory and introduced the field of quantum cryptography. *[Ch. 45]*

Jens "Jeb" Bergensten (b. 1979). Swedish video game programmer who became the lead designer of *Minecraft* in 2011 and serves as Mojang's Chief Creative Officer. *[Ch. 22]*

Robert Berger. American mathematician who proved the undecidability of the Domino Problem in 1964 by constructing the first aperiodic Wang tiles. *[Ch. 29]*

John Desmond Bernal (1901–1971). Irish crystallographer who pioneered the structural analysis of proteins and viruses, laying the foundations of molecular biology. *[Ch. 01]*

B. Andrei Bernevig (b. 1975). Romanian-American theoretical physicist who predicted the quantum spin Hall effect, enabling the experimental discovery of topological insulators. *[Ch. 11]*

Daniel Bernoulli (1700–1782). Swiss mathematician and physicist who authored *Hydrodynamica* and formulated Bernoulli's principle regarding fluid velocity and pressure. *[Ch. 28]*

Daniel J. Bernstein (b. 1971). American mathematician and cryptographer who designed Curve25519 and established code as protected speech in *Bernstein v. United States. [Ch. 18]*

Joseph A. Berry (b. 1941). American plant physiologist who co-developed the FvCB model, a biochemical framework describing photosynthetic carbon dioxide assimilation. *[Ch. 35]*

Joseph Bertrand (1822–1900). French mathematician who formulated Bertrand's Paradox, conjectured Bertrand's Postulate, and developed the Bertrand competition model in economics. *[Ch. 15]*

❁**Hans Bethe** (1906–2005). German-American theoretical physicist who formulated the theory of stellar energy production in 1939, winning the 1967 Nobel Prize. *[Ch. 10]*

Peter J. Bickel (b. 1940). American statistician who contributed to the theory of semiparametric models and co-authored the definitive example of Simpson's Paradox. *[Ch. 30]*

Eli Biham (b. 1960). Israeli cryptographer who co-introduced differential cryptanalysis with Adi Shamir, enabling the first practical attack on the Data Encryption Standard. *[Ch. 12]*

Jean-Baptiste Biot (1774–1862). French physicist who formulated the Biot-Savart Law with Félix Savart, establishing the fundamental relationship between electric current and magnetism. *[Ch. 13]*

George David Birkhoff (1884–1944). American mathematician who proved the Pointwise Ergodic Theorem and formulated the uniqueness theorem for the Schwarzschild metric.

N. David Birrell. British theoretical physicist who co-authored the influential monograph *Quantum Fields in Curved Space* with Paul Davies in 1982. *[Ch. 47]*

Colin R. Blyth (1922–2019). Canadian statistician who formally coined the term "Simpson's paradox" in the paper "On Simpson's Paradox and the Sure-Thing Principle." *[Ch. 30]*

Tom Bohman (b. unknown). American mathematician who utilized the triangle-free process to determine asymptotic bounds for off-diagonal Ramsey numbers. *[Ch. 09]*

❁**Niels Bohr** (1885–1962). Danish physicist who developed the Bohr model of the atom, positing that electrons orbit the nucleus at discrete energy levels. *[Ch. 15]*

Ludwig Boltzmann (1844–1906). Austrian physicist who established statistical mechanics and formulated the equation $S = k \log W$ linking entropy to probability. *[Ch. 28, 32, 34, 45]*

Enrico Bombieri (b. 1940). Italian mathematician who received the 1974 Fields Medal for contributions to analytic number theory, algebraic geometry, and analysis. *[Ch. 08]*

Jerry Bona (b. 1945). American mathematician who co-developed the Benjamin-Bona-Mahony equation, a model for long surface gravity waves of small amplitude. *[Ch. 07]*

Hermann Bondi (1919–2005). Austrian-British mathematician who co-developed the Steady State theory of the universe with Fred Hoyle and Thomas Gold in 1948. *[Ch. 43]*

Franz Bopp (1791–1867). German linguist who founded comparative grammar and historical linguistics by demonstrating the genetic relationship between Sanskrit and classical European languages. *[Ch. 05]*

Albert Bosma (b. unknown). Dutch astronomer who provided fundamental confirmation for dark matter halos by demonstrating that galactic rotation curves remain flat. *[Ch. 37]*

Frank P. Bowden (1903–1968). Australian physicist who demonstrated that the low friction of ice is caused by a water layer produced by frictional heating. *[Ch. 42]*

Robert Boyle (1627–1691). Anglo-Irish natural philosopher who wrote *The Sceptical Chymist* and formulated Boyle's Law describing gas pressure and volume. *[Ch. 28]*

Monika Bradač (b. unknown). Slovenian astrophysicist who led the mapping of dark matter in the Bullet Cluster, providing empirical evidence for particulate dark matter. *[Ch. 37]*

Leon Brillouin (1889–1969). French physicist who established the concept of Brillouin zones and co-developed the WKB approximation for the Schrödinger equation. *[Ch. 45]*

Malcolm Brown (1930–2017). British historian who co-authored the definitive book 'Christmas Truce: The Western Front December 1914' with Shirley Seaton. *[Ch. 38]*

Nicolaas de Bruijn (1918–2012). Dutch mathematician who developed Automath, the first computer language for checking proofs, and discovered De Bruijn sequences. *[Ch. 40]*

Viggo Brun (1885–1978). Norwegian mathematician who introduced "Brun's sieve" and proved the convergence of the sum of twin prime reciprocals. *[Ch. 08]*

Elisabeth Buck (unknown–2002). American biologist who verified the synchronous flashing of *Pteroptyx* fireflies in Southeast Asia, refuting the optical illusion hypothesis. *[Ch. 25]*

John Buck (1912–2005). American physiologist and entomologist who identified the mechanisms governing the synchronous flashing of Southeast Asian *Pteroptyx* fireflies. *[Ch. 25]*

Susanne von Caemmerer (b. 1953). Australian plant physiologist and mathematician who co-developed the FvCB model establishing the framework for C3 photosynthesis in 1980. *[Ch. 35]*

Annie Jump Cannon (1863–1941). American astronomer who developed the Harvard Classification Scheme and classified approximately 350,000 stars for the *Henry Draper Catalogue*. *[Ch. 27]*

Georg Cantor (1845–1918). German mathematician who founded set theory and introduced transfinite numbers to classify the sizes of infinite sets. *[Ch. 03, 17]*

Tom Cargill (b. unknown). American computer scientist at AT&T Bell Laboratories who formulated the "Ninety-Ninety Rule" and developed the "pi" debugger. *[Ch. 22]*

Sadi Carnot (1796–1832). French physicist who analyzed the theoretical limits of heat engines in *Reflections on the Motive Power of Fire*. *[Ch. 28]*

Sean Carroll (b. 1966). American theoretical physicist and philosopher who specializes in quantum foundations and communicates science through the *Mindscape* podcast. *[Ch. 02]*

Élie Cartan (1869–1951). French mathematician who developed the method of moving frames and established the use of exterior differential forms. *[Ch. 21]*

Hendrik Casimir (1909–2000). Dutch physicist who predicted the Casimir effect in 1948, an attractive force between uncharged parallel conducting plates. *[Ch. 47]*

Arthur Cayley (1821–1895). British mathematician who founded the theory of matrices and was the first to define the concept of a group. *[Ch. 07]*

Anders Celsius (1701–1744). Swedish astronomer and mathematician who proposed the centigrade temperature scale in 1742, now known as the Celsius scale. *[Ch. 28]*

David Chalmers (b. 1966). Australian philosopher and cognitive scientist who formulated the "hard problem of consciousness" in his 1996 book *The Conscious Mind*. *[Ch. 20, 50]*

Subrahmanyan Chandrasekhar (1910–1995). Indian-American astrophysicist who calculated the Chandrasekhar limit, establishing the maximum mass of stable white dwarf stars. *[Ch. 02]*

Jacques Charles (1746–1823). French inventor and scientist who piloted the first hydrogen balloon and formulated the gas principle known as Charles's Law. *[Ch. 28]*

Winston Churchill (1874–1965). British statesman who served as Prime Minister and led the United Kingdom against Nazi Germany during the Second World War. *[Ch. 09]*

Patricia Churchland (b. 1943). Canadian-American philosopher and founder of neurophilosophy who advocates for eliminative materialism and the replacement of "folk psychology" with neuroscience. *[Ch. 20]*

Paul Churchland (b. 1942). Canadian philosopher who advocates for eliminative materialism, arguing that "folk psychology" will be replaced by neuroscience. *[Ch. 20]*

Rudolf Clausius (1822–1888). German physicist and mathematician who formulated the Second Law of Thermodynamics and introduced the concept of "entropy." *[Ch. 28, 45]*

Donald D. Clayton (1935–2024). American theoretical astrophysicist who authored *Principles of Stellar Evolution and Nucleosynthesis* and predicted presolar grains in meteorites. *[Ch. 10]*

Douglas Clowe (b. unknown). American astrophysicist who provided direct empirical evidence for dark matter via the 2006 Bullet Cluster study. *[Ch. 37]*

Ronald H. Coase (1910–2013). British economist who received the Nobel Prize for introducing transaction costs and formulating the Coase Theorem. *[Ch. 30]*

William Coblentz (1873–1962). American physicist who established the foundations of infrared spectroscopy and quantified the efficiency of *Photinus pyralis* bioluminescence. *[Ch. 25]*

Paul Cohen (1934–2007). American mathematician who invented "forcing" in 1963 to prove the independence of the Continuum Hypothesis and the Axiom of Choice. *[Ch. 03]*

Jean-Baptiste Colbert (1619–1683). French statesman who served as Controller-General of Finances and systematized the economic policies of mercantilism under King Louis XIV.

Sidney Coleman (1937–2007). American theoretical physicist who formulated the Coleman-Mandula theorem and the Coleman-Weinberg mechanism, and derived the dynamics of false vacuum decay. *[Ch. 16]*

Francis Collins (b. 1950). American physician-geneticist who oversaw the successful sequencing of the human genome and directed the National Institutes of Health. *[Ch. 40]*

Mark Corrigan (Fictional). Socially anxious loan manager from *Peep Show* whose cynical internal monologues and dysfunctional friendship with Jeremy define the British sitcom's dark comedy. *[Ch. 09]*

Nicomo Cosca (Fictional). Charismatic mercenary captain from Joe Abercrombie's *First Law* series who leads the Thousand Swords with wit, treachery, and inebriation. *[Ch. 38]*

Harald Cramér (1893–1985). Swedish mathematician and statistician who established the Cramér-Rao bound and formalized statistical inference in *Mathematical Methods of Statistics*. *[Ch. 08]*

Francis Crick (1916–2004). British molecular biologist who co-discovered the double-helix structure of DNA in 1953 and shared the 1962 Nobel Prize. *[Ch. 49]*

John Dash (1923–2010). American experimental physicist who provided the physical basis for the "quasi-liquid layer" on ice surfaces, explaining ice friction and glacier motion. *[Ch. 42]*

Eustache Dauger (c. 1637–1703). French valet who is identified by historians as the candidate for the prisoner known as the "Man in the Iron Mask." *[Ch. 44]*

Paul C. W. Davies (b. 1946). British theoretical physicist and cosmologist who co-discovered the Fulling-Davies-Unruh effect and defined the Bunch-Davies vacuum state. *[Ch. 47]*

Bryce DeWitt (1923–2004). American theoretical physicist who pioneered quantum gravity by formulating the Wheeler-DeWitt equation and quantizing fields in curved spacetime. *[Ch. 14]*

Peter Debye (1884–1966). Dutch-American physicist and chemist who received the 1936 Nobel Prize for studies on molecular structure and dipole moments. *[Ch. 04, 31]*

Richard Dedekind (1831–1916). German mathematician who formalized real numbers via "Dedekind cuts" and introduced ideals to algebraic number theory. *[Ch. 03]*

Giuseppe Degrassi (b. 1959). Italian theoretical physicist who demonstrated the metastability of the electroweak vacuum using Higgs and top quark masses. *[Ch. 16]*

Stanislas Dehaene (b. 1965). French cognitive neuroscientist who characterized the neural basis of reading and numerical cognition, and developed the Global Neuronal Workspace theory.

Daniel Dennett (1942–2024). American philosopher and cognitive scientist who formulated the "intentional stance" and championed a materialist, evolutionary explanation of consciousness. *[Ch. 20]*

René Descartes (1596–1650). French philosopher and mathematician who formulated "Cogito, ergo sum" and developed the Cartesian coordinate system uniting geometry and algebra. *[Ch. 50]*

Jean-Pierre Desclaux (b. 1938). French physicist who developed computational codes proving that relativistic effects determine the yellow color of gold and liquidity of mercury. *[Ch. 01]*

Diophantus (c. 200–c. 284). Hellenistic mathematician who authored *Arithmetica* and pioneered the study of indeterminate equations known as Diophantine equations. *[Ch. 08]*

❁ **Paul Dirac** (1902–1984). British theoretical physicist who formulated the Dirac equation, reconciling quantum mechanics with special relativity and predicting antimatter. *[Ch. 01, 36]*

The Dogman (Fictional). Loyal scout and Named Man from Joe Abercrombie's *The First Law* trilogy known for his keen sense of smell, pragmatism, and comparative moral center among Northern warriors. *[Ch. 43]*

Jack Donaghy (Fictional). Conservative Vice President of East Coast Television portrayed by Alec Baldwin in *30 Rock*, known for his Six Sigma management style and mentorship of Liz Lemon. *[Ch. 14]*

Simon Donaldson (b. 1957). British mathematician who utilized Yang-Mills instantons to prove results regarding the intersection forms of smooth 4-manifolds. *[Ch. 24]*

Paul Drude (1863–1906). German physicist who developed the Drude model, applying kinetic theory to electrons to explain electrical and thermal conductivity. *[Ch. 04]*

Raphael Dubois (1849–1929). French physiologist who established the biochemical mechanisms of bioluminescence and coined the terms "luciferin" and "luciferase." *[Ch. 25]*

Richard Duda (b. unknown). American computer scientist who introduced the (ρ, θ) parameterization for the Hough transform and coauthored *Pattern Classification and Scene Analysis*. *[Ch. 41]*

Alexandre Dumas (1802–1870). French author who wrote major historical romances including *The Three Musketeers* and *The Count of Monte Cristo*. *[Ch. 44]*

Jörn Dunkel (b. unknown). German mathematician who argued that negative absolute temperatures are artifacts of the Boltzmann entropy definition. *[Ch. 28]*

Henri Dutrochet (1776–1847). French physician and botanist who discovered the phenomena of endosmosis and exosmosis and anticipated the unified cell theory. *[Ch. 31]*

Andy Dwyer (Fictional). Lovable but dimwitted shoe-shiner from *Parks and Recreation* who leads the band Mouse Rat and stars as "Johnny Karate", portrayed by Chris Pratt. *[Ch. 42]*

Freeman Dyson (1923–2020). British-American theoretical physicist who unified disparate quantum electrodynamics formulations and theorized the megastructure known as the "Dyson sphere." *[Ch. 24]*

Lisa Dyson (b. unknown). American physicist and entrepreneur who founded Kiverdi and Air Protein to produce sustainable goods using carbon-transformation technology. *[Ch. 34]*

Albrecht Dürer (1471–1528). German Renaissance artist and theorist who integrated geometry into art in the first German mathematics text, *Underweysung der Messung*. *[Ch. 29]*

Arthur Eddington (1882–1944). British astronomer who provided the first observational confirmation of General Relativity during the 1919 expedition to Príncipe. *[Ch. 06, 10, 34]*

George Efstathiou (b. 1955). British cosmologist who helped establish the standard model of Cold Dark Matter through pioneering computer simulations of cosmic structure. *[Ch. 02]*

Greg Egan (b. 1961). Australian science fiction author and mathematician who wrote *Permutation City* and devised improved upper bounds for superpermutations. *[Ch. 39]*

Paul Ehrenfest (1880–1933). Austrian theoretical physicist who bridged classical and quantum mechanics through the Ehrenfest theorem and the theory of adiabatic invariants. *[Ch. 24]*

Albert Einstein (1879–1955). German-born theoretical physicist who developed the theory of relativity and derived the mass-energy equivalence formula $E = mc^2$. *[Ch. 02, 06, 14, 36, 43]*

Modris Eksteins (b. 1943). Latvian-born Canadian historian who analyzed the cultural origins and impact of World War I in *Rites of Spring*. *[Ch. 38]*

Eratosthenes (c. 276 BC–c. 194 BC). Greek polymath who calculated the Earth's circumference and devised the "Sieve of Eratosthenes" for identifying prime numbers. *[Ch. 08]*

Paul Erdős (1913–1996). Hungarian mathematician who developed the probabilistic method in combinatorics and published approximately 1,500 papers with over 500 collaborators. *[Ch. 08, 09]*

M. C. Escher (1898–1972). Dutch graphic artist who visualized mathematical concepts including infinity, symmetry, and impossible objects through woodcuts and lithographs. *[Ch. 29]*

Louis Essen (1908–1997). British physicist who developed the first practical cesium atomic clock in 1955, enabling the redefinition of the second. *[Ch. 32]*

Euclid (c. 325 BCE–c. 265 BCE). Greek mathematician who authored the *Elements*, a treatise that systematized geometry through axioms and rigorous deduction. *[Ch. 08]*

Leonhard Euler (1707–1783). Swiss mathematician who founded modern mathematical analysis and standardized notations such as e, i, and π. *[Ch. 08, 21]*

Daniel Gabriel Fahrenheit (1686–1736). German physicist and instrument maker who invented the first reliable mercury thermometer and developed the Fahrenheit temperature scale. *[Ch. 28]*

Timothy R. Fallon (b. unknown). American evolutionary biochemist who elucidated the genomic pathways of bioluminescence by sequencing the North American firefly *Photinus pyralis*. *[Ch. 25]*

Michael Faraday (1791–1867). British scientist who discovered electromagnetic induction and the laws of electrolysis, establishing the foundations of classical electromagnetism and electrochemistry. *[Ch. 04, 42]*

Graham Farquhar (b. 1947). Australian plant physiologist who developed the Farquhar-von Caemmerer-Berry model of photosynthesis and methods for breeding drought-resistant wheat. *[Ch. 35]*

Lynn Faust (b. unknown). American naturalist who scientifically confirmed the existence of synchronous fireflies (*Photinus carolinus*) in the Great Smoky Mountains. *[Ch. 25]*

Solomon Feferman (1928–2016). American mathematician and philosopher who defined the Feferman-Schütte ordinal to measure the proof-theoretic strength of predicative analysis. *[Ch. 17]*

Mitchell Feigenbaum (1944–2019). American mathematical physicist who discovered universality in non-linear systems and calculated the Feigenbaum constants central to chaos theory. *[Ch. 48]*

Archduke Franz Ferdinand (1863–1914). Austro-Hungarian heir presumptive who was assassinated in Sarajevo in 1914, triggering the outbreak of World War I. *[Ch. 38]*

Richard P. Feynman (1918–1988). American theoretical physicist who developed the path integral formulation and received the 1965 Nobel Prize in Physics. *[Ch. 04, 13, 36]*

Alan Finkelstein (b. unknown). American biophysicist who elucidated the physics of water movement through lipid bilayers and membrane protein channels. *[Ch. 31]*

Armand Hippolyte Louis Fizeau (1819–1896). French physicist who conducted the first terrestrial measurement of the speed of light using a rotating toothed wheel apparatus. *[Ch. 13]*

Leopold Eliezer Zvi Flatto (b. unknown). American mathematician who authored the definitive monograph *Poncelet's Theorem* exploring connections to elliptic curves and billiard dynamics. *[Ch. 07]*

Ferdinand Foch (1851–1929). French general who served as Supreme Allied Commander and directed the offensive that forced German capitulation. *[Ch. 38]*

Jerry Fodor (1935–2017). American philosopher and cognitive scientist who developed the Language of Thought hypothesis and the theory of the Modularity of Mind. *[Ch. 20]*

Kent Ford (b. 1931). American astronomer who developed image tube spectrographs establishing the first robust observational evidence for dark matter. *[Ch. 37]*

Benjamin Fortson (b. unknown). American linguist and classicist who authored *Indo-European Language and Culture: An Introduction*, a standard reference in the field. *[Ch. 05]*

Joseph Fourier (1768–1830). French mathematician who developed the Fourier series and established heat propagation laws in *Théorie analytique de la chaleur*. *[Ch. 13]*

Ralph Fowler (1889–1944). British physicist who applied Fermi-Dirac statistics to white dwarf stars, explaining their stability through electron degeneracy pressure. *[Ch. 10]*

William A. Fowler (1911–1995). American nuclear physicist who won the 1983 Nobel Prize for his work on stellar nucleosynthesis. *[Ch. 10]*

Joseph von Fraunhofer (1787–1826). Bavarian optician and physicist who invented the diffraction grating and systematically mapped the dark absorption lines in the solar spectrum. *[Ch. 27]*

Sir John French (1852–1925). British Field Marshal who commanded the British Expeditionary Force on the Western Front during the first eighteen months of World War I. *[Ch. 38]*

Augustin-Jean Fresnel (1788–1827). French physicist and engineer who established the wave theory of light and invented the Fresnel lens for lighthouses. *[Ch. 13]*

Itzhak Fried (b. 1948). Israeli-American neurosurgeon who identified "concept cells" responding to specific individuals or objects using intracranial depth electrodes. *[Ch. 50]*

Harvey Friedman (b. 1948). American mathematician who initiated Reverse Mathematics and demonstrated "concrete incompleteness" by finding natural statements independent of Peano arithmetic. *[Ch. 17]*

Alexander Friedmann (1888–1925). Russian mathematician and physicist who derived the Friedmann equations from general relativity, predicting the expansion of the universe. *[Ch. 02, 43]*

Liang Fu (b. unknown). Chinese-American theoretical physicist who predicted three-dimensional topological insulators, identifying $Bi_{1-x}Sb_x$ alloys as the first material to exhibit this state. *[Ch. 11]*

Kenichi Fukui (1918–1998). Japanese theoretical chemist who became the first Asian Nobel laureate in Chemistry for developing the Frontier Molecular Orbital theory. *[Ch. 46]*

Stephen Fulling (b. 1945). American mathematical physicist who demonstrated that accelerating observers perceive a thermal bath of particles, known as the Fulling-Davies-Unruh effect. *[Ch. 47]*

Harry Furstenberg (b. 1935). American-Israeli mathematician who established the interaction between dynamics and number theory and proved Szemerédi's theorem using ergodic theory. *[Ch. 08]*

Tobias Fünke (Fictional). Failed actor and "never-nude" former psychiatrist from *Arrested Development*, known for unintentional double entendres and membership in the dysfunctional Bluth family. *[Ch. 12]*

Galileo Galilei (1564–1642). Italian astronomer and physicist who improved the telescope to provide empirical support for the Copernican heliocentric model. *[Ch. 06, 28]*

Rabban Gamliel (unknown–c. 114). Jewish Nasi of the Sanhedrin who centralized religious authority at Yavneh and formalized the Amidah prayer after the Second Temple's destruction. *[Ch. 26]*

George Gamow (1904–1968). Russian-American theoretical physicist who explained alpha decay via quantum tunneling and predicted cosmic microwave background radiation. *[Ch. 10]*

The Gang (b. 2005 (Show Debut)). Dysfunctional group of bar owners from *It's Always Sunny in Philadelphia* whose self-serving schemes and moral failures define the dark comedy sitcom. *[Ch. 50]*

Saadia Gaon (c. 882–942). Babylonian Jewish Gaon who pioneered Hebrew philology and authored *The Book of Beliefs and Opinions*, the first systematic synthesis of Jewish philosophy. *[Ch. 26]*

Martin Gardner (1914–2010). American writer who popularized recreational mathematics through his "Mathematical Games" column in *Scientific American* from 1956 to 1981. *[Ch. 15]*

Joseph Louis Gay-Lussac (1778–1850). French chemist and physicist who formulated the Law of Combining Volumes and the pressure-temperature relationship of gases. *[Ch. 28]*

Israel Gelfand (1913–2009). Soviet mathematician who made fundamental contributions to functional analysis, representation theory, and the theory of generalized functions. *[Ch. 07]*

Gelon (c. 266 BCE–c. 216 BCE). Syracusan co-regent and son of Hiero II who is the addressee of Archimedes' treatise *The Sand Reckoner*. *[Ch. 17]*

Daniel Genkin (b. unknown). Israeli computer scientist who co-discovered the Meltdown and Spectre vulnerabilities exposing design flaws in modern processors.

Sophie Germain (1776–1831). French mathematician and physicist who won the *Prix extraordinaire* for elasticity theory and formulated a partial proof for Fermat's Last Theorem. *[Ch. 13]*

Elbridge Gerry (1744–1814). American Founding Father and Vice President who enacted a redistricting bill that inspired the term "gerrymander." *[Ch. 30]*

Allan Gibbard (b. 1942). American philosopher and mathematician who proved Gibbard's theorem regarding voting manipulation and formulated norm-expressivism in *Wise Choices, Apt Feelings*. *[Ch. 09]*

🏅 **Walter Gilbert** (b. 1932). American physicist and biochemist who shared the 1980 Nobel Prize for developing the Maxam-Gilbert DNA sequencing method. *[Ch. 40]*

Michael van Ginkel (b. unknown). Dutch quantitative imaging researcher who proved the mathematical equivalence between the Hough transform and the Radon transform. *[Ch. 41]*

Thomas Gold (1920–2004). Austrian-American astrophysicist who co-proposed the Steady State theory and correctly identified pulsars as rotating neutron stars. *[Ch. 43]*

Dan Goldston (b. 1954). American mathematician who, with János Pintz and Cem Yıldırım, proved that gaps between consecutive primes can be arbitrarily smaller than average. *[Ch. 08]*

Anthony Gonzalez (b. unknown). American astronomer who co-authored the 2006 Bullet Cluster observation providing direct evidence for the existence of dark matter. *[Ch. 37]*

Chaim Goodman-Strauss (b. 1967). American mathematician who co-discovered the first aperiodic monotile in 2023, resolving a major open problem in tiling theory. *[Ch. 29]*

Paul Gordan (1837–1912). German mathematician known as the "King of Invariants" who proved the finite basis theorem for binary forms. *[Ch. 03]*

Marc H. Goroff (b. unknown). American physicist who proved with Augusto Sagnotti that pure Einstein gravity is not renormalizable at the two-loop level. *[Ch. 36]*

Richard Gott (b. 1947). American astrophysicist who demonstrated how cosmic strings theoretically allow time travel and formulated the statistical "Doomsday argument." *[Ch. 34]*

Thomas Graham (1805–1869). Scottish chemist who founded colloid chemistry and formulated Graham's Law regarding the diffusion and effusion rates of gases. *[Ch. 31]*

Ben Green (b. 1977). British mathematician who proved the Green-Tao theorem in 2004, showing that primes contain arbitrarily long arithmetic progressions. *[Ch. 08]*

Jakob Grimm (1785–1863). German philologist and mythologist who formulated Grimm's Law and co-edited *Kinder- und Hausmärchen* with his brother Wilhelm. *[Ch. 05]*

Mikhail Gromov (b. 1943). Russian-French mathematician who revolutionized geometry by introducing pseudoholomorphic curves and defining the concept of hyperbolic groups. *[Ch. 24]*

Alan Guth (b. 1947). American theoretical physicist and cosmologist who developed the theory of cosmic inflation to explain the early universe's exponential expansion. *[Ch. 16]*

Kurt Gödel (1906–1978). Austrian-American logician and mathematician who proved the Incompleteness Theorems, establishing that consistent arithmetic systems must contain undecidable propositions. *[Ch. 03, 14]*

Hadrian (76–138). Roman Emperor who consolidated imperial borders, commissioned Hadrian's Wall, rebuilt the Pantheon, and suppressed the Bar Kokhba revolt. *[Ch. 26]*

Thomas Hales (b. 1958). American mathematician who established the computer-assisted proof of the Kepler conjecture and led the Flyspeck verification project. *[Ch. 29]*

David B. Hall. American experimental physicist who collaborated with Bruno Rossi to provide the first direct verification of relativistic time dilation. *[Ch. 19]*

E. A. Hammel (1930–2022). American anthropologist and demographer who founded Berkeley's Department of Demography and coauthored the seminal *Science* paper on Simpson's Paradox. *[Ch. 30]*

G. H. Hardy (1877–1947). English mathematician recognized for his work in number theory, mentoring Srinivasa Ramanujan, and authoring *A Mathematician's Apology. [Ch. 08]*

Peter Hart (b. 1941). American computer scientist who co-invented the A* search algorithm, formalized the Hough transform, and coauthored *Pattern Classification and Scene Analysis. [Ch. 41]*

E. Newton Harvey (1887–1959). American physiologist who analyzed the biochemistry of bioluminescence and invented the centrifuge microscope to measure cellular viscosity. *[Ch. 25]*

Felix Hausdorff (1868–1942). German mathematician who founded modern topology and set theory by defining topological spaces in *Grundzüge der Mengenlehre. [Ch. 03]*

Stephen Hawking (1942–2018). British theoretical physicist who predicted that black holes emit thermal radiation, now known as Hawking radiation. *[Ch. 23, 47]*

Oliver Heaviside (1850–1925). English mathematician and physicist who reformulated Maxwell's field equations into vector calculus and established transmission line theory. *[Ch. 04, 39]*

G.W.F. Hegel (1770–1831). German philosopher who structured German Idealism by developing a dialectical framework in *The Phenomenology of Spirit. [Ch. 50]*

Erich Heilbronner (1921–2006). Swiss physical organic chemist who theoretically predicted Möbius aromaticity for cyclic conjugated systems containing $4n$ π-electrons. *[Ch. 46]*

⚙ **Werner Heisenberg** (1901–1976). German theoretical physicist who formulated the uncertainty principle and received the 1932 Nobel Prize for creating quantum mechanics. *[Ch. 36]*

Hermann von Helmholtz (1821–1894). German physician and physicist who formulated the law of conservation of energy and invented the ophthalmoscope. *[Ch. 10]*

Jan Baptist van Helmont (1579–1644). Flemish physician and chemist who founded pneumatic chemistry, identified carbon dioxide, and coined the term "gas." *[Ch. 35]*

Cris Luengo Hendriks (b. unknown). Digital image analysis researcher who is a principal developer of the DIPlib and DIPimage scientific image processing libraries. *[Ch. 41]*

John Herapath (1790–1868). English physicist and mathematician who formulated an early kinetic theory of gases proposing pressure arises from particle collisions. *[Ch. 28]*

Rabbi Herzog (1888–1959). Irish-Israeli religious leader who served as the Ashkenazi Chief Rabbi of Israel and resolved the International Date Line dispute. *[Ch. 26]*

David Hilbert (1862–1943). German mathematician who developed Hilbert spaces and presented twenty-three unsolved problems that directed mathematical research for decades. *[Ch. 03, 08, 17]*

Stefan Hilbert (b. unknown). German physicist who challenged standard entropy definitions by arguing negative absolute temperatures are inconsistent with thermodynamic principles. *[Ch. 28]*

Augustine of Hippo (354–430). North African theologian who profoundly influenced Western Christian thought through his major works *Confessions* and *The City of God*. *[Ch. 43]*

Eric Hobsbawm (1917–2012). British Marxist historian who authored a tetralogy documenting the "long nineteenth century" and "short twentieth century." *[Ch. 38]*

Michael Hobson (b. 1967). British astrophysicist who researches theoretical cosmology and co-authored the textbook *Mathematical Methods for Physics and Engineering*. *[Ch. 02]*

Jacobus van 't Hoff (1852–1911). Dutch physical chemist who received the inaugural Nobel Prize in Chemistry for discovering the laws of chemical dynamics and osmotic pressure. *[Ch. 31]*

Roald Hoffmann (b. 1937). American theoretical chemist who shared the 1981 Nobel Prize for co-formulating the Woodward-Hoffmann rules regarding pericyclic reactions. *[Ch. 46]*

Ron Holzman (b. unknown). Israeli mathematician who specializes in combinatorics and social choice, and co-proved the six-runner case of the Lonely Runner Conjecture. *[Ch. 09]*

Homer (c. 8th Century BCE). Ancient Greek poet who is traditionally credited with authoring the *Iliad* and the *Odyssey*, foundational works of Western literature. *[Ch. 38]*

Paul V. C. Hough (b. 1928). American physicist who invented the Hough transform in 1962, a feature extraction technique fundamental to computer vision. *[Ch. 41]*

Robin Houston (b. unknown). British mathematician and computer scientist who disproved the superpermutation conjecture and solved the sum of three cubes problem for 42. *[Ch. 39]*

Fritz Houtermans (1903–1966). German physicist who provided the first theoretical explanation of stellar nucleosynthesis with Robert Atkinson in 1929. *[Ch. 10]*

Fred Hoyle (1915–2001). British astronomer who developed the theory of stellar nucleosynthesis and coined the term "Big Bang" for the rival cosmology. *[Ch. 43]*

Edwin Hubble (1889–1953). American astronomer who proved that spiral nebulae are independent galaxies and provided evidence for the expanding universe. *[Ch. 02, 27]*

William Huggins (1824–1910). English astronomer who established astronomical spectroscopy and was the first to apply the Doppler effect to starlight. *[Ch. 27]*

Taylor L. Hughes (b. unknown). American theoretical physicist who co-developed the Bernevig-Hughes-Zhang model predicting the quantum spin Hall effect in mercury telluride. *[Ch. 11]*

Michael Hunkapiller (b. 1948). American biotechnologist who engineered the first automated fluorescent DNA sequencers, instruments that significantly accelerated the Human Genome Project.

Rabbi Dosa ben Hurkinos. Jewish Tanna of the *Mishnah* who urged Rabbi Yehoshua to accept Rabban Gamliel's decree to maintain unity. *[Ch. 26]*

Christiaan Huygens (1629–1695). Dutch physicist who formulated the wave theory of light in *Traité de la Lumière* and invented the pendulum clock. *[Ch. 13]*

Rabbi Eliezer ben Hyrcanus (c. 40 CE–c. 120 CE). Tannaitic sage who is central to the "Oven of Akhnai" narrative establishing that legal authority rests with consensus. *[Ch. 26]*

Constantine I (c. 272–337). Roman Emperor who became the first Christian ruler and founded Constantinople as the empire's new capital. *[Ch. 26]*

Hiero II (c. 308 BC–215 BC). Greek King of Syracuse who allied with Rome, established the *Lex Hieronica*, and commissioned works from Archimedes. *[Ch. 17]*

Hillel II (unknown–c. 365). Jewish Nasi of the Sanhedrin who established the fixed mathematical calendar around 359 CE, replacing observation-based methods. *[Ch. 26]*

Kaiser Wilhelm II (1859–1941). Last German Emperor who dismissed Bismarck and pursued aggressive policies that contributed to the outbreak of World War I.

Miki Imura (b. unknown). A independent Japanese designer and researcher who explores mathematical tilings, especially the Modulo Krinkle family of non-periodic tilings. *[Ch. 29]*

Jan Ingenhousz (1730–1799). Dutch physiologist who discovered the mechanism of photosynthesis in 1779 by showing that light enables green plant parts to release oxygen. *[Ch. 35]*

Herbert Ives (1882–1953). American physicist who directed the first public demonstration of long-distance television and confirmed time dilation via the Ives-Stilwell experiment. *[Ch. 25]*

Toru Iwatani (b. 1955). Japanese video game designer who created the arcade game *Pac-Man*, which was released by Namco in 1980. *[Ch. 22]*

Solid Jackson. Jazz slang term for "affirmation", personified as a minor character (a fisherman) in Terry Pratchett's *Jingo*; the name puns on 1940s jazz expressions. *[Ch. 28]*

Carl Gustav Jacob Jacobi (1804–1851). German mathematician who established the standard theory of elliptic functions in his treatise *Fundamenta nova theoriae functionum ellipticarum. [Ch. 07]*

He Jiankui (b. 1984). Chinese biophysicist who created the world's first gene-edited humans using CRISPR-Cas9 technology to confer resistance to HIV. *[Ch. 49]*

John Joly (1857–1933). Irish physicist and geologist who invented the steam calorimeter and estimated the Earth's age based on sodium accumulation. *[Ch. 42]*

Christine Jones (b. 1949). American astrophysicist who co-authored the 2006 Bullet Cluster study providing empirical evidence for the existence of dark matter. *[Ch. 37]*

Sir William Jones (1746–1794). British philologist who laid the foundations of comparative linguistics by proposing a common source for Sanskrit, Greek, and Latin. *[Ch. 05]*

Joshua (c. 14th–13th century BCE–c. 13th–12th century BCE). Israelite leader who succeeded Moses and led the conquest of Canaan as the central figure of the *Book of Joshua. [Ch. 26]*

Glen A. Rebka Jr. (1931–2015). American experimental physicist who co-conducted the Pound-Rebka experiment, confirming Einstein's gravitational redshift using the Mössbauer effect. *[Ch. 06]*

❁ **Raymond Davis Jr.** (1914–2006). American chemist and physicist who detected solar neutrinos, establishing the "solar neutrino problem" and sharing the 2002 Nobel Prize. *[Ch. 10]*

Étienne du Junca (unknown–1706). French Bastille major who recorded the arrival and death of the "Man in the Iron Mask" in his administrative journals. *[Ch. 44]*

Martin Kamen (1913–2002). American chemist who co-discovered the radioactive isotope Carbon-14, which became an essential tool for radiocarbon dating. *[Ch. 35]*

Charles L. Kane (b. 1963). American theoretical physicist who predicted the quantum spin Hall effect and pioneered the theory of topological insulators. *[Ch. 11]*

Immanuel Kant (1724–1804). German philosopher who synthesized rationalism and empiricism in his 1781 work, the *Critique of Pure Reason. [Ch. 24, 50]*

Craig Kaplan (b. unknown). Canadian computer scientist who co-discovered the first verified aperiodic monotiles, known as the "hat" and "spectre." *[Ch. 29]*

Avrohom Yeshaya Karelitz (Chazon Ish) (1878–1953). Haredi religious leader known as *Chazon Ish* who codified agricultural laws in Israel and defined the halakhic international dateline. *[Ch. 26]*

Edward Kasner (1878–1955). American mathematician who derived the Kasner metric and popularized the term "googol" in *Mathematics and the Imagination. [Ch. 17]*

Garry Kasparov (b. 1963). Russian chess grandmaster who became the youngest undisputed World Chess Champion and played historic matches against Deep Blue. *[Ch. 20]*

Alan R. Kay (b. unknown). American biologist who investigates the physiological roles of zinc and clarifies the mechanical and thermodynamic basis of osmosis. *[Ch. 31]*

Nathan Keller (b. 1981). Israeli mathematician and cryptographer who co-demonstrated an instant ciphertext-only cryptanalysis of the GSM encryption algorithms A5/1 and A5/2. *[Ch. 12]*

Charlie Kelly (b. c. 1976 (Fictional)). Illiterate janitor and wildcard from *It's Always Sunny in Philadelphia* who lives in squalor, huffs paint, and writes bizarre musicals while hopelessly pining for the Waitress. *[Ch. 27]*

Lord Kelvin (William Thomson) (1824–1907). British mathematical physicist and engineer who proposed the absolute temperature scale and formulated the laws of thermodynamics. *[Ch. 10, 28, 32, 42, 45]*

Johannes Kepler (1571–1630). German astronomer and mathematician who established the three laws of planetary motion, foundational to Newton's theory of universal gravitation. *[Ch. 29]*

Roy Kerr (b. 1934). New Zealand mathematician who discovered the Kerr metric, an exact solution to Einstein's field equations for rotating black holes. *[Ch. 23]*

Claus Kiefer (b. 1958). German theoretical physicist who investigates quantum gravity and black hole thermodynamics, and authored the textbook *Quantum Gravity*. *[Ch. 36]*

Matthew Kleban (b. unknown). American physicist who investigates observable implications of the multiverse, particularly cosmic bubble collisions in the Cosmic Microwave Background. *[Ch. 34]*

Felix Klein (1849–1925). German mathematician who formulated the Erlangen Program and transformed the University of Göttingen into a leading mathematical center. *[Ch. 14]*

Dan Kleitman (b. 1934). American mathematician who collaborated with Paul Erdős, simplified the Four Color Theorem, and advised on *Good Will Hunting*. *[Ch. 09]*

David Klenerman (b. 1959). British biophysical chemist who co-invented the sequencing-by-synthesis technology that reduced DNA sequencing costs by a million-fold.

Klaus von Klitzing (b. 1943). German physicist who discovered the integer quantum Hall effect and established the standard for electrical resistance R_K. *[Ch. 11]*

Donald Knuth (b. 1938). American computer scientist and mathematician who authored *The Art of Computer Programming* and created the TeX typesetting system. *[Ch. 17]*

Paul Kocher (b. 1973). American cryptographer who established the field of side-channel cryptanalysis and played a central role in discovering the Spectre vulnerability. *[Ch. 18]*

Mahito Kohmoto (b. 1954). Japanese theoretical physicist who co-derived the TKNN invariant, revealing the topological nature of the integer quantum Hall effect. *[Ch. 11]*

Andrey Kolmogorov (1903–1987). Russian mathematician who established the axiomatic foundations of probability theory in his 1933 monograph and formulated the KAM theorem. *[Ch. 48]*

Maurice Kraitchik (1882–1957). Russian-born Belgian mathematician who formulated the "Wallet Paradox" in *La Mathématique des Jeux*, a precursor to the Two Envelopes Problem. *[Ch. 15]*

Gerard Kuiper (1905–1973). Dutch-American astronomer who discovered the atmosphere of Saturn's moon Titan and hypothesized the existence of the Kuiper Belt. *[Ch. 27]*

Joseph-Louis Lagrange (1736–1813). Italian-French mathematician who transformed classical mechanics by grounding it in analytical principles in his treatise *Mécanique analytique*. *[Ch. 13]*

Leslie Lamport (b. 1941). American computer scientist who pioneered the theory of distributed systems and created the LaTeX document preparation system.

Rolf Landauer (1927–1999). German-American physicist who formulated Landauer's principle, demonstrating that the logical irreversibility of erasing information generates heat. *[Ch. 45]*

Eric Lander (b. 1957). American mathematician and geneticist who served as a principal leader of the Human Genome Project and founded the Broad Institute. *[Ch. 40]*

Pierre-Simon Laplace (1749–1827). French mathematician and astronomer who rigorously demonstrated the long-term stability of the solar system in *Mécanique Céleste. [Ch. 48]*

Anthony Lasenby (b. 1954). British astrophysicist at the University of Cambridge who researches the Cosmic Microwave Background and applies Geometric Algebra. *[Ch. 02]*

Rafał Latała (b. 1971). Polish mathematician who collaborated to verify the Gaussian Correlation Inequality by providing a streamlined exposition of Royen's proof. *[Ch. 33]*

César Lattes (1924–2005). Brazilian physicist who co-discovered the pion in 1947 using nuclear emulsions and detected the first artificially produced pions. *[Ch. 19]*

Comte de Lautréamont (1846–1870). French poet who wrote *Les Chants de Maldoror* under the pseudonym Comte de Lautréamont, influencing the Surrealists. *[Ch. 25]*

William Lawvere (1937–2023). American mathematician who introduced elementary topos theory and "Lawvere theories," redefining the connections between logic and geometry. *[Ch. 21]*

Tom Lehrer (b. 1928). American mathematician and musical satirist who composed and performed popular comic songs like "The Elements" during the 1950s and 1960s. *[Ch. 16]*

Gottfried Wilhelm Leibniz (1646–1716). German polymath who developed differential and integral calculus independently of Isaac Newton and invented the binary number system. *[Ch. 24]*

Robert B. Leighton (1919–1997). American physicist who co-authored *The Feynman Lectures on Physics*, discovered solar oscillations, and led Mariner imaging teams. *[Ch. 04]*

Georges Lemaître (1894–1966). Belgian priest and theoretical physicist who proposed the "Hypothesis of the Primeval Atom," foundation of the Big Bang. *[Ch. 02, 43]*

Andrew Lenard (1927–2020). Hungarian-born American mathematical physicist who co-authored the 1967 proof of the stability of matter with Freeman Dyson. *[Ch. 24]*

Sara Lewis (b. unknown). American biologist who researches firefly sexual selection and authored *Silent Sparks: The Wondrous World of Fireflies. [Ch. 25]*

Benjamin Libet (1916–2007). American neurophysiologist who demonstrated that the brain's readiness potential precedes conscious intention, challenging traditional views of free will. *[Ch. 50]*

Sophus Lie (1842–1899). Norwegian mathematician who created the theory of continuous transformation groups, known as Lie groups and Lie algebras. *[Ch. 21]*

Andrei Linde (b. 1948). Russian-American theoretical physicist who proposed chaotic inflation and formulated the theory of eternal inflation implying a multiverse. *[Ch. 34]*

Werner Lipp (b. unknown). Austrian systems security researcher who co-discovered the Meltdown and Spectre vulnerabilities exposing flaws in modern microprocessors. *[Ch. 18]*

Rabbi Yisrael Lipschitz (1782–1860). German rabbi who authored the *Tiferet Yisrael* commentary on the Mishnah and served as Chief Rabbi of Danzig. *[Ch. 26]*

John Edensor Littlewood (1885–1977). British mathematician who collaborated with G.H. Hardy for thirty-five years to advance number theory and mathematical analysis. *[Ch. 15]*

James Lloyd (b. 1933). American entomologist who discovered that female *Photuris* fireflies use bioluminescent aggressive mimicry to lure and devour males. *[Ch. 25]*

Sidney Loeb (1916–2008). American chemical engineer who co-invented the Loeb-Sourirajan membrane, rendering reverse osmosis a commercially viable desalination technology. *[Ch. 31]*

The Lorax. Environmental guardian from Dr. Seuss's *The Lorax* who "speaks for the trees" and warns against industrial greed destroying nature, becoming an icon of conservation. *[Ch. 35]*

Edward Lorenz (1917–2008). American mathematician and meteorologist who laid the foundations of chaos theory by discovering the "butterfly effect" in weather patterns. *[Ch. 48]*

Louvois (François-Michel le Tellier) (1641–1691). French Secretary of State for War who reorganized the army and founded the Hôtel des Invalides under King Louis XIV. *[Ch. 44]*

Frank De Luccia. American theoretical physicist who co-formulated the Coleman-De Luccia instanton to describe false vacuum decay within a gravitational framework. *[Ch. 16]*

Andrew D. Ludlow (b. unknown). American physicist at NIST who advances precision metrology by developing neutral ytterbium optical lattice clocks with record-setting stability. *[Ch. 32]*

Rabbi Isaac Luria (Ha'ari) (1534–1572). Sephardic Jewish mystic who founded Lurianic Kabbalah in Safed, transforming post-medieval Jewish thought. *[Ch. 26]*

Aleksandr Lyapunov (1857–1918). Russian mathematician who established the stability theory of dynamical systems and introduced Lyapunov functions and characteristic exponents. *[Ch. 48]*

M. H. Löb (1921–2006). German-born British mathematician who formulated Löb's Theorem, a formal result in provability logic related to Gödel's incompleteness theorems. *[Ch. 17]*

Martin H. Müser (b. unknown). German computational physicist and professor at Saarland University who researches tribology and contact mechanics across scales, and co-authored simulations revealing cold displacement-driven amorphization as the dominant ice lubrication mechanism. *[Ch. 42]*

Mac (b. 1977). Self-appointed head of security at Paddy's Pub from *It's Always Sunny in Philadelphia* who struggles with his sexuality, Catholic faith, and desperate need for his father's approval. *[Ch. 27]*

Maimonides (c. 1138–1204). Medieval Jewish philosopher and jurist who codified Jewish law in the *Mishneh Torah* and synthesized Aristotelian philosophy with the Torah. *[Ch. 43]*

Malachi. Hebrew prophet who is the attributed author of the *Book of Malachi* and the final prophet of the Hebrew Bible. *[Ch. 16]*

Gerald S. Manning (b. unknown). American physical chemist who formulated the counterion condensation theory explaining the electrostatic behavior of polyelectrolytes like DNA. *[Ch. 31]*

Maxim Markevitch (b. unknown). Russian-American astrophysicist who provided direct evidence for dark matter by analyzing the "Bullet Cluster" with *Chandra*. *[Ch. 37]*

Dariusz Matlak (b. unknown). Polish mathematician who, alongside Rafał Latała, verified and popularized Thomas Royen's proof of the Gaussian correlation inequality. *[Ch. 33]*

Ercole Antonio Mattioli (1640–1694). Italian diplomat who betrayed Louis XIV and is a prominent candidate for the identity of the "Man in the Iron Mask." *[Ch. 44]*

Allan Maxam (b. 1942). American molecular biologist who co-invented the Maxam-Gilbert DNA sequencing method, the first widely adopted technique for reading genetic code. *[Ch. 40]*

James Clerk Maxwell (1831–1879). Scottish mathematical physicist who formulated the classical theory of electromagnetic radiation, unifying electricity, magnetism, and light. *[Ch. 04, 13, 28, 45]*

David Francis Mayers (unknown–2023). British mathematician and physicist who provided numerical verification of relativistic orbital contraction in heavy atoms using the EDSAC computer. *[Ch. 01]*

James Maynard (b. 1987). British mathematician who won the 2022 Fields Medal for proving the Duffin-Schaeffer conjecture and refining bounds on prime gaps. *[Ch. 08]*

Pavlo "Pasha" Mazur (b. unknown). Ukrainian clinical embryologist who achieved the first live birth using pronuclear transfer to treat infertility caused by embryo arrest in 2017. *[Ch. 49]*

John McCarthy (1927–2011). American computer scientist who coined the term "artificial intelligence" and invented LISP, the standard programming language for AI. *[Ch. 20]*

Eugene J. Mele (b. 1950). American physicist who theoretically predicted topological insulators and identified the quantum spin Hall effect in graphene with Charles Kane. *[Ch. 11]*

Aaron ben Meïr. Gaon of Eretz Yisrael academy who instigated a major controversy regarding the Jewish calendar calculation in 921–922 against Babylonian authority. *[Ch. 26]*

Angelos Michaelides (b. unknown). British theoretical chemist who investigates the molecular mechanics of water and ice nucleation using computer simulations. *[Ch. 42]*

Hermann Minkowski (1864–1909). German mathematician who developed the four-dimensional concept of "Minkowski spacetime," providing the essential framework for general relativity. *[Ch. 14, 47]*

Shoukhrat Mitalipov (b. 1961). Kazakh-American biologist who successfully created human embryonic stem cells via somatic cell nuclear transfer in 2013.

Joni Mitchell (b. 1943). Canadian singer-songwriter and painter who blended folk and jazz on the albums *Blue* and *Court and Spark*.

Laurens Molenkamp (b. 1956). Dutch physicist who led the first experimental observation of the quantum spin Hall effect, confirming two-dimensional topological insulators. *[Ch. 11]*

William Morton (1819–1868). American dentist who conducted the first successful public demonstration of the use of diethyl ether for surgical anesthesia in 1846. *[Ch. 50]*

Jürgen Moser (1928–1999). German-American mathematician who completed the Kolmogorov-Arnold-Moser theorem regarding the stability of quasi-periodic motions in Hamiltonian systems. *[Ch. 48]*

Moses (c. 14th–13th Century BCE–c. 13th Century BCE). Hebrew prophet and central figure of the *Torah* who led the Exodus and received the Ten Commandments. *[Ch. 26]*

Kary Mullis (1944–2019). American biochemist who invented the Polymerase Chain Reaction technique in 1983, enabling rapid amplification of specific DNA sequences. *[Ch. 40]*

Javier P. Muniain. Mathematical physicist who co-authored the textbook *Gauge Fields, Knots and Gravity* with John Baez in 1994. *[Ch. 04]*

Joseph Myers (b. 1978). British mathematician and software developer who co-discovered the "hat" monotile, the first known single shape to tile the plane aperiodically. *[Ch. 29]*

August Möbius (1790–1868). German mathematician and astronomer who described the Möbius strip, a non-orientable surface with only one side and one boundary. *[Ch. 14]*

Barry Nalebuff (b. 1958). American economist and Yale professor who co-founded Honest Tea and co-authored *Co-opetition* to demonstrate strategic business frameworks. *[Ch. 15]*

John Napier (1550–1617). Scottish mathematician who invented logarithms to significantly simplify calculation and popularized the use of the decimal point. *[Ch. 21]*

Seth Neddermeyer (1907–1988). American physicist who co-discovered the muon and championed the implosion assembly method for plutonium nuclear weapons. *[Ch. 19]*

John von Neumann (1903–1957). Hungarian-American polymath who pioneered modern computer design and co-authored the foundational text *Theory of Games and Economic Behavior*. *[Ch. 03, 41]*

Amos Nevo (b. unknown). Israeli mathematician who researches ergodic theory, Lie groups, and the structure of lattice subgroups at the Technion. *[Ch. 07]*

Simon Newcomb (1835–1909). American astronomer who recalculated the motions of the solar system's major planets and established international standards for astronomical constants.

James Newman (1907–1966). American lawyer and editor who assisted in drafting the Atomic Energy Act of 1946 and compiled *The World of Mathematics*. *[Ch. 17]*

Joanna Newsom (b. 1982). American harpist and composer who creates polyrhythmic arrangements and dense lyrics on albums like *Ys* and *Have One on Me*. *[Ch. 16]*

Isaac Newton (1643–1727). English mathematician and physicist who formulated the laws of motion and universal gravitation in *Philosophiæ Naturalis Principia Mathematica. [Ch. 06, 13, 27]*

M. Peter Nightingale (b. unknown). Theoretical physicist who developed the Phenomenological Renormalization Group and co-authored the seminal TKNN paper. *[Ch. 11]*

M. den Nijs (b. unknown). Dutch theoretical physicist who co-authored the 1982 TKNN paper establishing the topological nature of the Quantum Hall Effect. *[Ch. 11]*

Logen Ninefingers. Exiled barbarian warrior from Joe Abercrombie's *The First Law* trilogy known as "The Bloody-Nine", struggling with a violent berserker alter-ego in the grimdark fantasy series. *[Ch. 22]*

Tomohiro Nishikado (b. 1944). Japanese game designer and engineer who created the arcade game *Space Invaders* at Taito in 1978. *[Ch. 22]*

Richard Nixon (1913–1994). American politician who served as the 37th U.S. president and normalized relations with China before resigning amidst the Watergate scandal. *[Ch. 21]*

Jean-Antoine Nollet (1700–1770). French clergyman and physicist who was the first to document the phenomenon of osmosis using pig bladders in 1748. *[Ch. 31]*

Pierre Nora (b. 1931). French historian and publisher who developed the concept of *lieux de mémoire* and directed the project *Les Lieux de mémoire. [Ch. 38]*

Lennart Norrby. Swedish chemist who demonstrated how relativistic effects cause the liquid state of mercury and the distinct color of gold. *[Ch. 01]*

Giuseppe Occhialini (1907–1993). Italian physicist who confirmed the existence of the positron and co-discovered the pion decay in cosmic rays. *[Ch. 19]*

J. Robert Oppenheimer (1904–1967). American theoretical physicist who led the Manhattan Project to develop the first nuclear weapons, becoming the "father of the atomic bomb." *[Ch. 23]*

Jeremiah Ostriker (b. 1937). American astrophysicist who co-formulated the Ostriker-Peebles criterion, providing theoretical evidence that spiral galaxies require dark matter halos for stability. *[Ch. 37]*

J. W. O'Connell. American researcher who co-authored the 1975 *Science* article providing a definitive real-world demonstration of Simpson's Paradox. *[Ch. 30]*

Thanu Padmanabhan (1957–2021). Indian theoretical physicist who formulated the thermodynamic perspective of gravity, proposing that gravity is an emergent phenomenon rather than a fundamental force. *[Ch. 02]*

Don Page (b. 1948). American theoretical physicist who formulated the "Page curve," a calculation central to the black hole information paradox. *[Ch. 34]*

Kenneth Parcell. Optimistically naive NBC page from *30 Rock* with ambiguous age and hints of immortality, revealed in the finale to be the ageless president of NBC. *[Ch. 06]*

Wolfgang Pauli (1900–1958). Austrian theoretical physicist who formulated the Pauli Exclusion Principle stating no two fermions can occupy the same quantum state. *[Ch. 36]*

M. G. Pavlides (b. unknown). Statistician who utilized geometric probability to quantify the likelihood of Simpson's Paradox in $2 \times 2 \times 2$ contingency tables. *[Ch. 30]*

John Peacock (b. 1956). British cosmologist who directed the 2dF Galaxy Redshift Survey to investigate the large-scale structure of the universe. *[Ch. 06]*

Karl Pearson (1857–1936). English mathematician who established the discipline of mathematical statistics and developed the Pearson product-moment correlation coefficient. *[Ch. 30]*

Jim Peebles (b. 1935). Canadian-American cosmologist who established the theoretical framework of physical cosmology and received the 2019 Nobel Prize. *[Ch. 37]*

Roger Penrose (b. 1931). British mathematical physicist who proved that black hole formation is a robust prediction of the general theory of relativity. *[Ch. 23, 29, 34]*

Arno Penzias (1933–2024). German-born American physicist who co-discovered the cosmic microwave background radiation, providing crucial evidence for the Big Bang theory. *[Ch. 43]*

Joshua ben Perachya (c. 2nd century BCE–c. 1st century BCE). Jewish sage and *Nasi* of the Sanhedrin who famously taught to judge every person favorably in *Pirkei Avot*. *[Ch. 26]*

Michael D. Perlman (b. unknown). American statistician who specializes in theoretical multivariate analysis, focusing on Wishart matrices, probability inequalities, and order-restricted inference. *[Ch. 30]*

Saul Perlmutter (b. 1959). American astrophysicist who shared the 2011 Nobel Prize for discovering that the expansion of the universe is accelerating. *[Ch. 02]*

Markus "Notch" Persson (b. 1979). Swedish programmer known as "Notch" who created *Minecraft* and sold it to Microsoft for $2.5 billion. *[Ch. 22]*

Wilhelm Pfeffer (1845–1920). German plant physiologist who constructed the "Pfeffer cell", the first semipermeable membrane osmometer, to measure osmotic pressure. *[Ch. 31]*

János Pintz (b. 1950). Hungarian mathematician who co-developed the GPY sieve method, demonstrating that prime gaps are arbitrarily smaller than average. *[Ch. 08]*

Kenneth S. Pitzer (1914–1997). American physical chemist who demonstrated that relativistic effects explain mercury's unique low melting point and liquid state. *[Ch. 01]*

Max Planck (1858–1947). German theoretical physicist who won the Nobel Prize for discovering energy quanta and laying the groundwork for quantum mechanics. *[Ch. 27, 36]*

Henri Poincaré (1854–1912). French mathematician and physicist who founded algebraic topology, formulated the Poincaré Conjecture, and laid the foundations of chaos theory. *[Ch. 07, 14, 33, 48]*

Siméon Denis Poisson (1781–1840). French mathematician and physicist who formulated the Poisson distribution and established key concepts in elasticity and potential theory. *[Ch. 13]*

Jean-Victor Poncelet (1788–1867). French mathematician and engineer who founded modern projective geometry through his seminal work *Traité des propriétés projectives des figures*. *[Ch. 07]*

Robert Pound (1919–2010). American physicist who provided the first precise terrestrial confirmation of gravitational redshift using the Mössbauer effect. *[Ch. 06, 28]*

Cecil Powell (1903–1969). British physicist who developed the photographic emulsion method and discovered the pion in cosmic rays in 1947. *[Ch. 19]*

John Henry Poynting (1852–1914). British physicist who formulated the Poynting vector in 1884, describing the magnitude and direction of electromagnetic energy flow. *[Ch. 04]*

Terry Pratchett (1948–2015). English author who wrote the *Discworld* series, comprising 41 satirical fantasy novels that parodied cultural and political themes. *[Ch. 23]*

Hillel Pratt (b. unknown). Israeli neurophysiologist who researches auditory and somatosensory evoked potentials and stimulus response in the auditory cortex. *[Ch. 05]*

The Preacher (Ecclesiastes). Philosophical voice of the biblical book *Ecclesiastes* who meditates on life's futility with "vanity of vanities, all is vanity", exploring meaning and mortality. *[Ch. 34]*

Huw Price (b. 1953). Australian philosopher of physics known for work on causation, the arrow of time, and pragmatism in science; Bertrand Russell Professor of Philosophy at Cambridge (2011–2020). *[Ch. 30]*

Joseph Priestley (1733–1804). English natural philosopher who independently discovered oxygen, which he termed "dephlogisticated air", and invented carbonated water. *[Ch. 35]*

Gavrilo Princip (1894–1918). Bosnian Serb revolutionary who assassinated Archduke Franz Ferdinand in Sarajevo in 1914, triggering the outbreak of World War I.

Edward Purcell (1912–1997). American physicist who shared the 1952 Nobel Prize for independently discovering nuclear magnetic resonance in condensed matter. *[Ch. 28]*

Pekka Pyykkö (b. 1941). Finnish physical chemist who demonstrated that relativistic effects determine the yellow color of gold and liquid state of mercury. *[Ch. 01]*

Stephen Quake (b. 1969). American bioengineer who pioneered microfluidic large-scale integration and invented non-invasive prenatal testing using cell-free fetal DNA.

Isidor Rabi (1898–1988). American physicist who received the Nobel Prize for developing the resonance method for recording magnetic properties of atomic nuclei. *[Ch. 19]*

Johann Radon (1887–1956). Austrian mathematician who developed the Radon transform, providing the theoretical basis for medical computed tomography. *[Ch. 41]*

Ilan Ramon (1954–2003). Israeli Air Force Colonel and first astronaut who died when Space Shuttle *Columbia* disintegrated during reentry in 2003. *[Ch. 26]*

Norman Ramsey (1915–2011). American physicist who invented the separated oscillatory fields method, which forms the basis of modern atomic clocks. *[Ch. 28, 32]*

Scott Randall (b. unknown). American astrophysicist who analyzed the Bullet Cluster to place robust constraints on the self-interaction cross-section of dark matter. *[Ch. 37]*

Terence Ranger (1929–2015). British historian of Zimbabwe who emphasized African agency and co-edited *The Invention of Tradition* with Eric Hobsbawm. *[Ch. 38]*

Lord Rayleigh (1842–1919). British physicist who discovered argon and formulated the theory of Rayleigh scattering explaining the blue color of the sky. *[Ch. 27]*

Agustín Rayo (b. 1973). Mexican philosopher and logician who defined "Rayo's number," a colossal integer derived using first-order set theory. *[Ch. 17]*

Thomas Reid (1710–1796). Scottish philosopher who founded the School of Common Sense and defended direct realism against the skepticism of David Hume. *[Ch. 50]*

Alan W. Rempel (b. unknown). American geophysicist who applies mathematical modeling to ice mechanics and co-authored "The physics of premelting" in *Reviews of Modern Physics*. *[Ch. 42]*

Judith Resnik (1949–1986). American electrical engineer and astronaut who was the second American woman in space and died in the 1986 *Challenger* disaster. *[Ch. 26]*

Marjorie Rice (1923–2017). American amateur mathematician who discovered four new families of convex pentagons that tile the plane using her own notation. *[Ch. 29]*

Frank Richards (1883–1961). British soldier and author who wrote *Old Soldiers Never Die*, a memoir documenting the 1914 Christmas Truce. *[Ch. 38]*

Bernhard Riemann (1826–1866). German mathematician who formulated Riemannian geometry and the Riemann Hypothesis regarding the distribution of prime numbers. *[Ch. 08]*

Adam Riess (b. 1969). American astrophysicist who shared the 2011 Nobel Prize for the discovery of the accelerating expansion of the universe. *[Ch. 02]*

Shimon the Righteous. Jewish High Priest of the Second Temple period who taught that the world rests on three pillars: Torah, Divine service, and loving-kindness. *[Ch. 26]*

Don Ringe (b. 1954). American linguist who specializes in Proto-Germanic reconstruction and applies computational cladistics to the study of language phylogeny. *[Ch. 05]*

Warren Robinett (b. 1951). American video game designer who developed the Atari title *Adventure* and created the first widely known "Easter egg." *[Ch. 22]*

Matthieu Rosenfeld (b. unknown). French mathematician and assistant professor at LIRMM, Université de Montpellier, who specializes in combinatorics on words and computer-assisted proofs, and proved the Lonely Runner Conjecture for eight and nine runners. *[Ch. 09]*

Bruno Rossi (1905–1993). Italian-American experimental physicist who invented the electronic coincidence circuit and initiated the field of extra-solar X-ray astronomy. *[Ch. 19]*

Hugo Rossi (b. 1935). American mathematician who specialized in complex analysis and co-authored the foundational text *Analytic Functions of Several Complex Variables. [Ch. 21]*

Jonathan Rothberg (b. 1963). American chemical engineer who pioneered massively parallel DNA sequencing and created the first handheld semiconductor-based ultrasound.

Thomas Royen (b. 1947). German statistician who proved the Gaussian Correlation Inequality in 2014, resolving a conjecture open since the 1950s. *[Ch. 33]*

Samuel Ruben (1913–1943). American chemist who co-discovered the radioactive isotope Carbon-14, which became fundamental to radiocarbon dating and metabolic tracing. *[Ch. 35]*

Vera Rubin (1928–2016). American astronomer who analyzed galaxy rotation curves, providing the first robust observational evidence for the existence of dark matter. *[Ch. 37]*

Donald Saari (b. 1940). American mathematician and economist who utilizes dynamical systems to analyze voting paradoxes and the Newtonian N-body problem. *[Ch. 09]*

Julius von Sachs (1832–1897). German botanist who founded modern experimental plant physiology and identified starch as the product of photosynthesis. *[Ch. 35]*

Marianna Safronova (b. unknown). American theoretical physicist who develops high-precision atomic clocks to search for dark matter and variations in fundamental constants. *[Ch. 32]*

Augusto Sagnotti (b. 1957). Italian theoretical physicist who demonstrated the non-renormalizability of pure Einstein gravity and developed the "orientifold" construction. *[Ch. 36]*

Bénigne Dauvergne de Saint-Mars (c. 1626–1708). French prison governor who served as the lifelong custodian of the unidentified captive known as the "Man in the Iron Mask." *[Ch. 44]*

Masahiro Sakurai (b. 1970). Japanese video game director best known for creating the *Kirby* and *Super Smash Bros.* franchises. *[Ch. 22]*

Matthew Sands (1920–2014). American physicist who served as a co-author of *The Feynman Lectures on Physics* with Richard Feynman and Robert Leighton. *[Ch. 04]*

Frederick Sanger (1918–2013). British biochemist who received two Nobel Prizes for determining the structure of insulin and developing methods for DNA sequencing. *[Ch. 40]*

Mark Satterthwaite (b. unknown). American economist who independently proved the Gibbard-Satterthwaite theorem, establishing that non-dictatorial voting systems are susceptible to strategic manipulation. *[Ch. 09]*

August Schleicher (1821–1868). German linguist who formulated the *Stammbaumtheorie* model of language evolution and pioneered the reconstruction of Proto-Indo-European. *[Ch. 05]*

Brian Schmidt (b. 1967). American-Australian astrophysicist who shared the 2011 Nobel Prize in Physics for discovering the accelerating expansion of the universe. *[Ch. 02]*

Erwin Schrödinger (1887–1961). Austrian physicist who formulated the wave equation and devised the "Schrödinger's cat" thought experiment to illustrate quantum superposition. *[Ch. 15, 36]*

Aaron Schurger (b. unknown). American cognitive neuroscientist who reinterpreted the "readiness potential" as stochastic neural noise accumulation, challenging Libet's evidence against free will. *[Ch. 50]*

Karl Schwarzschild (1873–1916). German physicist and astronomer who derived the Schwarzschild metric, the first exact solution to Einstein's field equations. *[Ch. 23]*

Julian Schwinger (1918–1994). American physicist who shared the 1965 Nobel Prize in Physics for developing renormalization in quantum electrodynamics. *[Ch. 47]*

Kurt Schütte (1909–1998). German mathematician focused on proof theory who independently identified the Feferman-Schütte ordinal Γ_0, characterizing the limits of predicative arithmetic. *[Ch. 17]*

Alexandru Scorpan (b. unknown). Mathematician specializing in geometric topology who authored *The Wild World of 4-Manifolds* and oversees production at Mathematical Sciences Publishers. *[Ch. 24]*

John Searle (b. 1932). American philosopher who formulated the "Chinese Room" thought experiment to challenge claims of strong artificial intelligence. *[Ch. 20]*

Angelo Secchi (1818–1878). Italian Jesuit astronomer who pioneered astronomical spectroscopy by creating the first comprehensive spectral classification system for stars. *[Ch. 27]*

Atle Selberg (1917–2007). Norwegian mathematician who developed the Selberg sieve, trace formula, and an elementary proof of the Prime Number Theorem. *[Ch. 08]*

Semirhage (Fictional). One of the Forsaken from Robert Jordan's *The Wheel of Time*, a former renowned healer who turned to the Shadow and became infamous for sadistic torture using the One Power. *[Ch. 48]*

Josef Sewald. German soldier who documented the spontaneous Christmas Truce of 1914 in letters serving as a definitive primary source. *[Ch. 38]*

Shammai (c. 50 BCE–c. 30 CE). Jewish sage who co-led the *Zugot* and founded the rigorous House of Shammai in opposition to Hillel. *[Ch. 26]*

⚗ **Dan Shechtman** (b. 1941). Israeli materials scientist who was awarded the 2011 Nobel Prize in Chemistry for the discovery of quasicrystals. *[Ch. 29]*

Shemaya (Unknown–c. late 1st century BCE). Jewish sage and *Nasi* of the Sanhedrin who constituted the fourth of the *Zugot* alongside Abtalion. *[Ch. 26]*

Shimon ben Shetach. Judean Pharisaic leader who restored Pharisaic authority and instituted the *Ketubah* to secure property rights for women. *[Ch. 26]*

⚗ **Osamu Shimomura** (1928–2018). Japanese organic chemist and marine biologist who discovered and isolated Green Fluorescent Protein, sharing the 2008 Nobel Prize in Chemistry.

Martin Silenus. Fictional character from the *Hyperion Cantos* series by Dan Simmons, first appearing in the 1989 novel *Hyperion*. *[Ch. 20]*

Edward H. Simpson (1922–2019). British statistician and civil servant who described Simpson's paradox in the 1951 paper "The Interpretation of Interaction in Contingency Tables." *[Ch. 30]*

Yakov Sinai (b. 1935). Russian-American mathematician who introduced Kolmogorov-Sinai entropy and analyzed "Sinai billiards" to prove the existence of chaos.

Ben Slater (b. unknown). British computational chemist who utilizes atomistic simulations to analyze defects in crystalline solids and the surface chemistry of ice. *[Ch. 42]*

Vesto Slipher (1875–1969). American astronomer who measured the radial velocities of spiral nebulae, providing the first evidence for the expanding universe. *[Ch. 27]*

David Smith (b. c. 1958). British amateur mathematician who discovered the "hat", the first known aperiodic monotile, resolving a major geometric tiling problem. *[Ch. 29]*

Hartland Snyder (1913–1962). American theoretical physicist who collaborated with J. Robert Oppenheimer to mathematically predict the implosion of stars into black holes. *[Ch. 23]*

Antigonus of Socho. Jewish sage of the Second Temple period who urged service to God out of love rather than reward. *[Ch. 26]*

Arnold Sommerfeld (1868–1951). German theoretical physicist who developed the Bohr-Sommerfeld model and introduced the fine-structure constant to explain spectral lines. *[Ch. 01, 04]*

Paul Sonnino (b. 1930). American historian specializing in Louis XIV who identified the "Man in the Iron Mask" as the valet Eustache Dauger. *[Ch. 44]*

Lorenzo Sorbo (b. unknown). Theoretical physicist who collaborated with Andreas Albrecht in 2004 to revitalize the discussion of the Boltzmann Brain paradox. *[Ch. 34]*

Srinivasa Sourirajan (1923–2022). Indian chemical engineer who co-invented the asymmetric cellulose acetate membrane with Sidney Loeb, making reverse osmosis practical. *[Ch. 31]*

Eddard Stark. Lord of Winterfell from George R.R. Martin's *A Song of Ice and Fire* whose unwavering honor and sense of duty lead to his tragic downfall in the political machinations of Westeros. *[Ch. 28]*

Alexei Starobinsky (1948–2023). Russian theoretical physicist who formulated one of the earliest models of cosmic inflation, known as Starobinsky inflation.

Paul Steinhardt (b. 1952). American theoretical physicist who introduced the concept of quasicrystals and co-proposed the cyclic model of the universe. *[Ch. 34]*

Sergey V. Sukhomlinov (b. unknown). Computational physicist at the Department of Materials Science and Engineering, Saarland University, who researches mechanochemistry and co-authored molecular dynamics simulations of displacement-driven amorphization in ice sliding. *[Ch. 42]*

Leonard Susskind (b. 1940). American theoretical physicist who is a founding father of string theory and formulated the holographic principle to address the black hole information paradox. *[Ch. 34]*

Leo Szilard (1898–1964). Hungarian-American physicist who conceived the nuclear chain reaction and initiated the Manhattan Project via the Einstein-Szilárd letter. *[Ch. 45]*

Judah ben Tabbai. Jewish scholar of the Second Temple period who served as one of the *Zugot* and established foundational judicial ethics. *[Ch. 26]*

David Tabor (1913–2005). British physicist who co-authored *The Friction and Lubrication of Solids* and showed that frictional heating creates a water film on ice. *[Ch. 42]*

Lincoln Taiz (b. unknown). American plant biologist who characterized the vacuolar H^+-ATPase and co-authored the foundational textbook *Plant Physiology and Development. [Ch. 35]*

M. Takata. Japanese physicist who experimentally verified the Babinet-Fresnel principle using ultrasonic waves, demonstrating that optical diffraction laws apply to acoustics. *[Ch. 13]*

Terence Tao (b. 1975). Australian-American mathematician who received the 2006 Fields Medal and proved that prime numbers contain arbitrarily long arithmetic progressions. *[Ch. 08]*

Alfred Tarski (1901–1983). Polish-American logician and mathematician who formalized the semantic conception of truth and co-discovered the Banach-Tarski paradox with Stefan Banach. *[Ch. 03]*

Joan Taylor (b. unknown). Australian independent mathematician who co-discovered the Socolar-Taylor tile, a hexagonal structure that stood as the leading approximation of an aperiodic monotile. *[Ch. 29]*

Max Tegmark (b. 1967). Swedish-American physicist who proposed the "Mathematical Universe Hypothesis" positing that the physical universe is a mathematical structure. *[Ch. 24]*

James Thomson (1822–1892). British physicist and engineer who theoretically predicted that the freezing point of water decreases with increasing pressure. *[Ch. 42]*

David J. Thouless (1934–2019). British theoretical physicist who pioneered the application of topology to condensed matter systems and co-developed the Kosterlitz-Thouless transition. *[Ch. 11]*

J.R.R. Tolkien (1892–1973). English writer and philologist who authored *The Hobbit* and *The Lord of the Rings*, establishing modern fantasy literature. *[Ch. 01]*

Robert Tomasulo (1934–2008). American computer architect who invented Tomasulo's algorithm, a dynamic scheduling technique enabling out-of-order execution in high-performance processors. *[Ch. 18]*

Giulio Tononi (b. 1960). Italian neuroscientist and psychiatrist who formulated Integrated Information Theory to investigate the physical substrates of subjective experience. *[Ch. 50]*

Charles H. Townes (1915–2015). American physicist whose work in quantum electronics led to the invention of the maser and the laser.

Tanupat Trakulthongchai (b. unknown). Mathematician who independently proved the Lonely Runner Conjecture for nine and ten runners in 2025, using a sieving method that refined

Rosenfeld's computer-assisted backtracking approach. *[Ch. 09]*

Alan Turing (1912–1954). British mathematician who laid the foundations of computer science and broke German Enigma codes during World War II. *[Ch. 20]*

Stephen Turner (b. unknown). American biophysicist and entrepreneur who invented zero-mode waveguides and founded Pacific Biosciences to pioneer single-molecule real-time sequencing.

Neil Turok (b. 1958). South African theoretical physicist who co-developed the cyclic model with Paul Steinhardt and founded the African Institute for Mathematical Sciences. *[Ch. 34]*

Mark Twain (1835–1910). American author who wrote *The Adventures of Tom Sawyer* and *Adventures of Huckleberry Finn*, utilizing vernacular speech to critique society. *[Ch. 13]*

John Tyndall (1820–1893). Irish physicist who discovered the Tyndall effect and proved water vapor and carbon dioxide absorb heat. *[Ch. 27]*

William G. Unruh (b. 1945). Canadian physicist who discovered the Unruh effect, demonstrating that accelerating observers perceive a vacuum as thermal particles. *[Ch. 47]*

Oswald Veblen (1880–1960). American mathematician who co-formulated the Veblen-Young axioms and played a definitive role in founding the Institute for Advanced Study. *[Ch. 17]*

Lars Vegard (1880–1963). Norwegian physicist who established "Vegard's Law" for solid solutions and identified nitrogen and oxygen in the aurora borealis. *[Ch. 31]*

Craig Venter (b. 1946). American biotechnologist who pioneered whole-genome shotgun sequencing and led the private effort to sequence the human genome. *[Ch. 40]*

Alexander Vilenkin (b. 1949). Soviet-born American cosmologist who introduced the theory of eternal inflation and the quantum creation of the universe from "nothing." *[Ch. 34]*

Leonardo da Vinci (1452–1519). Italian Renaissance polymath who painted the *Mona Lisa* and *The Last Supper* while studying anatomy and mechanics. *[Ch. 45]*

Giuseppe Vitali (1875–1932). Italian mathematician who constructed the Vitali set, proving the existence of non-measurable sets of real numbers. *[Ch. 03]*

Lucas J. van Vliet (b. 1966). Dutch Professor of Quantitative Imaging who developed recursive Gaussian derivative filters and structure tensors for image analysis. *[Ch. 41]*

Voltaire (1694–1778). French Enlightenment writer and philosopher who advocated for civil liberties and popularized the legend of the "Man in the Iron Mask." *[Ch. 44]*

S. S. Wainer (unknown–2021). British mathematical logician who developed the Wainer hierarchy to classify computable functions by growth rate using ordinal numbers. *[Ch. 17]*

Hao Wang (1921–1995). Chinese-American logician and philosopher who introduced Wang tiles to investigate the decision problem for tiling the plane. *[Ch. 29]*

John Waterston (1811–1883). Scottish physicist and civil engineer who independently developed the kinetic theory of gases in 1845, in a paper rejected as "nonsense." *[Ch. 28]*

James Watson (b. 1928). American molecular biologist who co-discovered the double helix structure of DNA with Francis Crick in 1953. *[Ch. 40]*

The Weakerthans (1997–2015). Canadian indie rock band led by John K. Samson who produced literary punk-folk albums like *Reconstruction Site*. *[Ch. 44]*

Brian Weber (b. unknown). American condensed matter physicist who used molecular dynamics to show ice friction is mediated by a mobile, disordered layer. *[Ch. 42]*

Karl Weierstrass (1815–1897). German mathematician cited as the "father of modern analysis" for establishing the strict epsilon-delta definition of limits. *[Ch. 03]*

Steven Weinberg (1933–2021). American theoretical physicist who formulated the electroweak theory unifying the weak force and electromagnetism, earning the Nobel Prize in Physics. *[Ch. 02, 36]*

Stanley Weintraub (1929–2019). American historian and biographer who authored approximately 50 books, including *Silent Night*, on military narratives and the Victorian era. *[Ch. 38]*

John Wettlaufer (b. unknown). American physicist and geophysicist who integrates theory and experiment to investigate soft condensed matter and the physics of ice. *[Ch. 42]*

Hermann Weyl (1885–1955). German mathematician who applied group theory to quantum mechanics and introduced gauge invariance in *Raum, Zeit, Materie*. *[Ch. 11]*

John Archibald Wheeler (1911–2008). American theoretical physicist who revitalized general relativity and popularized terms such as "black hole" and "wormhole." *[Ch. 06, 14, 23]*

Gerald Whitrow (1912–2000). British mathematician who demonstrated that life requires three spatial dimensions, providing a physical basis for the anthropic principle. *[Ch. 24]*

Wilhelm Wien (1864–1928). German physicist who derived the displacement law and received the 1911 Nobel Prize for laws governing heat radiation. *[Ch. 27]*

Eugene Wigner (1902–1995). Hungarian-American physicist who received the Nobel Prize for research concerning the structure of the atomic nucleus and elementary particles. *[Ch. 24]*

Andrew Wiles (b. 1953). British mathematician who proved Fermat's Last Theorem in 1995 by establishing the modularity theorem for semistable elliptic curves. *[Ch. 08]*

A. O. Williams (1913–2002). American physicist who calculated relativistic self-consistent fields for atomic structure and analyzed sound propagation in underwater environments. *[Ch. 01]*

Robert Wilson (b. 1936). American radio astronomer who co-discovered the cosmic microwave background radiation in 1964, providing critical evidence for the Big Bang. *[Ch. 43]*

Wit (b. unknown). Ancient immortal world-hopper from Brandon Sanderson's *Stormlight Archive* who serves as King's Wit on Roshar, wielding acerbic humor and cryptic wisdom while pursuing a millennia-spanning agenda. *[Ch. 18]*

Robert Burns Woodward (1917–1979). American organic chemist who received the Nobel Prize for synthesizing complex natural products and developed the Woodward-Hoffmann rules. *[Ch. 46]*

Allison Wu. Undergraduate researcher who authored an expository paper on the Banach-Tarski Paradox as part of the 2008 University of Chicago VIGRE REU program. *[Ch. 03]*

Louis XIV (1638–1715). French king who established a centralized absolute monarchy and reigned for 72 years, the longest in European history. *[Ch. 44]*

Itai Yanai (b. 1975). Israeli-American computational biologist and one of the pioneers of single-cell transcriptomics (CEL-Seq), co-author of *The Society of Genes*, and founding director of the Institute for Computational Medicine at NYU. *[Ch. 40]*

Rabbi Yehoshua (c. 40 CE–c. 130 CE). Jewish Tanna who served in the Second Temple and established the supremacy of the rabbinic court over astronomical observation. *[Ch. 26]*

Yose ben Yochanan. Jewish sage who served as *Av Beit Din* and formed the first of the *Zugot* with Yose ben Yoezer. *[Ch. 26]*

Yose ben Yoezer (unknown–c. 162 BCE). Jewish sage who served as the first Nasi of the Zugot and participated in the first recorded halachic dispute in the *Mishnah*. *[Ch. 26]*

Thomas Young (1773–1829). English polymath who demonstrated the wave nature of light and made key contributions to deciphering the Rosetta Stone. *[Ch. 13]*

Hideki Yukawa (1907–1981). Japanese theoretical physicist who predicted the existence of the meson to explain nuclear forces, becoming the first Japanese Nobel laureate. *[Ch. 19]*

G. Udny Yule (1871–1951). British statistician who described the Yule-Simpson effect and authored the textbook *An Introduction to the Theory of Statistics*. *[Ch. 30]*

Cem Yıldırım (b. 1961). Turkish mathematician who co-developed the GPY sieve method proving that small gaps between prime numbers occur infinitely often. *[Ch. 08]*

David ben Zakkai (unknown–c. 940). Babylonian Jewish Exilarch who asserted authority in the calendar dispute and engaged in a power struggle with Saadia Gaon. *[Ch. 26]*

Rabban Yochanan ben Zakkai (c. 30 BCE–c. 90 CE). Jewish sage who transformed Judaism into a Rabbinic tradition by establishing the academy at Yavneh after the Second Temple's destruction. *[Ch. 26]*

Dennis Zaritsky (b. 1964). American astronomer who co-authored the 2006 study on the Bullet Cluster providing direct empirical evidence for dark matter. *[Ch. 37]*

Eduardo Zeiger (b. unknown). American plant physiologist who co-authored the textbook *Plant Physiology* and established the role of blue light in stomatal regulation. *[Ch. 35]*

Yakov Zel'dovich (1914–1987). Soviet theoretical physicist who formulated the Sunyaev-Zel'dovich effect and predicted that rotating black holes amplify incident waves.

John Zhang (b. unknown). American fertility specialist who oversaw the birth of the world's first baby conceived via spindle nuclear transfer in 2016. *[Ch. 49]*

Shoucheng Zhang (1963–2018). American theoretical physicist who discovered the quantum spin Hall effect and predicted the first three-dimensional topological insulators. *[Ch. 11]*

Yitang Zhang (b. 1955). Chinese-American mathematician who proved the existence of a finite bound on the gaps between consecutive prime numbers. *[Ch. 08]*

Tamar Ziegler (b. 1971). Israeli mathematician specializing in ergodic theory who focuses on polynomial progressions in prime numbers and the Gowers norms. *[Ch. 08]*

Howard Zimmerman (1926–2012). American chemist who developed the Zimmerman-Traxler transition state model and formulated the Möbius-Hückel concept for pericyclic reactions. *[Ch. 46]*

Max Zorn (1906–1993). German-American mathematician who formulated Zorn's Lemma, a principle equivalent to the Axiom of Choice and essential for non-constructive proofs. *[Ch. 07]*

Valery Zukin (b. unknown). Ukrainian physician who led the team reporting the first birth using pronuclear transfer to treat infertility in 2017. *[Ch. 49]*

Fritz Zwicky (1898–1974). Swiss-American astrophysicist who inferred the existence of dark matter and coined the term "supernova" with Walter Baade. *[Ch. 31, 37]*

Hans Christian Ørsted (1777–1851). Danish physicist and chemist who discovered in 1820 that electric currents create magnetic fields, establishing the discipline of electromagnetism. *[Ch. 04]*

Bibliography

Scholarly Works

(1914). 'Christmas at the Front/An Interrupted Truce'. *Morning Post.* [Ch. 38]

Albert Einstein (1905). 'Zur Elektrodynamik bewegter Körper'. *Annalen der Physik* **322**(10), 891–921. [Ch. 1, 14, 19]

Albert Einstein (1907). Über das Relativitätsprinzip und die aus demselben gezogenen Folgerungen. Jahrbuch der Radioaktivität und Elektronik.

Albert Einstein (1916). Die Grundlage der allgemeinen Relativitätstheorie. Annalen der Physik. [Ch. 6]

Albert Einstein & Nathan Rosen (1935). The Particle Problem in the General Theory of Relativity. Physical Review.

Albrecht Dürer (1525). *Underweysung der messung mit dem zirckel und richtscheyt in Linien ebnen unnd gantzen corporen.* Hieronymus Andreae, Nuremberg. [Ch. 34]

Alexandru Scorpan (2005). *The Wild World of 4-Manifolds.* American Mathematical Society. [Ch. 24]

Allan Gibbard (1973). 'Manipulation of Voting Schemes: A General Result'. *Econometrica 41(4).* [Ch. 9]

Anaxagoras, c. 450 BCE. Fragment DK 59 A42, reported by various ancient sources including Diogenes Laërtius. [Ch. 10]

Annie Jump Cannon, Edward C. Pickering (1918-1924). *Henry Draper Catalog.* Harvard College Observatory. [Ch. 27]

Anonymous (1915). 'Letter from a doctor attached to the Rifle Brigade'. *The Times.* [Ch. 39]

Archimedes (c. 250 BCE). *The Sand Reckoner.* [Ch. 17]

Ashlock & Tillotson (1993). An automaton for game-theoretic representation. Proceedings of the 2nd Annual Conference on Evolutionary Programming.

Atila, A., Sukhomlinov, S. V., and Müser, M. H. (2024). *Phys. Rev. Lett.*, **133**, 236201. [Ch. 42]

Augustine of Hippo (c. 397). *Confessions.* Book XI. [Ch. 43]

Axelrod (1984). The Evolution of Cooperation. Basic Books.

B. T. Draine (2003). 'Interstellar Dust Grains'. *Annual Review of Astronomy and Astrophysics.* [Ch. 27]

Babinet (1837). Mémoire sur les couleurs des réseaux. Comptes Rendus de l'Académie des Sciences.

Ballard (1981). Generalizing the Hough transform to detect arbitrary shapes. Pattern Recognition.

Banach, S. and Tarski, A. (1924). *Fund. Math.* **6** [Ch. 3]

Bardeen, Carter & Hawking (1973). The Four Laws of Black Hole Mechanics. Communications in Mathematical Physics.

Barkan, E., Biham, E., Keller, N. (2008). Instant ciphertext-only cryptanalysis of GSM encrypted communication. *J. Cryptology* 21(3):392–429. [Ch. 12]

Baryshnikov, Y. (1997). Topological and discrete social choice: in a search of a theory. *Social Choice and Welfare*, **14**(2), 199–209. [Ch. 9]

Bassham, Benson & Calvin (1950). The path of carbon in photosynthesis. Journal of Biological Chemistry.

Benford (1938). The law of anomalous numbers. Proceedings of the American Philosophical Society.

Benjamin W. Fortson IV (2010). *Indo-European Language and Culture: An Introduction.* Wiley-Blackwell. [Ch. 5]

Bennett, C. H. (1982). *The Thermodynamics of Computation — A Review.* Int. J. Theor. Phys., 21, 905–940. [Ch. 45]

Bentley et al. (2008). *Nature* 456:53–59. [Ch. 40]

Berkson (1946). Limitations of the application of fourfold table analysis to hospital data. Biometrics Bulletin.

Bernevig, B. A., Hughes, T. L., and Zhang, S.-C. (2006). Quantum Spin Hall Effect and Topological Phase Transition in HgTe Quantum Wells. *Science,* **314**(5806), 1757–1761. [Ch. 11]

Bethe, H. A. (1939). Energy Production in Stars. *Phys. Rev.,* **55**, 434–456. [Ch. 10]

Bloch (1929). Über die Quantenmechanik der Elektronen in Kristallgittern. Zeitschrift für Physik.

Bogoliubov (1947). On the theory of superfluidity. Journal of Physics (USSR).

Boltzmann (1877). Über die Beziehung zwischen dem zweiten Hauptsatze der mechanischen Wärmetheorie und der Wahrscheinlichkeitsrechnung. Wiener Berichte. [Ch. 28, 34, 45]

Book of Joshua. Hebrew Bible. [Ch. 38]

Bopp (1816). Über das Conjugationssystem der Sanskritsprache in Vergleichung mit jenem der griechischen, lateinischen, persischen und germanischen Sprache. Andreäische Buchhandlung.

Boyle & Smith (1970). Charge-Coupled Semiconductor Devices. Bell System Technical Journal.

Brascamp & Lieb (1976). Best constants in Young's inequality, its converse, and its generalization to more than three functions. Advances in Mathematics.

C. J. Isham (1999). *Modern Differential Geometry for Physicists.* World Scientific. [Ch. 14]

Cantor (1891). Über eine elementare Frage der Mannigfaltigkeitslehre. Jahresbericht der Deutschen Mathematiker-Vereinigung.

Captain Josef Sewald, 17th Bavarian Regiment. Quoted in Stanley Weintraub, *Silent Night: The Story of the World War I Christmas Truce* (New York: Free Press, 2001). [Ch. 38]

Carroll, S. M. (2001). The cosmological constant. *Living Rev. Relativ.,* 4, 1. [Ch. 2]

Casimir (1948). On the attraction between two perfectly conducting plates. Proceedings of the Royal Netherlands Academy of Arts and Sciences.

Chandrasekhar (1931). The Maximum Mass of Ideal White Dwarfs. The Astrophysical Journal.

Christenson, J. H., Cronin, J. W., Fitch, V. L., and Turlay, R. (1964). Evidence for the 2π Decay of the K_2^0 Meson. *Phys. Rev. Lett.* **13**, 138–140. [Ch. 19]

Clausius (1850). Ueber die bewegende Kraft der Wärme. Annalen der Physik. [Ch. 42]

Clausius (1870). On a Mechanical Theorem Applicable to Heat. Philosophical Magazine. [Ch. 42]

Clayton, D. D. (1983). *Principles of Stellar Evolution and Nucleosynthesis.* University of Chicago Press. [Ch. 10]

Clowe, D., Bradač, M., Gonzalez, A. H., Markevitch, M., Randall, S. W., Jones, C., and Zaritsky, D. (2006). A Direct Empirical Proof of the Existence of Dark Matter. *ApJL*, **648**, L109-L113. [Ch. 37]

Coblentz, W.W. (1912). *A Physical Study of the Firefly.* Carnegie Institution of Washington, Publ. 164. [Ch. 25]

Compton (1923). A Quantum Theory of the Scattering of X-rays by Light Elements. Physical Review.

Condorcet (1785). Essai sur l'application de l'analyze à la probabilité des décisions rendues à la pluralité des voix. Imprimerie Royale.

D. G. Saari (1994). *Geometry of Voting.* Springer-Verlag. [Ch. 9]

Dahl & Nygaard (1966). SIMULA – an ALGOL-based simulation language. Communications of the ACM.

Daniel C. Dennett (1987). *The Intentional Stance.* MIT Press. [Ch. 20]

Daniel J. Bernstein (1999). 'Bernstein v. United States Department of Justice'. *9th Circuit, 176 F.3d 1132 (1999); withdrawn on rehearing en banc, 192 F.3d 1308.* [Ch. 17]

David Hilbert, "Über das Unendliche," *Mathematische Annalen* 95 (1926), pp. 161–190. [Ch. 1]

David Hilbert, as quoted in Calvin C. Clawson, *Mathematical Mysteries: The Beauty and Magic of Numbers* (New York: Basic Books, 1999), p. 258. [Ch. 1]

David J. Chalmers (1995). 'Facing Up to the Problem of Consciousness'. *Journal of Consciousness Studies.* [Ch. 1, 50]

David J. Chalmers (1996). *The Conscious Mind: In Search of a Fundamental Theory.* Oxford University Press. [Ch. 1, 20]

Davis, R., Harmer, D. S., and Hoffman, K. C. (1968). Search for Neutrinos from the Sun. *Phys. Rev. Lett.* **20**, 1205–1209. [Ch. 10]

de la Vallée Poussin, C.-J. (1896). Recherches analytiques sur la théorie des nombres premiers. Annales de la Société Scientifique de Bruxelles.

Debye, P. (1923). Physikalische Eigenschaften von Lösungen und Theorie der Elektrolyte. *Phys. Z.* 24:334–338. [Ch. 31]

Degrassi et al., *JHEP* **08**, 098 (2012). [Ch. 16]

Diels & Alder (1928). Synthesen in der hydroaromatischen Reihe. Justus Liebigs Annalen der Chemie.

Dirac, P. A. M. (1928). 'The Quantum Theory of the Electron'. *Proc. R. Soc. Lond. A* **117**, 610–624. [Ch. 1]

Dodelson, Scott (2003). *Modern Cosmology.* Academic Press. [Ch. 43]

Donaldson (1983). An application of gauge theory to four-dimensional topology. Journal of Differential Geometry.

Du Junca, Étienne (1690-1705). 'Registres des entrées et des sorties de la Bastille'. *Bibliothèque de l'Arsenal, Paris.* [Ch. 39]

Dubois (1887). Ueber die Magnetisirungsconstante des Wassers und den absoluten Werth der Susceptibilität des Eisens. Annalen der Physik und Chemie.

Duda, R. O., and Hart, P. E. (1972). Use of the Hough Transformation to Detect Lines and Curves in Pictures. *Commun. ACM* 15(1), 11–15. [Ch. 41]

Dunkel, J., Hilbert, S. (2014). *Consistent Thermostatistics Forbids Negative Absolute Temperatures.* Nat. Phys., 10, 67. [Ch. 28]

Dyson (1949). The S-Matrix in Quantum Electrodynamics. Physical Review. [Ch. 15]

Dyson, L., Kleban, M., Susskind, L. (2002). Disturbing implications of a cosmological constant. *JHEP* 0210:011. [Ch. 34]

Eccles & Jordan (1919). A trigger relay utilizing three-electrode thermionic vacuum tubes. The Radio Review.

Ecclesiastes 1:2, Hebrew Bible. [Ch. 34]

Eddington (1931). The End of the World: from the Standpoint of Mathematical Physics. Nature. [Ch. 34]

Edward Kasner and James R. Newman (1940). *Mathematics and the Imagination.* Simon & Schuster. [Ch. 17]

Egan, G. (2018). 'Superpermutations'. *gregegan.net/SCIENCE/Superpermutations/.* [Ch. 39]

Ehrenfest, P. (1917). *In what way does it become manifest in the fundamental laws of physics that space has three dimensions?* Proceedings of the Amsterdam Academy. [Ch. 24]

Eisenstein, D. J. et al. (2005). Detection of the Baryon Acoustic Peak. *ApJ* **633**, 560–574. [Ch. 2]

Elias (1955). Coding for noisy channels. IRE Convention Record.

Englert & Brout (1964). Broken Symmetry and the Mass of Gauge Vector Mesons. Physical Review Letters.

Erdős, P. (1934). 'A Theorem of Sylvester and Schur'. *J. London Math. Soc.* **9**, 282–288. [Ch. 8, 9]

Eric Hobsbawm, Terence Ranger (eds.) (1983). *The Invention of Tradition.* Cambridge University Press. [Ch. 38]

Eugene Wigner, "The Unreasonable Effectiveness of Mathematics in the Natural Sciences," *Communications in Pure and Applied Mathematics* 13, no. 1 (February 1960), pp. 1–14. [Ch. 24]

Exodus 10:21, Hebrew Bible. [Ch. 37]

Farquhar, G. D., von Caemmerer, S., Berry, J. A. (1980). A biochemical model of photosynthetic CO_2 assimilation in C_3 leaves. *Planta,* **149**, 78–90. [Ch. 35]

Feigenbaum, M. J. (1978). Quantitative universality for a class of nonlinear transformations. *Journal of Statistical Physics*, 19, 25–52. [Ch. 48]

Feldman, W. M. (1931). *Rabbinical Mathematics and Astronomy.* M. L. Cailingold. [Ch. 26]

Ferro, M. (1990). *The Great War 1914–1918.* Routledge. [Ch. 38]

Festinger (1957). A Theory of Cognitive Dissonance. Stanford University Press.

Feynman, R. P., Leighton, R. B., and Sands, M. (1964). *The Feynman Lectures on Physics.* Addison-Wesley. [Ch. 4]

Fick (1855). Ueber Diffusion. Annalen der Physik.

Finkelstein, A. (1987). Water Movement Through Lipid Bilayers, Pores, and Plasma Membranes. *Wiley.* [Ch. 31]

Frank Richards (1933). *Old Soldiers Never Die.* Faber and Faber Limited. [Ch. 38]

Friedmann (1922). Über die Krümmung des Raumes. Zeitschrift für Physik. [Ch. 2]

G. H. Hardy (1940). *A Mathematician's Apology.* Cambridge University Press. [Ch. 8]

G. Udny Yule (1911). *An Introduction to the Theory of Statistics.* C. Griffin and Company, Limited. [Ch. 30]

Gamow (1928). Zur Quantentheorie des Atomkernes. Zeitschrift für Physik.

Gauss (1809). Theoria motus corporum coelestium in sectionibus conicis solem ambientium. F. Perthes and I. H. Besser. [Ch. 33]

Geroch, R. (1967). *J. Math. Phys.*, **8.** [Ch. 14]

Gibbs (1876). On the Equilibrium of Heterogeneous Substances. Transactions of the Connecticut Academy of Arts and Sciences.

Golomb, S. W. (1955). Sequences with Randomness Properties. Glenn L. Martin Company.

Goroff, M. H., and Sagnotti, A. (1986). The ultraviolet behavior of Einstein gravity. *Nuclear Physics B.* [Ch. 36]

Gottfried Wilhelm Leibniz (1686). *Discourse on Metaphysics.* [Ch. 24]

Grimm, J. (1822). *Deutsche Grammatik.* Göttingen: Dieterich. [Ch. 5]

Gross, D. J. & Wilczek, F. (1973). Ultraviolet Behavior of Non-Abelian Gauge Theories. Physical Review Letters.

Gödel, K. (1931). 'Über formal unentscheidbare Sätze der Principia Mathematica und verwandter Systeme I'. *Monatshefte für Mathematik und Physik* **38**, 173–198. [Ch. 3, 21]

Gödel, K. (1938). The Consistency of the Axiom of Choice and of the Generalized Continuum-Hypothesis. *Proc. Natl. Acad. Sci.* **24**(12), 556–557. [Ch. 3]

H. Furstenberg (1955). 'On the Infinitude of Primes'. *American Mathematical Monthly.* [Ch. 8]

H. Poincaré (1890). 'Sur le problème des trois corps et les équations de la dynamique'. *Acta Mathematica.* [Ch. 48]

Hadamard, J. (1896). Sur la distribution des zéros de la fonction $\zeta(s)$ et ses conséquences arithmétiques. Bulletin de la Société Mathématique de France.

Harvey, E. Newton (1920). *The Nature of Animal Light.* J.B. Lippincott Company. [Ch. 17]

Harvey, E.N., and Stevens, K.L. (1928). The brightness of the light of the West Indian elaterid beetle, *Pyrophorus. J. Gen. Physiol.*, 12, 269–272. [Ch. 25]

Hegel, Georg Wilhelm Friedrich (1807). *Phänomenologie des Geistes.* Joseph Anton Goebhardt, Bamberg. [Ch. 50]

Heisenberg, W. (1927). 'Über den anschaulichen Inhalt der quantentheoretischen Kinematik und Mechanik'. *Z. Phys.* **43**, 172–198. [Ch. 1]

Henri Poincaré, *Calcul des Probabilités* (Paris: Gauthier-Villars, 1912), preface. [Ch. 7]

Henri Poincaré, *La Science et l'Hypothèse* (Paris: Flammarion, 1902). [Ch. 7]

Henri Poincaré, *Science et Méthode* (Paris: Flammarion, 1908). [Ch. 7]

Hermann Weyl (1918). *Raum, Zeit, Materie: Vorlesungen über allgemeine Relativitätstheorie.* Springer, Berlin. [Ch. 10]

Hermann Weyl, "Invariants," *Duke Mathematical Journal* 5 (1939), pp. 489–502. [Ch. 11]

Higgs, P. W. (1964). Broken Symmetries and the Masses of Gauge Bosons. Physical Review Letters.

Houston, R. (2014). *Obvious Does Not Imply True: The Minimal Superpermutation Conjecture Is False.* arXiv:1408.5108. [Ch. 39]

Hubble, E. (1929). A Relation between Distance and Radial Velocity among Extra-Galactic Nebulae. Proceedings of the National Academy of Sciences. [Ch. 2, 27, 43]

Hugo Rossi, "Mathematics Is an Edifice, Not a Toolbox," *Notices of the AMS* 43, no. 10 (October 1996), p. 1108. [Ch. 21]

Huw Price, "Retrieving the Baby: Reichenbach's Principle, Bell Locality, and Selection Bias," arXiv:2602.16985 (19 February 2026), Section 2, opening sentence. [Ch. 30]

Huygens, C. (1690). Traité de la Lumière. Pieter van der Aa. [Ch. 13, 24]

Hyslop, L. A. et al. (2016). Towards clinical application of pronuclear transfer. *Nature* **534**, 383–386. [Ch. 49]

International Bureau of Weights and Measures (BIPM) (2019). *The International System of Units (SI)*. BIPM. [Ch. 26]

International Meridian Conference (1884). International Conference Held at Washington for the Purpose of Fixing a Prime Meridian and a Universal Day. Gibson Bros. [Ch. 26]

Isaac Newton, *Principia Mathematica*, 2nd edition (1713), Definition VIII, Scholium. [Ch. 6]

Ives, H.E., and Coblentz, W.W. (1924). Photometric studies of luminous insects. *J. Opt. Soc. Am.*, 9(3), 217–236. [Ch. 25]

J.D. Bernal, lecture at Birkbeck College, University of London (1960). Published in Alan L. Mackay and Maurice Ebison (eds.), *The Harvest of a Quiet Eye* (London: Institute of Physics, 1977), p. 7. Original wording: "All that glisters may not be gold, but at least it contains free electrons." [Ch. 1]

James Clerk Maxwell, review in *Nature* 17 (7 February 1878), p. 278. [Ch. 34]

James D. Watson and Francis H.C. Crick, "Molecular Structure of Nucleic Acids," *Nature* 171 (25 April 1953), pp. 737–738. [Ch. 4]

James R. Newman (1956). *The World of Mathematics.* Simon and Schuster. [Ch. 4]

Jinek, M. et al. (2012). A Programmable Dual-RNA–Guided DNA Endonuclease in Adaptive Bacterial Immunity. *Science* **337**, 816–821. [Ch. 49]

John Archibald Wheeler. Earliest published version: "Space tells matter how to move; matter tells space how to curve," in C.W. Misner, K.S. Thorne, and J.A. Wheeler, *Gravitation* (San Francisco: W.H. Freeman, 1973), p. 5. The "spacetime" variant appears in J.A. Wheeler and K. Ford, *Geons, Black Holes, and Quantum Foam* (New York: Norton, 1998). [Ch. 6, 14]

John C. Baez, Javier P. Muniain (1994). *Gauge Fields, Knots and Gravity.* World Scientific Publishing Company. [Ch. 4]

John Keegan (1999). *The First World War.* A. Knopf. [Ch. 38]

John von Neumann, as recounted by Enrico Fermi to Freeman Dyson, 1953. Documented in Freeman Dyson, "A Meeting with Enrico Fermi," *Nature* 427 (22 January 2004), p. 297. [Ch. 1]

Jürgen Mayer, Khaled Khairy, and Jonathon Howard, "Drawing an Elephant with Four Complex Parameters," *American Journal of Physics* 78, no. 6 (June 2010), pp. 648–649. [Ch. 41]

Kahneman, D. & Tversky, A. (1979). Prospect Theory: An Analysis of Decision under Risk. Econometrica.

Kane, C. L. and Mele, E. J. (2005). $\mathbb{Z}_2$ Topological Order and the Quantum Spin Hall Effect. *Physical Review Letters*, **95**(14), 146802. [Ch. 11]

Kant, Immanuel (1781). *Critique of Pure Reason.* Johann Friedrich Hartknoch, Riga. [Ch. 50]

Kasianowicz, J. J., Brandin, E., Branton, D., & Deamer, D. W. (1996). Characterization of individual polynucleotide molecules using a membrane channel. Proceedings of the National Academy of Sciences.

Katori, H. (1999). Optical lattice clocks. Proceedings of the 6th Symposium on Frequency Standards and Metrology.

Kenneth J. Arrow (1963). *Social Choice and Individual Values.* John Wiley & Sons. [Ch. 9]

Kerr, R. P. (1963). 'Gravitational Field of a Spinning Mass as an Example of Algebraically Special Metrics'. *Phys. Rev. Lett.* **11**, 237–238. [Ch. 23]

Keynes, J. M. (1921). A Treatise on Probability. Macmillan and Co.

Kiefer, C. (2012). *Quantum Gravity* (3rd ed.). Oxford University Press. [Ch. 36]

Klein, F. (1882). Über Riemann's Theorie der Algebraischen Functionen und ihrer Integrale. B. G. Teubner. [Ch. 14]

Knuth, D. E. (1974). *Surreal Numbers.* Addison-Wesley.

Kolb, E. W., and Turner, M. S. (1990). *The Early Universe.* Addison-Wesley. [Ch. 43]

Kolmogorov, A. N. (1933). Grundbegriffe der Wahrscheinlichkeitsrechnung. Springer. [Ch. 15]

Kolmogorov, A. N. (1954). On Conservation of Conditionally Periodic Motions for a Small Change in Hamilton's Function. Doklady Akademii Nauk SSSR.

Kramers, H. A. (1930). Théorie générale de la rotation paramagnétique dans les cristaux. Proceedings of the Amsterdam Academy.

Kruskal, M. D. (1960). Maximal extension of Schwarzschild metric. Physical Review

Landauer, R. (1961). *Irreversibility and Heat Generation in the Computing Process.* IBM J. Res. Dev., 5(3), 183–191. [Ch. 45]

Langevin, P. (1911). L'évolution de l'espace et du temps. Scientia

Laplace, P. S. (1799). Traité de Mécanique Céleste. J.B.M. Duprat

Laplace, P. S. (1812). Théorie analytique des probabilités. Ve. Courcier [Ch. 33]

Laplace, P. S. (1814). Essai philosophique sur les probabilités. Courcier [Ch. 48]

Latała, R., Matlak, D. (2017). *Royen's Proof of the Gaussian Correlation Inequality.* In: Israel Seminar (GAFA) 2014–2016. Springer. [Ch. 33]

Lawvere, F. W. (1969). Adjointness in foundations. Dialectica

Lenard, A., and Dyson, F. J. (1967). Stability of matter I, II. *Journal of Mathematical Physics.* [Ch. 24]

Lense, J. & Thirring, H. (1918). Über den Einfluß der Eigenrotation der Zentralkörper auf die Bewegung der Planeten und Monde nach der Einsteinschen Gravitationstheorie. Physikalische Zeitschrift

Leonardo da Vinci, c. 1490s–1500s. From his notebooks. Various translations exist. [Ch. 5]

Leopold Flatto (2009). *Poncelet's Theorem.* American Mathematical Society. [Ch. 7]

Lincoln Taiz, Eduardo Zeiger, Ian Max Møller, Angus Murphy (2015). *Plant Physiology and Development, Sixth Edition.* Sinauer Associates, Inc. [Ch. 35]

Liouville, J. (1838). Note sur la théorie de la variation des constantes arbitraires. Journal de Mathématiques Pures et Appliquées

Lloyd, S. P. (1965). Least squares quantization in PCM. Bell Laboratories Technical Memorandum [Ch. 25]

Lohmann, K. (1929). Über die Pyrophosphatfraktion im Muskel. Naturwissenschaften

Lopez, J. V., Yuhki, N., Masuda, R., Modi, W., and O'Brien, S. J. (1994). Numt, a recent transfer and tandem amplification of mitochondrial DNA to the nuclear genome of the domestic cat. *J. Mol. Evol.* **39**, 174–190. [Ch. 49]

Lorenz, E. N. (1963). Deterministic nonperiodic flow. *Journal of the Atmospheric Sciences*, 20(2), 130–141. [Ch. 48]

Ludlow, A. D. et al. (2015). *Optical atomic clocks.* Rev. Mod. Phys. 87(2), 637–701. [Ch. 24, 32]

Lugaresi, E. et al. (1986). Fatal familial insomnia and dysautonomia with selective degeneration of thalamic nuclei. *N. Engl. J. Med.* **315**, 997–1003. [Ch. 50]

Lyapunov, A. M. (1892). The general problem of the stability of motion. Kharkov Mathematical Society

Lynn F. Faust, Paul A. Weston (2009). 'Degree-Day Prediction of Adult Emergence of Photinus carolinus (Coleoptera: Lampyridae)'. *Environmental Entomology.* [Ch. 25]

Lynn Frierson Faust (2017). *Fireflies, Glow-Worms, and Lightning Bugs.* University of Georgia Press. [Ch. 25]

M.C. Escher, "Approaches to Infinity," posthumously published essay (1973). Anthologized in Doris Schatschneider and Wallace Walker, eds. Full quote continues: "Long before there were people on the earth, crystals were already growing in the earth's crust." [Ch. 29]

M.P. Hobson, G.P. Efstathiou, and A.N. Lasenby, *General Relativity: An Introduction for Physicists* (Cambridge: Cambridge University Press, 2006). [Ch. 2]

Maimonides (c. 1170-1180). *Mishneh Torah.* [Ch. 43]

Maimonides (Moses ben Maimon), *The Guide for the Perplexed* (Moreh Nevukhim), c. 1190. Introduction, section on the definition of faith. Available in S. Pines translation (Chicago: University of Chicago Press, 1963). [Ch. 43]

Maki, Z., Nakagawa, M., & Sakata, S. (1962). Remarks on the unified model of elementary particles. Progress of Theoretical Physics

Malachi (450 BCE). *Book of Malachi.* [Ch. 16]

Malachi 3:19 (in Christian numbering 4:1), Hebrew Bible, c. 5th–4th century BCE. [Ch. 16]

Malcolm Brown (2005). *The Christmas Truce: The Western Front Comes to a Standstill.* [Ch. 38]

Manning, G.S., and Kay, A.R. (2023). The physical basis of osmosis. *J. Gen. Physiol.* 155:e202313332. [Ch. 31]

Martin Gardner (1964). *The Ambidextrous Universe: Left, Right, and the Fall of Parity.* Basic Books. [Ch. 46]

Martin Gardner (1982). *Aha! Gotcha: Paradoxes to Puzzle and Delight.* W. H. Freeman. [Ch. 15]

Masser, D. W. (1985). Lecture on the abc conjecture. Symposium on Diophantine Problems

Matiyasevich, Y. (1970). Enumerable sets are Diophantine. Doklady Akademii Nauk SSSR

Maurice Kraitchik (1953). *Mathematical Recreations.* Dover Publications. [Ch. 15]

Mauss, M. (1925). Essai sur le don. L'Année Sociologique [Ch. 38]

Maxwell, J. C. (1860). Illustrations of the dynamical theory of gases. Philosophical Magazine

Mayers, D. F. (1957). Relativistic Self-Consistent Field Calculation for Mercury. *Proc. R. Soc. Lond. A*, **241**, 93. [Ch. 1]

Maynard, J. (2015). Small gaps between primes. *Annals of Mathematics*, 181(1), 383–413. [Ch. 8]

Meissner, W. & Ochsenfeld, R. (1933). Ein neuer Effekt bei Eintritt der Supraleitfähigkeit. Naturwissenschaften

Meyer, H. (1899). Zur Theorie der Alkoholnarkose. Archiv für experimentelle Pathologie und Pharmakologie

Michael Atiyah, "Mathematics in the 20th Century: Geometry versus Algebra," *Mathematics Today* 37, no. 2 (2001), pp. 46–53. [Ch. 7]

Milgrom, M. (1983). A modification of the Newtonian dynamics as a possible alternative to the hidden mass hypothesis. The Astrophysical Journal

Minkowski, H. (1908). Raum und Zeit. Lecture at the 80th Assembly of German Natural Scientists and Physicians [Ch. 6, 47]

Mirollo, R. E. & Strogatz, S. H. (1990). Synchronization of pulse-coupled biological oscillators. SIAM Journal on Applied Mathematics.

Mishnaic sages (c. 200 CE). *Pirkei Avot.* Mishnah. [Ch. 26]

Misner, C. W., Thorne, K. S., Wheeler, J. A. (1973). *Gravitation.* W. H. Freeman and Company. [Ch. 6, 14]

Mitchell, P. (1961). Coupling of phosphorylation to electron and hydrogen transfer by a chemi-osmotic type of mechanism. Nature.

Modris Eksteins (1989). *Rites of Spring: The Great War and the Birth of the Modern Age.* Houghton Mifflin. [Ch. 38]

Moore, G. E. (1965). Cramming more components onto integrated circuits. Electronics.

Moritz Lipp, Michael Schwarz, Daniel Gruss, Thomas Prescher, Werner Haas, Anders Fogh, Jann Horn, Stefan Mangard, Paul Kocher, Daniel Genkin, Yuval Yarom, Mike Hamburg (2018). *Meltdown: Reading Kernel Memory from User Space.* 27th USENIX Security Symposium. [Ch. 18]

Myers, C. S. (1915). A contribution to the study of shell shock. The Lancet.

Möbius, A. F. (1865). Über die Bestimmung des Inhaltes eines Polyëders. Berichte über die Verhandlungen der Königlich Sächsischen Gesellschaft der Wissenschaften zu Leipzig.

N. D. Birrell and P. C. W. Davies (1982). *Quantum Fields in Curved Space.* Cambridge University Press. [Ch. 47]

Nalebuff, B. (1989). The Other Person's Envelope Is Always Greener. *J. Econ. Persp.*, **3**(1), 171–181. [Ch. 15]

Navarro, J. F., Frenk, C. S., and White, S. D. M. (1997). A Universal Density Profile from Hierarchical Clustering. *ApJ*, **490**, 493–508. [Ch. 37]

Newton, I. (1687). Philosophiæ Naturalis Principia Mathematica. Royal Society. [Ch. 6, 24]

Nicolas Léonard Sadi Carnot (1824). *Reflections on the Motive Power of Fire.* Bachelier, Paris. [Ch. 45]

Niels Bohr, as quoted in Ruth Moore, *Niels Bohr: The Man, His Science, & the World They Changed* (New York: Knopf, 1966), p. 140. [Ch. 15]

Nora, Pierre (1984). *Les Lieux de mémoire.* Gallimard, Paris. [Ch. 38]

Norrby, L. J. (1991). Why Is Mercury Liquid? Or, why do relativistic effects not get into chemistry textbooks? *J. Chem. Educ.*, **68**, 110. [Ch. 1]

Oesterlé, J. (1988). Nouvelles approches du "théorème" de Fermat. Séminaire Bourbaki [Ch. 33]

Oliver Heaviside (1893). *Electromagnetic Theory.* The Electrician Printing and Publishing Company, London. [Ch. 4]

Openshaw, S. (1984). The modifiable areal unit problem. Geo Books.

Originated with I.J. Good, address to Institute of Mathematical Statistics, 22 April 1971 (published in *The American Statistician*, June 1972). Ronald H. Coase refined it in a letter dated 17 October 1975: "If you torture the data long enough nature will confess.". [Ch. 40]

Osthoff, H. & Brugmann, K. (1878). Morphologische Untersuchungen auf dem Gebiete der indogermanischen Sprachen. S. Hirzel.

Overton, E. (1901). Studien über die Narkose. Gustav Fischer

P. W. Kasteleyn (1961). 'The statistics of dimers on a lattice: I. The number of dimer arrangements on a quadratic lattice'. *Physica 27, 1209-1225.* [Ch. 29]

Padmanabhan, T. (2003). Cosmological constant — the weight of the vacuum. *Phys. Rep.*, 380, 235. [Ch. 2]

Palermo, L., von Ranke, P. J., & da Silva, X. A. (1992). A spin-1/2 Ising model for a semi-infinite lattice in a transverse field. Physical Review B.

Pascal, B. (1670). Pensées. Guillaume Desprez.

Paul Gordan, as reported in: Max Noether, "Paul Gordan," *Mathematische Annalen* 75 (1914), pp. 1–41. The remark dates to approximately 1888, in response to Hilbert's proof of the finite basis theorem. [Ch. 1]

Paul Kocher, Jann Horn, Anders Fogh, Daniel Genkin, Daniel Gruss, Werner Haas, Mike Hamburg, Moritz Lipp, Stefan Mangard, Thomas Prescher, Michael Schwarz, Yuval Yarom (2019). *Spectre Attacks: Exploiting Speculative Execution.* IEEE Symposium on Security and Privacy. [Ch. 18]

Pauli, W. (1925). 'Über den Zusammenhang des Abschlusses der Elektronengruppen im Atom mit der Komplexstruktur der Spektren'. *Z. Phys.* **31**, 765–783. [Ch. 1]

Pauling, L. (1931). The Nature of the Chemical Bond. Journal of the American Chemical Society.

Pavlides, M. G., and Perlman, M. D. (2009). *J. Stat. Plan. Inference*, 139(1), 198–213. [Ch. 30]

Penrose, R. (1965). Gravitational Collapse and Space-Time Singularities. Physical Review Letters.

Penzias, A. A. & Wilson, R. W. (1965). A Measurement of Excess Antenna Temperature at 4080 Mc/s. The Astrophysical Journal.

Perlin, K. (1985). An Image Synthesizer. ACM SIGGRAPH Computer Graphics.

Persson, M. (2009). *Initial Geometry and Entity Source.* Mojang. [Ch. 22]

Peter Simkins (1991). *World War I: The Western Front.* [Ch. 4]

Pevzner et al. (2001). *PNAS* 98:9748–9753. [Ch. 40]

Planck Collaboration (2020). *A&A*, **641**, A6. [Ch. 43]

Planck, M. (1901). Ueber das Gesetz der Energieverteilung im Normalspectrum. Annalen der Physik.

Poincaré, H. (1885). Sur l'équilibre d'une masse fluide animée d'un mouvement de rotation. Acta Mathematica. [Ch. 7]

Politzer, H. D. (1973). Reliable Perturbative Results for Strong Interactions?. Physical Review Letters.

Pontecorvo, B. (1957). Mesonium and anti-mesonium. Zhurnal Eksperimental'noi i Teoreticheskoi Fiziki.

Popper, K. (1934). Logik der Forschung. Julius Springer.

Pound, R. V. & Rebka, G. A. Jr. (1960). Apparent Weight of Photons. Physical Review Letters. [Ch. 6, 32]

Poynting, J. H. (1884). On the Transfer of Energy in the Electromagnetic Field. Philosophical Transactions of the Royal Society of London.

Prusiner, S. B. (1982). Novel proteinaceous infectious particles cause scrapie. Science.

Purcell, E. M., Pound, R. V. (1951). *A Nuclear Spin System at Negative Temperature.* Phys. Rev., 81, 279. [Ch. 28]

Putnam, H. (1967). The Nature of Mental States. In Art, Mind, and Religion.

Pyykkö, P., Desclaux, J. P. (1979). Relativity and the periodic system of elements. *Acc. Chem. Res.*, **12** (8), 276–281. [Ch. 1]

Rabbi Eliezer ben Hyrcanus (attributed; medieval compilation) (1514). *Pirkei de-Rabbi Eliezer.* Constantinople. [Ch. 26]

Rabbi Yisrael Lipschitz, *Tiferet Yisrael* / Boaz on Mishnah Berakhot 1, c. 1830–1850. [Ch. 26]

Radon, J. (1917). Über die Bestimmung von Funktionen durch ihre Integralwerte längs gewisser Mannigfaltigkeiten. *Berichte der Sächsischen Akademie der Wissenschaften zu Leipzig*, 69, 262–277. [Ch. 41]

Radó, T. (1962). On non-computable functions. Bell System Technical Journal.

Ramsey, N. F. (1956). *Thermodynamics and Statistical Mechanics at Negative Absolute Temperatures.* Phys. Rev., 103, 20. [Ch. 28]

Rayleigh, Lord (1871). On the light from the sky, its polarization and color. Philosophical Magazine.

René Descartes (1637). *Discours de la méthode*, Part V. Leiden. [Ch. 20]

René Descartes (1641). *Meditations on First Philosophy.* Michel Soly. [Ch. 50]

René Descartes, *Principia Philosophiae* (Amsterdam: Elsevier, 1644), Part II. [Ch. 47]

Richard Nixon — Statement made in 1972 regarding inflation rates. Widely reported. [Ch. 13, 21]

Richard O. Duda, Peter E. Hart (1973). *Pattern Classification and Scene Analysis.* Wiley-Interscience. [Ch. 13]

Richard P. Feynman (1985). *QED: The Strange Theory of Light and Matter.* Princeton University Press. [Ch. 13]

Riemann, B. (1859). Ueber die Anzahl der Primzahlen unter einer gegebenen Grösse. Monatsberichte der Berliner Akademie. [Ch. 8]

Riess, A. G. et al. (1998). Observational Evidence from Supernovae for an Accelerating Universe and a Cosmological Constant. *Astron. J.* **116**, 1009–1038. [Ch. 2, 34, 43]

Ringe, D. (2006). *From Proto-Indo-European to Proto-Germanic.* Oxford University Press. [Ch. 5]

Robert L. Berger (1966). *The Undecidability of the Domino Problem.* Memoirs of the American Mathematical Society. [Ch. 29]

Robert M. Wald (1984). *General Relativity.* University of Chicago Press. [Ch. 23]

Robinson, J. (1950). Existential definability in arithmetic. Transactions of the American Mathematical Society. [Ch. 30]

Robinson, R. M. (1971). *Undecidability and Nonperiodicity for Tilings of the Plane.* Invent. Math. [Ch. 29]

Roger Penrose (2004). *The Road to Reality: A Complete Guide to the Laws of the Universe.* Jonathan Cape. [Ch. 29, 34]

Rossi, B., and Hall, D. B., *Physical Review*, 59(3), 223–228 (1941). [Ch. 19]

Royen, T. (2014). *A simple proof of the Gaussian correlation conjecture extended to multivariate gamma distributions.* Far East J. Theor. Stat. [Ch. 33]

Ruben, S., Randall, M., Kamen, M., and Hyde, J. L. (1941). Heavy Oxygen (^{18}O) as a Tracer in the Study of Photosynthesis. *J. Am. Chem. Soc.* **63**, 877–879. [Ch. 35]

Rybicki, G. B., and Lightman, A. P. (1979). *Radiative Processes in Astrophysics.* Wiley. [Ch. 27]

S Coleman and F De Luccia (1980). 'Gravitational effects on and of vacuum decay'. *Physical Review D.* [Ch. 16]

S.W. Hawking (1974). 'Black hole explosions?'. *Nature 248, 30-31.* [Ch. 23]

Safronova, M. S. et al. (2018). *Search for new physics with atomic clocks.* Rev. Mod. Phys. 90(2), 025008. [Ch. 32]

Sagnac, G. (1913). L'éther lumineux démontré par l'effet du vent relatif d'éther dans un interféromètre en rotation uniforme. Comptes Rendus de l'Académie des Sciences.

Sanger, F., Nicklen, S., & Coulson, A. R. (1977). DNA sequencing with chain-terminating inhibitors. Proceedings of the National Academy of Sciences.

Sara Lewis (2016). *Silent Sparks: The Wondrous World of Fireflies.* Princeton University Press. [Ch. 25]

Satterthwaite, M. A. (1975). Strategy-proofness and Arrow's conditions. *J. Econ. Theory* **10**(2), 187–217. [Ch. 9]

Schiffer, P. H. et al. (2018). The nematode *Acrobeloides nanus* reveals conserved phylum-specific expression patterns. *PNAS* **115**(6), 1269–1274. [Ch. 40]

Schleicher, A. (1861). Compendium der vergleichenden Grammatik der indogermanischen Sprachen. H. Böhlau.

Schrödinger, E. (1926). 'Quantisierung als Eigenwertproblem'. *Annalen der Physik* **384**(4), 361–376. [Ch. 1]

Schwarzschild, K. (1916). Über das Gravitationsfeld eines Massenpunktes nach der Einsteinschen Theorie. Sitzungsberichte der Königlich Preussischen Akademie der Wissenschaften.

Sean M. Carroll (2004). *Spacetime and Geometry: An Introduction to General Relativity.* Addison Wesley. [Ch. 2]

Searle, J. R. (1980). *Minds, Brains, and Programs.* Behavioral and Brain Sciences, 3(3), 417–457. [Ch. 1, 20]

Shannon, C. E. (1938). A Symbolic Analysis of Relay and Switching Circuits. Transactions of the American Institute of Electrical Engineers. [Ch. 22]

Shannon, C. E. & Weaver, W. (1949). The Mathematical Theory of Communication. University of Illinois Press.

Shechtman, D., Blech, I., Gratias, D., and Cahn, J. W. (1984). Metallic Phase with Long-Range Orientational Order and No Translational Symmetry. *Phys. Rev. Lett.* **53**, 1951–1953. [Ch. 29]

Silver, D. H. et al. (2013). ELOPER: elongation of paired-end reads. *Bioinformatics* **29**(11), 1465–1467. [Ch. 40]

Simpson, E. H. (1951). *J. R. Stat. Soc. B*, 13(2), 238–241. [Ch. 30]

Slater, B., and Michaelides, A. (2019). *Nat. Rev. Chem.*, **3**, 172. [Ch. 42]

Sobel, I. (1970). Camera Models and Machine Perception. Stanford University.

Soddy, F. (1913). The Radio-Elements and the Periodic Law. Nature.

Sonnino, Paul (2016). *The Search for the Man in the Iron Mask: A Historical Detective Story.* Rowman & Littlefield. [Ch. 39, 44]

Sophie Germain (1821). *Recherches sur la théorie des surfaces élastiques.* Académie des sciences. [Ch. 13]

Spergel, D. N. et al. (2003). First-Year WMAP Observations. *ApJS* **148**, 175–194. [Ch. 37]

Steinhardt, P. J. & Turok, N. (2002). A Cyclic Model of the Universe. Science.

Stephanopoulos, N. O. & McGhee, E. M. (2015). Partisan Gerrymandering and the Efficiency Gap. University of Chicago Law Review.

Stern, Sacha (2019). *The Jewish Calendar Controversy of 921/2 CE.* Brill. [Ch. 26]

Swadesh, M. (1955). Towards greater accuracy in lexicostatistic dating. International Journal of American Linguistics.

Szilard, L. (1929). *On the Decrease of Entropy in a Thermodynamic System by the Intervention of Intelligent Beings.* Z. Phys., 53, 840–856. [Ch. 45]

Tachibana, M. *et al.* (2009). Mitochondrial gene replacement in primate offspring and embryonic stem cells. *Nature* **461**, 367–372. [Ch. 49]

Tachibana, M. *et al.* (2013). Towards germline gene therapy of inherited mitochondrial diseases. *Nature* **493**, 627–631. [Ch. 49]

Tanya Lee Stone (2003). *Ilan Ramon: Israel's First Astronaut.* Millbrook Press. [Ch. 44]

Terraine, J. (1978). *To Win a War: 1918, the Year of Victory.* Sidgwick & Jackson. [Ch. 38]

The National Archives (UK) (1914). 'BEF Unit War Diaries, Dec 1914'. *The National Archives (UK).* [Ch. 38]

Thomas Reid (1785). *Essays on the Intellectual Powers of Man.* J. Bell, Edinburgh. [Ch. 50]

Thouless, D. J., Kohmoto, M., Nightingale, M. P., and den Nijs, M. (1982). Quantized Hall Conductance in a Two-Dimensional Periodic Potential. *Phys. Rev. Lett.* **49**, 405–408. [Ch. 11]

Timmins, G.S., et al. (2001). Firefly flashing is controlled by gating oxygen to light-emitting cells. *J. Exp. Biol.*, 204, 2795–2801. [Ch. 25]

Tomasulo, R. M. (1967). An Efficient Algorithm for Exploiting Multiple Arithmetic Units. IBM Journal of Research and Development.

Tononi, G. (2004). An information integration theory of consciousness. BMC Neuroscience.

Turing, A. M. (1936). On Computable Numbers, with an Application to the Entscheidungsproblem. Proceedings of the London Mathematical Society. [Ch. 17]

Turing, A. M. (1950). Computing Machinery and Intelligence. Mind. [Ch. 20]

Unruh, W. G. (1976). Notes on black-hole evaporation. Physical Review D.

van Ginkel, M., Luengo Hendriks, C. L., and van Vliet, L. J. (2004). *A short introduction to the Radon and Hough transforms and how they relate to each other.* Imaging Science and Technology Department, Delft University of Technology, The Netherlands. [Ch. 41]

Various (Tannaitic sages, school of Rabbi Akiva) (c. 200 CE). *Midrash Halakha.* [Ch. 5]

Vegard, L. (1908). Zur Theorie der osmotischen Erscheinungen. *Proc. Camb. Phil. Soc.* 15:13–23. [Ch. 31]

Vernam, G. S. (1926). Cipher Printing Telegraph Systems For Secret Wire and Radio Telegraphic Communications. Journal of the American Institute of Electrical Engineers.

Voltaire (1751). *Le Siècle de Louis XIV.* C. F. Henning. [Ch. 39, 44]

von Neumann, John, and Morgenstern, Oskar (1944). *Theory of Games and Economic Behavior.* Princeton University Press. [Ch. 3]

W. Deane Butcher, Translator's Preface to Stéphane Leduc, *The Mechanism of Life* (London: William Heinemann, 1911), p. vii. [Ch. 31]

Wallace, D. C., et al. (1988). Mitochondrial DNA mutation associated with Leber's hereditary optic neuropathy. Science.

Wang, H. (1961). Proving theorems by pattern recognition — II. Bell System Technical Journal.

Weber, B. et al. (2018). *J. Phys. Chem. Lett.*, **9**, 2838. [Ch. 42]

Weinberg, S. (1979). Ultraviolet divergences in quantum theories of gravitation. In S. W. Hawking and W. Israel (Eds.), *General Relativity: An Einstein Centenary Survey* (Cambridge University Press). [Ch. 36]

Weinberg, S. (1989). The cosmological constant problem. *Rev. Mod. Phys.*, 61, 1. [Ch. 2]

Weinberg, Steven (1995). *The Quantum Theory of Fields, Vol. I: Foundations.* Cambridge University Press. [Ch. 24]

Weintraub, Stanley (2001). *Silent Night: The Story of the World War I Christmas Truce.* Free Press. [Ch. 38]

Weyl, H. (1916). Über die Gleichverteilung von Zahlen mod. Eins. Mathematische Annalen. [Ch. 7]

Wiedemann, G., & Franz, R. (1853). Ueber die Wärme-Leitungsfähigkeit der Metalle. Annalen der Physik.

Wien, W. (1896). Ueber die Energieverteilung im Emissionsspectrum eines schwarzen Körpers. Annalen der Physik.

Wiles, A. (1995). Modular elliptic curves and Fermat's Last Theorem. Annals of Mathematics.

Williams, A. O. (1940). A Relativistic Self-Consistent Field for Cu^{+}. *Phys. Rev.*, **58**, 723. [Ch. 1]

Wills, C. (1967). The ecological relations of California voles. Ecology.

Wilson, K. G. (1971). Renormalization Group and Critical Phenomena. I. Renormalization Group and the Kadanoff Scaling Picture. Physical Review B.

Winston Churchill, speech to the House of Commons, 11 November 1947. Published in *Hansard*, vol. 444, col. 206–7. [Ch. 9]

Woodward, R. B., and Hoffmann, R. (1970). *The Conservation of Orbital Symmetry.* Verlag Chemie GmbH / Academic Press. [Ch. 46]

Wu, A. (2008). 'The Banach-Tarski Paradox'. *University of Chicago VIGRE REU.* [Ch. 3]

Yang, C. N., & Mills, R. L. (1954). Conservation of Isotopic Spin and Isotopic Gauge Invariance. Physical Review.

Yehudai Gaon (8th century). *Halakhot Pesukot.* [Ch. 26]

Zermelo, E. (1904). Beweis, daß jede Menge wohlgeordnet werden kann. Mathematische Annalen. [Ch. 3, 7]

Zhang, Y. (2014). Bounded gaps between primes. *Annals of Mathematics*, 179(3), 1121–1174. [Ch. 8]

Zimbardo, P. G. (1971). The Power and Pathology of Imprisonment. Congressional Record.

Zwicky, F. (1933). Die Rotverschiebung von extragalaktischen Nebeln. *Helv. Phys. Acta*, **6**, 110–127. [Ch. 37]

Literary, Film & Media Sources

"Let the Lord of Chaos rule." — Dark One's command, *Lord of Chaos* (1994) Prologue. Semirhage says it in *Knife of Dreams* (2005), Ch. 34. Robert Jordan, *The Wheel of Time.* [Ch. 48]

"The Gang" — *It's Always Sunny in Philadelphia*, Season 7, Episode 4, "Sweet Dee Gets Audited" (FX, 2011). [Ch. 50]

Albert Einstein — "Zeit ist das, was man an der Uhr abliest." From a 1926 discussion with Rudolf Carnap, as reported in Carnap's autobiography in *The Philosophy of Rudolf Carnap* (La Salle: Open Court, 1963), p. 37. [Ch. 6]

Albert Einstein, letter to Cornel Lanczos, 21 March 1942. Published in *Albert Einstein: The Human Side* (Princeton: Princeton University Press, 1979), eds. Helen Dukas and Banesh Hoffmann. [Ch. 6]

Andy Dwyer — *Parks and Recreation*, Season 6, Episode 1, "London, Part 1" (NBC, 2013). Fictional character played by Chris Pratt. [Ch. 42]

Brandon Sanderson (2010). *The Way of Kings (The Stormlight Archive, Book 1).* Tor Books. [Ch. 18]

Charlie Kelly — *It's Always Sunny in Philadelphia*, Season 5, Episode 3, "The Great Recession" (FX, 2009). [Epigraph, Ch. 27]

Cohen the Barbarian — Terry Pratchett, *The Last Hero* (London: Gollancz, 2001). Fictional character. Quote also used by @terryandrob to announce Pratchett's death, 12 March 2015. [Ch. 3]

Comte de Lautréamont (1869). *Les Chants de Maldoror.* Lacroix, Verboeckhoven et Cie. [Ch. 25]

Dan Simmons (1989). *Hyperion.* Doubleday. [Ch. 20]

Dave Barry, *Dave Barry Turns 50* (New York: Crown, 1998). [Ch. 4]

Douglas Adams, *The Hitchhiker's Guide to the Galaxy* (London: Pan Books, 1979), Chapter 1. [Ch. 14]

Dr. Seuss (Theodor Seuss Geisel), *The Lorax* (New York: Random House, 1971). [Ch. 35]

Eddard Stark — George R.R. Martin, *A Game of Thrones* (New York: Bantam, 1996), Chapter 1 ("Bran I"). [Ch. 28]

George R. R. Martin (1996). *A Game of Thrones.* Bantam Books. [Ch. 28]

Greg Egan (1994). *Permutation City.* Millennium Orion Publishing Group. [Ch. 39]

I.I. Rabi — "Who ordered that?" upon learning of the muon's discovery, 1936. Widely reported; see e.g., Robert P. Crease and Charles C. Mann, *The Second Creation* (New York: Macmillan, 1986). [Ch. 19]

Jack Donaghy — *30 Rock*, Season 1, Episode 19, "Corporate Crush" (NBC, 2007). Fictional character played by Alec Baldwin. [Ch. 14]

J.R.R. Tolkien, *The Fellowship of the Ring* (London: George Allen & Unwin, 1954), poem in Chapter 10 ("Strider"). [Ch. 1]

Jacob Grimm, Wilhelm Grimm (1812). *Kinder- und Hausmärchen.* Realschulbuchhandlung, Berlin. [Ch. 7]

Joanna Newsom, "Cosmia," *Ys* (Drag City, 2006). Song lyrics. [Ch. 16]

Kenneth Parcell — *30 Rock*, Season 5, Episode 19, "I Heart Connecticut" (NBC, 2011). [Ch. 6]

Lewis Carroll, *Alice's Adventures in Wonderland* (London: Macmillan, 1865), Chapter 2. [Ch. 25]

Logen Ninefingers — Joe Abercrombie, *The Blade Itself* (London: Gollancz, 2006). [Ch. 22]

Mac — *It's Always Sunny in Philadelphia*, Season 5, Episode 3, "The Great Recession" (FX, 2009). "That doesn't sound right, but I don't know enough about stars to dispute it." [Ch. 15]

Mark Twain, *Following the Equator* (Hartford: American Publishing Company, 1897), Chapter 15 epigraph. [Ch. 3]

Martin Silenus — Dan Simmons, *Hyperion* (New York: Doubleday, 1989), "The Poet's Tale.". [Ch. 20]

Nicomo Cosca — Joe Abercrombie, *Best Served Cold* (London: Gollancz, 2009). [Ch. 38]

Solid Jackson — Terry Pratchett, *Jingo* (London: Gollancz, 1997). [Ch. 28]

Super Hans — *Peep Show*, Series 3, Episode 2, "Sectioning" (Channel 4, 2005). Fictional character played by Matt King. [Ch. 9]

Terry Pratchett (1997). *Jingo.* Victor Gollancz. [Ch. 7]

The Dogman — Joe Abercrombie, *Before They Are Hanged* (London: Gollancz, 2007). Fictional character. [Ch. 43]

The Weakerthans, "Left and Leaving," *Left and Leaving* (G7 Welcoming Committee Records, 2000). Song lyrics by John K. Samson. [Ch. 44]

Tiffany Aching — Terry Pratchett, *A Hat Full of Sky* (London: Doubleday, 2004). [Ch. 46]

Tobias Funke — *Arrested Development*, Season 2, Episode 5, "Sad Sack" (Fox, 2004). [Ch. 12]

Tom Cargill — "Ninety-ninety rule." Reported in Jon Bentley, "Programming Pearls," *Communications of the ACM* 28, no. 9 (September 1985). [Ch. 22]

Wit (Hoid) to Shallan — Brandon Sanderson, *Oathbringer* (New York: Tor Books, 2017), Chapter 68, "Aim for the Sun." Fictional character. [Ch. 18]

About the Team

Alessandra Tosi was the managing editor for this book.

Annie Hine proofread this manuscript.

Jessica 'D. G.' Hayes provided illustrations and early reading of the manuscript.

Jeevanjot Kaur Nagpal designed the cover. The cover was produced in InDesign using the Fontin font. Cover image by Enny Silver and David H. Silver.

The author typeset the book in LaTeX and compiled the index.

The author produced the PDF, paperback and hardback editions.

Lucy Barnes was in charge of marketing.

This book was peer-reviewed by two anonymous referees. Experts in their field, these readers give their time freely to help ensure the academic rigour of our books. We are grateful for their generous and invaluable contributions.

www.ingramcontent.com/pod-product-compliance
Lightning Source LLC
LaVergne TN
LVHW052111210926
848482LV00025B/540
9781805118770